SOURCE BOOKS IN THE
HISTORY OF THE SCIENCES

GREGORY D. WALCOTT · *General Editor*

A SOURCE BOOK IN GREEK SCIENCE

SOURCE BOOKS IN THE HISTORY OF THE SCIENCES

GREGORY D. WALCOTT · *General Editor*

A SOURCE BOOK IN ASTRONOMY

HARLOW SHAPLEY · *Director, Harvard Observatory Harvard University*
AND HELEN E. HOWARTH · *Harvard University*

A SOURCE BOOK IN MATHEMATICS

DAVID EUGENE SMITH · *Columbia University*

A SOURCE BOOK IN PHYSICS

W. F. MAGIE · *Princeton University*

A SOURCE BOOK IN GEOLOGY

KIRTLEY F. MATHER · *Harvard University*
AND SHIRLEY L. MASON

A SOURCE BOOK IN GREEK SCIENCE

MORRIS R. COHEN · *College of the City of New York and University of Chicago* AND
I. E. DRABKIN · *College of the City of New York*

Endorsed by the American Philosophical Association, the American Association for the Advancement of Science, and the History of Science Society. Also by the American Anthropological Association, the Mathematical Association of America, the American Mathematica Society and the American Astronomical Society in their respective fields.

A SOURCE BOOK IN GREEK SCIENCE

By MORRIS R. COHEN, Ph.D.
Late Professor of Philosophy
College of the City of New York
and University of Chicago

and I. E. DRABKIN, Ph. D.
Department of Mathematics
College of the City of New York

FIRST EDITION

NEW YORK TORONTO LONDON

McGRAW-HILL BOOK COMPANY, INC.

1948

A SOURCE BOOK IN GREEK SCIENCE

SOURCE BOOKS IN THE HISTORY OF THE SCIENCES

General Editor's Preface

This series of Source Books aims to present the most significant passages from the works of the most important contributors to the major sciences during the last three or four centuries. So much material has accumulated that a demand for selected sources has arisen in several fields. Source books in philosophy have been in use for nearly a quarter of a century, and history, economics, ethics, and sociology utilize carefully selected source material. Recently, too, such works have appeared in the fields of psychology and eugenics. It is the purpose of this series, therefore, to deal in a similar way with the leading physical and biological sciences.

The general plan is for each volume to present a treatment of a particular science with as much finality of scholarship as possible from the Renaissance to the end of the nineteenth century. In all, it is expected that the series will consist of eight or ten volumes, which will appear as rapidly as may be consistent with sound scholarship.

In June, 1924, the General Editor began to organize the following Advisory Board:

Harold C. Brown	*Philosophy*	Stanford University
Morris R. Cohen	*Philosophy*	College of the City of N. Y.
Arthur O. Lovejoy	*Philosophy*	Johns Hopkins University
George H. Mead	*Philosophy*	University of Chicago
William P. Montague	*Philosophy*	Columbia University
Wilmon H. Sheldon	*Philosophy*	Yale University
Edward G. Spaulding	*Philosophy*	Princeton University
Joseph S. Ames	*Physics*	Johns Hopkins University
Frederick Barry	*Chemistry*	Columbia University
R. T. Chamberlin	*Geology*	University of Chicago
Edwin G. Conklin	*Zoology*	Princeton University
Harlow Shapley	*Astronomy*	Harvard University
David Eugene Smith	*Mathematics*	Columbia University
Alfred M. Tozzer	*Anthropology*	Harvard University

Each of the scientists on this board, in addition to acting in a general advisory capacity, is chairman of a committee of four or five men, whose business it is to make a survey of their special field and to determine the number of volumes required and the contents of each volume.

In December, 1925, the General Editor presented the project to the Eastern Division of the American Philosophical Association. After some discussion by the Executive Committee, it was approved and the philosophers of the board, with the General Editor as chairman, were appointed a committee to have charge of it. In November, 1927, the Carnegie Corporation of New York granted $10,000 to the American Philosophical Association as a revolving fund to help finance the series. In December, 1927, the American Association for the Advancement of Science approved the project, and appointed the General Editor and Profs. Edwin G. Conklin and Harlow Shapley a committee to represent that Association in cooperation with the Advisory Board. In February, 1928, the History of Science Society officially endorsed the enterprise. Endorsements have also been given by the American Anthropological Association, the Mathematical Association of America, the American Mathematical Society, and the American Astronomical Society within their respective fields.

The General Editor wishes to thank the members of the Advisory Board for their assistance in launching this undertaking; Dr. J. McKeen Cattell for helpful advice in the early days of the project and later; Dr. William S. Learned for many valuable suggestions; the several societies and associations that have given their endorsements; and the Carnegie Corporation for the necessary initial financial assistance.

In 1948 this series was extended to include *A Source Book in Greek Science* and *A Source Book in Medieval Science.*

A single volume, containing the most important contributions of the major sciences from 1900 to 1950, is planned for publication about 1960, and a similar volume each half century thereafter indefinitely.

GREGORY D. WALCOTT

LONG ISLAND UNIVERSITY
BROOKLYN, N. Y.
March, 1948

PREFACE

The notion that natural science began in the seventeenth century with Bacon, Galileo, and Descartes, or perhaps in the sixteenth with Copernicus, and that the Greeks were mere speculators and the medieval thinkers all sunk in theology and superstition is not merely an established popular error; it has become a basic dogma of modernistic philosophy and is even shared by some professional historians. This error is largely supported by the prevailing type of specialized education, which trains students of nature to look at things exclusively from the point of view of current conceptions and does not sufficiently equip them with philological or historical methods to investigate how the world appeared to man at other times. On the other hand, historians trained in the humanistic tradition often fail to discriminate justly between genuine science and folklore. In any case, the scientist's interest in the nature of things and the historian's interest in the development of man's critical knowledge are all too seldom united. Such a union of interests, however, can be furthered by the study of the history of science. It is to serve this study that the present series of Source Books was inaugurated.

While in those volumes of the Series that deal with the special sciences it was for practical purposes convenient to begin with the fifteenth century, there is an obvious need to supplement this material with extracts from ancient and medieval sources. The present volume makes available to students some of the most significant passages in what remains to us of Greek writings on science, including Latin contributions that reflect Greek methods and ideas. The plan of the Series calls for a similar volume dealing with medieval science.

The extracts here presented cannot take the place of a history. They cannot by themselves offer a systematic view of Greek science, nor can they indicate the place of science within the frame of Greek civilization as a whole. But no history of Greek science can convey an adequate picture unless the reader sees something of the original sources on which the historian relies. The presentation of these sources seems all the more desirable because the material on which we draw is often fragmentary and scattered in many diverse volumes.

There is, to be sure, some disadvantage in offering passages abstracted from their context. We have sought in part to overcome this disadvantage by the use of explanatory notes, but it is our hope that the reading of the extracts will ultimately lead serious students to seek a fuller knowledge of the original texts in more complete form.

There have been many outstanding contributions to the history of Greek science. To speak only of the past, the work of Meyer, though almost a century old, still remains the most comprehensive treatment of Greek botany. Hultsch, Zeuthen, Paul Tannery, Duhem, Heiberg, and Heath have illuminated Greek mathematics and astronomy. Diels has done much for the study of early Greek science. But the works of these men have appealed, for the most part, only to specialists. The present collection seeks to give some idea of the whole field of Greek scientific achievement, covering a period of about a thousand years. And it must not be forgotten that many Greek scientific treatises were written and read by cultivated people who did not regard themselves as specialists. These works should appeal today to those thoughtful readers who wish to achieve some understanding not only of the foundations of modern science, but of a vital element of the humanistic tradition.

In making the present collection we have had to face certain difficult questions. We have had to draw the line between what should be regarded as scientific material and that which would more properly be considered philosophic or speculative. Such a line must necessarily be arbitrary, since the Greeks themselves did not draw it very sharply. But in view of the general availability to English readers of Greek philosophic speculation in cosmology as well as on human affairs, it has seemed advisable to confine our material to that which would generally be regarded today as scientific in method, i.e., based, in principle, either on mathematics or on empirical verification.

Again, in considering technical material, e.g., in mathematics and astronomy, we have, as a rule, chosen to present the more elementary and fundamental ideas. We have felt that such a procedure, though it might not convey the details of the most advanced Greek achievement, would be more useful to the general reader.

By considering method rather than substantive results or truth according to modern standards, we have also tried to meet the difficult problem of how much attention to give theories that are now known to be false or even ridiculous. To leave out any reference to the doctrine of the four elements or to the atomism of Democritus would be to distort the picture of what Greek science really was and would certainly not be conducive to a better understanding of it. On the other hand, a source book of Greek science should not encroach on the field of Greek magic, superstition, and religion. No one can well deny that a good deal of what may be called "pseudo science," such as astrology and the like, can be found in the writings of such sober Greek scientists as Aristotle and Ptolemy. But it is well to remember that the intrusion of the occult can be found in modern writings such as Kepler's or Newton's and in the contributions to the early volumes of the transactions of the Royal Society down to the works of Lodge, Carrel, and Eddington in our own day. A difference of degree in this respect arises from the fact that, as modern science has become highly specialized and technically naturalistic in its experimental procedure, the author's personal views are more readily distinguished from the body of accepted science. But it cannot be supposed that even the most modern scientific work is completely free from erroneous views. The actual history of science as well as good logic shows that false theories, such as those of phlogiston, caloric, and "ether," far from necessarily hindering the growth of science, may sometimes help its advance by colligating phenomena, the connection between which would not otherwise have been considered.

The extent to which technological material should enter into a source book of Greek science is another difficult question. We have included chiefly material that seems either to have been a direct application of Greek scientific theory or to have contributed to its subsequent development.

In employing modern categories to classify the contents of this source book, we do not, of course, suggest that these were the categories of the ancient thinkers whose works are quoted. But we hope that this device will enable the modern reader concerned with a given special field to find more readily the ancient material that will interest him.

Wherever we have quoted previously published translations, we have given the name of the translator (further details may be found in the Acknowledgments, page xi); in the other cases the translation is the work of Dr. Drabkin. The notes to the various passages are quoted in each case from those of the translator, except where the notation [Edd.] specifically ascribes them to the editors.

NEW YORK, N. Y. MORRIS R. COHEN
I. E. DRABKIN

On January 28, 1947, shortly after our manuscript was made ready for the press, Professor Cohen died. The work had his final approval substantially as it now appears. But on numerous questions of detail that have arisen I have had to make the decisions without the benefit of his judgment. I can only hope that in so doing I have not seriously misrepresented his thought. In any case, the reader will understand that the responsibility for errors arising in this way is wholly mine.

I wish to acknowledge most gratefully our debt to all who so generously assisted us in our task. Professors Giorgio de Santillana and Benedict Einarson read the work in manuscript. Drs. Edward Rosen, Lee Lorch, and George Rosen and Professor Abraham Mazur examined portions of it in proof. Professors Abraham S. Halkin and S. Gerald Hindin answered inquiries dealing with their respective fields, Arabic and chemistry. Professor Gregory D. Walcott, the General Editor of the Series, in addition to handling countless details connected with the publication, read the proof sheets of the whole work. The efforts of these scholars have made it possible to eliminate many inaccuracies and to improve the text in many places.

I must also acknowledge my indebtedness to the Carnegie Foundation and to the Johns Hopkins University. As Carnegie Fellow in Greek and Roman Science at the Johns Hopkins University in 1941 and 1942 I was able to do a large part of my share of the present work. The fellowship also gave me the opportunity of studying at the university's Institute of the History of Medicine with such distinguished scholars as Henry E. Sigerist, Ludwig Edelstein, and Owsei Temkin.

At every stage of the progress of this work, from the preparation of the manuscript to the final proofreading and the making of the index, I have relied heavily on the knowledge, skill, and patient industry of my wife, Miriam Drabkin. My debt to her is far greater than I can readily express.

I. E. Drabkin

New York, N. Y.
May, 1948.

ACKNOWLEDGMENTS

The editors are deeply grateful to the following publishers for granting permission to quote from the designated works published by them.

EDWARD ARNOLD & CO.

Kenneth C. Bailey, *The Elder Pliny's Chapters on Chemical Subjects.* London, 1929–1932.

BULLETIN OF THE HISTORY OF MEDICINE.

Owsei Temkin and William L. Straus, Jr., "Galen's Dissection of the Liver and of the Muscles Moving the Forearm." *Bulletin of the History of Medicine* 19 (1946) 167-176.

THE CLARENDON PRESS, OXFORD.

Aristoxenus, *Elements of Harmonics.* Translated by H. S. Macran, Oxford, 1902.

Cyril Bailey, *Epicurus, the Extant Remains.* Oxford, 1926.

J. I. Beare, *Greek Theories of Elementary Cognition from Alcmaeon to Aristotle.* Oxford, 1906.

Thomas L. Heath, *Aristarchus of Samos.* Oxford, 1913.

Thomas L. Heath, *History of Greek Mathematics.* Oxford, 1921.

Lucretius, *On the Nature of Things.* Translated by Cyril Bailey, Oxford, 1921.

Charles Singer, *A Short History of Biology.* Oxford, 1931.

Charles Singer, *A Short History of Science.* Oxford, 1941.

The Oxford Translation of Aristotle. Edited by W. D. Ross.

De Audibilibus. Translated by T. Loveday and E. S. Forster, Oxford, 1913.

De Caelo. Translated by J. L. Stocks, Oxford, 1922.

De Generatione Animalium. Translated by Arthur Platt, Oxford, 1910.

De Incessu Animalium. Translated by A. S. L. Farquharson, Oxford, 1912.

De Memoria et Reminiscentia. Translated by J. I. Beare, Oxford, 1908.

De Partibus Animalium. Translated by William Ogle, Oxford, 1911.

De Plantis. Translated by E. S. Forster, Oxford, 1912.

De Sensu. Translated by J. I. Beare, Oxford, 1908.

De Somniis. Translated by J. I. Beare, Oxford, 1908.

Historia Animalium. Translated by D'Arcy Wentworth Thompson, Oxford, 1910.
Mechanica. Translated by E. S. Forster, Oxford, 1913.
Metaphysica. Translated by W. D. Ross, Oxford, 1928.
Meteorologica. Translated by E. W. Webster, Oxford, 1923.
Physica. Translated by R. P. Hardie and R. K. Gaye, Oxford, 1930.
Problemata. Translated by E. S. Forster, Oxford, 1927.
R. T. Gunther, *The Greek Herbal of Dioscorides*. Oxford, 1934. Reprinted by permission of the Executors.

ROBERT E. DENGLER.
Theophrastus. *De Causis Plantarum I*. Edited and translated by Robert E. Dengler, Philadelphia, 1927.

J. M. DENT & SONS, LTD., and
E. P. DUTTON & CO., INC.
Thomas L. Heath, *Greek Astronomy*. London and New York, 1932.

HARVARD UNIVERSITY PRESS, CAMBRIDGE, MASS.
Vitruvius, *On Architecture*. Translated by M. H. Morgan, Cambridge, Mass.: Harvard University Press, 1914.
The Loeb Classical Library, Cambridge, Mass.: Harvard University Press.
Athenaeus, *The Deipnosophists*. Translated by Charles B. Gulick, London, 1928.
Celsus, *On Medicine*. Translated by W. G. Spencer, London, 1935.
Sextus Julius Frontinus, *On the Water Supply of the City of Rome*. Translated by C. E. Bennett, London, 1925.
Galen, *On the Natural Faculties*. Translated by A. J. Brock, London, 1916.
The Greek Anthology. Translated by W. R. Paton, London, 1918.
Herodotus. Translated by A. D. Godley, London, 1921.
Hippocrates. Translated by W. H. S. Jones and E. T. Withington, London, 1923-1931. Selections from Epidemics; On Airs, Waters and Places; On Ancient Medicine; On Fractures; On Joints; On the Nature of Man; On Regimen in Acute Diseases; On Wounds in the Head; Prognostic; Regimen; Regimen in Health.
Plato, *Ion*. Translated by W. R. M. Lamb, London, 1925.
Plato, *Timaeus*. Translated by R. G. Bury, London, 1929.
Plutarch, *Lives*. Translated by Bernadotte Perrin, London, 1917.
Strabo, *Geography*. Translated by H. L. Jones, London, 1917.
Theophrastus, *An Enquiry into Plants*. Translated by Sir Arthur Hort, London, 1916.

Theophrastus, *Concerning Odours*. Translated by Sir Arthur Hort, London, 1916.

Theophrastus, *Concerning Weather Signs*. Translated by Sir Arthur Hort, London, 1916.

Journal of Chemical Education.

Earle Radcliffe Caley, "The Leyden Papyrus X." *Journal of Chemical Education* 3 (1926) 1149–1166.

Earle Radcliffe Caley, "The Stockholm Papyrus." Ibid. 4 (1927) 979–1002.

THE MACMILLAN COMPANY.

Sir Archibald Geikie and John Clarke, *Physical Science in the Time of Nero*. London, 1910.

G. M. Stratton, *Theophrastus and the Greek Physiological Psychology before Aristotle*. New York, 1917.

THE MACMILLAN COMPANY, CAMBRIDGE UNIVERSITY PRESS DEPARTMENT.

Aristotle, *De Anima*. Edited and translated by R. D. Hicks, Cambridge, 1907.

Thomas L. Heath, *The Works of Archimedes*. Cambridge, 1897.

Thomas L. Heath, *The Method of Archimedes*. Cambridge, 1912.

Thomas L. Heath, *Diophantus of Alexandria*. Second edition, Cambridge, 1910.

Thomas L. Heath, *The Thirteen Books of Euclid's Elements*. Cambridge, 1926.

UNIVERSITY OF MICHIGAN PRESS.

Nicomachus of Gerasa, *Introduction to Arithmetic*. Translated by M. L. D'Ooge with studies in Greek arithmetic by F. E. Robbins and L. C. Karpinski, New York, 1926.

TABLE OF CONTENTS

MATHEMATICS

The Greeks achieved greater success in pure mathematics than in any other branch of science. Not only did they lay the enduring foundations for the subsequent development of the subject, but they made notable and lasting contributions to its various divisions. Recent researches have increased our knowledge of Egyptian and Babylonian mathematics before and after the rise of the Greeks and have indicated remarkable skill in calculation and in the solution of various types of problem. But all the evidence indicates that the ideal of rigorously deductive proof, the method of developing a subject by a chain of theorems based on definitions, axioms, and postulates, and the constant striving for complete generality and abstraction are the specific contributions of the Greeks to mathematics.

From the modest beginnings at the opening of the sixth century B.C.[1] there is a rapid development in number theory and in geometry, culminating in the work of Euclid, Archimedes, and Apollonius. This development occupies, roughly, 350 years. Thereafter the discoveries are less striking, though there are, from time to time, great names, e.g., Hipparchus, Ptolemy, Pappus, and Diophantus. It is, however, not until the period from Descartes to our own day—also a period of about 350 years—that we have a comparable surge of creative activity in mathematics.

The rise of pure mathematics among the Greeks was accompanied by applications to astronomy and various branches of physics, particularly mechanics, hydrostatics, and optics.

A word on the fields of mathematics as they were developed by the Greeks may be appropriate here. Arithmetic—the theory of numbers, as distinguished from logistics, the art of calculating and of solving particular problems involving calculation—was prominent among the interests of the Pythagorean school. Contributions were made to the classification of numbers according to their composition, a system of representing numbers by points or figures was devised, and various types of proportion were analyzed. The discovery of incommensurability, probably in the fifth century B.C., immediately showed the inadequacy of the type of number theory that had been previously developed and led to a wider study of geometry and to the inclusion of number theory within the framework of geometry. It should be remembered, however, that almost throughout the whole Greek period interest in number also developed in the direction of mystical numerology.

It is in geometry that we find the greatest achievements of Greek mathematics. From the barest beginnings, probably transmitted in the first instance from Egypt, there is a steady flow of discovery, reaching its high-water mark in the work of Archimedes. The theory of proportion is generalized to include incommensurable magnitudes. The powerful "method of exhaustion" opens new avenues of investigation. In Euclid's *Elements* the achievements in elementary geometry are set forth in systematic form. New fields of geometrical discovery are cultivated by Archimedes and Apollonius. Spherical geometry and trigonometry, indispensable in astronomy and mathematical geography, are greatly advanced by Hipparchus in the second century B.C. and later by Menelaus and Ptolemy.

The field of "algebra," as we call it, is relatively neglected during this remarkable increase of geometrical knowledge, and it is only in the work of Diophantus, probably in the third

[1] Our knowledge of the early period is quite fragmentary and is usually based on accounts written many centuries later. Thus the ascription of particular discoveries to particular individuals should be read with reservations.

century A.D., that we have any systematic treatment of this branch among the Greeks. The Greeks never achieved a fruitful union of algebra and geometry such as came in modern times with the rise of analytical geometry. This is important for a proper estimate of Greek mathematics.

How profound was the influence of Greek mathematics on the resurgence of science in the sixteenth and seventeenth centuries is amply demonstrated by the works that were produced in these centuries.

In the selections given here an attempt has been made to trace some of the outstanding features of the Greek mathematical development. The availability to English readers of the late Sir Thomas L. Heath's *History of Greek Mathematics* and of Ivor Thomas's translation of Greek mathematical texts in the Loeb Classical Library has prompted the editors to devote less space to this field in order to give additional space to fields in which the material is not so readily available.

THE DIVISIONS OF MATHEMATICS PURE AND APPLIED

Proclus Diadochus, *Commentary on Euclid's Elements I*, pp. 38.1–42.8 (Friedlein)[1]

Such is the doctrine of the Pythagoreans and their fourfold division of mathematics.[2] But others, among them Geminus, prefer to divide mathematics in another way. They consider mathematics on the one hand as concerned with things conceived by the mind, and on the other hand as concerned with and applied to things perceived by the senses. By things conceived by the mind they mean those which the psyche makes objects of contemplation by itself, when it completely divorces itself from forms connected with matter. In that division of mathematics which has to do with things conceived by the mind they place the two most basic

---

[1] Proclus the Neoplatonist (A.D. 410–485) was born at Constantinople but spent most of his life at Athens, where he headed the Academy. Of his voluminous writings those extant include commentaries on Platonic dialogues, works on theology, astronomy, and physics, and the commentary on Euclid, *Elements* I. This latter work is very valuable for its information on the history of geometry. It contains numerous references to important works no longer extant. Thus, the selection given here is in part drawn from the work of Geminus (see p. 117) on mathematical theory. The discussion of pre-Euclidean geometry in the selection given below (p. 33) is based, directly or indirectly, on the *History of Geometry* by Eudemus, a pupil of Aristotle.

[2] I.e., arithmetic, music, geometry, and astronomy. The earliest extant reference to this division is by Archytas, a Pythagorean philosopher of the fourth century B.C. (see p. 35):

"They [the mathematicians] have given us clear knowledge about the speed of the stars, and their risings and settings, about geometry and numbers and spheric and, not least, music. For these studies seem to be sisters." (Diels, *Fragmente der Vorsokratiker* I⁵. 331. 5–8.)

A similar division is made in Plato, *Republic* 525 A–530 D, with a distinction between logistic and arithmetic and the addition of solid geometry (stereometry).

The fourfold division is repeated by various Neoplatonic philosophers, e.g., Nicomachus and Theon of Smyrna, and is continued in the medieval quadrivium.

The term "spheric" in this connection refers to spherical geometry in its special relation to the circular motions of the heavenly bodies.

and important branches, arithmetic and geometry. On the other hand, of that part of mathematics which devotes its attention to objects perceived by the senses they list six branches: mechanics, astronomy, optics, geodesy, canonics, and logistics.

They do not think, as some others do, that military science should be considered as one of the branches of mathematics, though, to be sure, it sometimes uses logistics, for example in the enumeration of companies,[1] and at other times geodesy, for example in the division and measurement of areas. Similarly, there is even more reason why neither history nor medicine may be considered a part of mathematics, even though historians often use mathematical theorems, as when they give the location of climata[2] or calculate the size and the diameter or circumference of cities.[3] Physicians, too, by similar methods clarify many of the problems relevant to their field. For the importance of astronomy in medicine is made clear by Hippocrates[4] and, in fact, by all who have discoursed on seasons and places. In the same way, then, the military tactician, without being a mathematician himself, uses the theorems of mathematics when, on occasion, he wishes to make the number of his troops appear as small as possible, and so arranges them in a circle,[5] or, again, when he wishes to make the number appear as large as possible, and so arranges them in a square or pentagon or some other polygon.

While these are the branches of mathematics as a whole, geometry, in its turn, is divided into the theory of the plane and stereometry.[6] For the study of points and lines cannot constitute a separate branch since the figures they form must be either plane or solid. Everywhere it is the task of geometry, in the case both of planes and solids, either to construct, or to compare or divide what has been constructed.

Similarly, arithmetic is divided into the study of linear, of plane, and of solid numbers.[7] For it treats of the kinds of numbers in and of them-

---

[1] Reading λόχων for λόγων (of Friedlein).

[2] The reference is to zones of latitude. See p. 171, below.

[3] Omitting καὶ διαμέτρους ἢ περιμέτρους (of Friedlein).

[4] See the treatise *On Airs, Waters, and Places* of the Hippocratic Collection.

[5] An application of the proposition that the area of a circle is greater than the area of any polygon of equal perimeter.

[6] I.e., solid geometry.

[7] One dot represents 1, two dots represent 2 (a linear number, for the dots define a line), three dots represent 3 (a plane number and triangular, ∴ , as are 6, ∴̇·, 10, 15, etc.), four dots represent 4 (a square number when all the dots are in the same plane, but a solid pyramidal number when one dot is outside the plane of the other three); so also there are oblong, pentagonal, hexagonal, cubic numbers, etc. The idea of representing numbers in this way, which tradition ascribed to Pythagoras, led to the discovery of many propositions in elementary number theory (see pp. 7–9).

selves, as they proceed from unity, the generation of plane numbers both similar and dissimilar,[1] and the progression to the third dimension.

Geodesy and logistics[2] are analogous to geometry and arithmetic, respectively, but are concerned with things perceived by the senses, not with numbers or figures as conceived by the mind. For it is not the business of geodesy to measure the cylinder or the cone, but to measure mounds as if they were cones, or wells as if they were cylinders. Geodesy deals, then, with straight lines, not as conceived by the mind, but as perceived by the senses, sometimes more sharply defined, as in the case of the rays of the sun, sometimes less sharply defined, as in the case of cords or a plumbline.

Similarly, one who employs logistics does not consider the properties of numbers by themselves, but always in connection with perceptible objects. Hence he gives to numbers a name after the objects that are being computed and thus speaks of *melites* and *phialites*.[3] He does not admit an [absolute] minimum unit, as does the arithmetician, yet he does use a minimum unit in relation to some subject matter. Thus one man serves the logistician as the unit of a crowd of men.

Again, optics and canonics are derived from geometry and arithmetic, respectively. The science of optics makes use of lines as visual rays and makes use also of the angles formed by these lines. The divisions of optics are: (*a*) the study which is properly called optics and accounts for illusions in the perception of objects at a distance, for example, the apparent convergence of parallel lines or the appearance of square objects at a distance as circular; (*b*) catoptrics, a subject which deals, in its entirety, with every kind of reflection of light and embraces the theory of images; (*c*) scenography (scene-painting), as it is called, which shows how objects at various distances and of various heights may so be represented in drawings that they will not appear out of proportion and distorted in shape.

The science of canonics[4] deals with the observed proportions of the notes of the musical scales and investigates the divisions of the mono-

---

[1] I.e., numbers of the form a^2 (square numbers) and ab (oblong and prolate numbers), respectively.

[2] Logistic was the art of calculating and of solving particular problems involving calculation, as distinguished from arithmetic, the science or theory of numbers. On this distinction see Plato, *Republic* 525 A, the scholium to *Charmides* 165 E, and T. L. Heath, *History of Greek Mathematics* I. 13–15.

[3] The reference is to problems involving, in the former case a number of apples (or of sheep), and in the latter case a number or weight of liquid measures (bowls). For examples of these problems see pp. 25–26.

[4] I.e., the science of musical intervals or "harmonics," as it was often called, as distinguished from what Plato calls "popular music," the art of singing or of playing instruments.

chord. It makes use, throughout, of sense perception and attributes greater importance to the ear than to the intellect, as Plato says.[1]

In addition to these there is the science called mechanics which is a division of the study of material objects perceived by the senses. The science of mechanics embraces: (*a*) the manufacture of engines useful in war, for example, the engines of defense which Archimedes is said to have constructed against the besiegers of Syracuse;[2] (*b*) the manufacture of wonderful devices, including those based on (1) air currents, e.g., devices such as Ctesibius and Hero[3] describe, (2) weights (lack of equilibrium producing motion, and equilibrium producing rest, according to the definition in the *Timaeus*),[4] (3) ropes and cables, by means of which the motion of living beings may be imitated; (*c*) the study of equilibrium, in general, and of so-called centers of gravity; (*d*) sphere construction, for depicting the revolutions of the heavenly bodies, a field in which Archimedes worked;[5] (*e*) in general, the whole subject of the kinetics of material bodies.

There remains astronomy, which is concerned with cosmic motions, the sizes and forms of the heavenly bodies, their illumination, their distances from the earth, and all other subjects of this sort. It makes wide use of sense perception but also has close connection with physical theory. Its branches are: (*a*) gnomonics, which is concerned with the measuring of the hours by the proper placing of gnomons; (*b*) meteoroscopy, which investigates the different elevations and distances of stars, and sets forth numerous other theorems of various sorts in the field of astronomy; (*c*) the science of dioptrics, which, with the use of the proper instruments,[6] investigates the positions[7] of the sun, moon, and the other stars.

The above is the account of the branches of mathematics that we have found in the ancient writers.

ARITHMETIC OR THE THEORY OF NUMBERS

The chief extant sources of our knowledge of Greek arithmetic are Books VII–X of Euclid's *Elements*, the *Introduction to Arithmetic* of Nicomachus (end of first century A.D.), and the *Arithmetic* of Diophantus (probably as late as the third century A.D.). There are, in addition, various commentaries, e. g., those of Theon of Smyrna (first half of the second century

---

[1] In *Republic* 531 B Plato criticizes empiricism in musical science (as opposed to the pure arithmeticism of the Pythagoreans). See p. 302, n. 5, below.

[2] See p. 316.

[3] See pp. 326–336.

[4] Cf. *Timaeus* 57 E ff., where, however, rest and motion are referred to equality and inequality in general.

[5] On the planetarium of Archimedes see p. 142.

[6] On the dioptra see pp. 139ff., 336ff. Dioptrics is not to be understood here in its later restricted sense as concerned with the phenomena of refraction.

[7] Reading ἐποχάς for ε̄ ἀποχάς (of Friedlein).

A.D.) and of Iamblichus (first half of fourth century A.D.), some material of arithmetical interest in Pappus's compilation made in the latter part of the third century A.D. (e. g., in Book III), as well as numerous arithmetical discussions in works on musical theory.

Though there is considerable question as to how much of our material goes back to the period before Euclid, it is clear that much of it does. We find, to give a single example, that a proof ascribed to Archytas is practically the same as that given by Euclid (see Boethius, *On Music* III. 11; Euclid, *Sectio Canonis* 3).

It should be pointed out that, while Euclid shows a purely mathematical interest in the rigorous demonstration of theorems about numbers, Nicomachus and Iamblichus are very largely influenced by the mystic numerology of the earlier Pythagoreans and of Plato.

The arithmetic of Diophantus is, as we shall see, unique in Greek mathematics. With its introduction of an elementary type of symbolism and its operational methods, it is akin to what we call "algebra."

A detailed account of the history of Greek systems of numeration is out of place here.[1] It may be noted, however, that an alphabetic system was used by the Greek mathematicians, as by the Phoenicians and Hebrews, and that the system was basically decimal,[2] though in the measurement of angles and in certain astronomical computations the Babylonian sexagesimal system was employed. It is impossible to say to what extent, if any, the Greek system of numeration, with its absence of zero, was an obstacle to progress in number theory. It will be seen, however, that multiplication and division involving extremely large numbers and cumbersome fractions, both decimal and sexagesimal, were carried through, and remarkably accurate methods of approximating the square roots and cube roots of non-square and non-cubic numbers were practiced.

Proportions

Archytas, Frag. 2 (Diels, *Fragmente der Vorsokratiker* I⁵. 334.16–335.13)[3]

There are three proportions in music, the arithmetic, the geometric, and the subcontrary or so-called harmonic. We have an arithmetic proportion when three terms are related with respect to excess, as follows: the first exceeds the second by as much as the second exceeds the third. In this proportion the ratio of the two larger terms is smaller, while the ratio of the two smaller terms is larger.[4] We have a geometric proportion when the first term is to the second as the second is to the third. The ratio of the two larger terms is equal to that of the two smaller. The subcontrary proportion, which we call harmonic, is that in which the terms are such that if the first exceeds the second by a certain part of the first,

[1] See T. L. Heath, *History of Greek Mathematics* I, ch. 2.

[2] "Why do all men, both foreigners and Greeks, count to ten? Why do they not count to some other number like 2, 3, 4, or 5, and then, beginning again, say 'one and five,' 'two and five,' etc., as they say 'one and ten' [ἕνδεκα], 'two and ten' [δώδεκα]? . . . Or is it that all men were born with ten fingers? Thus having, as it were, pebbles to the number of their fingers, they counted everything else according to this number." [Aristotle], *Problemata* XV. 3.

[3] See p. 35. The quotation is from a commentary on Ptolemy's *Harmonics* probably by Porphyrius the Neoplatonist, pupil of Plotinus in the third century.

[4] If $a - b = b - c$, then $a/b < b/c$ (where $a > b > c > 0$).

the second will exceed the third by the same part of the third.[1] In this proportion the ratio of the two larger terms is larger, while the ratio of the two smaller terms is smaller.[2]

Figured Numbers

Nicomachus, *Introduction to Arithmetic* II. 8–12. Translation of M. L. D'Ooge, New York, 1926[3]

Now a triangular number is one which, when it is analyzed into units, shapes into triangular form the equilateral placement of its parts in a plane. 3, 6, 10, 15, 21, 28, and so on, are examples of it. . . .[4]

The square is the next number after this, which shows us no longer 3, like the former, but 4, angles in its graphic representation, but is none the less equilateral. Take, for example, 1, 4, 9, 16, 25, 36, 49, 64, 81, 100. . . .[5]

This number also is produced if the natural series is extended in a line, increasing by 1, and no longer the successive numbers are added to the numbers in order, as was shown before,[6] but rather all those in alternate places, that is, the odd numbers. For the first, 1, is potentially the first

---

[1] If $a - b = (1/m)a$ and $b - c = (1/m)c, a/b > b/c$. [For $a/b = m/(m - 1)$ and $b/c = (m + 1)/m$, where m is a positive integer.]

[2] That these were the three fundamental means of the Pythagoreans is confirmed by Iamblichus in his commentary on Nicomachus' *Introduction to Arithmetic*, p. 100.19–25 (Pistelli):

"In antiquity, in the time of Pythagoras and the mathematicians of his school, there were only three means, the arithmetic, the geometric, and, third in order, that which was once called subcontrary, but was renamed harmonic by Archytas and Hippasus because it was seen to embrace harmonious and concordant ratios."

The basic musical consonances of the fourth, fifth, and octave (4:3, 3:2, and 2:1, respectively) are given by strings of lengths 3, 4, and 6. These numbers are in harmonic proportion (see pp. 294 ff., below).

Nicomachus, *Introduction to Arithmetic* II. 28 and Pappus, *Mathematical Collection* III. 23 give various additional means.

[3] Nicomachus of Gerasa probably flourished at the close of the first century A.D. His extant works are an *Introduction to Arithmetic* and a *Manual of Harmony*. There is, in addition, an extant metaphysical treatise on arithmetic which, if not by Nicomachus, seems to be based on one of his lost works. Other works ascribed to him, but no longer extant, deal with music, Platonic exegesis, and the life of Pythagoras. The interest of Nicomachus, even in his *Introduction to Arithmetic*, is not merely mathematical, but philosophical, and he is an important figure in the Pythagorean tradition of numerology. The elementary theorems on number that may be evolved from a consideration of figured numbers seem to go back to the early Pythagorean school. [Edd.]

[4] ·, ∴, ∴̣̇., ⁘, etc. The numbers are of the form $\frac{n}{2}(n + 1)$. [Edd.]

[5] ·, ∷, ⁞⁞⁞, etc. The numbers are of the form n^2. [Edd.]

[6] As in the case of triangular numbers, 1, 1 + 2, 1 + 2 + 3, etc. [Edd.]

square; the second, 1 plus 3, is the first in actuality; the third, 1 plus 3 plus 5, is the second in actuality. . . .[1]

The pentagonal number is one which likewise upon its resolution into units and depiction as a plane figure assumes the form of an equilateral pentagon. 1, 5, 12, 22, 35, 51, 70 and analogous numbers are examples. . . .[2]

The hexagonal, heptagonal, and succeeding numbers will be set forth in their series by following the same process, if from the natural series of numbers there be set forth series with their differences increasing by 1. . . .

To remind us, let us set forth rows of the polygons, written in parallel lines. . . .

Triangles	1	3	6	10	15	21	28	36	45	55
Squares	1	4	9	16	25	36	49	64	81	100
Pentagonals	1	5	12	22	35	51	70	92	117	145
Hexagonals	1	6	15	28	45	66	91	120	153	190
Heptagonals	1	7	18	34	55	81	112	148	189	235[3]

Theon of Smyrna, pp. 35.7–38.2 (Hiller)

It is a special property of square numbers that they are exactly divisible by 3 or leave a remainder of 1; again, that they are exactly divisible by 4 or leave a remainder of 1. And a square number either (1) leaves a remainder of 1 when divided by 3, and is exactly divisible by 4 (e.g., the number 4); or (2) leaves a remainder of 1 when divided by 4, and is exactly divisible by 3 (e.g., the number 9); or (3) is exactly divisible by both 3 and 4 (e.g., the number 36); or (4) is exactly divisible by neither 3 nor 4, but in each case leaves a remainder of 1 (e.g., the number 25).

---

[1] I.e., as we should say, $1 + 3 + 5 + \ldots + (2n - 1) = n^2$. [Edd.]

[2] The series is $1, 1 + 4, 1 + 4 + 7, 1 + 4 + 7 + 10$, etc., and the numbers are of the form $\frac{n}{2}(3n - 1)$. [Edd.]

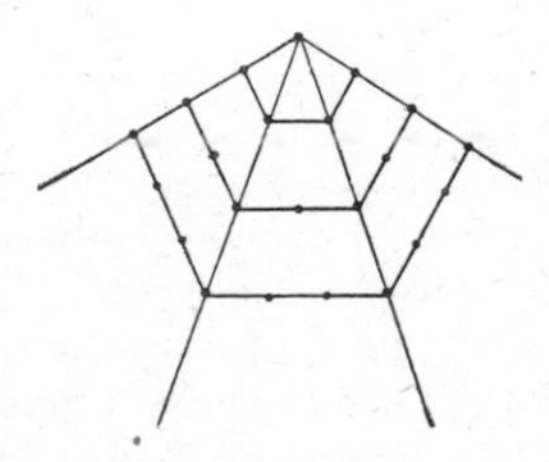

[3] Nicomachus gives various theorems based on this classification of numbers, e.g., every square after 1 is the sum of two consecutive triangles. $\frac{n}{2}(n - 1) + \frac{(n + 1)n}{2} = n^2$. Similarly, every pentagonal number is the sum of a triangular and a square number, every hexagonal number is the sum of a triangular and a pentagonal number, and so on. That is, every number greater than 1 in the table is the sum of the number immediately above it and the triangular number in the preceding column. [Edd.]

Iamblichus, *Commentary on Nichomachus' Introduction to Arithmetic*, p. 62.10–18 (Pistelli)

In the representation of polygonal numbers two sides in all cases remain the same, and are produced; but the additional sides are included by the application of the gnomon and a ways change.[1] There is one such additional side in the case of the triangle, two in the case of the square, three in the case of the pentagon, and so on indefinitely, the difference between the number of sides of the polygon and the number of sides which change being 2.

Plutarch, *Platonic Questions* 1003 F

Take any triangular number, multiply it by 8 and add 1. The result is a square number.[2]

---

[1] The gnomon is used in the theory of figured numbers as the difference between two consecutive numbers of the same type, e.g., two successive triangles, squares, or pentagonal numbers. The word designates the style of the sundial, and in fact any upright that casts a shadow. A carpenter's square to draw right angles is also called a gnomon; and in the theory of figured numbers, that is the shape of the gnomon for successive squares (see the second figure below). Whether for that reason or not, the word became general in the sense indicated.

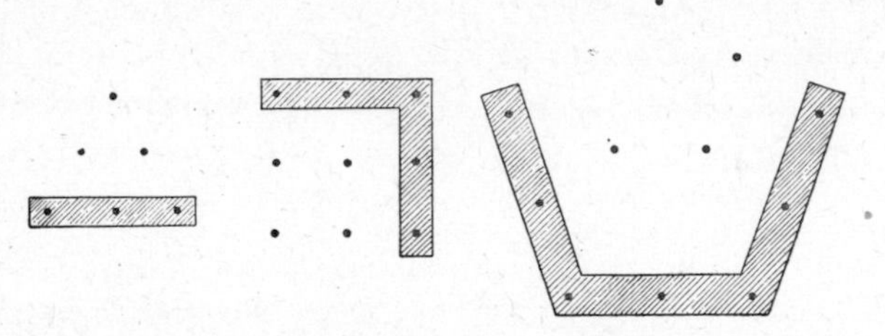

The gnomon is the number within the shaded area.

Iamblichus' proposition is that in an a-gonal number the number of sides that change is $a - 2$. It is readily shown that the successive gnomons of a polygon of a sides are $1, 1 + (a - 2), 1 + 2(a - 2), 1 + 3(a - 2)$, etc., and that the a-gonal number of side n is the sum of n terms in this series, or $n + \frac{1}{2}n(n-1)(a - 2)$. Diophantus ascribes this method of defining polygonal numbers to Hypsicles (second century B.C.).

[2] This theorem is also used in Diophantus, *Arithmetica* IV. 38. How far back it goes is doubtful, but it can easily be shown from a consideration of figured numbers. E. g., in the case of $5^2 = 8 \cdot 3 + 1$, we have four oblong numbers $2 \cdot 3$. And each oblong number of the form $n(n + 1)$ is twice a triangular number, as is obvious from a figure ⁘ (see p. 7, n. 4). The theorem is equivalent to $8 \cdot \frac{1}{2}n(n + 1) + 1 = (2n + 1)^2$. In the fragment on Polygonal Numbers Diophantus generalizes this theorem by proving that if P is any a-gonal number $8P(a - 2) + (a - 4)^2 =$ a square.

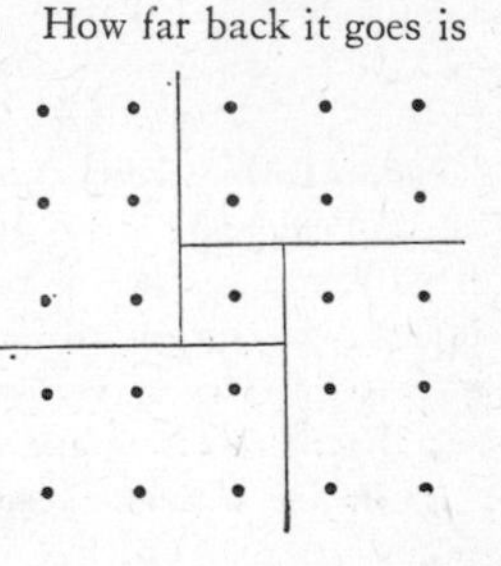

Perfect Numbers

Euclid, *Elements* VII, Definition 22; IX, Proposition 36. Translation of T. L. Heath, *The Thirteen Books of Euclid's Elements* II. 278, 421–424 (Cambridge, 1926)

Definition: A perfect number is that which is equal to its own parts.[1]

Proposition 36

If as many numbers as we please beginning from an unit be set out continuously in double proportion, until the sum of all becomes prime, and if the sum multiplied into the last make some number, the product will be perfect.

For let as many numbers as we please, *A*, *B*, *C*, *D*, beginning from an unit be set out in double proportion, until the sum of all becomes prime, let *E* be equal to the sum, and let *E* by multiplying *D* make *FG*;
I say that *FG* is perfect.

For, however many *A*, *B*, *C*, *D* are in multitude, let so many *E*, *HK*, *L*, *M* be taken in double proportion beginning from *E*;
therefore, *ex aequali*, as *A* is to *D*, so is *E* to *M*. [VII. 14][2]

Therefore the product of *E*, *D* is equal to the product of *A*, *M*. [VII. 19]

And the product of *E*, *D* is *FG*;
therefore the product of *A*, *M* is also *FG*.

Therefore *A* by multiplying *M* has made *FG*;
therefore *M* measures[3] *FG* according to the units in *A*.

And *A* is a dyad;

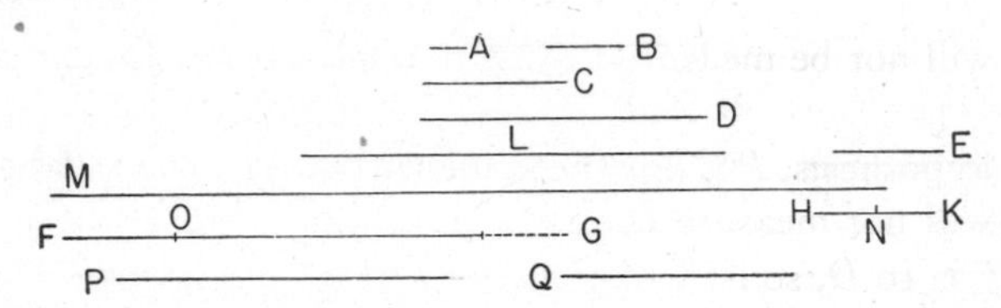

therefore *FG* is double of *M*.

---

[1] I.e., a number equal to the sum of its factors other than itself. Thus 6 (= 1 + 2 + 3) and 28 (= 1 + 2 + 4 + 7 + 14) are the first two perfect numbers. Euclid here proves that if $(2^n - 1)$, which is the sum of n terms of the series $1, 2, 2^2, \ldots 2^{n-1}$, is prime, then $2^{n-1}(2^n - 1)$ is a perfect number. This formula can be shown to include all even perfect numbers; it is not known whether any odd perfect numbers exist. Besides 6 and 28 the ancients knew of 496 and 8128 as perfect. Eight other perfect numbers are now known, the last being $2^{126}(2^{127} - 1)$. There is no extant reference to perfect numbers, in the sense here used, before Euclid. In early Pythagorean arithmetic, however, the numbers 3 and 10, because of certain of their properties, were called perfect. Cf. p. 12. [Edd.]

[2] If $a:b = b:c = c:d$, and $e:f = f:g = g:h\ (= a:b)$, then $a:d = e:h$. [Edd.]

[3] In line with the geometrical character of the Euclidean theory of numbers, this is the regular term for "divides without remainder." [Edd.]

But M, L, HK, E are continuously double of each other;
therefore E, HK, L, M, FG are continuously proportional in double proportion.

Now let there be subtracted from the second HK and the last FG the numbers HN, FO, each equal to the first E;
therefore, as the excess of the second is to the first, so is the excess of the last to all those before it. [IX. 35][1]

Therefore, as NK is to E, so is OG to M, L, KH, E.

And NK is equal to E;
therefore OG is also equal to M, L, HK, E.

But FO is also equal to E,
and E is equal to A, B, C, D and the unit.

Therefore the whole FG is equal to E, HK, L, M and A, B, C, D and the unit;
and it is measured by them.

I say also that FG will not be measured by any other number except A, B, C, D, E, HK, L, M and the unit.

For, if possible, let some number P measure FG,
and let P not be the same with any of the numbers A, B, C, D, E, HK, L, M.

And, as many times as P measures FG, so many units let there be in Q;
therefore Q by multiplying P has made FG.

But, further, E has also by multiplying D made FG;
therefore, as E is to Q, so is P to D. [VII. 19]

And, since A, B, C, D are continuously proportional beginning from an unit,
therefore D will not be measured by any other number except A, B, C. [IX. 13][2]

And, by hypothesis, P is not the same with any of the numbers A, B, C;
therefore P will not measure D.

But, as P is to D, so is E to Q;
therefore neither does E measure Q. [VII. Def. 20]

And E is prime;
and any prime number is prime to any number which it does not measure. [VII. 29]

Therefore E, Q are prime to one another.

But primes are also least, [VII. 21][3]

---

[1] See p. 23. [Edd.]

[2] If $1, a, a_2, \ldots a_n$ be a geometrical progression, and if a be prime, a_n will not be measured by any numbers except the preceding terms of the series (see Heath on Euclid IX. 13).

[3] That is, if a and b are prime to each other, the ratio $a:b$ is in its lowest terms (see Heath on Euclid VII. 21).

and the least numbers measure those which have the same ratio the same number of times, the antecedent the antecedent and the consequent the consequent; [VII. 20][1]
and, as E is to Q, so is P to D;
therefore E measures P the same number of times that Q measures D.

But D is not measured by any other number except A, B, C;
therefore Q is the same with one of the numbers A, B, C.

Let it be the same with B.

And, however many B, C, D are in multitude, let so many E, HK, L be taken beginning from E.

Now E, HK, L are in the same ratio with B, C, D;
therefore, *ex aequali*, as B is to D, so is E to L. [VII. 14]

Therefore the product of B, L is equal to the product of D, E. [VII. 19]

But the product of D, E is equal to the product of Q, P;
therefore the product of Q, P is also equal to the product of B, L.

Therefore, as Q is to B, so is L to P. [VII. 19]

And Q is the same with B;
therefore L is also the same with P;
which is impossible, for by hypothesis P is not the same with any of the numbers set out.

Therefore no number will measure FG except A, B, C, D, E, HK, L, M and the unit.

And FG was proved equal to A, B, C, D, E, HK, L, M and the unit; and a perfect number is that which is equal to its own parts; [VII. Def. 22]
therefore FG is perfect. Q.E.D.

Theon of Smyrna, p. 46.4–15 (Hiller)

Over-perfect (ὑπερτέλεοι) numbers are those the sum of whose factors is greater than the whole number, e.g., 12. For its half is 6, its third 4, its fourth 3, its sixth 2, and its twelfth 1. The sum of the factors is 16, which is greater than the original number 12.

Defective (ἐλλιπεῖς) numbers are those the sum of whose factors is less than the original number, e.g., 8. For its half is 4, its fourth 2, and its eighth 1. The same is true of 10, which the Pythagoreans called perfect for another reason. . . . The number 3 is also called perfect.

---

[1] That is, if a, b, c, and d are integers, and a/b is a fraction in its lowest terms, and $a{:}b = c{:}d$, then $a = (1/n)c$ and $b = (1/n)d$, where n is some integer (see Heath on Euclid VII. 20).

Friendly (Amicable) Numbers

Iamblichus, *Commentary on Nichomachus' Introduction to Arithmetic*, p. 35.1–7 (Pistelli)

They [the Pythagoreans] call a pair of numbers "friendly," endowing them with virtues and good qualities, e.g., 284 and 220, when the parts of each generate the other.[1] This accords with Pythagoras' definition of friendship. For when he was asked, "what is a friend?" he answered "another self." And this is clearly the case with these numbers.

Incommensurable Numbers

Aristotle, *Prior Analytics* I. 23 (41*a*23–27)

For all who draw a conclusion *per impossibile*[2] first infer, by syllogistic reasoning, a false proposition, and then demonstrate their original conclusion *ex hypothesi*, when something impossible results from the assumption of the contradictory [of this conclusion]. For example, they prove that the diagonal [of the square] is incommensurable [with the side] by showing that on the assumption that it is commensurable odd numbers are equal to even numbers.

The proof referred to by Aristotle is found in Euclid, *Elements* X, Appendix 27, a late interpolation there according to the editor, J. L. Heiberg. The proof may be briefly paraphrased as follows:

Let $ABCD$ be a square with diagonal AC and side AB.

If AC is commensurable with AB, the ratio $AC:AB$ may be represented by $m:n$, where m and n are integers prime to each other[3] and $m > 1$ since $AC > AB$.

Then $AC:AB = m:n$,

and $AC^2:AB^2 = m^2:n^2$.

But $AC^2 = 2AB^2$.

Therefore $m^2 = 2n^2$, whence m is even

and n *is odd.*

Since m is even, let $m = 2p$.

$$(2p)^2 = 2n^2,$$
$$4p^2 = 2n^2,$$
$$2p^2 = n^2.$$

Therefore n *is even.*

Since the assumption that AC is commensurable with AB leads to the impossible conclusion that the same number (n) is both odd and even, the assumption must be false.

---

[1] $284 = 1 + 2 + 4 + 5 + 10 + 11 + 20 + 22 + 44 + 55 + 110$ (the factors of 220), and $220 = 1 + 2 + 4 + 71 + 142$ (the factors of 284). In 1636, Fermat discovered that 17296 and 18416 are amicable numbers. Euler subsequently discovered 61 pairs of amicable numbers (T. L. Heath, *History of Greek Mathematics* I. 75).

[2] The reference is to proof by *reductio ad absurdum.*

[3] I.e., the fraction m/n is reduced to its lowest terms.

Therefore AC is not commensurable with AB.[1]

Plato, *Theaetetus* 147 D

Theodorus[2] was writing something for us about square roots, showing that the root of three [square] feet or five [square] feet is incommensurable in length with the one-foot length. And he took up the cases, one by one, until that of seventeen [square] feet. There, for some reason, he stopped. Now a thought occurred to us that since square roots of this kind seemed unlimited in number we should try to include them in one class by which we might designate them all.

---

[1] It is not certain how and when incommensurability and irrationality were first discovered. But the contrast between such a discovery and the mere approximation to the value of a surd is a contrast between night and day. (Even in the Old Babylonian period an approximation to $\sqrt{2}$ correct to six decimal places is said to have been attained.)

It has been suggested by K. von Fritz (*Annals of Mathematics*, 1945, p. 257) that Hippasus of Metapontum (*ca.* 450 B.C.) may have obtained the clue from a consideration of the side and diameter of the regular pentagon. Again, the first proofs probably lacked the refinement of that reproduced above. But in any case the discovery was of momentous significance in the history of mathematics. Having discovered the numerical relations of musical intervals and of other natural phenomena, the early Pythagoreans went on to assume the sufficiency of numbers to express all relations and considered basic reality to be number (cf. Philolaus Frag. 13 [Diels] and Aristotle, *Metaphysics* 1090*a*20–25.) That no whole number or rational fraction can describe the ratio, e.g., between the diagonal and the side of a square, or of a pentagon, thus offered a crucial difficulty. Tradition records the consternation of the Pythagoreans and their attempts to suppress the discovery. Later the proof of the incommensurability of $\sqrt{3}, \sqrt{5}, \ldots$ up to $\sqrt{17}$ indicated the existence of a whole class of incommensurables. And yet, lines describable as $\sqrt{2}$, $\sqrt{3}$, etc., could readily be constructed geometrically. This fact naturally caused mathematicians to prefer geometric methods, which could deal with continuous geometric magnitudes such as lines, areas, and the like. The discrete series of rational numbers had proved insufficient. This may in part account for the geometrical coloring that the theory of numbers receives in Plato, Euclid, and their successors. In fact, from the prevailing Greek viewpoint the subject of the incommensurable is part of geometry rather than of arithmetic and applies to the relation of magnitudes, not of numbers; from this viewpoint arithmetic proper deals only with rational numbers. By developing a theory of surds or irrationals as "real" numbers, modern mathematics has introduced continuity into arithmetic and has justified the previous practical use of the irrationals. The ancients arrived at somewhat similar results by elaborating a theory of proportion that was applicable to incommensurable as well as commensurable magnitudes. This development began soon after the discovery of incommensurability, was carried forward with great success by Theaetetus, and brought to perfection by Eudoxus. It may be noted that the geometric theory of proportion, in which geometric ratios take the place of "real" numbers, enables one to equate two variables whose limits are equal and is at the basis of the method of exhaustion so effectively used by Eudoxus and Archimedes (see pp. 45, 60).

[2] Theodorus of Cyrene, said to have been a pupil of Protagoras and a teacher of Plato, may have been the first to prove the incommensurability of $\sqrt{3}, \sqrt{5}, \ldots \sqrt{17}$. His method may have been analogous to that used for $\sqrt{2}$.

THE GENERAL THEORY OF PROPORTION AND THE THEORY OF INCOMMENSURABLE MAGNITUDES

Euclid, *Elements* V, Definitions 4, 5; X, Definitions and Propositions 1–3. Translation of T. L. Heath, *The Thirteen Books of Euclid's Elements*

V. Def. 4. Magnitudes are said to have a ratio to one another which are capable, when multiplied, of exceeding one another.

Def. 5. Magnitudes are said to be in the same ratio, the first to the second and the third to the fourth, when, if any equimultiples whatever be taken of the first and third, and any equimultiples whatever of the second and fourth, the former equimultiples alike exceed, are alike equal to, or alike fall short of, the latter equimultiples respectively taken in corresponding order.[1]

X. Definitions

1. Those magnitudes are said to be *commensurable* which are measured by the same measure, and those *incommensurable* which cannot have any common measure.

2. Straight lines are *commensurable in square* when the squares on them are measured by the same area, and *incommensurable in square* when the squares on them cannot possibly have any area as a common measure.

3. With these hypotheses, it is proved that there exist straight lines infinite in multitude which are commensurable and incommensurable respectively, some in length only, and others in square also, with an assigned straight line. Let then the assigned straight line be called *rational*, and those straight lines which are commensurable with it, whether in length and in square or in square only, *rational*, but those which are incommensurable with it *irrational*.[2]

---

[1] There is a tradition that this definition and the theory based on it, as developed in Book V, is the work of Eudoxus. Indicative of the importance of this definition is the fact that it is essentially the same as that used in modern mathematics (see T. L. Heath, *The Thirteen Books of Euclid's Elements* II. 125). On the crisis in mathematics that elicited the new theory, see p. 14, n. 1, above. Cf. the purely arithmetic definition (Euclid, *Elements* VII, Def. 21): "Numbers are in proportion when the first is the same multiple or part or parts of the second as the third is of the fourth."

As for Book X, much of the material was probably worked out by Theaetetus on the basis of an older definition of proportionality adapted so as to be applicable to incommensurable magnitudes. (The definition is referred to by Aristotle, *Topica* 158*b*32.) But in the form in which Euclid gives them, the proofs are probably patterned on those of Eudoxus. In fact some of the propositions are entirely dependent on this new definition of Eudoxus. [Edd.]

[2] . . . The relativity of the terms *rational* and *irrational* is well brought out in this definition. We may set out *any straight line* and call it rational, and it is then with reference to this assumed rational straight line that others are called *rational* or *irrational*.

We should carefully note that the signification of *rational* in Euclid is wider than in our

4. And let the square on the assigned straight line be called *rational* and those areas which are commensurable with it *rational*, but those which are incommensurable with it *irrational*, and the straight lines which produce them *irrational*, that is, in case the areas are squares, the sides themselves, but in case they are any other rectilineal figures, the straight lines on which are described squares equal to them.[1]

Proposition 1

Two unequal magnitudes being set out, if from the greater there be subtracted a magnitude greater than its half, and from that which is left a magnitude greater than its half, and if this process be repeated continually, there will be left some magnitude which will be less than the lesser magnitude set out.[2]

Let *AB*, *C* be two unequal magnitudes of which *AB* is the greater: I say that, if from *AB* there be subtracted a magnitude greater than its half, and from that which is left a magnitude greater than its half, and if this process be repeated continually, there will be left some magnitude which will be less than the magnitude *C*.

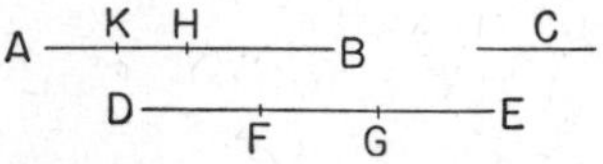

For *C* if multiplied will sometime be greater than *AB*.

[cf. V. Def. 4]

Let it be multiplied, and let *DE* be a multiple of *C*, and greater than *AB*;
let *DE* be divided into the parts *DF*, *FG*, *GE* equal to *C*,
from *AB* let there be subtracted *BH* greater than its half,
and, from *AH*, *HK* greater than its half,

---

terminology. With him, not only is a straight line commensurable *in length* with a rational straight line rational, but a straight line is rational which is commensurable with a rational straight line *in square only*. That is, if ρ is a rational straight line, not only is $(m/n)\rho$ rational, where m, n are integers and m/n in its lowest terms is not square, but $\sqrt{\frac{m}{n}} \cdot \rho$ is *rational* also. We should in this case call $\sqrt{\frac{m}{n}} \cdot \rho$ irrational. . . .

[In the course of Book X there is an elaborate classification of various types of irrationals called medials, apotomes, and binomials. The attempts of early algebraists to reduce all these types to algebra considerably influenced the development of the latter subject. Edd.]

[1] As applied to *areas*, the terms *rational* and *irrational* have, on the other hand, the same sense with Euclid as we should attach to them. . . .

[2] This proposition is used in the method of exhaustion (see, e.g., Euclid, *Elements* XII, Prop. 2, p. 46, below). Note, in this connection, Aristotle, *Physics* 266*b*2: "if continual additions be made to a finite quantity, any given quantity will be exceeded; similarly, if continual subtractions be made the remainder may be made less than any given quantity" (where, however, there is no provision against an infinite converging series), and 207*b*10 "a magnitude may be bisected endlessly," as well as the so-called Axiom of Archimedes "the excess by which the greater of (two) unequal areas exceeds the less can, by being added to itself, be made to exceed any given finite area" (see p. 75, below). [Edd.]

and let this process be repeated continually until the divisions in AB are equal in multitude with the divisions in DE.

Let, then, AK, KH, HB be divisions which are equal in multitude with DF, FG, GE.

Now, since DE is greater than AB,
and from DE there has been subtracted EG less than its half,
and, from AB, BH greater than its half,
therefore the remainder GD is greater than the remainder HA.

And, since GD is greater than HA,
and there has been subtracted, from GD, the half GF,
and, from HA, HK greater than its half,
therefore the remainder DF is greater than the remainder AK.

But DF is equal to C;
therefore C is also greater than AK.

Therefore AK is less than C.

Therefore there is left of the magnitude AB the magnitude AK which is less than the lesser magnitude set out, namely, C. Q.E.D.

And the theorem can be similarly proved even if the parts subtracted be halves.

Proposition 2

If, when the less of two unequal magnitudes is continually subtracted in turn from the greater, that which is left never measures the one before it, the magnitudes will be incommensurable.[1]

For, there being two unequal magnitudes AB, CD, and AB being the less, when the less is continually subtracted in turn from the greater, let that which is left over never measure the one before it;
I say that the magnitudes AB, CD are incommensurable.

For, if they are commensurable, some magnitude will measure them.

Let a magnitude measure them, if possible, and let it be E; let AB, measuring FD, leave CF less than itself, 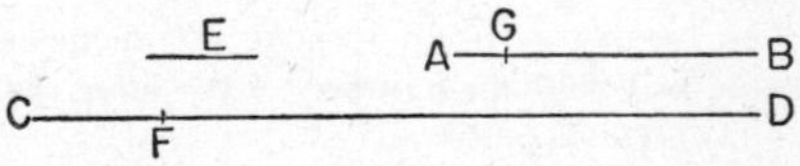
let CF measuring BG, leave AG less than itself,
and let this process be repeated continually, until there is left some magnitude which is less than E.

Suppose this done, and let there be left AG less than E.

Then, since E measures AB,

[1] This proposition states the test for incommensurable magnitudes, founded on the usual operation for finding the greatest common measure. The sign of the incommensurability of two magnitudes is that this operation never comes to an end, while the successive remainders become smaller and smaller until they are less than any assigned magnitude. . . .

while AB measures DF,
therefore E will also measure FD.

But it measures the whole CD also;
therefore it will also measure the remainder CF.

But CF measures BG;
therefore E also measures BG.

But it measures the whole AB also;
therefore it will also measure the remainder AG, the greater the less: which is impossible.

Therefore no magnitude will measure the magnitudes AB, CD;
therefore the magnitudes AB, CD are incommensurable. [X. Def. 1]

Therefore, etc.

Proposition 3

Given two commensurable magnitudes, to find their greatest common measure.[1]

Let the two given commensurable magnitudes be AB, CD of which AB is the less;
thus it is required to find the greatest common measure of AB, CD.

Now the magnitude AB either measures CD or it does not.

If then it measures it—and it measures itself also—AB is a common measure of AB, CD.

And it is manifest that it is also the greatest;
for a greater magnitude than the magnitude AB will not measure AB.

Next, let AB not measure CD.

Then, if the less be continually subtracted in turn from the greater, that which is left over will sometime measure the one before it, because AB, CD are not incommensurable; [cf. X. 2]

---

[1] This proposition for two commensurable *magnitudes* is, *mutatis mutandis*, exactly the same as VII. 2 for numbers. We have the process

b) a (p
$\underline{pb}$
c) b (q
$\underline{qc}$
d) c (r
$\underline{rd}$

where c is equal to rd and therefore there is no remainder.

It is then proved that d is a common measure of a, b; and next, by a *reductio ad absurdum*, that it is the *greatest* common measure, since any common measure must measure d, and no magnitude greater than d can measure d. . . .

[This procedure is sometimes called "the Euclidean algorithm." In this connection we may note, from VII, Prop. 1, the test for numbers prime to each other, viz., that in the process described above $rd \neq c$ and the process does not end until the remainder is unity. Edd.]

let *AB*, measuring *ED*, leave *EC* less than itself,
let *EC*, measuring *FB*, leave *AF* less than itself,
and let *AF* measure *CE*.

Since, then, *AF* measures *CE*,
while *CE* measures *FB*,
therefore *AF* will also measure *FB*.

But it measures itself also;
therefore *AF* will also measure the whole *AB*.

But *AB* measures *DE*;
therefore *AF* will also measure *ED*.

But it measures *CE* also;
therefore it also measures the whole *CD*.

Therefore *AF* is a common measure of *AB*, *CD*.

I say next that it is also the greatest.

For, if not, there will be some magnitude greater than *AF* which will measure *AB*, *CD*.

Let it be *G*.

Since then *G* measures *AB*,
while *AB* measures *ED*,
therefore *G* will also measure *ED*.

But it measures the whole *CD* also;
therefore *G* will also measure the remainder *CE*.

But *CE* measures *FB*;
therefore *G* will also measure *FB*.

But it measures the whole *AB* also,
and it will therefore measure the remainder *AF*, the greater the less: which is impossible.

Therefore no magnitude greater than *AF* will measure *AB*, *CD*;
therefore *AF* is the greatest common measure of *AB*, *CD*.

Therefore the greatest common measure of the two given commensurable magnitudes *AB*, *CD* has been found. Q.E.D.

Prime Numbers

Nicomachus, *Introduction to Arithmetic* I.13.2

The method of producing these numbers[1] is called a sieve by Eratosthenes, since we take the odd numbers mingled and indiscriminate and we separate out of them by this method of production, as if by some instrument or sieve, the prime and incomposite numbers by themselves,

[1] (1) Prime numbers, (2) composite numbers, and (3) a pair of numbers (e.g., 9 and 25), each composite, but having, relatively to each other, only unity as a common measure.

and the secondary and composite numbers by themselves, and we find separately those that are mixed.[1]

THE NUMBER OF PRIMES IS INFINITE

Euclid, *Elements* IX, Proposition 20. Translation of T. L. Heath, *The Thirteen Books of Euclid's Elements* II. 412

Prime numbers are more than any assigned multitude of prime numbers.[2]

Let A, B, C be the assigned prime numbers; I say that there are more prime numbers than A, B, C.

A——
B—— G————
C——
E————————D F

For let the least number measured by A, B, C be taken, and let it be DE; let the unit DF be added to DE.

Then EF is either prime or not.

First, let it be prime; then the prime numbers A, B, C, EF have been found, which are more than A, B, C.

Next let EF not be prime; therefore it is measured by some prime number. [VII. 31][3]

Let it be measured by the prime number G.

I say that G is not the same with any of the numbers A, B, C.

For, if possible, let it be so.

---

[1] Nicomachus goes on to explain the working of the sieve. All the odd numbers are set out in order and, beginning with 3, we note all the multiples of each. The method would not readily tell whether a very large number is prime, for we should have to try the multiples of all prime numbers up to the square root of the given number. The appended diagram, labeled "The Sieve of Eratosthenes," appears in manuscripts of this work.

3	5	7	9 (3 / 3)	11	13	15 (3 / 5)	17
19	21 (3 / 7)	23	25 (5 / 5)	27 (3 / 9)	29	31	33 (3 / 11)
35 (5 / 7)	37	39 (3 / 13)	41	43	45 (3 5 / 15 9)	47	49 (7 / 7)
51 (3 / 17)	53	55 (5 / 11)	57 (3 / 19)	59	61	63 (3 7 / 21 9)	65 (5 / 13)
67	69 (3 / 23)	71	73	75 (3 5 / 25 15)	77	79	81 (9 3 / 9 27)
83	85 (5 / 17)	87 (3 / 29)	89	91 (7 / 13)	93 (3 / 31)	95 (5 / 19)	97

[2] The proof of this important theorem that the number of prime numbers is infinite may be restated as follows:

Let a, b, and c be given prime numbers.

Consider the expression $(abc + 1)$.

If $(abc + 1)$ is prime, we have found an additional prime number.

If $(abc + 1)$ is not prime, it is exactly divisible by some prime. But this prime cannot be a, b, or c, for then it would divide exactly both abc and $(abc + 1)$, which is impossible. Therefore we have obtained a new prime, and this process may be continued indefinitely.

How the Greek theorems on prime, polygonal, and perfect numbers formed the foundation for the later development can best be seen in the appropriate sections of L. E. Dickson, *History of the Theory of Numbers.* [Edd.]

[3] Any composite number is measured by some prime number. (That the resolution of a composite number into primes can be done in only one way, the so-called "fundamental theorem of arithmetic," is not necessary to our proof but is essentially involved in VII. 30.) [Edd.]

Now A, B, C measure DE; therefore G also will measure DE.

But it also measures EF.

Therefore G, being a number, will measure the remainder, the unit DF: which is absurd.

Therefore G is not the same with any of the numbers A, B, C.

And by hypothesis it is prime.

Therefore the prime numbers A, B, C, G have been found which are more than the assigned multitude of A, B, C. Q.E.D.

Some Additional Theorems and Problems in the Theory of Numbers

RIGHT-ANGLED TRIANGLES WITH SIDES IN RATIONAL NUMBERS

Proclus Diadochus, *Commentary on Euclid's Elements I*, pp. 427.18–429.7 (Friedlein)

There are two kinds of right triangles, isosceles and scalene. Now in the case of isosceles right triangles we can never obtain [rational] numbers corresponding to the sides, for no square number is double another square, except as an approximation. Thus the square of 7 is double that of 5, less one. But that it is possible to find such numbers [i.e., rational numbers corresponding to the sides] in the case of scalene right triangles is clearly indicated by the theorem that the square on the hypotenuse is equal to the sum of the squares on the sides about the right angle. Such is the triangle in Plato's *Republic* with sides 3 and 4 and hypotenuse 5.[1] That is, the square of 5 is equal to the sum of the other two squares, for the square of 5 is 25, that of 3 is 9, and that of 4 is 16.

The theorem then is clear in the case of the numbers. Now certain methods of finding such triangles have been handed down, of which one is referred to Plato, the other to Pythagoras. The method of Pythagoras begins with odd numbers. It sets the given odd number as the lesser of the sides about the right angle. Then taking its square, subtracting one, and taking half the remainder, it sets the result down as the greater of the sides about the right angle. The hypotenuse is equal to this larger side plus 1.[2] E.g., this method takes 3, squares it, subtracts 1 from the resulting 9, takes half of 8, i.e., 4, and adds 1, obtaining 5. Thus there has been found a right-angled triangle with sides 3, 4, and 5.

The Platonic method begins with even numbers. It takes the given even number and sets it as one of the sides about the right angle. Now by squaring half this number and adding 1 to the result, it obtains the

---

[1] Cf. *Republic* 546 C.

[2] Beginning with any odd number, n, the Pythagorean formula gives n, $(n^2 - 1)/2$, and $(n^2 + 1)/2$ as the sides of the triangle. I.e., $n^2 + [(n^2 - 1)/2]^2 = [(n^2 + 1)/2]^2$.

hypotenuse, while by subtracting 1 from the square it obtains the other of the sides about the right angle.[1] For example, this method takes 4, squares half of it, making 4, subtracts 1 from 4 and makes 3, and adds 1 to 4 and makes 5, and obtains the same triangle as was produced by the other method.

Euclid, *Elements* X, Proposition 28, Lemma 1. Translation of T. L. Heath, *The Thirteen Books of Euclid's Elements* III. 63

To find two square numbers such that their sum is also square.

Let two numbers, AB, BC be set out, and let them be either both even or both odd.

Then since, whether an even number is subtracted from an even number, or an odd number from an odd number, the remainder is even, [IX. 24, 26]
therefore the remainder AC is even.

Let AC be bisected at D.

Let AB, BC also be either similar plane numbers, or square numbers, which are themselves also similar plane numbers.[2]

Now the product of AB, BC together with the square on CD is equal to the square on BD. [II. 6][3]

And the product of AB, BC is square, inasmuch as it was proved that, if two similar plane numbers by multiplying one another make some number, the product is square. [IX. 1][4]

Therefore two square numbers, the product of AB, BC, and the square on CD, have been found which, when added together, make the square on BD.

And it is manifest that two square numbers, the square on BD and the square on CD, have again been found such that their difference, the

---

[1] Beginning with any even number, $2n$, the Platonic formula gives $2n$, $n^2 - 1$, and $n^2 + 1$ as the sides of the triangle. I.e., $(2n)^2 + (n^2 - 1)^2 = (n^2 + 1)^2$.

On possible methods by which these formulae were obtained, see T. L. Heath, *History of Greek Mathematics* I. 80–81. The Greeks were not the first people who were interested in finding so-called "Pythagorean numbers." In Old Babylonian times there seems to have been a theoretic interest in such triads (see O. Neugebauer and A. Sachs, *Mathematical Cuneiform Texts*, pp. 37–41).

[2] a and b are similar plane numbers if $a = mn$ and $b = pq$, and $m:p = n:q$, where m, n, p, and q are integers. Obviously squares are similar. [Edd.]

[3] In algebraic form, if $AC = a$ and $BC = b$, $(a + b)b + (a/2)^2 = [(a/2) + b]^2$. In general, the product of two numbers plus the square of half their difference is equal to the square of half their sum. [Edd.]

[4] To revert to the example in n. 2, since $ab = mnpq$ and $mq = pn$, $ab = n^2p^2 = m^2q^2$, a square number. [Edd.]

product of AB, BC, is a square, whenever AB, BC are similar plane numbers.[1]

But when they are not similar plane numbers, two square numbers, the square on BD and the square on DC, have been found such that their difference, the product of AB, BC, is not square. Q.E.D.

THE SUMMATION OF A GEOMETRICAL SERIES

Euclid, *Elements* IX, Proposition 35. Translation of T. L. Heath, *The Thirteen Books of Euclid's Elements* II. 420–421

If as many numbers as we please be in continued proportion, and there be subtracted from the second and the last numbers equal to the first, then, as the excess of the second is to the first, so will the excess of the last be to all those before it.

Let there be as many numbers as we please in continued proportion, A, BC, D, EF, beginning from A as least, and let there be subtracted from BC and EF the numbers BG, FH, each equal to A; I say that, as GC is to A, so is EH to A, BC, D.

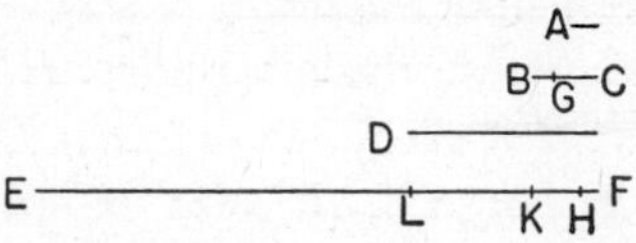

A
B G C
D
E L K H F

For let FK be made equal to BC, and FL equal to D.

Then, since FK is equal to BC,
and of these the part FH is equal to the part BG,
therefore the remainder HK is equal to the remainder GC.

And since, as EF is to D, so is D to BC, and BC to A, while D is equal to FL, BC to FK, and A to FH,
therefore, as EF is to FL, so is LF to FK, and FK to FH.

Separando, as EL is to LF, so is LK to FK, and KH to FH. [VII. 11, 13][2]

Therefore also, as one of the antecedents is to one of the consequents, so are all the antecedents to all the consequents; [VII. 12][3]

---

[1] Euclid's method of forming right-angled triangles in integral numbers . . . is as follows. Take two similar plane numbers, e.g., mnp^2, mnq^2, *which are either both even or both odd*, so that their difference is divisible by 2.

Now the product of the two numbers, or $m^2n^2p^2q^2$, is square [IX. 1], and, by II. 6,

$$mnp^2 \cdot mnq^2 + [(mnp^2 - mnq^2)/2]^2 = [(mnp^2 + mnq^2)/2]^2,$$

so that the numbers $mnpq$, $\frac{1}{2}(mnp^2 - mnq^2)$ satisfy the condition that the sum of their squares is also a square number.

It is also clear that $\frac{1}{2}(mnp^2 + mnq^2)$, $mnpq$ are numbers such that the *difference* of their squares is also a square.

[Euclid's formula is more general than those attributed by Proclus to Pythagoras and Plato and includes them. Cf. also Diophantus, *Arithmetica* II. 8 (p. 30, below). Edd.]

[2] These propositions prove, in effect, that if $a:b = c:d$, then $(a - c):(b - d) = a:b$, and $a:c = b:d$. [Edd.]

[3] This proposition proves, in effect, that if $a:b = c:d = e:f = \ldots$, then each ratio $= (a + c + e + \ldots):(b + d + f + \ldots)$. [Edd.]

therefore, as KH is to FH, so are EL, LK, KH to LF, FK, HF.

But KH is equal to CG, FH to A, and LF, FK, HF to D, BC, A; therefore, as CG is to A, so is EH to D, BC, A.

Therefore, as the excess of the second is to the first, so is the excess of the last to all those before it.[1] Q.E.D.

ON CUBIC NUMBERS

Nicomachus, *Introduction to Arithmetic* II. 20. 5

For when the successive odd numbers are set out in an endless series beginning with unity, observe that the first makes the potential cube, the sum of the next two makes the second cube, the sum of the next three following these makes the third cube, the sum of the four following these makes the fourth cube, the sum of the next five the fifth cube, and the sum of the next six the sixth cube, and so on indefinitely.[2]

---

[1] This proposition is perhaps the most interesting in the arithmetical Books since it gives a method, and a very elegant one, of *summing any series of terms in geometrical progression.*

Let $a_1, a_2, a_3, \ldots a_n, a_{n+1}$ be a series of terms in geometrical progression. Then Euclid's proposition proves that $(a_{n+1} - a_1):(a_1 + a_2 + \ldots + a_n) = (a_2 - a_1):a_1$.

For clearness' sake we will on this occasion use the fractional notation of algebra to represent proportions.

Euclid's method then comes to this.

Since
$$\frac{a_{n+1}}{a_n} = \frac{a_n}{a_{n-1}} = \ldots = \frac{a_2}{a_1},$$
we have, *separando*,
$$\frac{a_{n+1} - a_n}{a_n} = \frac{a_n - a_{n-1}}{a_{n-1}} = \ldots = \frac{a_3 - a_2}{a_2} = \frac{a_2 - a_1}{a_1},$$
whence, since, as one of the antecedents is to one of the consequents, so is the sum of all the antecedents to the sum of all the consequents, [VII. 12]
$$\frac{a_{n+1} - a_1}{a_n + a_{n-1} + \ldots + a_1} = \frac{a_2 - a_1}{a_1},$$
which gives $a_1 + a_2 + \ldots + a_n$, or S_n.

If, to compare the result with that arrived at in algebraical text-books, we write the series in the form
$$a, ar, ar^2, \ldots ar^{n-1} \quad (n \text{ terms}),$$
we have
$$\frac{ar^n - a}{S_n} = \frac{ar - a}{a}$$
or
$$S_n = \frac{a(r^n - 1)}{r - 1}.$$

[Archimedes performs the equivalent of summing the infinite geometric series $1 + \frac{1}{4} + (\frac{1}{4})^2 + \ldots$ (*Quadrature of a Parabola* 22). Edd.]

[2] That is, $1 = 1^3$
$3 + 5 = 2^3$
$7 + 9 + 11 = 3^3$, etc.

It is not known whether Nicomachus (or his predecessors) arrived at the formulae:
$$[n(n-1)+1] + [n(n-1)+3] + \ldots + [n(n-1)+2n-1] = n^3,$$
and $1^3 + 2^3 + 3^3 + \ldots + n^3 = (n^2/4)(n+1)^2$, though in the case of squares Archimedes (*On Spirals* 10) had performed the equivalent of finding $1^2 + 2^2 + 3^2 + \ldots + n^2 = (n/6)(n+1)(2n+1)$.

ALGEBRA

The great triumphs of Greek mathematics were predominantly in geometry, and, in fact, the advances in the theory of numbers and arithmetic were made in close connection with geometry. But both in pre-Greek times, as Egyptian papyri and Babylonian cuneiform tablets attest, and in Greek times, as we see from the arithmetical epigrams of the Greek anthology, from problems such as the Cattle Problem of Archimedes, and from those in Hero's *Metrica*, the algebraic solution of equations was a subject of study. Recent research has, in fact, indicated a highly developed technique among the Babylonians for the solution of many types of equation long before the beginnings of Greek mathematical thought. The work of Diophantus, who flourished probably in the third century A.D., represents the highest development in this branch of Greek mathematics, though it is impossible to say how much of it is drawn from predecessors, Greek and non-Greek. Of Diophantus' works only the *Arithmetica* is extant in substantial part (six of the original thirteen books). Here we have a collection of problems leading to equations determinate and indeterminate in one or more unknowns of degree as high, in one case, as the sixth. Generally it is required to find numbers such that various linear, quadratic, or cubic expressions containing them are squares or cubes, etc.

Diophantus' method is best seen in the actual problems. In general, however, it may be said that rather than adhere to a single method Diophantus employs artifices and devices that best fit the particular problem. The result is a great tribute to the author's resourcefulness but is not entirely satisfactory in point of generality.

An outstanding feature of Diophantus' work is the use of symbols for the unknown and its powers and for subtraction. Diophantus confines himself to positive rational solutions, integral or fractional; surds and "imaginary" roots are excluded, as are negative solutions. For his solution of a mixed quadratic equation Diophantus seems to have a formula akin to our quadratic formula based on completing the square (cf. Hero, *Geometrica* 21). Indeed this type of formula probably underlies the procedures used as early as the Old Babylonian period.

The work of Diophantus was studied by Greek and Arab mathematicians of medieval times and was the subject of renewed interest in the Renaissance and later. Of the numerous commentaries on Diophantus two may be mentioned, that of the famous Hypatia, daughter of Theon of Alexandria, at the beginning of the fifth century, and that of Fermat in the seventeenth. Diophantine Analysis is the name given today to the branch of mathematics that is concerned with the investigation of certain types of indeterminate equations.

Preceding the problems from Diophantus, we have given here, by way of illustration, a few epigrams of the Greek Anthology and the Cattle Problem of Archimedes. It may also be noted that a type of indeterminate analysis is involved in the finding of right triangles having sides in whole numbers (see p. 21), which exercised the Greeks from early Pythagorean times. The notes appended to Heath's paraphrase of Diophantus are based on the same author's account of Diophantus in *History of Greek Mathematics* II. ch. 20. In the selection of problems the aim has been to include many different types of equation.

Some Mathematical Problems in the Greek Anthology

Greek Anthology XIV, 3, 12, 49, 126, 130, 144. Translation of W. R. Paton (London, 1918)[1]

3

Cypris thus addressed Love, who was looking down-cast: "How, my child, hath sorrow fallen on thee?" And he answered: "The Muses stole

[1] There are some 44 epigrams in the form of mathematical problems found in the great collection of about 6,000 epigrams known as the Greek Anthology. Most of the mathematical epigrams seem to have been collected by Metrodorus, a grammarian who may have lived as late as the fifth or sixth century A.D. The problems, however, are of a type that was

and divided among themselves, in different proportions, the apples I was bringing from Helicon, snatching them from my bosom. Clio got the fifth part, and Euterpe the twelfth, but divine Thalia the eighth. Melpomene carried off the twentieth part, and Terpsichore the fourth, and Erato the seventh; Polyhymnia robbed me of thirty apples, and Urania of a hundred and twenty, and Calliope went off with a load of three hundred apples. So I come to thee with lighter hands, bringing these fifty apples that the goddesses left me."[1]

12

Croesus the king dedicated six bowls weighing six minae, each one drachma heavier than the other.[2]

49

Make me a crown weighing sixty minae, mixing gold and brass, and with them tin and much-wrought iron. Let the gold and bronze together form two-thirds, the gold and tin together three-fourths, and the gold and iron three-fifths. Tell me how much gold you must put in, how much brass, how much tin, and how much iron, so as to make the whole crown weigh sixty minae.[3]

---

common at a much earlier time. Some are set in a form involving the distribution of apples and the determination of the weight of bowls and thus throw light on a type of problem referred to by Plato (see p. 4, above). In fact half of the epigrams lead to simple equations in one unknown and are no different from the type found in Egyptian mathematics long before the Greek period (e.g., in the Rhind papyrus). Others lead to easy simultaneous equations, and there are two cases of indeterminate equations of the first degree. [Edd.]

[1] The original number of apples, 3360, is obtained by the solution of

$$x/5 + x/12 + x/8 + x/20 + x/4 + x/7 + 30 + 120 + 300 + 50 = x. \quad \text{[Edd.]}$$

[2] Since one mina = 100 drachmas, the solution is 97½ drachmas, 98½ drachmas, etc. [Edd.]

[3] The problem leads to four simultaneous linear equations in four unknowns:

$$\begin{aligned} x + y &= 40, \\ x + z &= 45, \\ x + w &= 36, \\ x + y + z + w &= 60, \end{aligned}$$

whence the gold is found to be 30½, the brass 9½, the tin 14½, and the iron 5½ minas. A formula for solving a set of equations of the type involved in this problem was said to have been set forth by a Thymaridas, whose identity and date are matters of controversy. The formula is given by Iamblichus, who goes on to discuss extensions of Thymaridas' method to other types of equations.

In modern symbols the rule would be stated:

if
$$x + x_1 + x_2 + \ldots + x_{n-1} = s,$$
and
$$\begin{aligned} x + x_1, &= a_1, \\ x + x_2 &= a_2, \\ &\ldots \\ x + x_{n-1} &= a_{n-1}, \end{aligned}$$
then
$$x = \frac{(a_1 + a_2 + \ldots + a_{n-1}) - s}{n - 2}.$$

See Iamblichus, *Commentary on Nicomachus' Introduction to Arithmetic*, pp. 62–68 (Pistelli). [Edd.]

126

This tomb holds Diophantus. Ah, how great a marvel! The tomb tells scientifically the measure of his life. God granted him to be a boy for the sixth part of his life, and adding a twelfth part to this, He clothed his cheeks with down; He lit him the light of wedlock after a seventh part, and five years after his marriage He granted him a son. Alas! late-born wretched child; after attaining the measure of half his father's life, chill Fate took him. After consoling his grief by this science of numbers for four years he ended his life.[1]

130

Of the four spouts one filled the whole tank in a day, the second in two days, the third in three days, and the fourth in four days. What time will all four take to fill it?[2]

144

A. How heavy is the base I stand on together with myself! *B.* And my base together with myself weighs the same number of talents. *A.* But I alone weigh twice as much as your base. *B.* And I alone weigh three times the weight of yours. [3]

The Cattle Problem of Archimedes[4]

Archimedes, *Opera* II. 528–532 (Heiberg)

A problem that Archimedes solved and, in a letter to Eratosthenes of Cyrene, sent to those who were working on such problems in Alexandria.

---

[1] The solution of the simple equation, $x/6 + x/12 + x/7 + 5 + x/2 + 4 = x$, gives $x = 84$. That is, Diophantus' boyhood lasted fourteen years, his youthhood seven, he married at thirty-three, became a father at thirty-eight, and lost his son four years before his own death at the age of eighty-four. It is impossible to say how accurately this gives the facts about Diophantus' life. [Edd.]

[2] $12/25$ of a day. A hardy perennial of the schoolroom. [Edd.]

[3] This problem leads to no determinate solution, for the four unknowns are connected by three equations,

$$s_1 + b_1 = s_2 + b_2,$$
$$s_1 = 2b_2,$$
$$s_2 = 3b_1.$$

The first statue weighs $4/3$ as much as the second; the base of the second statue weighs twice as much as that of the first. [Edd.]

[4] This problem appears in the form of a poem of 44 verses (in elegiac couplets). We have given it as an example of indeterminate analysis among the Greeks. Some have doubted not only that Archimedes composed the epigram, but that he propounded or solved the problem in the form in which it is here given, and we cannot, at any rate, be sure that this precise problem precedes Diophantus. But it is certain that a cattle problem of such difficulty as to become proverbial was connected with Archimedes' name long before the time of Dio-

Compute the number of cattle of the Sun, O stranger, and if you are wise apply your wisdom and tell me how many once grazed on the plains of the island of Sicilian Thrinacia, divided into four herds by differences in the color of their skin—one milk-white, the second sleek and dark-skinned, the third tawny-colored, and the fourth dappled.

In each herd there was a great multitude of bulls, and there were these ratios. The number of white bulls, mark well, O stranger, was equal to one-half plus one-third the number of dark-skinned, in addition to all the tawny-colored; the dark-skinned bulls were equal to one-fourth plus one-fifth the number of dappled, in addition to all the tawny-colored. The number of dappled bulls, observe, was equal to one-sixth plus one-seventh the white, in addition to all the tawny-colored.

Now for the cows there were these conditions: the number of white cows was exactly equal to one-third plus one-fourth of the whole dark-skinned herd; the number of dark-skinned cows, again, was equal to one-fourth plus one-fifth of the whole dappled herd, bulls included; the number of dappled cows was exactly equal to one-fifth plus one-sixth of the whole tawny-colored herd as it went to pasture; and the number of tawny-colored cows was equal to one-sixth plus one-seventh of the whole white herd.

Now if you could tell me, O stranger, exactly how many were the cattle of the Sun, not only the number of well-fed bulls, but the number of cows as well, of each color, you would be known as one neither ignorant nor unskilled in numbers, but still you would not be reckoned among the wise. But come now, consider these other facts, too, about the cattle of the Sun.

---

phantus (cf. Cicero, *Ad Atticum* XII. 4; XIII. 28; Scholium to Plato, *Charmides* 165 E, in which "logistics" is said to investigate the problem called by Archimedes the "Cattle Problem," as well as *melite* and *phialite* numbers [see pp. 4, 25–26, above]).

Our problem connects eight unknowns by means of seven equations. If W, X, Y, Z represent the number of white, dark-skinned, tawny, and dappled bulls, respectively, and w, x, y, z the number of cows of the respective colors, we have

$$W = (\tfrac{1}{2} + \tfrac{1}{3})\,X + Y,$$
$$X = (\tfrac{1}{4} + \tfrac{1}{5})\,Z + Y,$$
$$Z = (\tfrac{1}{6} + \tfrac{1}{7})\,W + Y,$$
$$w = (\tfrac{1}{3} + \tfrac{1}{4})\,(X + x),$$
$$x = (\tfrac{1}{4} + \tfrac{1}{5})\,(Z + z),$$
$$z = (\tfrac{1}{5} + \tfrac{1}{6})\,(Y + y),$$
$$y = (\tfrac{1}{6} + \tfrac{1}{7})\,(W + w).$$

In addition, $W + X$ = a square number,

$Y + Z$ = a triangular number (i.e., of the form $[n(n + 1)/2]$. See p. 7).

The smallest numbers that satisfy all these conditions—for there is an infinite number of solutions—are staggering. More than 200,000 digits would be required to give the number of bulls or cows of each herd. For the method of solution and bibliographical references see T. L. Heath, *The Works of Archimedes* 319–326 (Cambridge, 1897).

When the white bulls were mingled with the dark-skinned, their measure in length and depth was equal as they stood unmoved,[1] and the broad plains of Thrinacia were all covered with their number. And, again, when the tawny-colored bulls were joined with the dappled ones they stood in perfect triangular form beginning with one and widening out, without the addition or need of any of the bulls of other colors.

Now if you really comprehend this problem and solve it giving the number in all the herds, go forth a proud victor, O stranger, adjudged, mark you, all-powerful in this field of wisdom.

PROBLEMS FROM DIOPHANTUS, *Arithmetica*

Paraphase of T. L. Heath, *Diophantus of Alexandria* (2d. ed. Cambridge, 1910)

I.7.[2] From the same [required] number to subtract two given numbers, so as to make the remainders have to one another a given ratio.[3]

---

[1] Since the bulls are longer than broad, it has been argued that if the *figure* formed is to be square, the *number* $(W + X)$ must be rectangular (i.e., the product of two unequal factors) rather than square. This makes the problem somewhat easier. But apart from the fact that there is no requirement that the bulls all face the same way, "measure" here may mean "number," in which case πλίνθου, the reading of the manuscripts, would have special force, for though the *number* would be square, the *shape* would be a plinth. All this is conjectural, as is the reading πλήθους adopted for the translation.

[2] Though the Diophantine problems are given in Heath's paraphrase, the reader must not conclude that the substitution of algebraic symbols for ordinary language had gone as far as the paraphrase might suggest. To guard against this conclusion we give now a literal translation of the first problem. In connection with this translation it should be noted that x is used for the unknown quantity, the Greek text using an abbreviation or symbol which varies in various manuscripts and is of doubtful origin. The sign for minus in the manuscripts is ⋔. In other problems additional devices of symbolic representation are employed. Though the Greek system of alphabetic numeration may seem very cumbersome in comparison with ours, a glance at some of the complicated calculations required in the problems will serve to offset the impression that this was an insuperable obstacle to the solution of problems involving large numbers.

"I.7. From the same number to subtract two given numbers so that the remainders will have a given ratio to one another.

Let the numbers to be subtracted from the same number be 100 and 20, and let the larger remainder be three times that of the smaller.

Let the required number be $1x$. If I subtract from it 100, the remainder is $1x - 100$ units; if I subtract from it 20, the remainder is $1x - 20$ units. Now the larger remainder will have to be three times the smaller. Therefore three times the smaller will be equal to the larger. Now three times the smaller is $3x - 300$ units; and this is equal to $1x - 20$ units.

Let the deficiency be added in both cases. $3x$ equals $1x + 280$ units. If we subtract equals from equals, $2x$ equals 280 units, and x is 140 units.

Now as to our problem. I have set the required number as $1x$; it will therefore be 140 units. If I subtract from it 100, the remainder is 40; and if I subtract from it 20, the remainder is 120. And the larger remainder is three times the smaller." [Edd.]

[3] Equation of the first degree in one unknown, $x - a = m(x - b)$.

Given numbers 100, 20, given ratio 3:1.

Required number x. Therefore $x - 20 = 3(x - 100)$, and $x = 140$.

I.13. To divide a given number thrice into two numbers such that one of the first pair has to one of the second pair a given ratio, and the second of the second pair to one of the third pair another given ratio, and the second of the third pair to the second of the first pair another given ratio.[1]

Given number 100, ratio of greater of first parts to lesser of second 3:1, of greater of second to lesser of third 2:1, and of greater of third to lesser of first 4:1.

x lesser of third parts.

Therefore greater of second parts $= 2x$, lesser of second $= 100 - 2x$, greater of first $= 300 - 6x$.

Hence lesser of first $= 6x - 200$, so that greater of third $= 24x - 800$.

Therefore $25x - 800 = 100$, $x = 36$, and the respective divisions are (84, 16), (72, 28), (64, 36).

I.37. To find two numbers in a given ratio and such that the square of the lesser also has to the sum of both a given ratio.[2]

Given ratios 3:1 and 2:1.

Lesser number x, which is found to be 8.

The numbers are 8, 24.

II.8. To divide a given square number into two squares.[3]

Given square number 16.

x^2 one of the required squares. Therefore $16 - x^2$ must be equal to a square.

Take a square of the form $(mx - 4)^2$, m being any integer and 4 the number which is the square root of 16, *e.g.*, take $(2x - 4)^2$, and equate it to $16 - x^2$.

---

[1] Determinate system of equations of the first degree,

$$\left.\begin{array}{c} x_1 + x_2 = y_1 + y_2 = z_1 + z_2 = a \\ \\ x_1 = my_2,\ y_1 = nz_2,\ z_1 = px_2 \end{array}\right\} \begin{array}{l} x_1 > x_2 \\ y_1 > y_2 \\ z_1 > z_2 \end{array}$$

[2] Determinate system of equations reducible to the first degree,

$$y = mx$$
$$x^2 = n(x + y)$$

[3] It is to this proposition that Fermat appended his famous note, in which he enunciates what is known as the "last theorem" of Fermat. The text of the note is as follows:

"On the other hand it is impossible to separate a cube into two cubes, or a biquadrate into two biquadrates, or generally *any power except a square into two powers with the same exponent.* I have discovered a truly marvelous proof of this, which however the margin is not large enough to contain." [I.e., $x^n + y^n = z^n$ has no solution in positive integers for $n>2$. A complete proof of this theorem is still lacking. Edd.]

Therefore $4x^2 - 16x + 16 = 16 - x^2$,
or $5x^2 = 16x$, and $x = 16/5$.

The required squares are therefore $256/25$, $144/25$.[1]

IV. 1. To divide a given number into two cubes such that the sum of their sides is a given number.[2]

Given number 370, given sum of sides 10.

Sides of cubes $5 + x$, $5 - x$, satisfying one condition.

Therefore[3] $30x^2 + 250 = 370$, $x = 2$,
and the cubes are 7^3, 3^3, or 343, 27.

I. 22. To find three numbers such that, if each give to the next following a given fraction of itself, in order, the results after each has given and taken may be equal.[4]

Let first give $1/3$ of itself to second, second $1/4$ of itself to third, third $1/5$ of itself to first.

Assume first to be a number of x's divisible by 3, say $3x$, and second to be a number of *units* divisible by 4, say 4.

Therefore second after giving and taking becomes $x + 3$.

Hence the first also after giving and taking must become $x + 3$; it must therefore have taken $x + 3 - 2x$, or $3 - x$; $3 - x$ must therefore be $1/5$ of third, or third $= 15 - 5x$.

Lastly, $15 - 5x - (3 - x) + 1 = x + 3$,
or $13 - 4x = x + 3$, and $x = 2$.

The numbers are 6, 4, 5.

---

[1] By Diophantus' method a^2 may be divided into $(4a^2m^2)/(m^2 + 1)^2$ and $a^2(m^2 - 1)^2/(m^2 + 1)^2$, a result obtained by letting $(mx - a)^2 = a^2 - x^2$. Note, however, that Diophantus himself does not give the general formula but confines himself to a special case. Yet general propositions in the theory of numbers are asserted by Diophantus in the course of his work, e.g., that if $2n + 1$ (n an integer) is the sum of two squares, n must not be odd (*Arithmetica* V. 9), that the difference of two cubes may (using fractions) be transformed into the sum of two cubes (*Arithmetica* V. 16), and many others. The proof, extension, and discussion of Diophantus' propositions marked the rise of modern number theory. [Edd.]

[2] Determinate system reducible to equations of second degree,

$$s^3 + t^3 = a$$
$$s + t = b$$

(Diophantus puts $s = b/2 + x$, $t = b/2 - x$. The numbers a, b are so chosen that $(1/3b)(a - b^3/4)$ is a square.)

[3] Adding the cubes of $(5 + x)$ and $(5 - x)$. [Edd.]

[4] System of equations, apparently indeterminate, but really reduced, by arbitrary assumptions, to determinate equations of the first degree,

$$u - (1/m)u + (1/p)z = y - (1/n)y + (1/m)u = z - (1/p)z + (1/n)y.$$ (The value of y is assumed.)

III. 19. To find four numbers such that the square of their sum *plus* or *minus* any one singly gives a square.[1]

Since, in any right-angled triangle, (sq. on hypotenuse) $\pm$ (twice product of perps.) = a square, we must seek four right-angled triangles [in rational numbers] having the same hypotenuse, or we must find a square which is divisible into two squares in four different ways; and "we saw how to divide a square into two squares in an infinite number of ways." [II. 8][2]

Take right-angled triangles in the smallest numbers (3, 4, 5) and (5, 12, 13); and multiply the sides of the first by the hypotenuse of the second and *vice versa*.

This gives the triangles (39, 52, 65) and (25, 60, 65); thus 65^2 is split up into two squares in *two* ways.

Again, 65 is "naturally" divided into two squares in two ways, namely into $7^2 + 4^2$ and $8^2 + 1^2$ "which is due to the fact that 65 is the product of 13 and 5, each of which numbers is the sum of two squares."

Form now a right-angled triangle[3] from 7, 4. The sides are $(7^2 - 4^2, 2 \cdot 7 \cdot 4, 7^2 + 4^2)$ or (33, 56, 65).

Similarly, forming a right-angled triangle from 8, 1, we obtain $(2 \cdot 8 \cdot 1, 8^2 - 1^2, 8^2 + 1^2)$ or 16, 63, 65.

Thus 65^2 is split into two squares in *four* ways.

Assume now as the sum of the numbers $65x$ and

---

[1] Indeterminate analysis of the second degree,

$$(y_1 + y_2 + y_3 + y_4)^2 \pm y_1 = \begin{cases} t^2 \\ t'^2 \end{cases}$$

$$(y_1 + y_2 + y_3 + y_4)^2 \pm y_2 = \begin{cases} u^2 \\ u'^2 \end{cases}$$

$$(y_1 + y_2 + y_3 + y_4)^2 \pm y_3 = \begin{cases} v^2 \\ v'^2 \end{cases}$$

$$(y_1 + y_2 + y_3 + y_4)^2 \pm y_4 = \begin{cases} w^2 \\ w'^2 \end{cases}$$

(Diophantus finds . . . four different rational right-angled triangles with the same hypotenuse, namely (65,52,39), (65,60,25), (65,56,33), (65,63,16), or, what is the same thing, a square which is divisible into two squares in four different ways; this will solve the problem, since, if h, p, b be the three sides of a right-angled triangle, $h^2 \pm 2pb$ are both squares.

Put therefore $y_1 + y_2 + y_3 + y_4 = 65x$,

$$\text{and } y_1 = 2.39.52\,x^2,\ y_2 = 2.25.60x^2,$$

$$y_3 = 2.33.56\,x^2,\ y_4 = 2.16.63x^2;$$

this gives $12768x^2 = 65x$, and $x = \frac{65}{12768}$.)

[2] See p. 30, above. [Edd.]

[3] If there are two numbers p, q, to "form a right-angled triangle" from them means to take the numbers $p^2 + q^2$, $p^2 - q^2$, $2pq$. These are the sides of a right-angled triangle, since $(p^2 + q^2)^2 = (p^2 - q^2)^2 + (2pq)^2$.

as first number $2 \cdot 39 \cdot 52x^2 = 4056x^2$,
as second number $2 \cdot 25 \cdot 60x^2 = 3000x^2$,
as third number $2 \cdot 33 \cdot 56x^2 = 3696x^2$,
as fourth number $2 \cdot 16 \cdot 63x^2 = 2016x^2$,
the coefficients of x^2 being four times the areas of the four right-angled triangles respectively.

The sum $12768x^2 = 65x$, and $x = \frac{65}{12768}$.

The numbers are

$$\frac{17136600}{163021824}, \frac{12675000}{163021824}, \frac{15615600}{163021824}, \frac{8517600}{163021824}.$$

VI.1. To find a (rational) right-angled triangle such that the hypotenuse *minus* each of the sides gives a cube.[1]

Let the required triangle be formed from x, 3.

Therefore hypotenuse $= x^2 + 9$, perpendicular $= 6x$, base $= x^2 - 9$.

Thus $x^2 + 9 - (x^2 - 9) = 18$ should be a cube, but it is not.

Now $18 = 2 \cdot 3^2$; therefore we must replace 3 by m, where $2 \cdot m^2$ is a cube; and $m = 2$.

We form, therefore, a right-angled triangle from x, 2, namely $(x^2 + 4, 4x, x^2 - 4)$; and one condition is satisfied.

The other gives $x^2 - 4x + 4 =$ a cube; therefore $(x - 2)^2$ is a cube, or $x - 2$ is a cube $= 8$, say.

Thus $x = 10$, and the triangle is (40, 96, 104).

GEOMETRY

The History of Geometry to the Time of Euclid

Proclus Diadochus, *Commentary on Euclid's Elements I*, pp. 64.7–70.18 (Friedlein)[2]

We must next speak of the origin of geometry in the present world cycle. For, as the remarkable Aristotle tells us, the same ideas have repeatedly come to men at various periods of the universe. It is not, he goes on to say, in our time or in the time of those known to us that the sciences have first arisen, but they have appeared and again disappeared, and will continue to appear and to disappear, in various cycles, of which the number both past and future is countless. But since we must speak of the origin of the arts and sciences with reference to the present world cycle, it was,

[1] This is one of a series of problems of constructing right-angled triangles with sides in rational numbers and satisfying various other conditions. If hypotenuse is h and sides p and q, $h - p = u^3$, $h - q = v^3$.
(Form a right-angled triangle from x, m, so that $h = x^2 + m^2$, $p = 2mx$, $q = x^2 - m^2$; thus $h - q = 2m^2$, and as this must be a cube, we put $m = 2$; therefore $h - p = x^2 - 4x + 4$ must be a cube, or $x - 2 =$ a cube, say n^3, and $x = n^3 + 2$.)

[2] On Proclus see p. 2 above. Note also p. 1, n. 1.

we say, among the Egyptians that geometry is generally held to have been discovered. It owed its discovery to the practice of land measurement. For the Egyptians had to perform such measurements because the overflow of the Nile would cause the boundary of each person's land to disappear.[1] Furthermore, it should occasion no surprise that the discovery both of this science and of the other sciences proceeded from utility, since everything that is in the process of becoming advances from the imperfect to the perfect. The progress, then, from sense perception to reason and from reason to understanding is a natural one. And so, just as the accurate knowledge of numbers originated with the Phoenicians through their commerce and their business transactions, so geometry was discovered by the Egyptians for the reason we have indicated.

It was Thales,[2] who, after a visit to Egypt, first brought this study to Greece. Not only did he make numerous discoveries himself, but laid the foundations for many other discoveries on the part of his successors, attacking some problems with greater generality and others more empirically. After him Mamercus,[3] the brother of the poet Stesichorus, is said to have

---

[1] Cf. Herodotus II. 109: "This king [Sesostris, *ca.* 1300 B.C.] moreover (so they said) divided the country among all the Egyptians by giving each an equal square parcel of land, and made this his source of revenue, appointing the payment of a yearly tax. And any man who was robbed by the river of a part of his land would come to Sesostris and declare what had befallen him; then the king would send men to look into it and measure the space by which the land was diminished, so that thereafter it should pay the appointed tax in proportion to the loss. From this, to my thinking, the Greeks learnt the art of measuring land; the sunclock and the sundial, and the twelve divisions of the day, came to Hellas not from Egypt but from Babylonia." (Translation of A. D. Godley, London, 1921.)

Note the different emphasis in Aristotle, *Metaphysics* 981*b*20–25: "Hence when all such inventions were already established, the sciences which do not aim at giving pleasure or at the necessities of life were discovered, and first in the places where men first began to have leisure. That is why the mathematical arts were founded in Egypt; for there the priestly caste was allowed to be at leisure." (Translation of W. D. Ross, Oxford, 1928.)

The debt of Greek geometry to Egypt is implied also in passages of Plato, Hero of Alexandria, Diodorus Siculus, Strabo, and others.

[2] See p. 92. Several geometric propositions are associated with the name of Thales in Proclus's commentary, e.g., that a diameter bisects a circle, that the base angles of an isosceles triangle are equal, that when two lines intersect the vertical angles are equal, that two triangles are congruent when two angles and a side of one are equal, respectively, to two angles and the side of the other. Proclus tells us, on the authority of Eudemus, that Thales found the distance of ships at sea by using this last theorem. Diogenes Laertius preserves a statement of Pamphila that Thales first inscribed a right triangle in a circle, though Diogenes notes another tradition, that the discoverer was Pythagoras. How much Thales had to do with these propositions is the subject of controversy.

In this connection it must be remembered that what constitutes an acceptable demonstration is not the same in every period. Just as early attempts at demonstrations must have differed considerably from the later canonical proofs, so the modern mathematician cannot in every case be satisfied with Euclid's proofs.

[3] Other readings are Mamertius and Ameristus.

embraced the study of geometry, and in fact Hippias of Elis[1] writes that he achieved fame in that study.

After these Pythagoras[2] changed the study of geometry, giving it the form of a liberal discipline, seeking its first principles in ultimate ideas, and investigating its theorems abstractly and in a purely intellectual way. It was he who discovered the subject of proportions[3] and the construction of the cosmic figures.[4] After him Anaxagoras[5] of Clazomenae devoted himself to many of the problems of geometry, as did Oenopides[6] of Chios, who was a little younger than Anaxagoras. Plato in *The Rivals*[7] mentions them both as having achieved repute in mathematics. After them Hippocrates of Chios[8] the discoverer of the quadrature of the lune, and Theodorus of Cyrene[9] gained fame in geometry. For Hippocrates is the first man on record who also composed Elements.

Plato, who lived after Hippocrates and Theodorus, stimulated to a very high degree the study of mathematics and of geometry in particular because of his zealous interest in these subjects. For he filled his works with mathematical discussions, as is well known, and everywhere sought to awaken admiration for mathematics in students of philosophy.

At this time lived also Leodamas of Thasos, Archytas of Tarentum,[10] and Theaetetus of Athens,[11] all of whom increased the number of theorems

---

[1] Hippias flourished in the latter half of the fifth century B.C. He may have discovered the quadratrix, a curve that may be used for squaring the circle or trisecting an angle (see p. 57).

[2] See the special note on Pythagorean geometry, p. 41.

[3] Following a conjecture of Diels. The reading of the manuscripts, referring to the subject of irrationals, can hardly stand.

[4] See p. 43.

[5] See p. 93. Plutarch says (*De Exilio* 17.607 E) that Anaxagoras while in prison wrote on the squaring of the circle.

[6] There is a tradition that Oenopides discovered the obliquity of the ecliptic and calculated a Great Year of 59 years. (According to Heath, *History of Greek Mathematics* I. 175, Oenopides was concerned with the problem of finding the least integral number of complete years that would contain an exact number of lunar months.) The problems of drawing a perpendicular to a line from a point outside the line and of constructing at a given point in a given line an angle equal to a given angle are attributed to Oenopides by Proclus (*Commentary on Euclid's Elements I*, pp. 283.7; 333.5).

[7] *Erastae* 132 A–B.

[8] See p. 54.

[9] See p. 14.

[10] The work of Archytas, a contemporary of Plato, is known only through fragments and references in the works of others. He found a remarkable solution of the problem of doubling the cube (p. 63), wrote on arithmetic and musical theory (p. 286), and seems to have invented mechanical contrivances (p. 335). His proof that between two numbers n and $n + 1$ no rational geometric mean may be inserted is repeated by Euclid, *Sectio Canonis* 3.

[11] Theaetetus, another contemporary of Plato, made important investigations in connection with irrationals and the geometry of the five regular solids (see also pp. 14–15, 43).

and made progress towards a more scientific ordering thereof. Neoclides and his pupil, Leon, who were younger than Leodamas, made many additions to the work of their predecessors. As a result the Elements composed by Leon are more carefully worked out both in the number and in the usefulness of the propositions proved. Leon also first discussed the *diorismi* (distinctions), that is the determination of the conditions under which the problem posed is capable of solution, and the conditions under which it is not.[1]

A little younger than Leon was Eudoxus of Cnidus,[2] an associate of Plato's school, who first increased the number of the general theorems, as they are called, and added three new proportions to the three then known. Using the method of analysis he greatly extended the theory of the section, a subject which had originated with Plato. Amyclas of Heraclea, a friend of Plato, Menaechmus,[3] a pupil of Eudoxus who had also been associated with Plato, and Dinostratus,[4] a brother of Menaechmus, brought the whole of geometry to an even higher degree of perfection. Theudius of Magnesia achieved a reputation for excellence both in mathematics and in the other parts of philosophy, for his Elements were excellently arranged and many of the (theretofore) limited[5] propositions were put in more general form. Furthermore, Athenaeus of Cyzicus, a contemporary, distinguished himself in mathematics generally and in geometry in particular. These men spent their time together in the Academy and collaborated in their investigations. Hermotimus of Colophon extended the work that had been done by Eudoxus and Theaetetus, made many discoveries in the Elements, and put together some material on the subject of Loci. Philippus of Mende[6] who,

---

[1] There is a passage in Plato's *Meno* (86 E–87 B) which indicates that Leon was not the discoverer of *diorismi*.

[2] The importance of Eudoxus for the history of mathematics can hardly be exaggerated. In perfecting a theory of proportion that included incommensurable as well as commensurable magnitudes he laid the foundation for the method of exhaustion. This enabled the Greek mathematicians to solve problems that are essentially problems in infinitesimal analysis. Archimedes, whose work represents the highest development in this field among the Greeks, refers to Eudoxus' discoveries (see p. 70).

The precise meaning of Proclus's statements is obscure. Is the "section" that of solids by planes, or the so-called "golden section" (p. 50) ? On these and the other questions see T. L. Heath, *History of Greek Mathematics* I. 323–326. On Eudoxus' theory of concentric spheres see p. 101, below.

[3] Menaechmus used the parabola and the rectangular hyperbola in his solutions of the Delian problem. The tradition that he was the discoverer of the conic sections arose from this fact and the reference of Eratosthenes (?) to the "triads" of Menaechmus (see p. 66).

[4] According to Pappus, Dinostratus made use of the quadratrix in connection with the problem of squaring the circle.

[5] Reading μερικῶν.

[6] Medma, according to another reading. He is often identified with Philip of Opus, the astronomer and mathematician, whose name is sometimes associated with the publication of Plato's *Laws* and the authorship of the *Epinomis*.

as a pupil of Plato, was inspired by him to the study of mathematics, conducted investigations along lines suggested by Plato, and in particular set before himself researches which he thought would contribute to the philosophy of Plato. Now those who have written histories trace the development of the science of geometry down to Philippus.[1]

Euclid,[2] who was not much younger than Hermotimus and Philippus, composed Elements, putting in order many of the theorems of Eudoxus, perfecting many that had been worked on by Theaetetus, and furnishing with rigorous proofs propositions that had been demonstrated less rigorously by his predecessors. Euclid lived in the time of the first Ptolemy, for Archimedes, whose life overlapped the reign of this Ptolemy too,[3] mentions Euclid. Furthermore, there is a story that Ptolemy once asked Euclid whether there was any shorter way to a knowledge of geometry than by the study of the Elements. Whereupon Euclid answered that there was no royal road to geometry. He is, then, younger than Plato's pupils and older than Eratosthenes and Archimedes, who, as Eratosthenes somewhere remarks, were contemporaries.

By choice Euclid was a follower of Plato and connected with this school of philosophy. In fact he set up as the goal of the Elements as a whole the construction of the so-called Platonic figures.[4]

There are, in addition, many other mathematical works by Euclid, written with remarkable accuracy and scientific insight, such as the *Optics*, the *Catoptrics*, works on the *Elements of Music*, and the book *On Divisions*.[5]

---

[1] The discussion thus far is probably based, directly or indirectly, on the *History of Geometry* by Eudemus, a pupil of Aristotle.

[2] Euclid, of whose life very little is known other than what is contained in the account given by Proclus, flourished about 300 B.C. and is one of the outstanding figures not only of the golden age of Greek mathematics but of all mathematical history. Though his *Elements* represent the consummation of three hundred years of Greek mathematics, Euclid's original contribution was by no means insignificant. This work, which in one form or another has served as a textbook for twenty-two hundred years and has passed through countless editions, is in fifteen books, of which all but the last two are probably by Euclid. Besides the *Elements* Euclid wrote various works in pure and applied mathematics of which the *Data*, *Phaenomena*, and *Optics* are extant in Greek; the book on *Divisions of Figures* is extant in Arabic and Latin translations. The *Porisms*, *Conics*, *Pseudaria*, and *Surface Loci* have been lost. An extant *Catoptrics*, attributed to Euclid, is almost certainly not his. Of the two extant theoretical books on music attributed to Euclid, the *Sectio Canonis* may be his, but the *Introductio Harmonica* is almost certainly not. There is, however, no reason to doubt Proclus's statement that Euclid wrote on music. The correctness of the attribution to Euclid of various other works extant in Arabic translations from the Greek is very doubtful.

[3] The reference is to the fact that most of Archimedes' life fell in the succeeding reigns. Ptolemy I ruled from 306 to 283 B.C., Archimedes lived from 287 to 212, and Euclid flourished about 300.

[4] I.e., the five regular polyhedra, the subject of the last book (XIII) of Euclid's *Elements* (see p. 43 below). One is hardly justified in speaking of this as the goal of the whole work.

[5] See pp. 257, 261. For the work *On Divisions* see T. L. Heath, *History of Greek Mathematics* I. 425–430.

But he is most to be admired for his *Elements of Geometry* because of the choice and arrangement of the theorems and problems made with regard to the elements. For he did not include all that he might have included, but only those theorems and problems which could fulfill the functions of elements. . . . If you seek to add or subtract anything, are you not unwittingly cast adrift from science and carried away toward falsehood and ignorance?

Now there are many things which seem to be grounded in truth and to follow from scientific principles but actually are at variance with these principles and deceive the more superficial. It was for this reason that Euclid set forth methods for intelligent discrimination in such matters, too. With these methods not only shall we be able to train beginners in this study to detect fallacies, but we shall be able to escape deception ourselves. The work in which he gives us this preparation he entitled *Pseudaria*.[1] In it he recounts in order various types of fallacies, training us to understand each type by all sorts of theorems, setting the true alongside the false, and linking the refutation of the fallacy with empirical examples. This book, therefore, is for the purification and training of the understanding, while the *Elements* contain the complete and irrefutable guide to the scientific study of the subject of geometry.

Analysis and Synthesis in Geometry

Pappus of Alexandria, *Mathematical Collection* VII. 1–3.[2] Translation of T. L. Heath *History of Greek Mathematics* II. 400–401

The so-called Ἀναλυόμενος [Treasury of Analysis] is, to put it shortly, a special body of doctrine provided for the use of those who, after finishing the ordinary Elements, are desirous of acquiring the power of solving prob-

---

[1] The *Pseudaria* is not extant.

[2] Pappus of Alexandria lived in the latter part of the third and the first part of the fourth century A.D. In addition to a work on geography and commentaries on works of Euclid, of which only fragments exist, and a commentary on Ptolemy's *Almagest*, of which a considerable portion is extant, Pappus wrote a treatise on mathematics, *Collectio Mathematica*, which has survived substantially intact. The work seems to have been intended as a handbook to be used in connection with and as a supplement to the study of the great geometers of the Hellenistic period. It contains many problems, theorems, lemmas, and solutions not elsewhere found, and collects material that had previously been scattered among scores of treatises. With the loss of so many works of Greek mathematics Pappus becomes a source of the greatest importance for the modern historian.

The *Collectio* deals with such matters as the history of the Delian Problem (the problem of finding two mean proportionals, in continued proportion, between two given straight lines); the inscribing of the five regular solids in a sphere; the properties of the spiral, the conchoid, and the quadratrix, in connection with the problems of squaring the circle and trisecting an angle; problems in isoperimetry, with analogous extensions in solid geometry; various mechanical applications of geometry; and, perhaps most important of all, the lemmas and discussions in connection with the works constituting the so-called "Treasury of Analysis."

lems which may be set them involving [the construction of] lines, and it is useful for this alone. It is the work of three men, Euclid, the author of the Elements, Apollonius of Perga, and Aristaeus the elder, and proceeds by way of analysis and synthesis.

Analysis, then, takes that which is sought as if it were admitted and passes from it through its successive consequences to something which is admitted as the result of synthesis: for in analysis we admit that which is sought as if it were already done (γεγονός) and we inquire what it is from which this results, and again what is the antecedent cause of the latter, and so on, until by so retracing our steps we come upon something already known or belonging to the class of first principles, and such a method we call analysis as being solution backwards (ἀνάπαλιν λύσιν).

But in *synthesis*, reversing the process, we take as already done that which was last arrived at in the analysis and, by arranging in their natural order as consequences what before were antecedents, and successively connecting them one with another, we arrive finally at the construction of what was sought; and this we call synthesis.

Now analysis is of two kinds, the one directed to searching for the truth and called *theoretical*, the other to finding what we are told to find and called *problematical*. (1) In the *theoretical* kind we assume what is sought as if it were existent and true, after which we pass through its successive consequences, as if they too were true and established by virtue of our hypothesis, to something admitted: then (*a*) if that something admitted is true, that which is sought will also be true and the proof will correspond in the reverse order to the analysis, but (*b*) if we come upon something admittedly false, that which is sought will also be false. (2) In the *problematical* kind we assume that which is propounded as if it were known, after which we pass through its successive consequences, taking them as true, up to something admitted: if then (*a*) what is admitted is possible and obtainable, that is, what mathematicians call *given*, what was originally proposed will also be possible, and the proof will again correspond in the reverse order to the analysis, but if (*b*) we come upon something admittedly impossible, the problem will also be impossible.[1]

< . . . So much, then, for the definition of analysis and synthesis. Of the books already mentioned the list of those forming the Ἀναλυόμενος is as follows:>[2]

---

[1] This statement could hardly be improved upon except that it ought to be added that each step in the chain of inference in the analysis must be *unconditionally convertible*; that is, when in the analysis we say that, if A is true, B is true, we must be sure that each statement is a necessary consequence of the other, so that the truth of A equally follows from the truth of B. This, however, is almost implied by Pappus when he says that we inquire, not what it is (namely, B) which follows from A, but what it is (B) from which A follows, and so on.

[For an example of analysis and synthesis see the following selection. Edd.]

[2] The material within < > is inserted to complete the translation. [Edd.]

Euclid's *Data*, one Book, Apollonius's *Cutting-off of a ratio*, two Books, *Cutting-off of an area*, two Books, *Determinate Section*, two Books, *Contacts*, two Books, Euclid's *Porisms*, three Books, Apollonius's *Inclinations* or *Vergings* (νεύσεις), two Books, the same author's *Plane Loci*, two Books, and *Conics*, eight Books, Aristaeus's *Solid Loci*, five Books, Euclid's *Surface-Loci*, two Books, Eratosthenes's *On Means*, two Books. There are in all thirty-three Books, the contents of which up to the *Conics* of Apollonius I have set out for your consideration, including not only the number of the propositions, the *diorismi*[1] and the cases dealt with in each Book, but also the lemmas which are required; indeed I have not, to the best of my belief, omitted any question arising in the study of the Books in question.[2]

AN EXAMPLE OF ANALYSIS AND SYNTHESIS

Pappus, *Mathematical Collection* VII, Proposition 108 (p. 836.24 [Hultsch])[3]

Given two points D, E within a given circle. To draw lines from D and E meeting the circumference at A, so that if these lines are produced to meet the circumference at B and C, BC will be parallel to DE.

ANALYSIS: Suppose the construction done. Draw BF tangent to the circle at B. Then

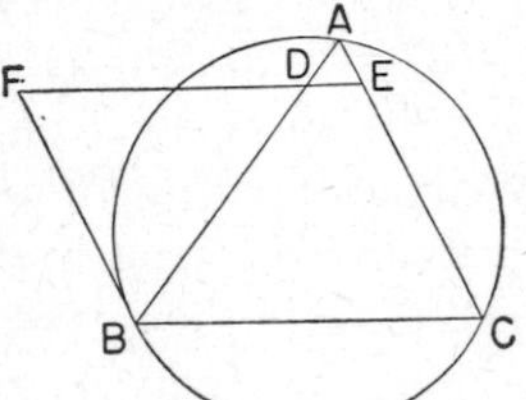

$\angle FBA = \angle C = \angle AED.$

$\therefore B, F, A,$ and E lie on the same circle.

$\therefore$ rectangle $BD.DA$ = rectangle $FD.DE$.

But rectangle $BD.DA$ is given [Euclid, *Data* 92] (for ADB is drawn from a given point D to a circle given in position).

---

[1] "The *diorismos* is a statement in advance as to when, how, and in how many ways the problem will be capable of solution" according to a gloss on this passage, which has intruded itself into the text after the definitions of analysis and synthesis. See p. 36. [Edd.]

[2] Aristaeus preceded Euclid in the fourth century B.C. His work here referred to is a treatise on conic sections, which were called "solid loci" to distinguish them from those that could be constructed with straight edge and compass, the so-called "plane loci." Apollonius of Perga, who followed Archimedes, toward the end of the third century B.C., brought the subject of conics to its highest development. Besides his *Conica* his treatises, including those referred to here, recorded extensions and discoveries of great importance in the geometry of line and circle as well as in conic sections. His researches in mathematical astronomy included the theory of epicycles and eccentric circles in connection with the planetary system (see p. 128). [Edd.]

[3] Analysis was the most general method used by the Greek geometers for the solution of more difficult problems (cf. Proclus, *Commentary on Euclid's Elements I*, p. 242.14–17). Much of the difficulty in studying Greek geometry is occasioned by the fact that, having discovered a solution analytically, the geometer removes all trace of the analysis and leaves only the synthetic proof with no clue as to the method employed. That is why examples such as those in Pappus, Books IV and VII, where both the *analysis* and *synthesis* are given, are of such interest. In the *Method* of Archimedes we have, similarly, the method of discovery preserved alongside the synthetic proof.

∴ $FD.DE$ is given.

But DE is given.

∴ FD is also given.

And D is given.

∴ F is also given.

Now FB has been drawn tangent from a given point F to a circle given in position.

∴ FB is given in position.

Now the circle, too, is given in position.

∴ point B is given.

Again, point D is given.

∴ BD is given in position.

And, since the circle is given in position, point A is given.

But both D and E are given.

∴ lines DA and AE are given in position.

SYNTHESIS: The synthesis of our problem is as follows:

A. *Construction*:

Let ABC be the circle given in position and D, E the two given points. Let any line ADB be drawn.[1]

Make the rectangle $ED.DF$ equal to the rectangle $AD.DB$.[2]

From F draw BF tangent to the circle.

Draw CEA.

B. *Proof*:

Since $\angle FBA = \angle AED$ (for points A, F, B, and E lie on the same circle), and also

$\angle FBA = \angle C$ (for FB is a tangent and BA a secant),

$\therefore \angle AED = \angle C$,

and BC is parallel to DE. Q.E.D.

NOTE ON PYTHAGOREAN GEOMETRY

Of the geometrical work associated by tradition with Pythagoras and the early Pythagoreans, the following may be noted:

1. Eudemus (Proclus, *Commentary on Euclid's Elements I*, p. 379.2) ascribed the discovery of the theorem of the sum of the angles of a triangle to the Pythagoreans. Their proof as he gives it involves knowledge of the theory of parallels, for it is the familiar proof by drawing a line through one vertex parallel to the opposite side.

2. Tradition ascribed to Pythagoras the discovery of the theorem (Euclid, *Elements* I. 47) that the square on the hypotenuse of a right-angled triangle is equal to the sum of the squares on the other two sides. How far Pythagoras or the early Pythagoreans went toward a general proof of the theorem is doubtful.

---

[1] The meaning is that *any* chord through D will enable us to obtain the product $AD.DB$. The precise location of A is determined only after F and B have been found.

[2] I.e., construct DF a fourth proportional to ED, AD, and DB.

3. Of special importance is the so-called method of "application of areas"—i.e., the construction of a rectangle or parallelogram equal in area to a given area and having as base a given line, or a segment thereof, or the line produced. This method, whose invention is ascribed to the Pythagoreans, was later extended, probably by Eudoxus, and is important in the Greek geometrical algebra, being equivalent to the solution of the general quadratic equation, as set forth in Euclid. It is also a forerunner of the theory of conic sections.

Cf. Plutarch, *On the Epicurean Life* 1094 B:

"And Pythagoras sacrificed an ox upon discovering this theorem . . . whether it was the theorem that the square on the hypotenuse is equal to the sum of the squares about the right angle, or the problem about the application (παραβολή) of the area."

Plutarch, *Convivial Questions* 720 A:

"For among the most essentially geometrical theorems or rather problems is the following: given two figures, to apply a third equal to the one and similar to the other. Upon this discovery, they say, Pythagoras made a sacrifice, for this is without doubt much more profound and scientific than the theorem in which he proved that the square on the hypotenuse is equal to the sum of the squares on the sides containing the right angle."

Proclus Diadochus, *Commentary on Euclid Elements I*, pp. 419.15–420.12:

"These things are ancient, says Eudemus, and they are discoveries of the Muse of the Pythagoreans—the application of areas (παραβολὴ τῶν χωρίων), their exceeding (ὑπερβολή) and their falling short (ἔλλειψις). It was from the Pythagoreans that more recent geometers took these names and applied them to the so-called conic curves, calling one of them the *parabola*, another the *hyperbola*, and the third the *ellipse*, while those wonderful men of old saw what was signified by these names in connection with the construction in a plane of areas upon a finite straight line.

"For when a straight line is set out and you place the given area along the whole of this line they say that you are then *applying* (παραβάλλειν) that area; but when you make the length of the area greater than that of the given straight line, you are then making it *exceed* (ὑπερβάλλειν), and when you make it less, so that after the area is drawn a portion of the given straight line protrudes, you are making it fall short (ἐλλείπειν). Thus in the sixth book Euclid refers to the *exceeding* and the *falling short* but here [I, Prop. 44] he had need of *application*, since he wished to apply to the given straight line an area equal to the given triangle, so that we might have not merely the construction of a parallelogram equal to the given triangle, but its application to a finite straight line." [1]

4. There is considerable doubt as to when the irrational was discovered. The evidence seems to point not to Pythagoras himself, but to some Pythagorean, possibly Hippasus of Metapontum, toward the middle of the fifth century (see pp. 14, 296). Indeed, those (e.g., E. Frank) who minimize the scientific achievements of the early Pythagoreans are inclined to put the discovery even later, about 400 B.C.

The Pythagorean development of "side" and "diameter" numbers may have been used as a means of approximating the value of $\sqrt{2}$. This subject, too, has important bearings in connection with Greek geometrical algebra and indeterminate analysis.

Cf. Proclus, *Commentary on Plato's Republic* II. 27.11–17:

"The Pythagoreans set forth this elegant theorem about the diameters and sides, that the diameter when added to the side of which it is the diameter becomes a side, and the

---

[1] Various propositions of Euclid's *Elements*, Book II, develop this Pythagorean method and are equivalent to the geometrical solution of such equations as $x^2 + ax = b^2$. Extensions in Book VI, due perhaps to Eudoxus, are equivalent to the solution of the most general type of quadratic, with *diorismi* to exclude imaginary or negative solutions.

side when doubled and added to its own diameter becomes a diameter. And this is proved by him [Euclid] geometrically on the basis of the proposition in the second book."

There follows a statement of Euclid, *Elements* II. 10, that if AB is bisected at C and produced to D, then $AD^2 + BD^2 = 2\,(AC^2 + CD^2)$.

A C B D

Now if AC is x and BD is y, we have

$$(2x + y)^2 + y^2 = 2x^2 + 2(x + y)^2$$

or

$$(2x + y)^2 - 2(x + y)^2 = 2x^2 - y^2.$$

That is, if values be assigned to y, x which will satisfy one of the equations $2x^2 - y^2 = \pm 1$, the corresponding values of $(2x + y)$ and $(x + y)$ when substituted for y and x, respectively, will satisfy the other.

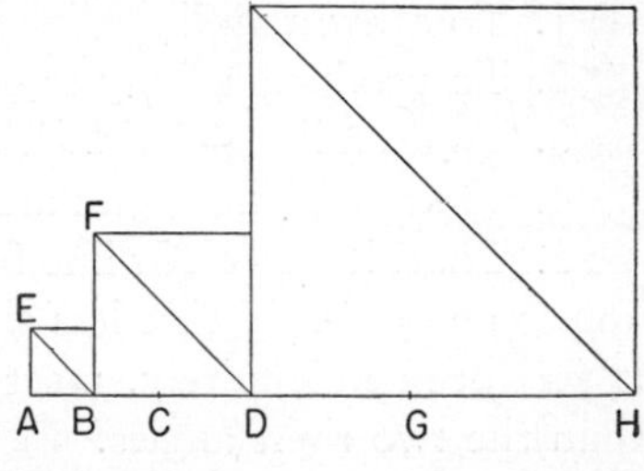

Commencing with $y = 1$ and $x = 1$, we have the successive pairs of integral diameter and side numbers (3,2) (7,5) (17,12) (41,29) . . . (Cf. Theon of Smyrna, pp. 42.10–44.17 [Hiller].) The pairs of numbers, written as fractions, form increasingly close approximations to $\sqrt{2}$, since they are rational approximations to the sides and hypotenuses (diameters) of increasing isosceles right triangles. This is seen from the diagram consisting of successive squares with $AB = BC$, $CD = EB$, $DG = BD$, $GH = FD$, etc.

5. The tradition of the discovery of the five regular polyhedra,[1] the only possible regular polyhedra (i.e., tetrahedron, cube, octahedron, icosahedron, and dodecahedron), is somewhat confused. It seems that the early Pythagoreans investigated their construction.[2] But the working out of mathematical proofs in this connection is probably due to Theaetetus. The subject of the regular solids is later considered in great detail and with great rigor by Euclid, *Elements* XIII. We may quote here the scholiast on Euclid, *Elements* XIII, who writes (V, 654 [Heiberg]):

"In this book, the thirteenth, the five so-called Platonic figures are described. They are not Plato's, however, but three of them, the cube, pyramid, and dodecahedron, belong to the Pythagoreans, while the octahedron and icosahedron are due to Theaetetus. They were connected with the name of Plato because he treats of them in the *Timaeus*."

The regular polyhedra were given cosmic significance in early Pythagorean philosophy and were connected with the theory of the elements.[3] The tetrahedron was connected with fire, the cube with earth, the octahedron with air, and the icosahedron with water. Aristotle, developing a Pythagorean idea, connected the dodecahedron with the fifth element, the aether of the celestial sphere.

It has in recent years been argued that the tradition, as we have it, gives too much credit to the early Pythagoreans and too little to the Ionians in connection with the development of mathematics before Euclid. And there has also been a tendency to place certain discoveries

---

[1] Archimedes investigated so-called "semiregular solids," i.e., those having regular polygons, not all of the same kind, as faces. See Pappus, *Mathematical Collection* V.

[2] Iamblichus, *Life of Pythagoras* 18.88, attributes the discovery of the dodecahedron to Hippasus. Hippasus no doubt made investigations, but the canonical proofs as we know them probably belong to a later generation.

[3] See Philolaus in Diels, *Frag. d. Vors.* I^5. 314.12–15, Aëtius, *Placita* II. 6.5, and the *locus classicus*, Plato, *Timaeus* 53 C–55 C. These solids are called "Platonic bodies" (see p. 37).

in mathematics, acoustics, and astronomy later than they had previously been thought. Some of the difficulty on this latter point might be avoided if, in interpreting the tradition, we remember what has often been pointed out, that the requirements for a mathematical construction or proof may vary from age to age, and indeed in early Greek mathematics probably varied from generation to generation.

Elementary Geometry

POSTULATES

Euclid, *Elements* I, Postulates. Translation of T. L. Heath, *The Thirteen Books of Euclid's Elements* I. 154

Let the following be postulated:[1]

1. To draw a straight line from any point to any point.
2. To produce a finite straight line continuously in a straight line.
3. To describe a circle with any centre and distance.
4. That all right angles are equal to one another.
5. That, if a straight line falling on two straight lines make the interior angles on the same side less than two right angles, the two straight lines, if produced indefinitely, meet on that side on which are the angles less than the two right angles.

The Parallel Postulate

Proclus Diadochus, *Commentary on Euclid's Elements I*, pp. 191.21–193.9 (Friedlein). Translation of T. L. Heath, *The Thirteen Books of Euclid's Elements* I. 202–203[2]

This [fifth postulate] ought even to be struck out of the Postulates altogether; for it is a theorem involving many difficulties which Ptolemy, in a certain book, set himself to solve, and it requires for the demonstration of it a number of definitions as well as theorems. And the converse of it is actually proved by Euclid himself as a theorem. It may be that some would be deceived and would think it proper to place even the assumption in question among the postulates as affording, in the lessening of the two right angles, ground for such an instantaneous belief that the straight lines converge and meet. To such as these Geminus correctly replied that we have learned from the very pioneers of this science not to have any regard to mere plausible imaginings when it is a question of the reason-

---

[1] The postulates, in connection with the definitions and axioms, constitute the principles of Euclidean geometry. The first three are postulates of construction. The fourth is required in order that the fifth may be applied. The fifth, which is probably Euclid's own attempt to found a sound theory of parallels, has given rise to the greatest controversy. Not only were unsuccessful efforts made to prove the postulate, but non-Euclidean geometries were developed in modern times on the basis of a denial of the postulate. [Edd.]

[2] The controversy on Euclid's fifth postulate about parallel lines began in antiquity. Attempts to prove the postulate or to substitute another definition of parallel lines were made by Posidonius, Geminus, Ptolemy, Proclus, and others. [Edd.]

ings to be included in our geometrical doctrine. For Aristotle says that it is as justifiable to ask scientific proofs of a rhetorician as to accept mere plausibilities from a geometer; and Simmias is made by Plato to say that he recognizes as quacks those who fashion for themselves proofs from probabilities. So in this case the fact that, when the right angles are lessened, the straight lines converge is true and necessary; but the statement that, since they converge more and more as they are produced, they will sometime meet is plausible but not necessary, in the absence of some argument showing that this is true in the case of straight lines. For the fact that some lines exist which approach indefinitely, but yet remain non-secant (ἀσύμπτωτοι), although it seems improbable and paradoxical, is nevertheless true and fully ascertained with regard to other species of lines.[1] May not then the same thing be possible in the case of straight lines which happens in the case of the lines referred to? Indeed, until the statement in the Postulate is clinched by proof, the facts shown in the case of other lines may direct our imagination the opposite way. And, though the controversial arguments against the meeting of the straight lines should contain much that is surprising, is there not all the more reason why we should expel from our body of doctrine this merely plausible and unreasoned (hypothesis)?

It is then clear from this that we must seek a proof of the present theorem, and that it is alien to the special character of postulates. But how it should be proved, and by what sort of arguments the objections taken to it should be removed, we must explain at the point where the writer of the Elements is actually about to recall it and use it as obvious. It will be necessary at that stage to show that its obvious character does not appear independently of proof, but is turned by proof into matter of knowledge.

THE METHOD OF EXHAUSTION

Euclid, *Elements* XII, Proposition 2. Translation of T. L. Heath, *The Thirteen Books of Euclid's Elements* III. 371–373

Circles are to one another as the squares on the diameters.

Let *ABCD*, *EFGH* be circles, and *BD*, *FH* their diameters; I say that, as the circle *ABCD* is to the circle *EFGH*, so is the square on *BD* to the square on *FH*.

For, if the square on *BD* is not to the square on *FH* as the circle *ABCD* is to the circle *EFGH*,
then, as the square on *BD* is to the square on *FH*, so will the circle *ABCD* be either to some less area than the circle *EFGH*, or to a greater.

---

[1] The reference is to curves like the hyperbola and conchoid, which have asymptotes. [Edd.]

First, let it be in that ratio to a less area *S*.

Let the square *EFGH* be inscribed in the circle *EFGH*;

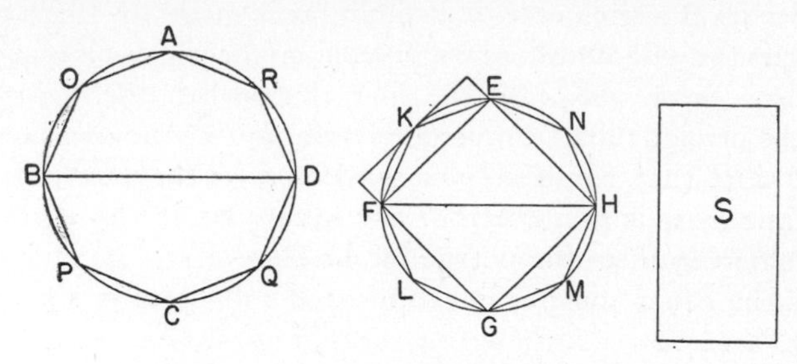

then the inscribed square is greater than the half of the circle *EFGH*, inasmuch as, if through the points *E*, *F*, *G*, *H* we draw tangents to the circle, the square *EFGH* is half the square circumscribed about the circle, and the circle is less than the circumscribed square;
hence the inscribed square *EFGH* is greater than the half of the circle *EFGH*.

Let the circumferences *EF*, *FG*, *GH*, *HE* be bisected at the points *K*, *L*, *M*, *N*,
and let *EK*, *KF*, *FL*, *LG*, *GM*, *MH*, *HN*, *NE* be joined;
therefore each of the triangles *EKF*, *FLG*, *GMH*, *HNE* is also greater than the half of the segment of the circle about it, inasmuch as, if through the points *K*, *L*, *M*, *N* we draw tangents to the circle and complete the parallelograms on the straight lines *EF*, *FG*, *GH*, *HE*, each of the triangles *EKF*, *FLG*, *GMH*, *HNE* will be half of the parallelogram about it,
while the segment about it is less than the parallelogram;
hence each of the triangles *EKF*, *FLG*, *GMH*, *HNE* is greater than the half of the segment of the circle about it.[1]

Thus, by bisecting the remaining circumferences and joining straight lines, and by doing this continually, we shall leave some segments of the circle which will be less than the excess by which the circle *EFGH* exceeds the area *S*.

For it was proved by the first theorem of the tenth book that, if two unequal magnitudes be set out, and if from the greater there be subtracted a magnitude greater than the half, and from that which is left a greater than the half, and if this be done continually, there will be left some magnitude which will be less than the lesser magnitude set out.

[1] In order to show that by doubling the number of sides of the inscribed polygon indefinitely, the area of the circle will be "exhausted," Euclid seeks to show that the difference in area between any polygon in the series and the circle is less than half the difference between the preceding polygon and the circle. [Edd.]

[1] Let segments be left such as described, and let the segments of the circle *EFGH* on *EK*, *KF*, *FL*, *LG*, *GM*, *MH*, *HN*, *NE* be less than the excess by which the circle *EFGH* exceeds the area *S*.

Therefore the remainder, the polygon *EKFLGMHN*, is greater than the area *S*.

Let there be inscribed, also, in the circle *ABCD* the
polygon *AOBPCQDR* similar to the polygon *EKFLGMHN*;
therefore, as the square on *BD* is to the square on *FH*, so is the
polygon *AOBPCQDR* to the polygon *EKFLGMHN*. [XII. 1][2]

But, as the square on *BD* is to the square on *FH*, so also is the circle *ABC* to the area *S*;
therefore also, as the circle *ABCD* is to the area *S*, so is the polygon *AOBPCQDR* to the polygon *EKFLGMHN*; [V. 11][3]
therefore, alternately, as the circle *ABCD* is to the polygon inscribed in it, so is the area *S* to the polygon *EKFLGMHN*. [V. 16][4]

---

[1] From this point on the proof is given in modern notation by Heath as follows: "Now let X, X' be the areas of the circles, d, d' their diameters, respectively.

Then, if $X:X' \neq d^2:d'^2$,

$$d^2:d'^2 = X:S,$$

where S is some area either greater or less than X'.

I. Suppose $S < X'$.

Continue the construction of polygons in X' until we arrive at one which leaves over segments together less than the excess of X' over S, i.e. a polygon such that

$$X' > (\text{polygon in } X') > S.$$

Inscribe in the circle X a polygon similar to that in X'.

Then (polygon in X) : (polygon in X') $= d^2:d'^2$ [XII.1] $= X:S$, by hypothesis;
and, alternately, (polygon in X)$:X =$ (polygon in X')$:S$.

But (polygon in X) $< X$;
therefore (polygon in X') $< S$.

But, by construction, (polygon in X') $> S$:
which is impossible.

Hence S cannot be *less* than X' as supposed.

II. Suppose $S > X'$.

Since $d^2:d'^2 = X:S$,
we have, inversely, $d'^2:d^2 = S:X$.

Suppose that $S:X = X':T$,
whence, since $S > X'$, $X > T$. [V. 14]

Consequently $d'^2:d^2 = X':T$,
where $T < X$.

This can be proved impossible in exactly the same way as shown in Part I.

Hence S cannot be greater than X' as supposed.

Since then S is neither greater nor less than X', $S = X'$, and therefore $d^2:d'^2 = X:X'$. . . ."

[2] Similar polygons inscribed in circles are to one another as the squares on the diameters. [Edd.]

[3] This proposition states, in effect, that if $a:b = c:d$, and $c:d = e:f$, then $a:b = e:f$. [Edd.]

[4] This proposition states, in effect, that if $a:b = c:d$, then $a:c = b:d$. [Edd.]

But the circle *ABCD* is greater than the polygon inscribed in it; therefore the area *S* is also greater than the polygon *EKFLGMHN*.

But it is also less:
which is impossible.

Therefore, as the square on *BD* is to the square on *FH*, so is not the circle *ABCD* to any area less than the circle *EFGH*.

Similarly we can prove that neither is the circle *EFGH* to any area less than the circle *ABCD* as the square on *FH* is to the square on *BD*.

I say next that neither is the circle *ABCD* to any area greater than the circle *EFGH* as the square on *BD* is to the square on *FH*.

For, if possible, let it be in that ratio to a greater area *S*.

Therefore, inversely, as the square on *FH* is to the square on *DB*, so is the area *S* to the circle *ABCD*.

But, as the area *S* is to the circle *ABCD*, so is the circle *EFGH* to some area less than the circle *ABCD*;
therefore also, as the square on *FH* is to the square on *BD*, so is the circle *EFGH* to some area less than the circle *ABCD*: [V. 11]
which was proved impossible.

Therefore, as the square on *BD* is to the square on *FH*, so is not the circle *ABCD* to any area greater than the circle *EFGH*.

And it was proved that neither is it in that ratio to any area less than the circle *EFGH*;
therefore, as the square on *BD* is to the square on *FH*, so is the circle *ABCD* to the circle *EFGH*

Therefore, etc.[1] Q.E.D.

THE "PYTHAGOREAN" THEOREM

Euclid, *Elements* I, Proposition 47. Translation of T. L. Heath, *The Thirteen Books of Euclid's Elements* I. 349–350

In right-angled triangles the square on the side subtending the right angle is equal to the squares on the sides containing the right angle.

Let *ABC* be a right-angled triangle having the angle *BAC* right;

I say that the square on *BC* is equal to the squares on *BA*, *AC*.

For let there be described on *BC* the square *BDEC*, and on *BA*, *AC* the squares *GB*, *HC*; [I. 46]
through *A* let *AL* be drawn parallel to either *BD* or *CE*, and let *AD*, *FC* be joined.

---

[1] This form of the proof may be due to Eudoxus (see p. 36). Note that Archimedes, in applying the method of exhaustion (e.g., in *Measurement of a Circle*, Prop. I, p. 60, below), proved the second part by *circumscribing* successive polygons. It was thus unnecessary for him to reduce the second case to the first. Eudemus attributed a proof of this proposition to Hippocrates of Chios (see p. 55). The nature of his proof is not known. [Edd.]

Then, since each of the angles *BAC*, *BAG* is right, it follows that with a straight line *BA*, and at the point *A* on it, the two straight lines *AC*, *AG* not lying on the same side make the adjacent angles equal to two right angles;

therefore *CA* is in a straight line with *AG*. [I. 14]

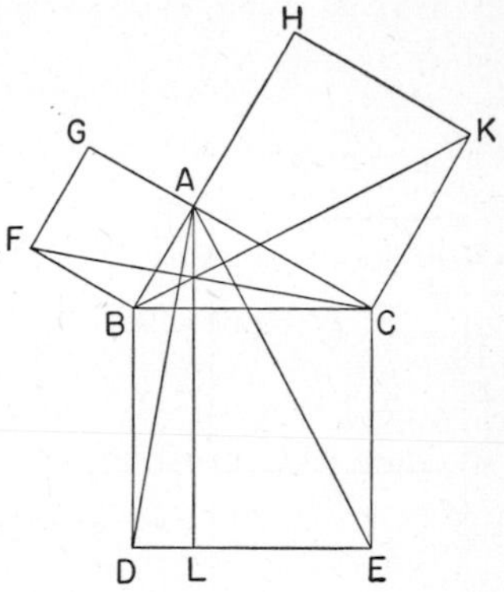

For the same reason

BA is also in a straight line with *AH*.

And, since the angle *DBC* is equal to the angle *FBA*:

for each is right:

let the angle *ABC* be added to each;

therefore the whole angle *DBA* is equal to the whole angle *FBC*. [*C.N.* 2]

And, since *DB* is equal to *BC*, and *FB* to *BA*,

the two sides *AB*, *BD* are equal to the two sides *FB*, *BC* respectively;

and the angle *ABD* is equal to the angle *FBC*;

therefore the base *AD* is equal to the base *FC*,

and the triangle *ABD* is equal to the triangle *FBC*. [I. 4]

Now the parallelogram *BL* is double of the triangle *ABD*, for they have the same base *BD* and are in the same parallels *BD*, *AL*. [I. 41]

And the square *GB* is double of the triangle *FBC*,

for they again have the same base *FB* and are in the same parallels *FB*, *GC*. [I. 41]

But the doubles of equals are equal to one another.

Therefore the parallelogram *BL* is also equal to the square *GB*.

Similarly, if *AE*, *BK* be joined,

the parallelogram *CL* can also be proved equal to the square *HC*;

therefore the whole square *BDEC* is equal to the two squares *GB*, *HC*. [*C.N.* 2]

And the square *BDEC* is described on *BC*, and the squares *GB*, *HC* on *BA*, *AC*.

Therefore the square on the side *BC* is equal to the squares on the sides *BA*, *AC*.

Therefore etc.[1] Q.E.D.

---

[1] This is the famous "windmill" proof of the so-called "Pythagorean theorem." The tradition connecting Pythagoras with this theorem is not clear, and it may be doubted that he had arrived at a general proof. In any case, the proof as given by Euclid is probably an original proof for the purpose of including the theorem in Book I rather than postponing it to the books on proportion.

Whether or not the Pythagoreans had a general proof, their work in the theory of numbers indicates an acceptance of its truth (see pp. 21–23). [Edd.]

THE GOLDEN SECTION

Euclid, *Elements* II, Proposition 11. Translation of T. L. Heath, *The Thirteen Books of Euclid's Elements* I. 402

To cut a given straight line so that the rectangle contained by the whole and one of the segments is equal to the square on the remaining segment.[1]

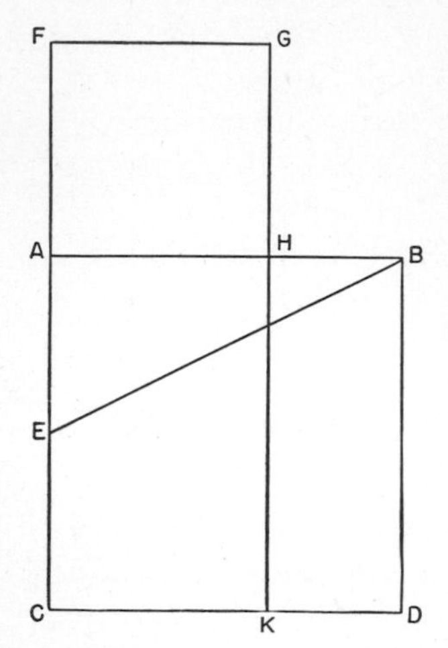

Let AB be the given straight line;
thus it is required to cut AB so that the rectangle contained by the whole and one of the segments is equal to the square on the remaining segment.

For let the square $ABDC$ be described on AB; [I. 46]
let AC be bisected at the point E, and let BE be joined;
let CA be drawn through to F, and let EF be made equal to BE;
let the square FH be described on AF, and let GH be drawn through to K.

I say that AB has been cut at H so as to make the rectangle contained by AB, BH equal to the square on AH.

For, since the straight line AC has been bisected at E, and FA is added to it,

the rectangle contained by CF, FA together with the square on AE is equal to the square on EF. [II. 6]

But EF is equal to EB;

therefore the rectangle CF, FA together with the square on AE is equal to the square on EB.

But the squares on BA, AE are equal to the square on EB, for the angle at A is right;

therefore the rectangle CF, FA together with the square on AE is equal to the squares on BA, AE.

Let the square on AE be subtracted from each;

---

[1] This is essentially the problem of dividing a line into extreme and mean ratio: the smaller segment is to the larger as the larger is to the whole line. The construction given is held by Heath to be pre-Euclidean; an alternative method is given later by Euclid (VI. 30).

The solution of the problem is necessary in order to construct the regular pentagon, and hence the regular dodecahedron, one of the five so-called "Platonic figures."

The ratio of the larger segment to the whole line is $(\sqrt{5}-1)/2$ (approximately .618). A rectangle of unit length with height equal to $(\sqrt{5}-1)/2$ was thought to have the most beautiful proportions. Because of such aesthetic considerations the names "golden section" and "golden rectangle" came into use in modern times. [Edd.]

therefore the rectangle *CF*, *FA* which remains is equal to the square on *AB*.

Now the rectangle *CF*, *FA* is *FK*, for *AF* is equal to *FG*;
and the square on *AB* is *AD*;

therefore *FK* is equal to *AD*.

Let *AK* be subtracted from each;

therefore *FH* which remains is equal to *HD*.

And *HD* is the rectangle *AB*, *BH*, for *AB* is equal to *BD*;
and *FH* is the square on *AH*;

therefore the rectangle contained by *AB*, *BH* is equal to the square on *HA*;

therefore the given straight line *AB* has been cut at *H* so as to make the rectangle contained by *AB*, *BH* equal to the square on *HA*. Q.E.F.

THE HORN-ANGLE

Euclid, *Elements* III, Proposition 16. Translation of T. L. Heath, *The Thirteen Books of Euclid's Elements* II. 37–39

The straight line drawn at right angles to the diameter of a circle from its extremity will fall outside the circle, and into the space between the straight line and the circumference another straight line cannot be interposed; further the angle of the semicircle is greater, and the remaining angle less, than any acute rectilineal angle.

Let *ABC* be a circle about *D* as centre and *AB* as diameter;
I say that the straight line drawn from *A* at right angles to *AB* from its extremity will fall outside the circle.

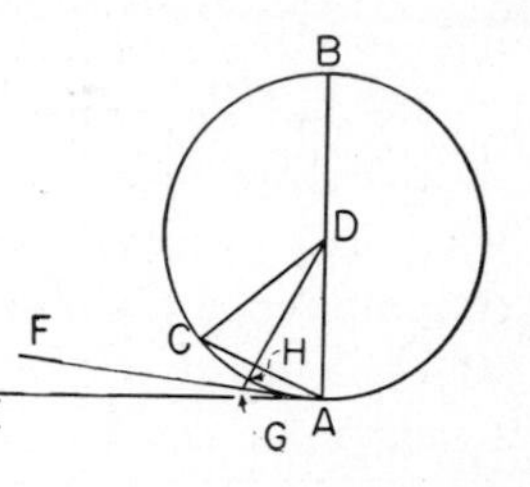

For suppose it does not, but, if possible, let it fall within as *CA*, and let *DC* be joined.

Since *DA* is equal to *DC*,

the angle *DAC* is also equal to the angle *ACD*. [I. 5]

But the angle *DAC* is right;

therefore the angle *ACD* is also right:

thus, in the triangle *ACD*, the two angles *DAC*, *ACD* are equal to two right angles: which is impossible. [I. 17]

Therefore the straight line drawn from the point *A* at right angles to *BA* will not fall within the circle.

Similarly we can prove that neither will it fall on the circumference;

therefore it will fall outside.

Let it fall as *AE*;

I say next that into the space between the straight line *AE* and the circumference *CHA* another straight line cannot be interposed.

For, if possible, let another straight line be so interposed, as *FA*, and let *DG* be drawn from the point *D* perpendicular to *FA*.

Then, since the angle *AGD* is right,

and the angle *DAG* is less than a right angle,

AD is greater than *DG*. [I. 19]

But *DA* is equal to *DH*;

therefore *DH* is greater than *DG*, the less than the greater: which is impossible.

Therefore another straight line cannot be interposed into the space between the straight line and the circumference.

I say further that the angle of the semicircle contained by the straight line *BA* and the circumference *CHA* is greater than any acute rectilineal angle,

and the remaining angle contained by the circumference *CHA* and the straight line *AE* is less than any acute rectilineal angle.

For, if there is any rectilineal angle greater than the angle contained by the straight line *BA* and the circumference *CHA*, and any rectilineal angle less than the angle contained by the circumference *CHA* and the straight line *AE*, then into the space between the circumference and the straight line *AE* a straight line will be interposed such as will make an angle contained by straight lines which is greater than the angle contained by the straight line *BA* and the circumference *CHA*, and another angle contained by straight lines which is less than the angle contained by the circumference *CHA* and the straight line *AE*.

But such a straight line cannot be interposed;

therefore there will not be any acute angle contained by straight lines which is greater than the angle contained by the straight line *BA* and the circumference *CHA*, nor yet any acute angle contained by straight lines which is less than the angle contained by the circumference *CHA* and the straight line *AE*.[1]

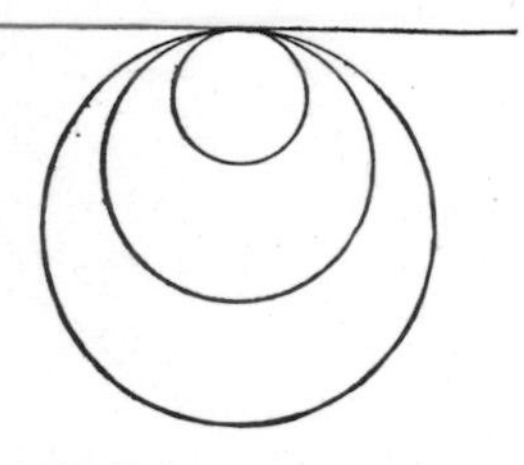

[1] The nature of the angle between the diameter and the circumference and of the so-called "horn-angle," the angle between the tangent and the circumference or between two tangent circumferences, was the subject of widespread discussion in antiquity. The angle between the diameter and the circumference seems to decrease, while the angle between the tangent and the circumference seems to increase as the circles become smaller. And yet the former angle always remains greater, and the latter angle smaller than any acute rectilineal angle.

This paradox may well have exercised men like Zeno of Elea, Protagoras, Democritus, and Eudoxus. Not only were concepts of tangency and of angle involved, but, like so many other paradoxes of the Greeks, the nature of continuity itself.

The controversy continued in the Middle Ages and in modern times. Among those who dealt with the problem were Campanus in the thirteenth century, Cardan, Peletier, and Vieta in the sixteenth, and Galileo and Wallis in the seventeenth. For a summary of recent studies

PORISM. From this it is manifest that the straight line drawn at right angles to the diameter of a circle from its extremity touches the circle.

Q.E.D.

THE THREE FAMOUS GEOMETRICAL PROBLEMS OF THE GREEKS

Not very long after the beginnings of Greek geometry, three problems—the squaring of the circle, the doubling of the cube, and the trisection of an angle—began to occupy the mathematicians. Unable to solve these problems by the use merely of ruler and compass—the impossibility of such solutions was definitely proved only in the nineteenth century—the Greeks attacked the problems by the use of conics and certain higher plane curves. At the same time, proceeding in another direction, they transformed the problems into problems of obtaining increasingly accurate approximations to irrational and transcendental numbers.

1. THE SQUARING OF THE CIRCLE

The squaring of the circle was first viewed as the problem of constructing a rectilinear area equal to that of a given circle.

Cf. Proclus Diadochus, *Commentary on Euclid's Elements I*, pp. 422.24–423.5 (Friedlein).

It was, I think, this problem [i.e., Euclid, *Elements* I. 45—to construct, in a given rectilinear angle, a parallelogram equal to a given rectilinear figure] which led the ancients to investigate the squaring of the circle. For if a parallelogram may be found equal to every rectilinear figure, it is worth investigating whether it is not also possible to prove rectilinear figures equal to circular. Now Archimedes proved that every circle is equal to a right-angled triangle of which one of the sides about the right angle is equal to the radius of the circle, and the base is equal to the perimeter.

a. *Early Attempts*

One of the earliest attempts at solution was that of Antiphon, the Sophist, a contemporary of Socrates, who proceeded by continually doubling the number of sides of an inscribed regular polygon. Here we have the germ of the method of exhaustion and the method of approximation used by Archimedes. If, as the commentators indicate, Antiphon believed he would ultimately make the polygon coincide with the circle, he may have been expressing the opinion of Sophists like Protagoras, who denied the existence of pure mathematical lines, tangencies, etc. In any case, as Aristotle says (*Physics* 185*a*15), "while it is the task of the geometer to refute the quadrature by segments it is not the task of the geometer to refute the quadrature of Antiphon," for Antiphon proceeds from hypotheses that are contrary to those of geometry.[1]

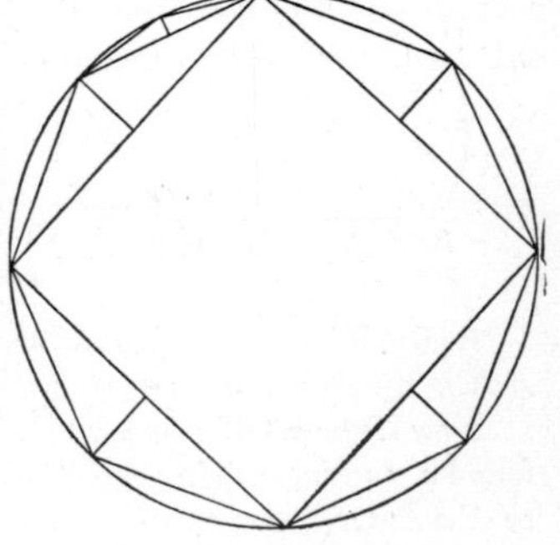

Cf. the attempted solution of Bryson (*fl. ca.* 400 B.C.) as given by Alexander, *Comm. Arist. Soph. Elench.*, p. 90.10–21 (Wallies):

---

on the horn angle (including that formed by non-circular arcs) see E. Kasner, "The Recent Theory of the Horn Angle," *Scripta Mathematica* 11 (1945) 263–267. [Edd.]

[1] See Themistius, *Commentary on Aristotle's Physics*, pp. 3.30–4.7 (Schenkl), and Simplicius, *Commentary on Aristotle's Physics*, pp. 54.20–55.24 (Diels).

But Bryson's quadrature of the circle is a piece of captious sophistry for it does not proceed on principles proper to geometry but on principles of more general application. He circumscribes a square about the circle, inscribes another within the circle, and constructs a third square between[1] the first two. He then says that the circle between the two squares [i.e., the inscribed and the circumscribed] and also the intermediate square are both smaller than the outer square and larger than the inner, and that things larger and smaller than the same things, respectively, are equal. Therefore, he says, the circle has been squared. But to proceed in this way is to proceed on general and false assumptions, general because these assumptions might be generally applicable to numbers, times, spaces, and other fields, but false because both 8 and 9 are smaller and larger than 10 and 7 respectively, but are not equal to each other.

b. *Hippocrates of Chios*

Somewhat before Bryson, Hippocrates of Chios had succeeded in squaring certain special cases of lunes, i.e., he had found that the area between the arcs of two intersecting circles with convexity on the same side was, in certain special cases, equal to a rectilinear figure. Now some, among them apparently Aristotle, took it that Hippocrates believed that he had solved the problem of squaring the circle. At any rate this is the occasion of our direct knowledge of Hippocrates' work. For in commenting on the reference in Aristotle to a "quadrature by segments" (p. 53, above) the commentators Themistius, Philoponus, and Simplicius indicate that the reference is to Hippocrates.

The most elaborate account is given by Simplicius (*Commentary on Aristotle's Physics*, pp. 58–69 [Diels]), in two parts, the first presumably based on the commentator Alexander, the second, much more important, expressly quoting Eudemus's *History of Geometry*.

In the first part Simplicius gives the simple proof that if a semicircle ABC be described on the side of an inscribed square as diameter, the shaded lune $ABC = \triangle ACD$. (I)

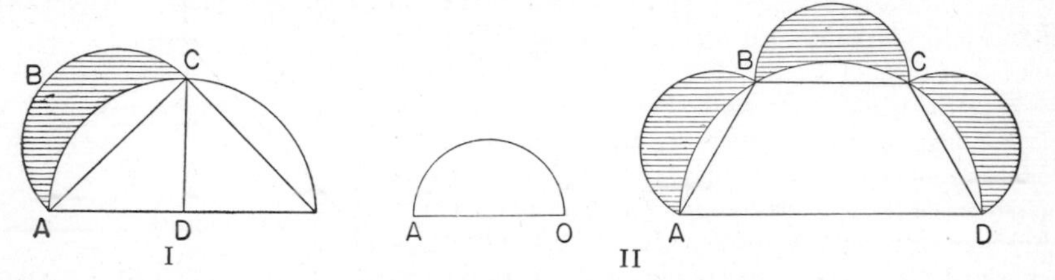

Again, if semicircles be described on three of the sides of a regular hexagon, the trapezium $ABCD$ equals the three shaded lunes plus semicircle AO (where $AO = \frac{1}{2}AD$). (II)

Now if the rectilinear figure equal to the lunes be subtracted from the trapezium we have found a rectilinear figure equal to semicircle AO, etc. This is the argument which, according to Alexander (*op. cit.*, p. 60.18), Aristotle was refuting. The fallacy in it is obvious, Simplicius tells us, for the lune squared in case I is not the same as the lune in case II.

After considerable discussion, Simplicius proceeds (p. 60.22):

[1] Precisely how the third square is constructed is not indicated. Since the area of the smaller square is $2r^2$ (where r is the radius of the circle) and of the larger is $4r^2$, it is clear that, whether the third square is a geometric or an arithmetic mean between the other two, it will be considerably smaller than the circle. It has been held that Bryson's method was to

But Eudemus in his *History of Geometry* says that Hippocrates did not prove the quadrature of the lune with reference to the side merely of a square [i.e., case I above], but generally, as one would say. For if every lune has its outer circumference either equal to, greater than, or less than a semicircle, and if Hippocrates squares the lune having its outer circumference equal to, greater than, and less than a semicircle, he would seem to have established a general proof. I shall set forth verbatim what Eudemus said, adding for the sake of clearness some few things recalled from Euclid's *Elements*. This I do because of the abbreviated style of Eudemus, who in the ancient manner gave his proofs in concise form.

There follows (pp. 61–68) the oldest Greek mathematical passage of any length that is extant, one that has been much studied for the history of mathematical terminology and method. We translate a few lines (p. 61.1–9) and then summarize the rest of the argument.

The quadratures of lunes which were held to belong to an unusual type of problem because of their relation to the circle were first written about by Hippocrates and seemed to be given correctly. We shall therefore consider them at length and go through them. Now he made it his leading principle, and set it forth as the first of the propositions useful for these quadratures, that similar segments of circles have to each other the same ratio as the square of their bases. And this he demonstrated by showing that the squares of diameters have the same ratio as their circles.[1]

Eudemus then describes three special cases in which Hippocrates found a lune equal to a rectilinear figure, viz., the cases in which the similar segments compared are in the ratio of 1:2 (case III); 1:3 (case IV); 2:3 (case V). In case III the outer circumference of the lune is equal to a semicircle, in case IV it is greater than and in case V it is less than a semicircle. Case III is, of course, the equivalent of case I.

CASE III. Lune $ABCEA = \triangle ABC$ (seg. AC is similar to seg. AB. AB and BC are sides of an inscribed square. Arc ABC is a semicircle.)

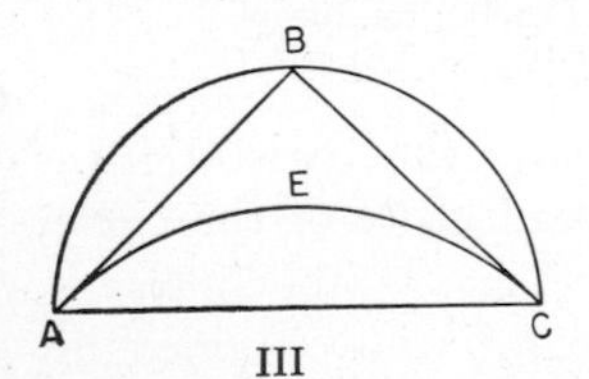

III

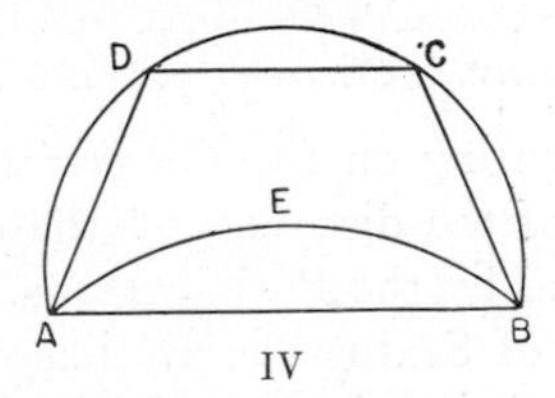

IV

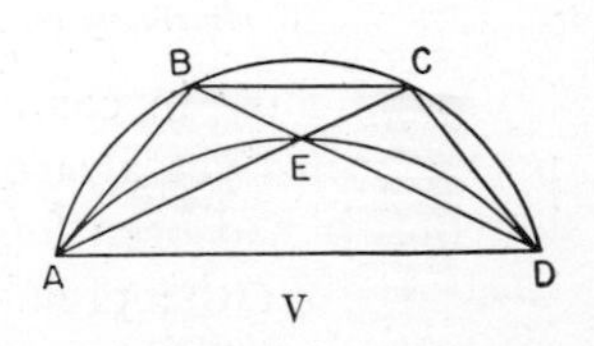

V

CASE IV. Lune $ADCBEA$ = trapezium $ABCD$ (seg. AEB is similar to segments AD, DC, and CB. $AD = DC = CB = AB/\sqrt{3}$. Arc $ADCB$ is shown to be greater than a semicircle.)

CASE V. Lune $ABCDEA$ = pentagon $ABCDEA$ (seg. AED is similar to segments AB, BC, and CD. $AB = BC = CD = \sqrt{2}AE/\sqrt{3} = \sqrt{2}ED/\sqrt{3}$. Arc $ABCD$ is shown to be less than a semicircle. The construction could be performed by the method of applica-

---

use inscribed and circumscribed polygons and keep multiplying the number of sides (a procedure by which Archimedes arrived at his approximation of π), but just how he finally constructed the intermediate polygon that was to be equal to the circle in area is not clear.

[1] See pp. 45–48.

tion of areas of Euclid, II.6. This may have been discovered by the Pythagoreans before the time of Hippocrates, but it is possible that he used mechanical means for the construction.)

Now in addition to squaring the three types of lunes in cases III, IV, and V, Hippocrates found that the sum of a lune and circle was equal to the sum of two rectilinear figures in a special case.

Lune $ABCDA$ + small circle = $\triangle ABC$ + small hexagon. (Arc ADC is similar to arc AB and to the other arcs of the two hexagons. 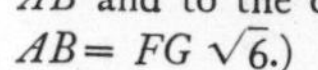 $AB = FG\sqrt{6}$.)

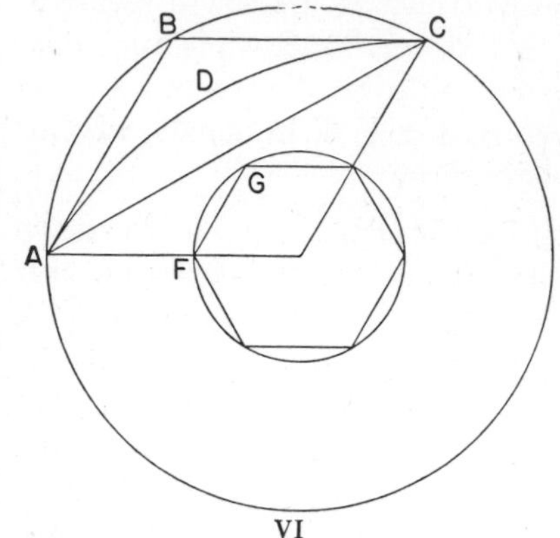

VI

It has been thought that Hippocrates' fallacy consisted in saying that because he had been able to square a (lune + circle) (case VI) and each kind of lune (cases III, IV, and V), he could, by a simple subtraction, express the area of the circle as a constructible rectilinear figure.

Of course the lune in case VI is different from any of the lunes in cases III, IV, and V, and so there is here no general solution to the problem of squaring the circle. Now there is considerable doubt whether Hippocrates committed the fallacy in question, or some similar type of fallacy, and if so, whether knowingly, and if not, how he came to be misinterpreted. These matters need not delay us. But it should be noted that he solved (cases III, IV, V) three of the five cases of lunes squarable by plane methods, i.e., by use of ruler and compass. The other cases seem not to have been discovered until the eighteenth century. On this whole matter, see T. L. Heath, *History of Greek Mathematics* I. 183–201 and W. D. Ross, *Aristotle's Physics*, p. 463 (Oxford, 1936).

c. *Higher Curves*

Curves other than those constructible with rule and compass were used to solve the three classic problems as well as various others. Some of these curves were used for the solution of more than one of the problems.

Simplicius, *Commentary on Aristotle's Physics*, p. 60.7–18 (Diels); (cf. Simplicius, *Commentary on Aristotle's Categories*, p. 192.16–25 [Kalbfleisch])

Iamblichus[1] in commenting on the *Categories* says that Aristotle was probably not yet aware of the discovery of the quadrature of the circle, but that it had been found by the Pythagoreans. This, Iamblichus says, is clear from the proofs of Sextus the Pythagorean,[2] who obtained his method of proof from a tradition that goes back to early times. And later, he says, Archimedes used the spiral,[3] Nicomedes[4] the curve specially called

[1] Iamblichus the Neoplatonist flourished at the beginning of the fourth century A.D. His extant works deal with various aspects of Pythagorean mathematics and philosophy.

[2] Sextus probably lived at the time of Augustus. The early Pythagoreans, so far as we know, did not succeed in solving the problem by the use of higher curves.

[3] See extract below. With the spiral the length of the circumference of the circle could be found and hence the area.

[4] Nicomedes flourished in the latter half of the third century B.C. He is best known for his discovery of the conchoid (sometimes called "cochloid"), of which one kind was used for the trisection of an angle and the duplication of the cube.

the quadratrix,[1] Apollonius[2] a curve which he himself called "sister of the cochloid" but which is the same as the curve of Nicomedes, and Carpus a curve which he simply calls "a curve generated by double motion." [3] And many others, says Iamblichus, solved the problem by various constructions. But it seems that all these effected the construction involved in this proposition by mechanical means.[4]

The Spiral

Archimedes, *On Spirals*, Introduction. Translation of T. L. Heath, *The Works of Archimedes*, p. 154 (Cambridge, 1897)

If a straight line of which one extremity remains fixed be made to revolve at a uniform rate in a plane until it returns to the position from which it started, and if, at the same time as the straight line revolves, a point move at a uniform rate along the straight line, starting from the fixed extremity, the point will describe a spiral in the plane. I say then that the area bounded by the spiral and the straight line which has returned to the position from which it started is a third part of the circle described with the fixed point as centre and with radius the length traversed by the point along the straight line during the one revolution. And, if a straight line touch the spiral at the extreme end of the spiral, and another straight line be drawn at right angles to the line which has revolved and resumed its position from the fixed extremity of it, so as to meet the tangent, I say that the straight line so drawn to meet it is equal to the circumference of the circle.[5]

The Quadratrix

Pappus, *Mathematical Collection* IV. 30–32 (pp. 250.33–258.19 [Hultsch])

For the squaring of the circle a certain curve was used by Dinostratus and Nicomedes and some other later geometers. This curve took its name from its property, for it was called by them the quadratrix. It is generated as follows.

---

[1] The quadratrix may have been discovered by Hippias of Elis (latter part of the fifth century B.C.) for the purpose of trisecting the angle. Whether it was used by him for the quadrature of the circle, as it was by Nicomedes, Dinostratus (brother of Menaechmus), and others, is not clear.

[2] See p. 76. What curve is referred to is not known.

[3] The date of Carpus is uncertain, but he may have lived in the first or second century A.D. The curve referred to is likewise uncertain.

[4] As opposed to the so-called "plane" methods of ruler and compass.

[5] Since the area of a circle is half the product of its radius and circumference, this rectification is a solution of the problem of quadrature. The theorem referred to is a special case of Archimedes' general theorem about the subtangent corresponding to a tangent at any turn of the spiral. His method of construction in this work is criticized by Pappus. See T. L. Heath, *History of Greek Mathematics* II. 68.

Construct a square $ABCD$.

With A as center describe arc BED.

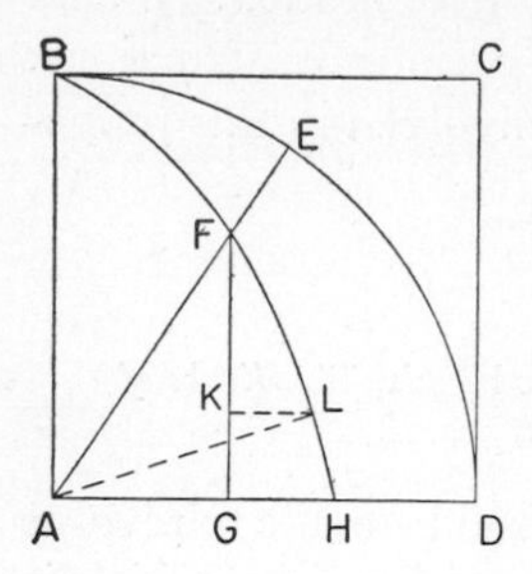

Now let AB be moved in such a way that A remains fixed and B is carried along arc BED.

Again, let BC, always remaining parallel to AD, follow point B as it is carried along BA. That is, *in equal time* let BA move with uniform motion through angle BAD [i.e., let B describe the arc BED] and BC move past line AB [i.e., let B traverse AB].[1] Clearly both AB and BC will coincide with line AD at the same moment. Now while this motion is taking place, lines BC and BA will intersect at some point that keeps changing along with the change in the position of the lines. And in the space between lines BA, AD and arc BED a curve, such as BFH, is described, concave in the same sense [as arc BED]. This curve is held to be useful for finding a square equal to a given circle. Its chief property is this. If any straight line, such as AFE, be drawn to the arc, BA will be to FG as the whole arc [BED] is to arc ED. For this is clear from the method of generating the curve.[2]

[There follow two criticisms of the quadratrix made by Sporus, a contemporary of Pappus. (1) The whole construction is a *petitio principii*, for in order to make the two motions terminate together, the ratio of AB to arc BED must be known; (2) the construction does not enable us to find point H, for at the end of their motions AB and BC coincide with AD and do not intersect. And it is precisely point H that is needed for solving the problem of quadrature.]

But we must first consider the problem [i.e., the quadrature of the circle] that is solved by this curve.

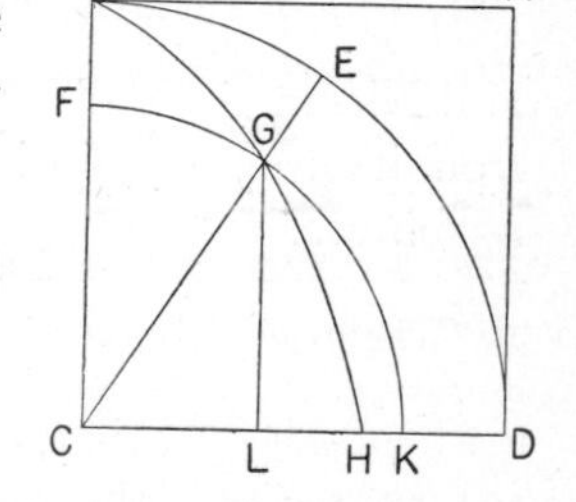

If $ABCD$ is a square, BED the arc of a circle with center C, BGH the quadratrix, as above described, it may be proved that

$$\text{arc } DEB : BC = BC : CH.$$

For if this is not the case, then [the last term will be] greater or less than CH.

I. Suppose, if possible, that this term is greater than CH, i.e., CK.

[1] That is, as AB starts to rotate about A, BC begins to move toward AD, the motion in each case being uniform and ending at the same time. The point B has a double function, as a point on AB and as a point on BC.

[2] That is, the rate of change of $\angle EAD$ is the same as that of FG. This property makes the curve useful for the division of an angle in any ratio, and, in fact, the quadratrix may first have been used by its reputed discoverer, Hippias, for the trisection of an angle.

If, in the figure, it is desired to divide $\angle EAD$ in any ratio, say in the ratio of FG to GK, lay off GK on FG; draw KL perpendicular to FG, meeting the quadratrix at L. Draw AL. Then $\angle EAD : \angle LAD = FG : GK$.

With C as center, draw arc FGK cutting the quadratrix at G. Draw the perpendicular GL. Draw CG and produce it to E.

Now since [by our assumption] arc $BED:BC = BC$ (or CD)$:CK$,

and $CD:CK =$ arc DEB:arc FGK

(for circumferences of circles have the same ratio as their diameters),

$\therefore$ arc $FGK = BC$.

But since, by the property of the quadratrix,

$BC:GL =$ arc BED:arc ED

$=$ arc FGK:arc GK,

and it was proved that arc $FGK = BC$,

it follows that arc $GK = GL$, which is absurd.

$\therefore$ it is not true that arc $BED:BC = BC$: a magnitude greater than CH.

II. And I say that neither is it true for a magnitude less than CH.

For, if possible, let there be such a magnitude, CK. With center C draw arc FMK. Draw KG perpendicular to CD cutting the quadratrix at G.

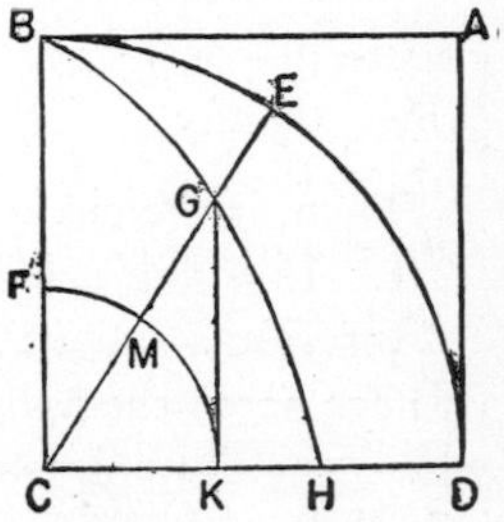

Draw CG and produce it to E.

As above, we shall prove that

arc $FMK = BC$,

and that $BC:GK =$ arc BED:arc ED

$=$ arc FMK:arc MK.

Clearly, therefore, arc $MK = GK$, which is absurd.[1]

$\therefore$ it is not true that arc

$BED:BC = BC$: a magnitude less than CH.

But this was also proved for a magnitude greater than CH.

$\therefore$ arc $BED:BC = BC:CH$.

Now it is also clear that a straight line taken as a third proportional to CH and BC will be equal to arc BED, and four times this line will be equal to the circumference of the whole circle.

Now since we have found a straight line equal to the circumference of the circle, it is clear that we can easily construct a square equal to the circle itself. For the product of the perimeter of the circle and its radius is double the area of the circle, as Archimedes proved.[2]

d. *The Area of a Circle*

Archimedes, *Measurement of a Circle*, Proposition I. Paraphrase of T. L. Heath, *The Works of Archimedes*, pp. 91–93[3]

The area of any circle is equal to a right-angled triangle in which one of the

---

[1] The demonstration assumes the proposition that for an acute angle the arc in circular measure is less than the tangent.

[2] *Measurement of a Circle*, Proposition I.

[3] The short treatise *Measurement of a Circle* is the source of the famous approximation $3\frac{10}{71} < \pi < 3\frac{1}{7}$. The first proposition is a typical illustration of the method of exhaustion. [Edd.]

sides about the right angle is equal to the radius, and the other to the circumference, of the circle.

Let *ABCD* be the given circle, *K* the triangle described.

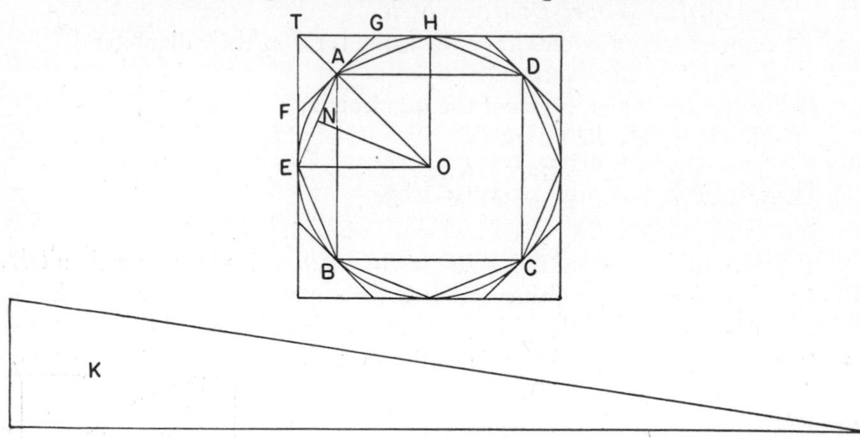

Then, if the circle is not equal to *K*, it must be either greater or less.

I. If possible, let the circle be greater than *K*.

Inscribe a square *ABCD*, bisect the arcs *AB*, *BC*, *CD*, *DA*, then bisect (if necessary) the halves, and so on, until the sides of the inscribed polygon whose angular points are the points of division subtend segments whose sum is less than the excess of the area of the circle over *K*.[1]

Thus the area of the polygon is greater than *K*.

Let *AE* be any side of it, and *ON* the perpendicular on *AE* from the centre *O*.

Then *ON* is less than the radius of the circle and therefore less than one of the sides about the right angle in *K*. Also the perimeter of the polygon is less than the circumference of the circle, i.e., less than the other side about the right angle in *K*.

Therefore the area of the polygon is less than *K*;[2] which is inconsistent with the hypothesis.

Thus the area of the circle is not greater than *K*.

II. If possible, let the circle be less than *K*.

Circumscribe a square, and let two adjacent sides, touching the circle in *E*, *H*, meet in *T*. Bisect the arcs between adjacent points of contact and draw the tangents at the points of bisection. Let *A* be the middle point of the arc *EH*, and *FAG* the tangent at *A*.

[1] Euclid had proved (*Elements* XII, 2) that this was possible and Archimedes elsewhere (*On the Sphere and Cylinder* I, 6) proves the corresponding proposition, also used here, for the *circumscribed* polygon. Cf. Euclid, *Elements* X, 1 (quoted above, p. 16). [Edd.]

[2] For the area of the polygon is equal to the product of its perimeter (which is less than that of the circle) by *ON* (which is less than the radius of the circle). [Edd.]

Then the angle *TAG* is a right angle.
Therefore $TG > GA$,
$> GH$.

It follows that the triangle *FTG* is greater than half the area *TEAH*.[1]

Similarly, if the arc *AH* be bisected and the tangent at the point of bisection be drawn, it will cut off from the area *GAH* more than one-half.

Thus, by continuing the process, we shall ultimately arrive at a circumscribed polygon such that the spaces intercepted between it and the circle are together less than the excess of *K* over the area of the circle.[2]

Thus the area of the polygon will be less than *K*.

Now, since the perpendicular form *O* on any side of the polygon is equal to the radius of the circle, while the perimeter of the polygon is greater than the circumference of the circle, it follows that the area of the polygon is greater than the triangle *K*;[3] which is impossible.

Therefore the area of the circle is not less than *K*.

Since then the area of the circle is neither greater nor less than *K*, it is equal to it.

Approximations to π

Archimedes, *Measurement of a Circle*, Proposition 3

The circumference of any circle is greater than three times the diameter, and the excess is less than a seventh part of the diameter but more than ten seventy-firsts.[4]

Hero of Alexandria, *Metrica* I. 26 (p. 66.13–19 [Schöne])

Archimedes also shows in his book on *Plinthides and Cylinders* that the ratio of the circumference of any circle to its diameter is greater than 211875:67441,

---

[1] For, since $\triangle TGA > \triangle GHA$,
$\triangle TGA > \frac{1}{2} \triangle TAH$,
and $\triangle FTG > \frac{1}{2} (\triangle TAH + \triangle TAE)$,
$> \frac{1}{2}$ area *TEAH*. [Edd.]

[2] See p. 48. [Edd.]

[3] For the area of the polygon is equal to the product of its perimeter (which is greater than that of the circle) by *OA* (which is a radius of the circle). [Edd.]

[4] Archimedes arrives at the famous approximation $3\frac{10}{71} < \pi < 3\frac{1}{7}$ by showing (1) that the perimeter of a circumscribed polygon of 96 sides (and, therefore, *a fortiori* the circumference of the circle) is less than $3\frac{1}{7}$ times the diameter, and (2) that the perimeter of an inscribed polygon of 96 sides (and, therefore, *a fortiori* the circumference of the circle) is greater than $3\frac{10}{71}$ times the diameter. Obviously, by continuing the process of doubling the number of sides (i.e., halving the central angle) the accuracy of the approximation may be increased.

The involved calculation required for the proof of this theorem is set forth in Eutocius' commentary, and only the principal steps are given by Archimedes.

The initial assumption of $\frac{265}{153} < \sqrt{3} < \frac{1351}{780}$ has given rise to numerous conjectures as to ancient methods of approximation. See T. L. Heath, *History of Greek Mathematics* II. 51–52, 323–326.

and less than 197888:62351. But since these figures are inconvenient for measurements, they are reduced to the smallest numbers, 22:7.[1]

Ptolemy, *Almagest* VI. 7 (p. 513.1–5 [Heiberg])

The ratio [of the circumference to the diameter] remains constant, $3 + \frac{8}{60} + \frac{30}{(60)^2} : 1$. This ratio is almost exactly the mean between the values $3\frac{10}{71}$ and $3\frac{1}{7}$ which Archimedes was content to give.[2]

Eutocius, *Commentary on Archimedes' Measurement of a Circle* III. 258 (Heiberg)

Now we must observe that Apollonius of Perga in his *Ocytocion* demonstrated this with other numbers, obtaining a closer approximation [than Archimedes did]. Now while Apollonius' approximation is more accurate, it is not useful for Archimedes' purpose. For Archimedes, as we have said, sought in this book [*Measurement of a Circle*] to find an approximation for practical purposes.[3]

2. THE DUPLICATION OF THE CUBE

The problem of constructing a cube double a given cube in volume was the subject of investigation beginning at least as early as the fifth century B.C. It was soon reduced to the problem of inserting two mean proportionals between two given lines. This problem cannot be solved by ruler and compass, requiring as it does the construction of $\sqrt[3]{2}$, but is solved in various ways with conics, higher plane curves, and mechanical devices. Eutocius in his commentary on Archimedes' treatise preserves twelve solutions attributed to Plato, Hero of Alexandria,[4] Philo of Byzantium, Apollonius, Diocles, Pappus, Sporus, Menaechmus (two solutions), Archytas, Eratosthenes, and Nicomedes.

In giving Eratosthenes' solution Eutocius quotes a letter purporting to be by Eratosthenes and containing a history of the problem. The letter, which we give here, is generally considered not to be genuine, though the solution itself (which is also ascribed to Eratosthenes by Pappus) and the concluding epigram are held to be by Eratosthenes.

Eratosthenes, Letter to Ptolemy Euergetes, in Eutocius, *Commentary on Archimedes' Sphere and Cylinder*, pp. 88–96 [Heiberg]

Eratosthenes to King Ptolemy, greetings.

The story goes that one of the ancient tragic poets represented Minos having a tomb built for Glaucus, and that when Minos found that the

---

[1] Of the work of Archimedes here referred to nothing is known. The figures, as given by the editor, are not so accurate as the approximation $3\frac{10}{71} < \pi < 3\frac{1}{7}$, and certain emendations have been suggested. The original figures evidently gave a closer approximation than $3\frac{1}{7}$.

[2] Ptolemy's approximation gives $\pi = 3.14166\ldots$ It might have been obtained from the table of chords (p. 83 below). The perimeter of an inscribed regular polygon of 360 sides =

$$360 \text{ crd } 1° = 360 \cdot \frac{1 + \frac{2}{60} + \frac{50}{(60)^2}}{120} = 3 + \frac{8}{60} + \frac{30}{(60)^2}.$$

[3] Of the *Ocytocion*, "means of swift birth," nothing is known. Eutocius goes on to mention other approximations. There is also evidence in the Indian mathematician Āryabhaṭṭa (fifth century A.D.), whose sources may have been Greek, of the use of $\pi = 3.1416$.

[4] The solution of Hero of Alexandria, which is practically the same as those of Apollonius of Perga and Philo of Byzantium, is given on p. 322.

tomb measured a hundred feet on every side, he said: "Too small is the tomb you have marked out as the royal resting place. Let it be twice as large. Without spoiling the form quickly double each side of the tomb."

This was clearly a mistake. For if the sides are doubled the surface is multiplied fourfold and the volume eightfold.

Now geometers, too, sought to find a way to double the given solid without altering its form. This problem came to be known as the duplication of the cube, for, given a cube, they sought to double it. Now when all had sought in vain for a long time, Hippocrates of Chios first discovered that if a way can be found to construct two mean proportionals in continued proportion between two given straight lines, the greater of which is double the lesser, the cube will be doubled.[1] So that his difficulty was resolved into another no less perplexing.

Some time later certain Delians, they say, seeking by order of the oracle to double an altar, fell into the same difficulty.[2] And so they sent representatives to ask the geometers of Plato's school in the Academy to find the solution for them. These geometers zealously tackled the problem of finding two mean proportionals between two given lines. And Archytas of Tarentum is said to have obtained a solution with semicylinders,[3] while Eudoxus used so-called "curved lines."[4] All who solved the problem succeeded in finding the deductive proof, but they were not able to dem-

---

[1] On Hippocrates of Chios, see pp. 35, 54.

If $a:x = x:y = y:b$, it follows that $a^3/x^3 = a/b$.

Therefore, if $b = 2a$, $x^3 = 2a^3$, i.e., x is the side of a cube of volume double the cube of side a.

Cf. Proclus, *Eucl.* p. 212.24 (Friedlein):

"Reduction is the passing from one problem or theorem to another. If the latter is known or solved, the former will also become clear. Thus after investigating the duplication of the cube, they transferred their investigation to another problem, on which the former depends, i.e., the finding of two mean proportionals. From that time on the problem they investigated was how, given two straight lines, two mean proportionals could be found. They say that Hippocrates of Chios first used the method of reduction in the case of difficult problems."

There is doubt that Hippocrates first used the method of reduction, which is, in a sense, implicit in any series of theorems. But he first made this specific reduction.

[2] In a fragment of Eratosthenes' *Platonicus* quoted by Theon of Smyrna, p. 2.3–12 (Hiller), the story is somewhat differently told.

"Eratosthenes in his work entitled *Platonicus* says that, when the god announced to the Delians [through the oracle] that in order to be freed from a plague they must construct an altar double the existing one, their architects were sorely perplexed in trying to find how a solid could be made double another solid, and came to Plato to ask about it. Thereupon he told them that the god had given this oracle to the Delians not because he wanted an altar double the size, but to reproach and rebuke the Greeks for their neglect of mathematics and their disregard for geometry."

[3] Archytas' remarkable solution involves the determination of the point of intersection of three surfaces of revolution, viz., a right cone, a cylinder, and a torus of inner diameter zero.

[4] The nature of the "curved lines" and of the whole solution of Eudoxus is doubtful.

onstrate the construction in a practical and useful way, with the exception of Menaechmus[1] (though he accomplished this only to a very small degree and with difficulty).

Now I have discovered an easy method of finding, by the use of an instrument, not only two but as many mean proportionals as desired between two given lines. With this discovery we shall be able to convert into a cube any given solid whose surfaces are parallelograms, or to change it from one form to another, and, again, to construct a solid of the same form as the given solid but larger, i.e., preserving the similarity. And we shall also be able to apply this in constructing altars and temples. We shall be able, furthermore, to convert our liquid and dry measures,[2] the metretes and the medimnus, into a cube, and from the side of this cube to measure the capacity of other vessels in terms of these measures. My method will also be useful for those who wish to increase the size of catapults and ballistas. For, if the throw is to be increased,[3] all the elements of these engines, the thicknesses, lengths, and the sizes of the openings, wheel casings, and cables must be increased in proportion. But this cannot be done without finding the mean proportionals. I have described below for you the demonstration and the method of construction of my device.

Let two unequal straight lines, *AE* and *DT*, be given, between which it is required to find two mean proportionals in continued proportion. Let *AE* be perpendicular to a line *ET*. Erect upon *ET* three [equal] parallelograms, *AZ*, *ZI*, and *IT*, in order. Draw diagonals *AZ*, *LH*, and

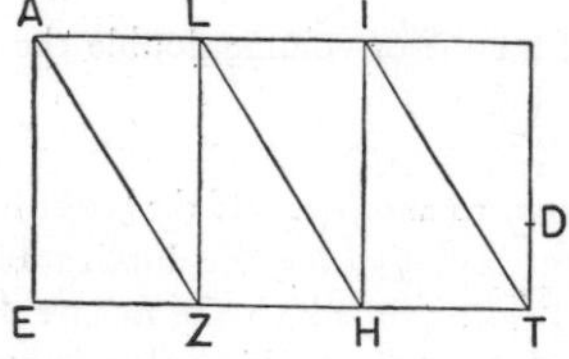

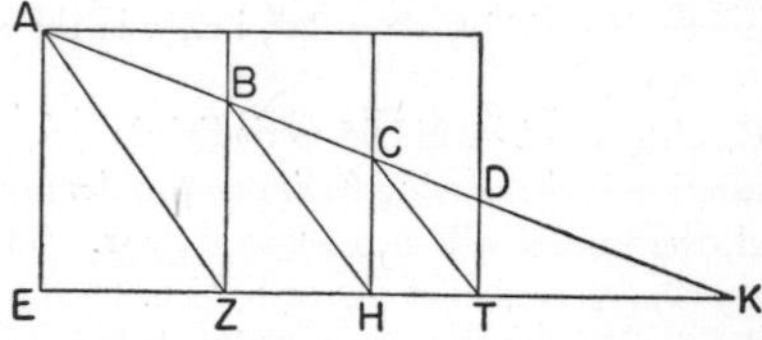

IT. These diagonals will be parallel. Now while the middle parallelogram *ZI* remains fixed, let parallelograms *AZ* and *IT* be pushed so that *AZ*

[1] The solutions of Menaechmus involve the determination of the intersection of two conics (two parabolas in the first case, a parabola and a hyperbola in the second).

In modern notation it is easy to see that

if $a:x = x:y = y:b$,

$x^2 = ay$, $y^2 = bx$, and $xy = ab$.

The intersection of any two of these curves will give the required solution corresponding to $x = \sqrt[3]{a^2b}$, $y = \sqrt[3]{ab^2}$. The first two curves are parabolas and the third is a rectangular hyperbola.

Some have considered Menaechmus the discoverer of the conic sections, but the names of the sections are due to Apollonius (see pp. 66, 78).

[2] I.e., the standard containers for these measures.

[3] The reference seems to be to the length of the trajectory of the missile. Elsewhere (pp. 319, 325, below) the size of the various elements of the engine is made to depend on the weight of the missile to be hurled.

moves above and IT below the middle parallelogram, as in the second figure [page 64], until A, B, C, and D[1] lie along the same straight line. Draw this line and let it intersect ET, produced, at K.

$AK:KB = EK : KZ$ (since AE and BZ are parallel),
and $AK:KB = ZK : KH$ (since AZ and BH are parallel).
$\therefore AK:KB = EK : KZ = KZ : KH$.
Again, since $BK : KC = ZK : KH$ (since BZ and CH are parallel),
and $BK : KC = KH : KT$ (since BH and CT are parallel),
$\therefore BK : KC = ZK : KH = KH : KT$.
But $EK : KZ = ZK : KH$.
$\therefore EK : KZ = ZK : KH = HK : KT$.
But $EK : KZ = AE : BZ$,
and $ZK : KH = BZ : CH$,
and $HK : KT = CH : DT$.
$\therefore AE : BZ = BZ : CH = CH : DT$.

That is, two mean proportionals, BZ and CH, have been found between AE and DT.

Such, then, is the demonstration on geometrical surfaces. But to find the two mean proportionals by an instrument, construct a frame of wood, ivory, or bronze having three equal flat surfaces as thin as possible. Let the middle surface be fixed, and the other two move along grooves; the size and shape of the surfaces may vary as desired.[2] For the proof is not affected.

In order that the required lines may be obtained more accurately, care must be taken that when the surfaces are brought together all parts remain parallel,[3] and fit one another snugly without gaps.

The instrument in bronze is placed on the votive monument beneath the crown of the column and is held fast with lead. Under the instrument the proof is set down concisely with a diagram and after this an epigram. These have been copied below for you, so that you may have them just as they are on the votive column. Of the two figures the second is engraved on the column.[4]

"Given two straight lines, to find two mean proportionals in continued proportion. Let AE and DT be given. I draw together the surfaces in

[1] B and C are not specifically defined, but from the description it is clear that B is the intersection of the movable side LZ of the first parallelogram with the fixed diagonal LH of the second, and C is the intersection of the fixed side IH of the second parallelogram with the movable diagonal IT of the third. The first and third parallelograms must so be moved that B and C are collinear with A and D, which, of course, themselves move as the parallelograms are brought together. When points B and C are found the problem is solved, as BZ and CH are the required mean proportionals.

[2] I.e., not only the size of the rectangles, but the ratio of length to breadth may vary. Again, right triangles may be substituted for rectangles.

[3] I.e., to their original position.

[4] What follows purports to have been copied from the monument.

the instrument until points A, B, C, and D are all in a straight line. Consider these points as they are in the second[1] diagram [page 64].

$AK:KB = EK:KZ$ (since AE and BZ are parallel),

and $AK:KB = ZK:KH$ (since AZ and BH are parallel).

$\therefore EK:KZ = KZ:KH = AE:BZ = BZ:CH.$

Similarly we shall be able to show that

$$ZB:CH = CH:DT.$$

$\therefore AE$, BZ, CH, and DT are in continued proportion.

That is, two mean proportionals between the two given lines have been found.

Now if the given lines are not equal to AE and DT, by taking AE and DT proportional to the given lines we shall obtain the means between AE and DT and then transfer the results to the given lines. Thus we shall have done what was required. And if it is required to find more mean proportionals, we shall achieve our purpose in each case by constructing one more surface on the instrument than the number of means required. The proof is the same in this case.

[2] If, my friend, you seek to make from a small cube a cube twice as large, and readily convert any solid form into another, here is your instrument. You can, then, measure a fold, or a grain pit, or the broad hollow of a well, if between two rulers you find means the extreme ends of which converge.[3] Do not seek the cumbersome procedure with Archytas' cylinders, or to make the three Menaechmian sections of the cone; seek not the type of curved line described by god-fearing Eudoxus. For with these plates of mine you could readily construct ten thousand means beginning with a small base.

You are a happy father, Ptolemy, because you enjoy youth with your son and have yourself given him all that is precious to the Muses and to kings. May he hereafter, heavenly Zeus, receive the scepter from your hand: so may it come to pass. And let whoever sees this votive column say: 'This is an offering of Eratosthenes of Cyrene'."

3. THE TRISECTION OF AN ANGLE

When geometers found that they could not in general trisect an angle, or, in fact, divide it in an arbitraty ratio by plane methods, they had recourse to other constructions. Prominent was the so-called "verging" (νεῦσις), where a line must be drawn through a given point in such a way that its intercept between two given lines which it intersects is a given length. The problem of trisection was reduced to a construction of this type for the solution of which the conchoid (or cochloid) of Nicomedes was specifically invented. Pappus also gives solu-

[1] This word must be deleted unless the whole clause be taken as an interpolation on the part of the writer of the letter, and not part of the inscription.

[2] The rest is the epigram in nine elegaic couplets.

[3] This is obscure. Perhaps the reference is to the fact that the ends of the parallels converge toward K.

tions to the problem of trisection by means of conics without recourse to a νεῦσις. It is quite possible, too, that the quadratrix was used for this purpose as early as the fifth century. See p. 58, n. 2.

Pappus of Alexandria, *Mathematical Collection* IV. 57–59 (pp. 270–272 [Hultsch])

When ancient geometers desired to divide a given rectilinear angle into three equal parts, they were baffled for the following reason. There are, we say, three types of problem in geometry, the so-called "plane," "solid," and "linear" problems. Those that can be solved with straight line and circle are properly called "plane" problems, for the lines by which such problems are solved have their origin in a plane. Those problems that are solved by the use of one or more sections of the cone are called "solid" problems. For it is necessary in the construction to use the surfaces of solid figures, that is to say, of cones. There remains the third type, the so-called "linear" problem. For the construction in these cases curves other than those already mentioned are required, curves having a more varied and forced origin and arising from more irregular surfaces and from complex motions. Of this character are the curves discovered in the so-called "surface loci"[1] and numerous others even more involved discovered by Demetrius of Alexandria in his *Treatise on Curves* and by Philo of Tyana[2] from the interweaving of plectoids and of other surfaces of every kind. These curves have many wonderful properties. More recent writers have indeed considered some of them worthy of more extended treatment, and one of the curves is called "the paradoxical curve" by Menelaus.[3] Other curves of the same type are spirals, quadratrices, cochloids, and cissoids.

Now it is considered a serious type of error for geometers to seek a solution to a plane problem by conics or linear curves and, in general, to seek a solution by a curve of the wrong type. Examples of this are to be found in the problem of the parabola in the fifth book of Apollonius's *Conics*,[4] and in the use of a solid νεῦσις with respect to a circle in Archimedes' work on the spiral.[5] For in the latter case it is possible without the use of anything solid to prove Archimedes' theorem, viz., that the circumference of the circle traced at the first turn is equal to the straight line drawn at right angles to the initial line and meeting the tangent to the spiral.[6]

---

[1] The precise meaning of this term is doubtful. On the lost work of Euclid entitled *Surface Loci* see T. L. Heath, *History of Greek Mathematics* I. 439.

[2] Demetrius and Philo are otherwise unknown.

[3] P. Tannery has conjectured that the curve referred to is the intersection of a sphere and a cylinder tangent to it internally and of diameter equal to the radius of the sphere (see T. L. Heath, *History of Greek Mathematics* II. 261). On Menelaus, see p. 82.

[4] The reference may be to Apollonius's use of a rectangular hyperbola where a circle would give the same points of intersection (see T. L. Heath, *op. cit.*, II. 167).

[5] The reference is to the construction in Archimedes, *On Spirals* 8 and 9 (see T. L. Heath, *op. cit.*, II. 68).

[6] See p. 57, n. 5, above.

In view of the existence of these different classes of problem, the geometers of the past who sought by planes to solve the aforesaid problem of the trisection of an angle, which is by its nature a solid problem, were unable to succeed. For they were as yet unfamiliar with the conic sections and were baffled for that reason. But later with the help of the conics they trisected the angle using the following νεῦσις for the solution.[1]

Solution by a "Verging" (νεῦσις)

Pappus, *ibid.*, pp. 274–276

Now that this has been proved,[2] the given rectilinear angle may be trisected as follows.

Let ABC be an acute angle, and from any point [of AB] draw the perpendicular AC. Complete parallelogram FC and produce FA. Now, since FC is a rectangular parallelogram, let line ED be inserted between AE and AC, verging toward B and equal to twice AB.[3] It has been shown above that this is possible.

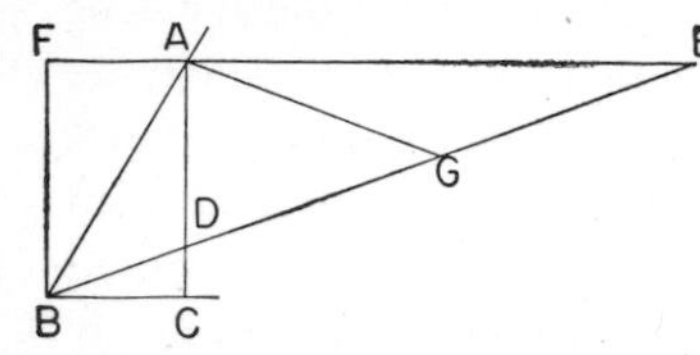

Now I say that $\angle EBC = \frac{1}{3}\angle ABC$.

For bisect DE at G and draw AG.

$DG = GA = GE.$

$\therefore DE = 2AG = 2AB.$

$\therefore AB = AG.$

$\therefore \angle ABD = \angle AGD.$

But $\angle AGD = 2\angle AED = 2\angle DBC.$

$\therefore \angle ABD = 2\angle DBC.$

And if we bisect $\angle ABD$, we shall have trisected $\angle ABC$.

---

[1] This solution may be shown to be equivalent to the solution of a cubic equation. A "plane problem," on the contrary, depends on equations of the first or second degree, only.

[2] I.e., how to insert a line so that the intercept cut off on it by two other lines is equal to a given length, the required line to verge to (i.e., when produced, to pass through) a given point. This can be made to depend on the intersection of a circle and a hyperbola, or by the special curve, the conchoid, invented for this purpose by Nicomedes in the third century B.C. The conchoid is produced by pivoting a straight line PB about a fixed point B in such a way that that part of PB which lies between a fixed line AC (called the "ruler") and the curve is always constant. By adding the constant distance PD to any radius vector (BD) from B to AC, the conchoid may be "constructed" by points.

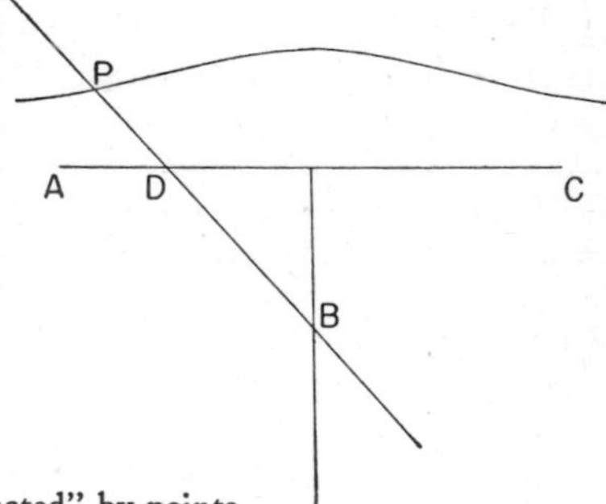

[3] I.e., let a line be drawn through B so that its intercept DE between AC and AE is equal to $2AB$. E is the point where FA produced meets a conchoid having B as pivot point, AC as ruler, and $2AB$ as constant distance. If BE is drawn, ED will meet the required conditions.

Mechanical and Geometric Methods Contrasted

Archimedes, *Method*, Introduction and Proposition 1.[1] Translation and paraphrase of T. L. Heath (Cambridge, 1912)

"Archimedes to Eratosthenes greeting.

I sent you on a former occasion some of the theorems discovered by me, merely writing out the enunciations and inviting you to discover the proofs, which at the moment I did not give. The enunciations of the theorems which I sent were as follows.

1. If in a right prism with a parallelogrammic base a cylinder be inscribed which has its bases in the opposite parallelograms,[2] and its sides [i.e., four generators] on the remaining planes (faces) of the prism, and if through the centre of the circle which is the base of the cylinder and (through) one side of the square in the plane opposite to it a plane be drawn, the plane so drawn will cut off from the cylinder a segment which is bounded by two planes and the surface of the cylinder, one of the two planes being the plane which has been drawn and the other the plane in which the base of the cylinder is, and the surface being that which is between the said planes; and the segment cut off from the cylinder is one sixth part of the whole prism.

2. If in a cube a cylinder be inscribed which has its bases in the opposite parallelograms[3] and touches with its surface the remaining four planes (faces), and if there also be inscribed in the same cube another cylinder which has its bases in other parallelograms and touches with its surface the remaining four planes (faces), then the figure bounded by the surfaces

---

[1] The *Method*, a work of Archimedes long thought lost but discovered in 1906, provides an insight into a powerful type of analysis used by that mathematician. By thinking of a plane figure as consisting of a set of lines and conceiving these lines as balanced by corresponding lines of a figure of known area, we may discover the area of the first figure. Analogously we may discover the volume of solids. But this method of equilibrium is, in Archimedes' judgment, merely a heuristic device, and does not constitute a proof. The actual proof is rigorously geometrical and is accomplished with the aid of the method of exhaustion and *reductio ad absurdum*.

Archimedes does not speak of infinitesimal elements between the lines, in the case of plane figures, or between the planes, in the case of solids, but his method seems to involve such a concept. At any rate we have here, as well as in other treatises, something very much like the process of integration. In this Archimedes may be called a precursor of the inventors of the calculus, though his Greek successors did not, so far as we know, pursue this aspect of his work.

Of the passage given, the part within quotation marks is a translation, the rest a reduction to modern notation and phraseology. Material in square brackets, other than that which is obviously explanatory, represents a conjectural restoration where the manuscript is illegible. [Edd.]

[2] The parallelograms apparently are *squares*.

[3] I.e., squares.

of the cylinders, which is within both cylinders, is two-thirds of the whole cube.[1]

Now these theorems differ in character from those communicated before; for we compared the figures then in question, conoids and spheroids and segments of them, in respect of size, with figures of cones and cylinders: but none of those figures have yet been found to be equal to a solid figure bounded by planes; whereas each of the present figures bounded by two planes and surfaces of cylinders is found to be equal to one of the solid figures which are bounded by planes. The proofs then of these theorems I have written in this book and now send to you. Seeing moreover in you, as I say, an earnest student, a man of considerable eminence in philosophy, and an admirer [of mathematical inquiry], I thought fit to write out for you and explain in detail in the same book the peculiarity of a certain method, by which it will be possible for you to get a start to enable you to investigate some of the problems in mathematics by means of mechanics. This procedure is, I am persuaded, no less useful even for the proof of the theorems themselves; for certain things first became clear to me by a mechanical method, although they had to be demonstrated by geometry afterwards because their investigation by the said method did not furnish an actual demonstration. But it is of course easier, when we have previously acquired, by the method, some knowledge of the questions, to supply the proof than it is to find it without any previous knowledge. This is a reason why, in the case of the theorems the proof of which Eudoxus was the first to discover, namely, that the cone is a third part of the cylinder, and the pyramid of the prism, having the same base and equal height, we should give no small share of the credit to Democritus, who was the first to make the assertion with regard to the said figure[2] though he did not prove it.[3] I am myself in the position of having first made the discovery of the

---

[1] The geometrical proof of the first of these propositions appears at the end of the present treatise, though in fragmentary form; that of the second proposition is lost, though the form which the proof took may be conjectured with reasonable probability. [Edd.]

[2] The problems of the pyramid and the cone are of the same type. [Edd.]

[3] Cf. the question put by Democritus (Plutarch, *De Communibus Notitiis*, p. 1079 E):

"If a cone is cut by a plane parallel to its base, how are we to think of the surfaces of the sections? Are they equal or unequal? For if they are unequal they will make the cone uneven, for the cone will have many step-like indentations and roughnesses. And if the surfaces are equal, the sections will be equal and the cone will obviously have the properties of the cylinder, for it will consist of equal circles. Yet at the same time it will [being a cone] consist of unequal circles. Now this is absurd."

Heath conjectures that Democritus, who had thought of solids as made up of an indefinitely large number of indefinitely thin plane sections, could easily have noted that the three pyramids into which a prism is divided satisfy, in pairs, a test of equality based on such indefinite division into thin sections. The extension to pyramids of polygonal base and then to the cone would be easy, though it is clear that Democritus could not furnish a rigorous

theorem now to be published [by the method indicated], and I deem it necessary to expound the method partly because I have already spoken of it[1] and I do not want to be thought to have uttered vain words, but equally because I am persuaded that it will be of no little service to mathematics; for I apprehend that some, either of my contemporaries or of my successors, will, by means of the method when once established, be able to discover other theorems in addition, which have not yet occurred to me.

First then I will set out the very first theorem which became known to me by means of mechanics, namely, that

Any segment of a section of a right-angled cone (i.e., a parabola) is four-thirds of the triangle which has the same base and equal height,

and after this I will give each of the other theorems investigated by the same method. Then, at the end of the book, I will give the geometrical [proofs of the propositions] . . . .

[I premise the following propositions which I shall use in the course of the work.]

1. If from [one magnitude another magnitude be subtracted which has not the same centre of gravity, the centre of gravity of the remainder is found by] producing [the straight line joining the centres of gravity of the whole magnitude and of the subtracted part in the direction of the centre of gravity of the whole] and cutting off from it a length which has to the distance between the said centres of gravity the ratio which the weight of the subtracted magnitude has to the weight of the remainder.

[*On the Equilibrium of Planes*, I. 8]

2. If the centres of gravity of any number of magnitudes whatever be on the same straight line, the centre of gravity of the magnitude made up of all of them will be on the same straight line. [Cf. *ibid.*, I. 5]

3. The centre of gravity of any straight line is the point of bisection of the straight line. [Cf. *ibid.*, I. 4]

4. The centre of gravity of any triangle is the point in which the straight lines drawn from the angular points of the triangle to the middle points of the (opposite) sides cut one another. [*Ibid.*, I. 13, 14]

5. The centre of gravity of any parallelogram is the point in which the diagonals meet. [*Ibid.*, I. 10]

6. The centre of gravity of a circle is the point which is also the centre [of the circle].

7. The centre of gravity of any cylinder is the point of bisection of the axis.

---

proof of these propositions.

Problems and paradoxes of the infinite and the infinitesimal were prominent in Greek thought, but in view of the predominantly philosophic character of the ancient debates we have felt that the topic, in general, lies outside the scope of the present work. [Edd.]

[1] Cf. Preface to *Quadrature of Parabola*. [See p. 74, below. Edd.]

8. The centre of gravity of any cone is [the point which divides its axis so that] the portion [adjacent to the vertex is] triple [of the portion adjacent to the base].

[All these propositions have already been] proved.[1] [Besides these I require also the following proposition, which is easily proved:

If in two series of magnitudes those of the first series are, in order, proportional to those of the second series and further] the magnitudes [of the first series], either all or some of them, are in any ratio whatever [to those of a third series], and if the magnitudes of the second series are in the same ratio to the corresponding magnitudes [of a fourth series], then the sum of the magnitudes of the first series has to the sum of the selected magnitudes of the third series the same ratio which the sum of the magnitudes of the second series has to the sum of the (correspondingly) selected magnitudes of the fourth series. [*On Conoids and Spheroids*, Prop. 1.]"

Proposition 1

Let ABC be a segment of a parabola bounded by the straight line AC and the parabola ABC, and let D be the middle point of AC. Draw the straight line DBE parallel to the axis of the parabola and join AB, BC.

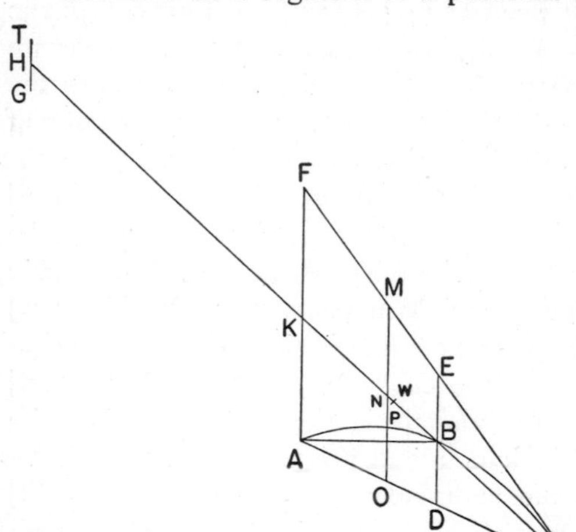

Then shall the segment ABC be $\frac{4}{3}$ of the triangle ABC.

From A draw AKF parallel to DE, and let the tangent to the parabola at C meet DBE in E and AKF in F. Produce CB to meet AF in K, and again produce CK to H, making KH equal to CK.

Consider CH as the bar of a balance, K being its middle point.

Let MO be any straight line parallel to ED, and let it meet FC, CK, AC in M, N, O and the curve in P.

Now since CE is a tangent to the parabola and CD the semi-ordinate,

$$EB = BD;$$

"for this is proved in the Elements [of Conics]."[2]

---

[1] The problem of finding the centre of gravity of a cone is not solved in any extant work of Archimedes . . . .

[2] The important works on conics before the time of Archimedes were those of Aristaeus and Euclid (see pp. 40, 75). [Edd.]

Since FA, MO are parallel to ED, it follows that

$$FK = KA, MN = NO.$$

Now, by the property of the parabola, "proved in a lemma,"

$MO:OP = CA:AO$ [Cf. *Quadrature of Parabola*, Prop. 5]

$= CK:KN$ [Euclid, VI. 2]

$= HK:KN.$

Take a straight line TG equal to OP, and place it with its centre of gravity at H, so that $TH = HG$; then, since N is the centre of gravity of the straight line MO,

and $$MO:TG = HK:KN,$$

it follows that TG at H and MO at N will be in equilibrium about K.

[*On the Equilibrium of Planes*, I. 6, 7]

Similarly, for all other straight lines parallel to DE and meeting the arc of the parabola, (1) the portion intercepted between FC, AC with its middle point on KC and (2) a length equal to the intercept between the curve and AC placed with its centre of gravity at H will be in equilibrium about K.

Therefore K is the centre of gravity of the whole system consisting (1) of all the straight lines as MO intercepted between FC, AC and placed as they actually are in the figure and (2) of all the straight lines placed at H equal to the straight lines as PO intercepted between the curve and AC.

And, since the triangle CFA is made up of all the parallel lines like MO, and the segment CBA is made up of all the straight lines like PO within the curve,

it follows that the triangle, placed where it is in the figure, is in equilibrium about K with the segment CBA placed with its centre of gravity at H.[1]

Divide KC at W so that $CK = 3KW$;

then W is the centre of gravity of the triangle ACF; "for this is proved in the books on equilibrium." [Cf. *On the Equilibrium of Planes* I. 15]

Therefore $\triangle ACF$: (segment ABC) $= HK:KW$

$= 3:1.$

Therefore segment $ABC = \frac{1}{3} \triangle ACF.$

But $\triangle ACF = 4 \triangle ABC.$[2]

Therefore segment $ABC = \frac{4}{3} \triangle ABC.$

"Now the fact here stated is not actually demonstrated by the argument used; but that argument has given a sort of indication that the conclusion is true. Seeing then that the theorem is not demonstrated,

[1] Note the important assumption which, in modern terms, is that the sum of the moments of each part of a figure, acting where that part is, is equal to the moment of the whole figure considered as concentrated at its center of gravity. [Edd.]

[2] For $\triangle ABC = \frac{1}{2} \triangle KAC$ (since $CB = BK$)

$= \frac{1}{4} \triangle ACF.$ [Edd.]

but at the same time suspecting that the conclusion is true, we shall have recourse[1] to the geometrical demonstration which I myself discovered and have already published."

Archimedes, *Quadrature of the Parabola*, Preface.[2] Translation of T. L. Heath, *The Works of Archimedes*, pp. 233–234

Archimedes to Dositheus greeting.

When I heard that Conon, who was my friend in his lifetime, was dead, but that you were acquainted with Conon and withal versed in geometry, while I grieved for the loss not only of a friend but of an admirable mathematician, I set myself the task of communicating to you, as I had intended to send to Conon, a certain geometrical theorem which had not been investigated before but has now been investigated by me, and which I first discovered by means of mechanics[3] and then exhibited by means of geometry. Now some of the earlier geometers tried to prove it possible to find a rectilineal area equal to a given circle and a given segment of a circle;[4] and after that they endeavoured to square the area bounded by the section of the whole cone and a straight line, assuming lemmas not easily conceded, so that it was recognised by most people that the problem was not solved.[5] But I am not aware that any one of my predecessors has attempted to square the segment bounded by a straight line and a section of a right-angled cone [a parabola],[6] of which problem I have now discovered the solution. For it is here shown that every segment bounded by a straight line and a section of a right-angled cone [a parabola] is four-thirds of the triangle which has the same base and equal height with the segment, and for the demonstration of this property the following lemma is assumed: that the excess by which the greater of (two) unequal areas exceeds the less can, by being added to

---

[1] Some hold the meaning to be that the geometrical proof will actually be given at the end of the present work, though others dispute this. [Edd.]

[2] In this treatise Archimedes proves that a segment of a parabola is equal to four-thirds the triangle having the same base and altitude as the segment. In the preface, quoted below, he claims the discovery of the proof. There is a proof based on mechanics in the *Method* (quoted above), and there are two proofs in the *Quadrature of the Parabola*, one based on mechanics, the other purely geometrical, both of them confirmed by the method of exhaustion. Other theorems proved by the method of exhaustion are cited in the preface. [Edd.]

[3] The reference is to the *Method* as well as to the use of mechanical means in the first part of the present treatise. [Edd.]

[4] On the quadrature of the circle see pp. 53–62, above. [Edd.]

[5] Nothing is known of attempts to square the ellipse or segments thereof (which seems to be the meaning here) before Archimedes. [Edd.]

[6] Archimedes still adhered to the view of the conic sections as made on a right circular cone by a plane perpendicular to an element. The section is a parabola, ellipse, and hyperbola, when the vertex angle of the cone is right, acute, and obtuse, respectively (see p. 76). [Edd.]

itself, be made to exceed any given finite area.[1] The earlier geometers have also used this lemma; for it is by the use of this same lemma that they have shown that circles are to one another in the duplicate ratio of their diameters, and that spheres are to one another in the triplicate ratio of their diameters, and further that every pyramid is one third part of the prism which has the same base with the pyramid and equal height; also, that every cone is one third part of the cylinder having the same base as the cone and equal height they proved by assuming a certain lemma similar to that aforesaid. And, in the result, each of the aforesaid theorems has been accepted no less than those proved without the lemma. As therefore my work now published has satisfied the same test as the propositions referred to, I have written out the proof and send it to you, first as investigated by means of mechanics, and afterwards too as demonstrated by geometry. Prefixed are, also, the elementary propositions in conics which are of service in the proof. Farewell.

Conic Sections

From the time when Menaechmus, a younger contemporary of Plato, first used conic sections for the solution of the problem of the duplication of the cube (pp. 64, 66), the investigation of these curves proceeded at a rapid pace. Aristaeus and Euclid in their works on conics (or "solid loci") seem to have proved propositions dealing with the fundamental properties of conics and their ordinates, axes, tangents, etc. The focus-directrix property and the classification of conics by what we call the "eccentricity" may have been known as early as the time of Aristaeus and Euclid, though the first explicit enunciation is in Pappus. Archimedes in various works deals with conics and the solids generated by their revolution about an axis. Some of his most elegant theorems, in which he performs the equivalent of the operation of integration, deal with areas, volumes, and centers of gravity in connection with various conics and the solids derived from them.

Though the standard treatises of Euclid and Aristaeus on the elements of conics are not extant we have, almost in its entirety, the work that marks the culmination of this branch

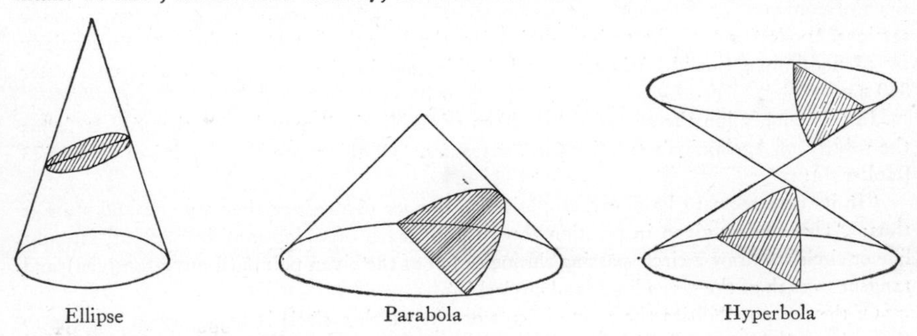

of Greek mathematics, the treatise of Apollonius of Perga. Of the eight books, the first four are extant in Greek, the next three in Arabic; the eighth is lost. The subject is taken up

[1] This lemma, upon which is based the method of exhaustion, was probably introduced by Eudoxus, though it is usually known as the Axiom of Archimedes. Cf. Euclid, *Elements* V, Def. 4, and X, Prop. 1 (pp. 15–17). [Edd.]

from the very elements, but goes beyond earlier treatises in the fullness and generality of treatment and in the new branches that are developed.

It may be noted here that the names "section of a right-angled cone," "section of an acute-angled cone," and "section of an obtuse-angled cone" were in general use before Apollonius, who seems to have made the terms "parabola," "ellipse," and "hyperbola" standard for these curves. That is, in the early view, the cone was obtained by the revolution of a right triangle about one of its sides and the section was considered to be made by a plane perpendicular to a generating element. Not that mathematicians like Archimedes were unaware of the possibility of obtaining conics by sections other than at right angles to a generator and with other than right cones. But the names indicate the original method employed. The particular kind of conic section obtained depended then on whether the vertex angle of the cone was right, acute, or obtuse. (See figures on preceding page.)

Apollonius of Perga, *Treatise on Conic Sections*, Preface to Books I and IV. Translation of T. L. Heath, *A History of Greek Mathematics* II. 128–131[1]

General Preface

Apollonius to Eudemus, greeting. . . .

Of the eight books [of my treatise on conic sections] the first four form an elementary introduction. The first contains the modes of producing the three sections[2] and the opposite branches (of the hyperbola), and the fundamental properties subsisting in them, worked out more fully and generally than in the writings of others. The second book contains the properties of the diameters and the axes of the sections as well as the asymptotes, with other things generally and necessarily used for determining limits of possibility (διορισμοί);[3] and what I mean by diameters and axes respectively you will learn from this book. The third book contains many remarkable theorems useful for the syntheses of solid loci[4]

[1] Apollonius of Perga flourished in the latter part of the third and early part of the second century B.C. Apart from the *Conics* eleven other works are attributed to him. But only one, *On the Cutting-off of a Ratio*, is extant (in an Arabic version); in the other cases descriptions of contents, or fragments, or mere titles have come down. In the theory of conic sections Apollonius brought to its culmination the branch that had been developed chiefly by Menaechmus, Euclid, Aristaeus, and Archimedes in the century and a quarter after 350 B.C.

In this connection the so-called "Problem of Apollonius" may be mentioned. It forms the subject of Apollonius's treatise *On Tangencies*. Of this work Pappus writes (p. 644.23 [Hultsch]):

"In it there seem to be many propositions, but we can reduce them to one and state it thus: 'There being given in position three things, each of which may be a point, straight line or circle, to draw a circle passing through each of the given points (if any are given) and tangent to each of the given lines [and circles].' "

Of the ten possibilities the very difficult one in which a circle is to be drawn tangent to three given circles was solved by Apollonius with ruler and compass alone. In modern times Vieta and Newton also contributed "plane" solutions. [Edd.]

[2] I.e., the parabola, ellipse, and hyperbola. [Edd.]

[3] See p. 36. [Edd.]

[4] I.e., loci depending on the section by a plane of a solid figure such as a cone or a cylinder. [Edd.]

and for *diorismi*; the most and prettiest of these theorems are new, and it was their discovery which made me aware that Euclid did not work out the synthesis of the locus with respect to three and four lines,[1] but only a chance portion of it, and that not successfully; for it was not possible for the said synthesis to be completed without the aid of the additional theorems discovered by me. The fourth book shows in how many ways the sections of cones can meet one another and the circumference of a circle; it contains other things in addition, none of which have been discussed by earlier writers, namely, the questions in how many points a section of a cone or a circumference of a circle can meet [a double-branch hyperbola, or two double-branch hyperbolas can meet one another].

The rest of the books are more by way of surplusage (περιουσιαστικώτερα): one of them deals somewhat fully with *minima* and *maxima*,[2] another with equal and similar sections of cones, another with theorems of the nature and determinations of limits, and the last with determinate conic problems. But of course, when all of them are published, it will be open to all who read them to form their own judgement about them, according to their own individual tastes. Farewell.

Preface to Book IV

Apollonius to Attalus,[3] greeting.

Some time ago I expounded and sent to Eudemus of Pergamum the first three books of my conics which I have compiled in eight books, but, as he has passed away, I have resolved to dedicate the remaining books to you because of your earnest desire to possess my works. I am sending you on this occasion the fourth book. It contains a discussion of the question, in how many points at most is it possible for sections of cones to meet one another and the circumference of a circle, on the assumption that they do not coincide throughout, and further in how many points at most a section of a cone or the circumference of a circle can meet the hyperbola with two branches, [or two double-branch hyperbolas can meet one

[1] The three-line locus is, in its simplest form, the locus of a point such that the product of its distance to two given straight lines is equal to the square of its distance to a third straight line. The four-line locus is the locus of a point such that the product of its distance to two given straight lines is equal to the product of its distance to two other given straight lines. The problem is made more general by requiring that the lines drawn from the point to the given lines be at any given angles, and the products in any given ratio (not necessarily equal). The loci are conics. The analogous two-line locus was shown to be a straight line by Apollonius in his lost work on *Plane Loci* (Pappus, p. 664). On these and related problems see T. L. Heath, *Apollonius of Perga*, Introduction, ch. 5 and 6; *History of Greek Mathematics* II. 402; see also p. 79, below. [Edd.]

[2] E.g., the longest and shortest lines that can be drawn from a given point to a conic. [Edd.]

[3] King Attalus I of Pergamum (ruled 241–197 B.C.). The last five books of Apollonius's *Conics* are dedicated to him. [Edd.]

another]; and, besides these questions, the book considers a number of others of a similar kind. Now the first question Conon expounded to Thrasydaeus, without, however, showing proper mastery of the proofs, and on this ground Nicoteles of Cyrene, not without reason, fell foul of him. The second matter has merely been mentioned by Nicoteles, in connexion with his controversy with Conon, as one capable of demonstration; but I have not found it demonstrated either by Nicoteles himself or by any one else. The third question and others akin to it I have not found so much as noticed by any one. All the matters referred to, which I have not found anywhere, required for their solution many and various novel theorems, most of which I have, as a matter of fact, set out in the first three books, while the rest are contained in the present book. These theorems are of considerable use both for the syntheses of problems and for *diorismi*. Nicoteles indeed, on account of his controversy with Conon, will not have it that any use can be made of the discoveries of Conon for the purpose of *diorismi*; he is, however, mistaken in this opinion, for, even if it is possible, without using them at all, to arrive at results in regard to limits of possibility, yet they at all events afford a readier means of observing some things, e.g., that several or so many solutions are possible, or again that no solution is possible; and such foreknowledge secures a satisfactory basis for investigations, while the theorems in question are again useful for the analysis of *diorismi*. And, even apart from such usefulness, they will be found worthy of acceptance for the sake of the demonstrations themselves, just as we accept many other things in mathematics for this reason and for no other.

Pappus, *Mathematical Collection*, pp. 672–674 (Hultsch)

Apollonius supplemented the four books of Euclid's *Conics*, and having added four, produced a work of eight books on *Conics*. Now Aristaeus who wrote a treatise, still extant, on solid loci in five books, in continuation of Euclid's *Conics*, [and the other predecessors of Apollonius] called the conics (1) the section of an acute-angled cone, (2) the section of a right-angled cone, (3) the section of an obtuse-angled cone. But since all three curves can be obtained in the case of each of these three types of cones by cutting them in different ways, Apollonius, it seems, did not see on what basis his predecessors assigned names, calling one a section of an acute-angled cone, though it might also be the section of a right-angled or obtuse-angled cone, calling the second a section of a right-angled cone, though it might well be a section of an acute- or an obtuse-angled cone, and calling the third a section of an obtuse-angled cone, though it might be a section of an acute- or a right-angled cone. And so he adopted new names, giving to the sections which had previously been known as sections of acute-angled, right-angled, and obtuse-angled cones, the names ellipse, parabola, and hyperbola, respectively. Each name referred to a specific property of the curve in question, for a certain area applied to a given line in the case of the section of the acute-angled cone is deficient (ἐλλείπει) by a rectangle, in the case of the section of an obtuse-angled cone exceeds (ὑπερβάλλει) by a rectangle, and in the case of the section of a right-angled cone is neither deficient nor exceeds.[1]

THE PROBLEM OF PAPPUS

Pappus, *Mathematical Collection*, p. 680.12–30 (Hultsch)

But in the case of more than six lines,[2] they no longer can say "if the figure represented by the product of four lines has a given ratio to that represented by the product of the others," for there is no figure represented by more than three dimensions.[3] And yet certain very recent writers permitted themselves to interpret such things, but when they referred to the product of a rectangle by a square or a rectangle, what they said was entirely without meaning. Yet they might have expressed and indicated their meaning generally through the use of compound ratios, both in the case of the previous propositions and in the case of those now under discussion, in the following way: If from a given point lines are drawn making given angles with lines given in position, and if the product of the ratios of the first of the lines so drawn to the second, and the third to the fourth,

---

[1] That is, in the case of the parabola, the square of the ordinate is equal to the product of the parameter and the abscissa, in the case of the hyperbola exceeds this product, and in the case of the ellipse falls short of this product. For the geometrical representation of these areas and lines, see T. L. Heath, *History of Greek Mathematics* II. 138. Cf. p. 42, above.

Geminus, quoted by Eutocius in the introduction to the latter's commentary on the *Conics* of Apollonius, has a similar discussion of the older and newer terminology. But Eutocius—he can hardly be quoting Geminus here—refers the names parabola, hyperbola, and ellipse to other considerations, viz., that in the section of an obtuse-angled cone by a plane at right angles to it, the sum of the vertex angle of the cone and the angle made by the cutting plane *exceeds* two right angles, in the case of an acute-angled cone, this sum *falls short of* two right angles, and in the case of the parabola the cutting plane is always parallel to an element of the cone.

It may be noted that the terms for the solids generated by the revolution of conics about their axes are analogous to those of the conic sections. Thus in Archimedes, who uses the older terminology, the terms "right-angled conoid," "obtuse-angled conoid," and "spheroid" are used for paraboloid, hyperboloid, and ellipsoid, respectively. The spheroid is "oblong" (παραμᾶκες) or "flat" (ἐπιπλατύ), according as the ellipse is revolved on its major or minor axis.

[2] Pappus had spoken of three- and four-line loci (see p. 77, n. 1, above), and also of five- and six-line loci (p. 678. 26 [Hultsch]). He goes on to generalize in the passage given here.

[3] This passage is generally considered to show how difficult it was even for the best Greek mathematicians to accept the algebraic approach to geometrical problems. But is not Pappus rather indicating that the approach *must* be algebraic, and objecting to the use of a terminology based on three-dimensional geometry for a problem of this type?

and of the fifth to the sixth, and of the seventh to a given straight line (if there are seven lines, but if there are eight, then the ratio of the seventh to the eighth) is also given, the locus of the point will be a curve given in position. And similarly, for any number of lines, odd or even.

But though these propositions follow naturally after that on the four-line locus, they have not succeeded in synthesizing any of them to the extent of discovering the locus.[1]

An Anticipation of Guldin's Theorem

Pappus, *Mathematical Collection*, p. 682.7–20 (Hultsch)

If solids are generated by a complete revolution of plane figures about an axis, the ratio of the volumes of these solids is the product of the ratio of the areas of the plane figures by the ratio of the straight lines drawn at equal angles to the axes of rotation from the centers of gravity of the plane figures.

In the case of solids generated by an incomplete revolution, the ratio of the volumes is the product of the ratio of the areas of the plane figures by the ratio of the lengths of the arcs described by the centers of gravity of these plane figures. Clearly, the ratio of these arcs is itself the product of the ratio of the straight lines (drawn at equal angles to the axes of rotation from the centers of gravity of the plane figures) by the ratio of the angles described about the axes of revolution by the extremities of these lines (i.e., by the centers of gravity of the plane figures).[2] These propositions, which are essentially one, involve numerous theorems of various kinds dealing with lines, surfaces, and solids, all proved together and by a single demonstration, and including those never before proved as well as those already proved, e.g., the propositions proved in the twelfth book of the *Elements*.

---

[1] I.e., if the lines drawn from a given point at given angles to n given lines are $p_1, p_2, p_3, \ldots p_n$, and $\frac{p_1}{p_2} \cdot \frac{p_3}{p_4} \cdots \frac{p_{n-1}}{p_n} = r$ (a given ratio), to find the locus of the point. (Where n is odd, the last ratio in the first member of the equation is p_n/a, a being a fixed length.)

This is called the Problem of Pappus, and in various forms has engaged the attention of geometers since the seventeenth century, when Descartes, Fermat, and Roberval busied themselves with it.

[2] Some have doubted the authenticity of the passage but it is probably ancient. If V_1, V_2 are the volumes compared, A_1, A_2 the areas, p_1, p_2 the perpendiculars drawn from the center of gravity to the axis, and q_1, q_2 the measure of the angles, we have

$$\frac{V_1}{V_2} = \frac{A_1}{A_2} \cdot \frac{p_1}{p_2} \cdot \frac{q_1}{q_2}.$$

This is essentially the theorem commonly attributed to Guldin, who may have obtained it from Pappus. Pappus's proof is not extant, and Guldin's is unsatisfactory. A proof based on the method of indivisibles was given by Cavalieri in the seventeenth century.

On the Sagacity of Bees in Building Their Cells

Pappus, *Mathematical Collection*, pp. 304–308 (Hultsch). Translation of T. L. Heath, *History of Greek Mathematics* II. 389

It is of course to men that God has given the best and most perfect notion of wisdom in general and of mathematical science in particular, but a partial share in these things he allotted to some of the unreasoning animals as well. To men, as being endowed with reason, he vouchsafed that they should do everything in the light of reason and demonstration, but to the other animals, while denying them reason, he granted that each of them should, by virtue of a certain natural instinct, obtain just so much as is needful to support life. This instinct may be observed to exist in very many other species of living creatures, but most of all in bees. In the first place their orderliness and their submission to the queens who rule in their state are truly admirable, but much more admirable still is their emulation, the cleanliness they observe in the gathering of honey, and the forethought and housewifely care they devote to its custody. Presumably because they know themselves to be entrusted with the task of bringing from the gods to the accomplished portion of mankind a share of ambrosia in this form, they do not think it proper to pour it carelessly on ground or wood or any other ugly and irregular material; but, first collecting the sweets of the most beautiful flowers which grow on the earth, they make from them, for the reception of the honey, the vessels which we call honeycombs, (with cells) all equal, similar and contiguous to one another, and hexagonal in form. And that they have contrived this by virtue of a certain geometrical forethought we may infer in this way. They would necessarily think that the figures must be such as to be contiguous to one another, that is to say, to have their sides common, in order that no foreign matter could enter the interstices between them and so defile the purity of their produce. Now only three rectilineal figures would satisfy the condition, I mean regular figures which are equilateral and equiangular; for the bees would have none of the figures which are not uniform. . . . There being then three figures capable by themselves of exactly filling up the space about the same point, the bees by reason of their instinctive wisdom chose for the construction of the honeycomb the figure which has the most angles, because they conceived that it would contain more honey than either of the two others.

Bees, then, know just this fact which is of service to themselves, that the hexagon is greater than the square and the triangle and will hold more honey for the same expenditure of material used in constructing the different figures. We, however, claiming as we do a greater share in wisdom than bees, will investigate a problem of still wider extent, namely, that, of all equilateral and equiangular plane figures having an equal perimeter, that which has the greater number of angles is always greater, and the greatest plane figure of all those which have a perimeter equal to that of the polygons is the circle.[1]

TRIGONOMETRY

The Greek geometers before the time of Hipparchus had proved propositions both in plane and in spherical geometry that might be translated into equivalent trigonometric terms. Thus Euclid proves the equivalent of $\tan \alpha/\tan \beta > \alpha/\beta$ when α and β are acute angles and $\alpha > \beta$ (see p. 260), and Aristarchus of Samos makes approximations to what are really trigonometric ratios by the use of the relation $\sin \alpha/\sin \beta < \alpha/\beta < \tan \alpha/\tan \beta$ for $\pi/2 > \alpha > \beta > 0$. Aristarchus does not give a proof, presumably because such a proof was simple and well known. Earlier Greek spherical geometry similarly investigated the relative size of arcs but, so far as we know, did not arrive at methods for the construction of accurate tables that would give the numerical value of any arc.

This achievement seems to belong to Hipparchus. The development was undoubtedly motivated by the requirements of a mathematical astronomy that was exclusively based on circular motion. But whereas we have only fragmentary indications of Hipparchus's work in this field, we are fortunate to possess Ptolemy's Table of Chords, with a detailed account of its construction.

The table is used to solve not only plane triangles but spherical triangles as well. This is made possible by the application of propositions in spherical geometry that are taken up by Ptolemy, but were already part of the standard treatment of the subject. These propositions are first found in systematic form in the third book of Menelaus's *Sphaerica*, though they were probably known to Hipparchus. This treatise, written sometime before the *Almagest*, probably toward the end of the first century A.D., is extant in an Arabic version. It takes up the geometry of the spherical triangle and proves the famous Theorem of Menelaus.

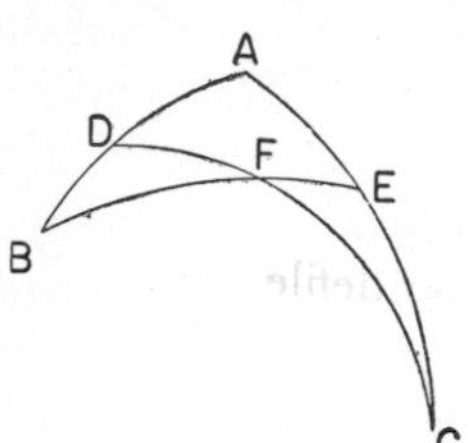

The theorem may be stated in modern terms as follows. If on a sphere two arcs of great circles, ADB and AEC, are intersected by two other arcs of great circles, DFC and BFE, which meet at F, each of the four arcs being less than a semicircle, then

$$\frac{\sin CE}{\sin AE} = \frac{\sin CF}{\sin FD} \cdot \frac{\sin DB}{\sin BA}.$$

The formulae necessary for the solution of spherical triangles may be deduced from this theorem. This is what Ptolemy does in the course of his treatise.

A word may be appropriate here on Ptolemy's method of constructing his Table of Chords, as given in *Almagest* I. 10. It should first be noted, however, that instead of the trigonometric ratios, sines, cosines, etc., Ptolemy, probably following the practice of Hipparchus, used "chords" corresponding to various arcs or central angles. The circle is divided into 360

[1] This charming preface introduces the subject of isoperimetry. A special treatise had been written on this subject by Zenodorus at some unknown previous date. After taking up the problems of plane figures, Pappus goes on to consider the proposition that of all solids having equal surface the sphere has the greatest volume. He succeeds in proving the sphere greater than any of the five *regular* solids of equal surface, and greater than the cone or cylinder of equal surface. Though he then goes on to describe the thirteen semiregular solids of Archimedes, he does not compare the volume of the sphere with that of any of these semiregular or of any irregular solids of surface equal to that of the sphere. A completely general proof was not given until the nineteenth century. [Edd.]

degrees, each degree into 60 minutes, each minute into 60 seconds; to the diameter, i.e., the chord of 180°, is assigned the value 120, and the problem is to compute the length of each chord from ½° to 180° by half degrees. The connection between a table of chords and a table of sines is immediately seen from the equation crd $x = 2 \sin (x/2)$. It was the Hindu and Arab mathematicians who later made the change from chords to sines.

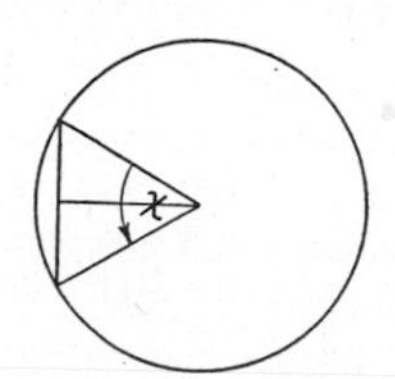

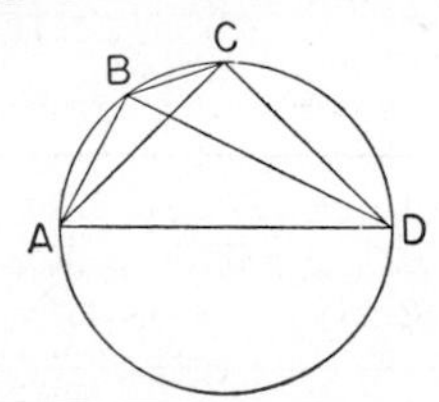

The basic theorem is the so-called "Theorem of Ptolemy," that the product of the diagonals of a quadrilateral inscribed in a circle is equal to the sum of the products of the opposite sides, i.e.,

$$AC \cdot BD = AB \cdot CD + BC \cdot AD.$$

Now by making one of the sides of the inscribed quadrilateral the diameter of the circle, Ptolemy develops the following formulae:

(1) $\text{crd } (a - b) \cdot \text{crd } 180° = \text{crd } a \cdot \text{crd } (180° - b) - (\text{crd } b) \cdot \text{crd } (180° - a).$

By an application of the relation between chords and sines noted above, we observe that (1) yields the equivalent of our formula, $\sin (x - y) = \sin x \cos y - \cos x \sin y$.

(2) $\left(\text{crd} \frac{a}{2}\right)^2 = \frac{1}{2} (\text{crd } 180°) \left\{(\text{crd } 180°) - [\text{crd } (180° - a)]\right\}.$

This is equivalent to our formula for the half angle, $\sin^2(x/2) = \frac{1}{2} (1 - \cos x)$.

(3) $(\text{crd } 180°) \cdot (\text{crd } [180° - (a + b)]) = [\text{crd } (180° - a)] [\text{crd } (180° - b)] - (\text{crd } a)(\text{crd } b).$

This is equivalent to our formula, $\cos (x + y) = \cos x \cos y - \sin x \sin y$.

Now being easily able to calculate the chords representing the sides of regular inscribed polygons of 3, 4, 5, 6, and 10 sides, i.e., crd 120°, 90°, 72°, 60°, and 36°,[1] Ptolemy applies formula (1), above, to obtain crd 12° (i.e., crd [72° − 60°]), and then applies formula (2) to obtain successively, crd 6°, 3°, 1½°, and ¾°.

Now to obtain crd 1°, Ptolemy uses the relation $\sin \alpha / \sin \beta < \alpha / \beta$ for $\pi/2 > \alpha > \beta > 0$, which he proves at this point. He then proceeds:

$$\text{crd } 1° : \text{crd } \tfrac{3}{4}° < 1 : \tfrac{3}{4},$$
$$\text{and crd } 1\tfrac{1}{2}° : \text{crd } 1° < 1\tfrac{1}{2} : 1.$$
$$\therefore \tfrac{4}{3} \text{ crd } \tfrac{3}{4}° > \text{crd } 1° > \tfrac{2}{3} \text{ crd } 1\tfrac{1}{2}°.$$

Knowing (crd ¾°) = $0^p47'8''$ and (crd 1½°) = $1^p34'15''$, he is able to get the very close approximation, crd 1° = $1^p\,2'50''$.

Then, applying formula (2), above, he is able to approximate crd ½°, and with the addition formula (3), above to fill out the table from ½° − 180°, by steps of half degrees.

The third column is a table of proportional parts and gives the amount by which the length of the chord is augmented for every minute of arc over the last half- or full-degree value in the table.

Since crd $x = 2 \sin (x/2)$, the table is equivalent to a table of sines from ¼ to 90° by quarter degrees. Conversion into decimals shows the table to be accurate, in general, to about

[1] These generally involve the extraction of the square root of surds. In his commentary on this chapter of the *Almagest*, Theon of Alexandria gives, as an example, the complete calculation of $\sqrt{4500}$, for which Ptolemy merely set down the result $67 + 4/60 + 55/(60)^2$. This example is important for an understanding of Greek methods of calculation in the sexagesimal system. See T. L. Heath, *History of Greek Mathematics* I. 61.

five places, sometimes more. E.g., crd 120° (= 2 sin 60° = $\sqrt{3}$) is $103^p55'23''$. Since the radius 60^p is reduced to unity, 2 sin 60° works out to 1.7320509, correct to six decimals. The value of π computed from the table is 3.14166. . . .

The following is the portion of Ptolemy's Table of Chords (*Almagest* I. 11) corresponding to arcs of 116° to 124½°.

Arcs	Chords			Prop. Parts			Arcs	Chords			Prop. Parts		
116	101	45	57	0	33	11	120½	104	11	2	0	31	4
116½	102	2	33	0	32	57	121	104	26	34	0	30	49
117	102	19	1	0	32	43	121½	104	41	59	0	30	35
117½	102	35	22	0	32	29	122	104	57	16	0	30	21
118	102	51	37	0	32	15	122½	105	12	26	0	30	7
118½	103	7	41	0	32	0	123	105	27	30	0	29	52
119	103	23	44	0	31	46	123½	105	42	26	0	29	37
119½	103	39	37	0	31	32	124	105	57	14	0	29	23
120	103	55	23	0	31	18	124½	106	11	55	0	29	8

Application of Trigonometry

Ptolemy, *Almagest* II. 3

Given the length of the longest day, to determine the latitude of a place, and conversely.[1]

A. Now let it be required, given the length of the longest day, to find the elevation of the pole [i.e., the latitude], that is, arc BZ of the meridian.

On the same figure as before,

$$\frac{\text{crd 2 arc } ET}{\text{crd 2 arc } AT} = \frac{\text{crd 2 arc } EH}{\text{crd 2 arc } HB} \cdot \frac{\text{crd 2 arc } BZ}{\text{crd 2 arc } ZA}.$$

$$\begin{aligned} \text{But 2 arc } ET &= 37°30', \text{ and crd 2 arc } ET = 38^p34'22'', \\ \text{2 arc } TA &= 142°30', \text{ and crd 2 arc } AT = 113^p37'54'', \\ \text{Again, 2 arc } EH &= 60°, \quad \text{and crd 2 arc } EH = 60^p, \\ \text{2 arc } HB &= 120°, \quad \text{and crd 2 arc } HB = 103^p55'23''. \end{aligned}$$

$$\therefore \frac{\text{crd 2 arc } BZ}{\text{crd 2 arc } ZA} = \frac{38^p34'22''}{113^p37'54''} \cdot \frac{103^p55'23''}{60^p}$$

$$= \frac{70^p33'}{120^p}, \text{ very nearly.}$$

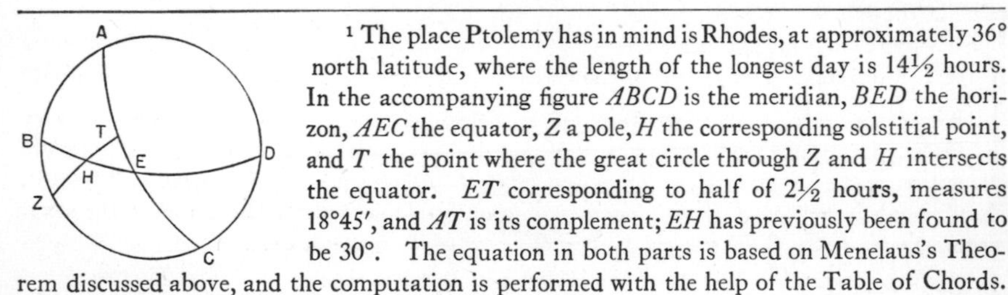

[1] The place Ptolemy has in mind is Rhodes, at approximately 36° north latitude, where the length of the longest day is 14½ hours. In the accompanying figure $ABCD$ is the meridian, BED the horizon, AEC the equator, Z a pole, H the corresponding solstitial point, and T the point where the great circle through Z and H intersects the equator. ET corresponding to half of 2½ hours, measures 18°45′, and AT is its complement; EH has previously been found to be 30°. The equation in both parts is based on Menelaus's Theorem discussed above, and the computation is performed with the help of the Table of Chords.

$$\text{Now crd } 2 \text{ arc } ZA = 120^p.$$
$$\therefore \text{crd } 2 \text{ arc } BZ = 70^p 33',$$
$$2 \text{ arc } BZ = 72°1',$$
$$\text{and } BZ = 36°, \text{ very nearly.}$$[1]

B. Now conversely, let arc BZ, representing, on the same figure, the elevation of the pole, be given. Suppose it is determined by observation to be 36°. Let it be required to find the difference between the shortest or longest day and the day of the equinox, i.e., 2 arc ET.

By the same theorem

$$\frac{\text{crd } 2 \text{ arc } ZB}{\text{crd } 2 \text{ arc } BA} = \frac{\text{crd } 2 \text{ arc } ZH}{\text{crd } 2 \text{ arc } HT} \cdot \frac{\text{crd } 2 \text{ arc } TE}{\text{crd } 2 \text{ arc } EA}.$$

But 2 arc ZB = 72°, and crd 2 arc ZB = $70^p 32' 3''$,
2 arc BA = 108°,[2] and crd 2 arc BA = $97^p\ 4'56''$.
Again, 2 arc ZH = 132°17′20″,[3] and crd 2 arc ZH = $109^p 44' 53''$,
2 arc HT = 47°42′40″,[4] and crd 2 arc HT = $48^p 31' 55''$.

$$\therefore \frac{\text{crd } 2 \text{ arc } TE}{\text{crd } 2 \text{ arc } EA} = \frac{70^p 32' 3''}{97^p 4' 56''} \cdot \frac{48^p 31' 55''}{109^p 44' 53''}$$
$$= \frac{31^p 11' 26''}{97^p 4' 56''}$$
$$= \frac{38^p 34'}{120^p}, \text{ very nearly.}$$

$$\text{But crd } 2\ EA = 120^p.$$
$$\therefore \text{crd } 2 \text{ arc } TE = 38^p 34',$$
$$\text{and } 2 \text{ arc } TE = 37°30', \text{ very nearly.}$$

This represents 2½ equinoctial hours. Q.E.D.

SOME NUMERICAL PROBLEMS IN MENSURATION

Hero of Alexandria, *Metrica* I. 8, III. 20, II. 12[5]

I. 8. There is a general method whereby, if the three sides of any triangle are given, it is possible to find the area without finding the altitude.

---

[1] I.e., a place where the length of the longest day is 14½ hours is at latitude 36°. The equation, in modern terms, is $\cos(\pi a/24) = -\tan\phi \tan\omega$, where ϕ is the latitude of the place, ω the obliquity of the ecliptic, and a the length of the longest day in hours. The maximum value of a in this formula is 24, the length of the longest day at the arctic and antarctic circles. A more complicated formula is necessary for the higher latitudes. The discussion does not take account of (1) the fact that the sun is not a point of light, and (2) the effect of atmospheric refraction in lengthening the time between apparent sunrise and apparent sunset.

[2] I.e., 2(90° − 36°).

[3] I.e., 2(90° − 23°51′20″ [obliquity of the ecliptic]).

[4] I.e., double the obliquity of the ecliptic.

[5] In his *Metrica* Hero gives us a series of numerical problems in finding the areas of all sorts of plane figures, including the area of an ellipse and of a parabolic segment, and the volumes of various solids, including the cone, cylinder, frustum of a pyramid and of a cone, segment of a sphere. The collection is of great interest, showing the practical application

For example, let the sides of the triangle be of 7, 8, and 9 units.

$$7+8+9=24,$$
$$24/2=12,$$
$$12-7=5,$$
$$12-8=4,$$
$$12-9=3,$$
$$12\times 5=60,$$
$$60\times 4=240,$$
$$240\times 3=720,$$
$$\text{Area of triangle}=\sqrt{720}.$$[1]

Now since 720 has no rational square root, we shall obtain a very close approximation to $\sqrt{720}$ as follows:

The nearest [perfect integral] square to 720 is 729.

$$\sqrt{729}=27,$$
$$\frac{720}{27}=26\tfrac{2}{3},$$
$$27+26\tfrac{2}{3}=53\tfrac{2}{3},$$
$$\frac{53\frac{2}{3}}{2}=26\tfrac{1}{2}\tfrac{1}{3},$$
$$\therefore \sqrt{720}=\text{approximately } 26\tfrac{1}{2}\tfrac{1}{3}.$$

For $(26\frac{1}{2}\frac{1}{3})^2 = 720\frac{1}{36}$, the difference being only $\frac{1}{36}$ of a unit.

Now if we desire that the difference be less than $\frac{1}{36}$ we commence with the number just obtained, $720\frac{1}{36}$ instead of with 729, and by proceeding in the same way we shall find that the difference will be much less than $\frac{1}{36}$.[2]

---

of geometry in contrast to the purely theoretical approach of the great Greek books on geometry. There are also in the *Metrica* some passages which have a theoretical interest, for they seem to indicate systematic methods of approximation to square and cube roots of non-square and non-cubic numbers. Since this is a vexed problem in the history of Greek mathematics we have taken two cases, $\sqrt{720}$ and $\sqrt[3]{100}$, from Hero. In addition we give an example of the solution of numerical problems in geometry.

[1] This is an application of the so-called Heronian formula, Area $=\sqrt{s(s-a)(s-b)(s-c)}$, where a, b, and c are the sides and s half the perimeter. The geometrical proof is given both in the *Metrica* and in the *Dioptra* of Hero, but there is strong evidence that we owe the formula to Archimedes.

[2] The procedure is as follows:

Let $x=\sqrt{A}$.

If x_1 is the first approximation, i.e., the square root of the nearest perfect square to A, Let $A/x_1 = x_1'$.

Now, the second approximation is the arithmetic mean of x_1 and x_1'.

(1) That is $x_2=\dfrac{x_1+x_1'}{2}$.

Similarly let $\dfrac{A}{x_2}=x_2'$,

(2) and $x_3=\dfrac{x_2+x_2'}{2}$, and so on.

Hero, in the example given, does not go beyond equation (1). If the process is repeated

III. 20. Let there be a pyramid, with base $ABCD$ of any form, and vertex at point E, with side AE of 5 units. It is required to cut the pyramid by a plane parallel to the base in such a way that the pyramid cut off at the vertex is, let us say, four times the remaining solid. Let $ZHTK$ be the section made by the cutting plane. Thus AZ will be a side of the solid $ABCDZHTK$.

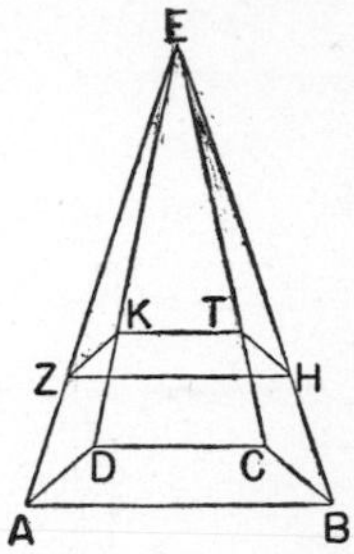

$$\frac{\text{pyramid } ABCDE}{\text{pyramid } ZTHKE} = \frac{5}{4}.$$

Now pyramids are to each other as the cubes of homologous sides.

$$\therefore \frac{AE^3}{EZ^3} = \frac{5}{4}.$$

But $AE^3 = 125$.

$\therefore EZ^3 = 100$.

We must therefore find $\sqrt[3]{100}$ by approximation.

$\sqrt[3]{100}$ = very nearly $4\frac{9}{14}$, as we shall show below.

Thus if EZ be taken as $4\frac{9}{14}$ and the pyramid be cut at point Z by a plane parallel to the base, the problem is solved.

The synthesis is as follows.

$$5^3 = 125.$$

Since the ratio of the two parts is 4:1, we have

$$4 + 1 = 5,$$
$$125 \times 4 = 500,$$
$$500/5 = 100,$$
$$\sqrt[3]{100} = 4\tfrac{9}{14} = EZ.$$

We shall now show how to find $\sqrt[3]{100}$.

Take the cubic numbers nearest to 100, both greater and less, i.e., 125 and 64.

$$125 - 100 = 25,$$
$$100 - 64 = 36,$$
$$5 \times 36 = 180,$$
$$180 + 100 = 280,$$
$$180/280 = 9/14.$$

Adding $\frac{9}{14}$ to the cube root of the smaller cubic number [64] we have $4\frac{9}{14}$. This will be the approximation to $\sqrt[3]{100}$.[1]

---

we get, from equation (2), $x_3 = 26\frac{1609}{1932}$, of which the square is $720\frac{1}{3732624}$.

For a discussion of this method in connection with Archimedes' approximation to $\sqrt{3}$ (p. 61, above), see T. L. Heath, *History of Greek Mathematics* II. 325.

[1] Precisely what general formula is applied in this procedure is not entirely clear. It has been suggested that the formula is $\sqrt[3]{A} \sim a + \frac{(a+1)d_1}{(a+1)d_1 + ad_2}$, where a^3 and $(a+1)^3$ are the nearest perfect cubes below and above A, and d_1 and d_2 the differences between A and a^3, and A and $(a+1)^3$, respectively. For a discussion see T. L. Heath, *op. cit.*, II. 341.

II. 12. Let it be required to measure the segment of a sphere, having a base of diameter 12 and height 2.[1]

Now Archimedes proves that the ratio of a segment of a sphere to a cone with same base and equal height is equal to the ratio of the sum of the radius of the sphere and the height of the remaining segment, to the height of the remaining segment.[2]

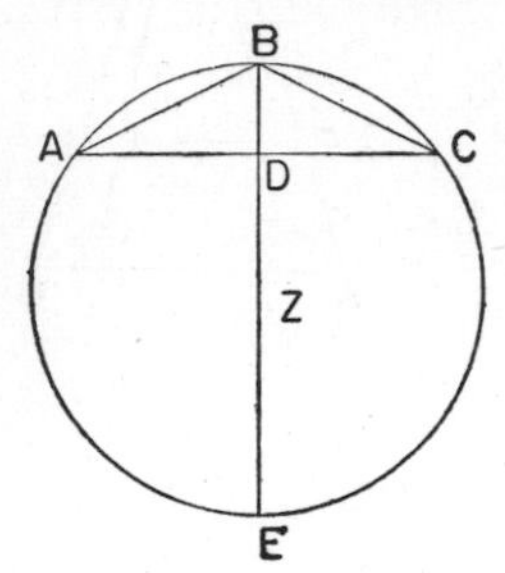

Let the segment in question be ABC, with altitude BD. Let Z be the center of the sphere.

$$\therefore \frac{\text{segment } ABC}{\text{cone } ABC} = \frac{DE + EZ}{DE}.$$

Since AC is given, AD and AD^2 are given.

That is, $BD \cdot DE$ is given.

But since BD is given, DE is also given, as is the whole of BE.

$\therefore EZ$ is given, and also $DE + EZ$.

But since DE is given, the ratio of the cone, with base equal to the diameter of circle AC and height BD, to the segment of the sphere is also given.

$\therefore$ The segment of the sphere is given.

We shall on the basis of this analysis proceed as follows:

$$\left(\frac{12}{2}\right)^2 = 36,$$
$$\frac{36}{2} = 18,$$
$$18 + 2 = 20,$$
$$\frac{20}{2} = 10,$$
$$10 + 18 = 28,$$
$$2 \times 2 = 4,$$
$$4 \times 4 = 16,$$
$$16 \times 28 = 448,$$
$$448 \times 11/14 = 352,$$
$$\frac{352}{3} = 117\tfrac{1}{3}.$$

This will be the volume of the segment.

[1] The reference evidently is to a segment of one base.

[2] Archimedes, *On the Sphere and the Cylinder* II, Prop. 2, Corollary.

ASTRONOMY AND MATHEMATICAL GEOGRAPHY

An interest in the heavenly bodies and their motion was not uncommon among the peoples of antiquity. Practical concerns of navigation, agriculture, and timekeeping generally provided the original motive. The Egyptians and Babylonians had made astronomical observations and noted various astronomical cycles for purposes of the calendar long before the rise of Greek science.

The great contribution of the Greeks was in the formulation of geometrical systems to represent the motions of the heavenly bodies. The period from Thales to Plato is, generally speaking, one in which speculative cosmology predominates, though certain discoveries which tradition ascribes to this period, e.g., the source of the moon's light, the cause of eclipses, the sphericity of the earth, and the obliquity of the ecliptic, are of prime scientific importance. For this period, however, one must bear in mind that the evidence is fragmentary and none too reliable (see p. 1, n. 1).

In the time of Plato a series of remarkable developments was initiated. Plato had proposed the problem of showing how the apparently irregular motions of the planets might be derived from the composition of certain uniform circular motions. Eudoxus sought to solve this problem with his system of concentric spheres. Callippus elaborated this system, and Aristotle converted it from a purely geometrical to a mechanical structure. Difficulties inherent in it led to the rise of systems involving epicycles and eccentric circles. Apollonius, Hipparchus, and Ptolemy were preeminent in this development.

Along with the formulation of these great astronomical systems there were other important developments. The hypothesis that night and day are due to the rotation not of the heavens but of the earth was set forth by Heraclides of Pontus, a pupil of Plato. Though the evidence is not certain, Heraclides may also have taught that Venus and Mercury revolve around the sun, one step in the direction of a complete heliocentric system. Such a heliocentric system seems to have been put forward as a hypothesis by Aristarchus of Samos, who thus in the third century B.C. anticipated Copernicus. Increasingly precise and systematic observations were made possible by the improved methods and instruments of Hipparchus. To Hipparchus, too, is generally ascribed the discovery of the precession of the equinoxes.

Geometry and trigonometry were applied to the problem of the size of the earth and the sizes and distances of the sun and the moon. With advances in astronomical observation there were corresponding advances in dealing with the problems of mathematical geography, the determination of latitude and longitude, and map making. In this branch the work of Hipparchus and Ptolemy represents the highest stage reached by the Greeks.

It should be noted that paralleling the rise of Greek astronomy there developed a highly refined mathematical astronomy among the Babylonians. Yet the question of the mutual influence between the geometrical methods of the Greeks and the numerical methods of the Babylonians remains largely unanswered.

We have selected passages from two different kinds of Greek writings, the one technical, like Ptolemy's *Almagest* and *Geography*, the other more popular or elementary, like Geminus's *Introduction* or Strabo's *Geography*.

THE SCOPE OF ASTRONOMY CONTRASTED WITH THAT OF PHYSICS

Simplicius, *Commentary on Aristotle's Physics*, pp. 291.21–292.31 (Diels). Translation of T. L. Heath, *Aristarchus of Samos*, pp. 275–276 (Oxford, 1913)

Alexander carefully quotes a certain explanation by Geminus taken from his summary of the *Meteorologica* of Posidonius.[1] Geminus's comment, which is inspired by the views of Aristotle, is as follows:

"It is the business of physical inquiry to consider the substance of the heaven and the stars, their force and quality, their coming into being and their destruction, nay, it is in a position even to prove the facts about their size, shape, and arrangement; astronomy, on the other hand, does not attempt to speak of anything of this kind, but proves the arrangement of the heavenly bodies by considerations based on the view that the heaven is a real *κόσμος*, and further, it tells us of the shapes and sizes and distances of the earth, sun, and moon, and of eclipses and conjunctions of the stars, as well as of the quality and extent of their movements. Accordingly, as it is connected with the investigation of quantity, size, and quality of form or shape, it naturally stood in need, in this way, of arithmetic and geometry. The things, then, of which alone astronomy claims to give an account it is able to establish by means of arithmetic and geometry. Now in many cases the astronomer and the physicist will propose to prove the same point, e.g., that the sun is of great size or that the earth is spherical, but they will not proceed by the same road. The physicist will prove each fact by considerations of essence or substance, of force, of its being better that things should be as they are, or of coming into being and change; the astronomer will prove them by the properties of figures or magnitudes, or by the amount of movement and the time that is appropriate to it. Again, the physicist will in many cases reach the cause by looking to creative force; but the astronomer, when he proves facts from external conditions, is not qualified to judge of the cause, as when, for instance,

[1] Posidonius (*ca.* 135–*ca.* 51 B.C.), a notable Stoic philosopher and polymath, was born in Apamea in Syria, traveled extensively, and died at Rome. He was interested not only in philosophy but *inter alia* in astronomy, descriptive and mathematical geography, meteorology, and history. Of his voluminous written work only fragments survive. He exerted wide influence on Roman thought (e.g., in Cicero, Lucretius, and Seneca) and on that of the Middle Ages. Reference is made elsewhere to his determination of the size of the earth and the distance and size of the sun (pp. 115, 149).

Geminus (see pp. 2, 117) may have been a pupil of Posidonius. At any rate he wrote an exposition (which was later epitomized) of the *Meteorologica* of Posidonius. The passage here quoted may, as its last sentence indicates, be by Posidonius himself. It was transmitted by Alexander of Aphrodisias in Caria, leader of the Peripatetic School in the early part of the third century A.D. and so highly esteemed a commentator of the works of Aristotle as to receive the title "The Exegete." [Edd.]

he declares the earth or the stars to be spherical; sometimes he does not even desire to ascertain the cause, as when he discourses about an eclipse; at other times he invents by way of hypothesis, and states certain expedients by the assumption of which the phenomena will be saved. For example, why do the sun, the moon, and the planets appear to move irregularly? We may answer that, if we assume that their orbits are eccentric circles or that the stars describe an epicycle, their apparent irregularity will be saved; and it will be necessary to go further and examine in how many different ways it is possible for these phenomena to be brought about, so that we may bring our theory concerning the planets into agreement with that explanation of the causes which follows an admissible method. Hence we actually find a certain person, Heraclides of Pontus,[1] coming forward and saying that, even on the assumption that the earth moves in a certain way, while the sun is in a certain way at rest, the apparent irregularity with reference to the sun can be saved. For it is no part of the business of an astronomer to know what is by nature suited to a position of rest, and what sort of bodies are apt to move, but he introduces hypotheses under which some bodies remain fixed, while others move, and then considers to which hypotheses the phenomena actually observed in the heaven will correspond. But he must go to the physicist for his first principles, namely, that the movements of the stars are simple, uniform, and ordered, and by means of these principles he will then prove that the rhythmic motion of all alike is in circles, some being turned in parallel circles, others in oblique circles." Such is the account given by Geminus, or Posidonius in Geminus, of the distinction between physics and astronomy, wherein the commentator is inspired by the views of Aristotle.[2]

---

[1] This passage has been relied on to credit Heraclides of Pontus, a pupil of Plato and Aristotle, with the honor of being the first to anticipate the Copernican theory. It is generally held, however, that the name is a late interpolation here, and that by "a certain person" Geminus (or Posidonius) referred to Aristarchus of Samos, who is generally credited with first suggesting the heliocentric hypothesis (p. 107). Heraclides, to be sure, taught the daily rotation of the earth and may have held that Mercury and Venus revolved around the sun (see pp. 105-107). [Edd.]

[2] In modern terms this distinction is in the main equivalent to that between celestial kinematics (i.e., the theory of the motions of the heavenly bodies) and astrophysics. Astronomy in the meaning of our passage is a purely mathematical science. It is to be noted, however, that in choosing hypotheses which will "save the phenomena" the astronomer is guided by basic requirements which, in the ancient view, are derived from physics, e.g., the requirements that motions be circular and uniform.

In connection with the distinction between physics and astronomy given here, another observation may be made. When the Greek astronomers attempted to resolve the observed motions of the heavenly bodies into component motions of various spheres, some considered the spheres as purely geometrical constructs or devices for the representation of the observed motion, whereas others attributed physical reality to the whole mechanism. Among the former was Eudoxus, among the latter Aristotle, who, from the viewpoint of the present

SOME EARLY THEORIES

There is among the Pre-Socratics much astronomical and cosmological speculation that we should today consider as philosophical rather than scientific. Though well aware that a sharp distinction here is modern rather than ancient, we have considered it preferable for our purpose to indicate the tradition of scientific contributions. This tradition is, for the most part, known to us through scanty fragments, from discussion in later writers, and from a considerable sum of doxographical material stemming in the main from Theophrastus's *Physical Opinions*. Extreme caution is necessary in dealing with all these sources, especially with accounts ascribing particular discoveries to this or that individual. For a full discussion of this early period of Greek astronomy, see T. L. Heath, *Aristarchus of Samos*, Part I.

Thales

Herodotus I. 74. Translation of T. L. Heath, *Greek Astronomy*, p. 1 (London, 1932)

When, in the sixth year, they [the Lydians and the Medes] encountered one another, it so fell out that, after they had joined battle, the day suddenly turned into night. Now that this transformation of day (into night) would occur was foretold to the Ionians by Thales of Miletus, who fixed as the limit of time this very year in which the change actually took place.[1]

Theon of Smyrna, p. 198.16–18 (Hiller). Translation of T. L. Heath, *op. cit.*, p. 2

Eudemus relates in his *Astronomies* . . . that Thales was the first to discover the eclipse of the sun and the fact that the sun's period with respect to the solstices is not always the same.[2]

Anaximander

Hippolytus, *Refutation of All Heresies* I. 6.3. Translation of T. L. Heath, *op. cit.*, p. 6

The earth is poised aloft supported by nothing, and remains where it is because of its equidistance from all other things. Its form is rounded, circular, like a stone pillar; of its plane surfaces one is that on which we stand, the other is opposite.[3]

---

passage, was thus combining astronomical with physical theory. (Cf., however, *Metaphysics*, Λ.8, p. 102, below.) The two approaches persisted through ancient and medieval astronomy. [Edd.]

[1] The eclipse in question, a total eclipse of the sun, may have been that which occurred May 28, 585 B.C. There is general agreement that Thales had no means of making sure predictions of eclipses, though some have held that he could approximate the time of a *possible* eclipse by using a Babylonian cycle of lunations. The visibility of the eclipse at a given place, its path, and its nature, whether total or partial, depended on factors beyond Thales' knowledge. Moreover, his use of the Babylonian cycle is now considered highly improbable.

In various forms the story of Thales' discovery is referred to by Diogenes Laertius, Cicero, Pliny, Clement of Alexandria, and Eusebius. On the question whether Thales discovered the cause of eclipses, the ancient tradition is itself doubtful (see p. 94). [Edd.]

[2] The reference seems to be to the fact that the period from the summer solstice to the winter solstice is approximately 184½ days, whereas that from the winter solstice to the summer solstice is approximately 180½ days. [Edd.]

[3] Anaximander of Miletus, like Thales, flourished in the first half of the sixth century

Pythagoras

Theon of Smyrna, p. 150.12–18 (Hiller). Translation of T. L. Heath, *op. cit.*, pp. 11–12[1]

The impression of variation in the movement of the planets is produced by the fact that they appear to us to be carried through the signs of the zodiac in certain circles of their own, being fastened in spheres of their own and moved by their motion, as Pythagoras was the first to observe, a certain varied and irregular motion being thus grafted, as a qualification, upon their simply and uniformly ordered motion in one and the same place [i.e., that of the daily rotation from east to west].[2]

Anaxagoras

Hippolytus, *Refutation of All Heresies* I. 8.6–10.[3] Translation of John Burnet, *Early Greek Philosophy* (London, 1930)

6. The sun and the moon and all the stars are fiery stones carried round by the rotation of the aether. Under the stars are the sun and moon, and also certain bodies which revolve with them, but are invisible to us.

B.C. The conception of the earth as a cylinder freely suspended without support in space is a step toward the notion of the earth's sphericity. Elsewhere Anaximander gives the ratio of the depth to the breadth of the cylinder as one-third. He is also said to have made a map of the inhabited portion of the earth (see p. 153). [Edd.]

[1] Pythagoras of Samos flourished in the latter half of the sixth century B.C. Tradition connects him with progress in arithmetical and geometrical theory and with the origin of the science of acoustics. See pp. 43, 294. He was the founder of a religious brotherhood in Magna Graecia. Our sources seem often to confound Pythagoras's own contribution with that of his followers. In opposition to the view that Pythagoras first put forward the doctrine of the sphericity of the earth, see, e.g., W. A. Heidel, *The Frame of the Ancient Greek Maps* (New York, 1937). [Edd.]

[2] There is much discussion as to the correctness of the attribution of this discovery to Pythagoras. It may be noted that Pythagoras is not mentioned in the passage of Aëtius (II. 16.2–3): "Some mathematicians hold that the planets move from west to east, contrary to the motion of the fixed stars. With this Alcmaeon, too, agrees."

Now there is every likelihood that the distinction referred to between the motion of planets and fixed stars could be deduced on the basis of any extended record of observations. For the fact that the axis of daily rotation is not the same as the axis of planetary motion along the zodiac makes it impossible for planetary motion to be conceived as resulting merely from diurnal motions at a slower rate than that of the fixed stars. But the view—whoever it was who first conceived it—that the apparently complex motion of the planets might be viewed as the resultant of a group of uniform circular motions, was of momentous importance for Greek astronomy. For it was a step toward those geometrical systems which are the essence of the Greek contribution to astronomy. [Edd.]

[3] Anaxagoras of Clazomenae, near Smyrna, flourished in the middle of the fifth century B.C. He correctly ascribed the light of the moon to the sun and gave a substantially correct explanation of solar and lunar eclipses. His doctrine of substances and the power of Mind (*Nous*) in effecting transformations and his treatment of the heavenly bodies as not essentially different from terrestrial bodies are especially noteworthy. [Edd.]

7. We do not feel the heat of the stars because of the greatness of their distance from the earth; and, further, they are not so warm as the sun, because they occupy a colder region. The moon is below the sun, and nearer us.

8. The sun surpasses the Peloponnesos in size. The moon has not a light of her own, but gets it from the sun.[1] The course of the stars goes under the earth.

9. The moon is eclipsed by the earth screening the sun's light from it, and sometimes, too, by the bodies below the moon coming before it. The sun is eclipsed at the new moon, when the moon screens it from us.[2] Both the sun and the moon turn in their courses owing to the repulsion of the air. The moon turns frequently, because it cannot prevail over the cold.

10. Anaxagoras was the first to determine what concerns the eclipses and the illumination of the sun and moon. And he said the moon was of earth, and had plains and ravines in it. The Milky Way was the reflexion of the light of the stars that were not illuminated by the sun. Shooting stars were sparks, as it were, which leapt out owing to the motion of the heavenly vault.

Aristotle, *Meteorologica* 345*a*25–29. Translation of E. W. Webster (Oxford, 1923)

Anaxagoras, Democritus, and their schools say that the milky way is the light of certain stars. For, they say, when the sun passes below the earth some of the stars are hidden from it. Now the light of those on which the sun shines is invisible, being obscured by the rays of the sun. But the milky way is the peculiar light of those stars which are shaded by the earth from the sun's rays.

Oenopides

Theon of Smyrna, p. 198 (Hiller)[3]

Eudemus in his *Astronomies* tells us that Oenopides first discovered

[1] We have for this the far better authority of Plato, *Cratylus* 409 A: "As he [Anaxagoras] recently said, that the moon has its light from the sun." [Edd.]

[2] Cf. Aëtius II.29.6: "Anaxagoras, . . . in agreement with the mathematicians, holds that the moon undergoes obscuration each month through conjunction with the sun, by which it is illuminated; and that the moon is eclipsed when it falls within the shadow of the earth, as the earth comes between both the stars [the sun and the moon], that is, when the moon is blocked off [from the sun]." [Edd.]

[3] On Oenopides, see p. 35. As is the case with most references to astronomical discovery in the early period, there is considerable conflict both in the sources and in their interpretation. Once the planetary and solar motions came to be considered as the resultant of more than one component motion, the notion of an axis oblique to that of the diurnal motion followed easily. The amount of obliquity could be approximately derived from a comparison of the shadows cast by the sun at noon at a given place on the longest and shortest days of the year.

The reference to a Great Year recalls a problem frequently considered by the Greek

the obliquity[1] of the zodiac circle and the duration of the Great Year. . . . (Others discovered) that the axis of the fixed stars and that of the planets are separated from each other by the angle which is subtended by the side of [an inscribed regular] pentadecagon, i.e., 24°.

PYTHAGOREAN ASTRONOMY

The Central Fire

Aristotle, *De Caelo* II. 13. Translation of J. L. Stocks (Oxford, 1922)

It remains to speak of the earth, of its position, of the question whether it is at rest or in motion, and of its shape.

I. As to its *position* there is some difference of opinion. Most people—all, in fact, who regard the whole heaven as finite—say it lies at the centre. But the Italian philosophers known as Pythagoreans take the contrary view.[2] At the centre, they say, is fire, and the earth is one of the stars, creating night and day by its circular motion about the centre.[3] They further construct another earth in opposition to ours to which they give the name counter-earth. . . .

II. As to the position of the earth, then, this is the view which some advance, and the views advanced concerning its *rest or motion* are similar. For here too there is no general agreement. All who deny that the earth lies at the centre think that it revolves about the centre, and not the earth only but, as we said before, the counter-earth as well. Some of them even

---

astronomers, viz., the period required (1) for the sun and moon or (2) for the sun, moon, and the five planets known to the Greeks to return to the same position in the heavens. With changing calculations of the length of the month and year there were various estimates of the length of these Great Years. Cycles containing an exact number of days, lunar months, and solar years were also investigated by the Babylonians and introduced at various times into Greece. The tradition is that the Great Year of Oenopides was 59 years, but there is some doubt as to his methods. See Diels, *Frag. d. Vors.* I^5. 394.

[1] Reading *λόξωσιν* for *διάζωσιν* (Hiller).

[2] The theory of a central fire and a counterearth was developed by the Pythagoreans, probably at the end of the fifth century B.C. Philolaus, a Pythagorean of that period, is associated with this theory by ancient authors. We have here the first instance of a non-geocentric system. A central fire occupies the center of the universe; the earth and a counter-earth revolve around the central fire. Since in so revolving the earth always presents the same hemisphere to the central fire, the hemisphere that is not inhabited, it follows that the inhabitants of the earth never see the fire. For the same reason they do not see the counter-earth, which revolves with the earth around the central fire but inside the earth's orbit. [Edd.]

[3] In presenting the same face to the central fire in its daily revolution, the earth must necessarily rotate once each day around its axis. It is this that makes day and night (whether the sun reflects only the central fire or all the fire of the universe; for in Philolaus's system the sun has no light of its own). Now it is very doubtful that the Pythagoreans accepted the implications of their theory and considered the sphere of the fixed stars as motionless. It has been suggested that the motion they assigned to this sphere was so slow as to be imperceptible. [Edd.]

consider it possible that there are several bodies so moving, which are invisible to us owing to the interposition of the earth. This, they say, accounts for the fact that eclipses of the moon are more frequent than eclipses of the sun: for in addition to the earth each of these moving bodies can obstruct it. Indeed, as in any case the surface of the earth is not actually a centre but distant from it a full hemisphere, there is no more difficulty, they think, in accounting for the observed facts on their view that we do not dwell at the centre, than on the common view that the earth is in the middle.[1] Even as it is, there is nothing in the observations to suggest that we are removed from the centre by half the diameter of the earth. Others, again, say that the earth, which lies at the centre, is "rolled," and thus in motion, about the axis of the whole heaven. So it stands written in the *Timaeus*.[2]

The Harmony of the Spheres

Alexander of Aphrodisias, *Commentary on Aristotle's Metaphysics*, p. 542*a*5–18 (Brandis). Translation of T. L. Heath, *Greek Astronomy*

They (the Pythagoreans) said that the bodies which revolve round the centre have their distances in proportion, and some revolve more quickly, others more slowly, the sound which they make during this motion being deep in the case of the slower, and high in the case of the quicker; these sounds then, depending on the ratio of the distances, are such that their combined effect is harmonious. . . . Thus, the distance of the sun from the earth being, say, double the distance of the moon, that of Aphrodite triple, and that of Hermes quadruple, they considered that there was some arithmetical ratio in the case of the other planets as well, and that the movement of the heaven is harmonious. They said that those bodies move most quickly which move at the greatest distance, that those bodies move most slowly which are at the least distance, and that the bodies at intermediate distances move at speeds corresponding to the sizes of their orbits.[3]

---

[1] As T. L. Heath, following Schiaparelli, indicates, the Pythagoreans would hold "that parallax is as negligible in one case as in the other." That is, they would have to hold that the earth's distance from the center is very small compared to that of the other heavenly bodies, and the radius of its orbit not many times greater than the radius of the earth itself (*Aristarchus*, pp. 100–101). [Edd.]

[2] On this vexed passage and the many problems it raises see T. L. Heath, *Aristarchus of Samos*, pp. 174–178, where the conclusion is reached that Plato does not in the *Timaeus* affirm the axial rotation of the earth or, in fact, any other motion, but is deliberately misrepresented here by Aristotle. For a different view see F. M. Cornford, *Plato's Cosmology*, p. 130 (London, 1937). [Edd.]

[3] On Alexander of Aphrodisias, see p. 90. All sorts of conjectures have been made as to the precise details of the Pythagorean system of the harmony of the spheres. Indeed there is good evidence that at various times various systems were put forward by the Pythag-

Cf. also the following passages on Pythagorean astronomy. The translation is that of T. L. Heath, *Greek Astronomy*.

Simplicius, *Commentary on Aristotle's De Caelo*, p. 512 (Heiberg)

The earth, then, being like one of the stars, moves round the centre and, according to its position with reference to the sun, makes night and day.

Aëtius, *Placita*

II.7.7. [Philolaus holds that] the middle is naturally first in order, and round it ten divine bodies move as in a dance, the heaven and < after the sphere of fixed stars> [1] the five planets, after them the sun, under it the moon, under the moon the earth and under the earth the counter-earth; after all these comes the fire which is placed like a hearth around the centre.

III.11.3. Philolaus the Pythagorean places the fire in the middle, for this is the Hearth of the All; second to it he puts the counter-earth, and third the inhabited earth, which is placed opposite to, and revolves with, the counter-earth; this is the reason why those who live in the counter-earth are invisible to those who live in our earth.

III.13.1–2. Others maintain that the earth remains at rest. But Philolaus the Pythagorean held that it revolves round the fire in an oblique circle, in the same way as the sun and moon.

II.20.12. Philolaus the Pythagorean holds that the sun is transparent like glass, and that it receives the reflection of the fire in the universe, and transmits to us both light and warmth . . . .

NOTE ON PLATO'S ASTRONOMY

We do not deal at length with the astronomy of Plato for the reason that what he has to say in the Dialogues is for the most part philosophical meditation rather than what we should call systematic science. A few points, however, may be noted.

1. We find Plato setting the very problem that is at the basis of a science of astronomy as opposed to a mere description or observation of the heavens.

Thus Simplicius writes (*Commentary on Aristotle's De Caelo*, p. 488.18 [Heiberg]): "Eudoxus of Cnidus, as Eudemus recounts in the second book of his *History of Astronomy* and as Sosigenes repeats on the authority of Eudemus, is said to have been the first of the Greeks to deal with this type of hypothesis. For Plato, Sosigenes says, set this problem for students of astronomy: 'By the assumption of what uniform and ordered motions can the apparent motions of the planets be accounted for?' "

---

oreans. But the general principle underlying these systems—as well as that in Plato's *Republic*—is the attempt to link planetary motions and distances with the notes of the musical scale. It is a manifestation of the Pythagorean and Platonic emphasis on number as the basis of physical theory. [Edd.]

[1] The words within < > translate a conjectural restoration of the text by Diels. See his *Fragmente der Vorsokratiker* I^5. 403. [Edd.]

This problem, investigated by all the Greek astronomers since the time of Eudoxus, was essentially the problem of combining various uniform circular motions in such a way that from their resultant the observed motions of the planets, with their progressions, retrogressions, and stationary points, could be deduced and predicted. It was in answer to this problem that the theories of concentric spheres, the heliocentric theory of Aristarchus, and the theories of eccentric circles and of epicycles were framed. It was, in a sense, the same problem that the modern systems from that of Copernicus on have sought to solve. There have been changes in the conditions of the problem (e.g., the abandonment of the requirement of circular motion by Kepler), and the creators of systems have not always viewed their systems merely as hypothetical constructs that would "save the phenomena." But the type of inquiry has been along the lines first suggested, if our tradition is correct, by Plato.

2. From a very obscure passage in the *Timaeus* (40 B), coupled with a passage in Aristotle's *De Caelo*, as well as some subsequent references in ancient sources, some have sought to credit Plato with the theory of the earth's daily rotation on its axis. This attribution is generally considered unsound at present. (See p. 96, n. 2, above.)

3. There are numerous passages bearing on Plato's astronomy, particularly in the *Republic*, *Timaeus*, *Laws*, and, if it can be held to reflect Platonic views, the *Epinomis*. These passages deal with astronomy in education and with astronomy as the mathematical study not only of the motions of the heavenly bodies, but, in a sense, of pure motion divorced from objects of actual observation; some passages deal, more in poetic and mythical than in scientific vein, with the creation of the universe, the harmony of the spheres, recurrent cycles of Great Years, and similar themes. But there is always—and this is in keeping with Plato's Pythagorean outlook in these matters—the emphasis on number, on the basic uniformity that underlies apparent irregularity, and on circular motion and spherical form as the most perfect of their kind. Without making substantive contributions to astronomy, Plato, by his influence on others, contributed to the development of the mathematical astronomy that was to be one of the crowning achievements of Greek science.

ELEMENTARY NOTIONS IN ASTRONOMY

Euclid, *Phaenomena*, Preface.[1] Translation of T. L. Heath, *Greek Astronomy*

Since the fixed stars are always seen to rise from the same place and to set at the same place,[2] and those which rise at the same time are seen always to rise at the same time, and those which set at the same time always to set at the same time, and these stars in their courses from rising to setting remain always at the same distances from one another, while this can only happen with objects moving with circular motion, when the eye (of the observer) is equally distant in all directions from the circumference, as is proved in the Optics, we must assume that the (fixed) stars move circularly, and are fastened in one body, while the eye is equidistant from the

[1] We have in Euclid's *Phaenomena* a treatise on traditional spherics, or spherical geometry (see p. 2), with special reference to elementary astronomy. The work, however, is purely geometrical and there is no attempt to formulate a system of planetary, lunar, and solar motions. We have therefore placed this passage before those dealing with such systems. We have seen (p. 82) how, with the development of spherical trigonometry from spherics, a powerful instrument for precise calculation and the testing and refining of hypotheses was put into the hands of Greek astronomers. [Edd.]

[2] See pp. 122–125, below. [Edd.]

circumferences of the circles. But a certain star is seen between the Bears which does not change from place to place, but turns about in the position where it is. And, since this star appears to be equidistant in all directions from the circumferences of the circles in which the rest of the stars move, we must assume that the circles are all parallel, so that all the fixed stars move in parallel circles having for one pole the aforesaid star.

Now some of the stars are seen neither to rise nor to set because they are borne on circles which are high up and are called "always visible" circles. These stars are those which come next to the visible pole and reach as far as the arctic circle.[1] And, of these stars, those nearer the pole move on smaller circles, and those on the arctic circle on the greatest circle, the latter stars appearing actually to graze the horizon.

But all the stars which follow on these towards the south are all seen to rise and to set because their circles are not wholly above the earth, but part of them is above, and the remainder below, the earth. And of the segments of the several circles that are above the earth, that appears larger which is nearer to the greatest of the always-visible circles, while of the segments under the earth that which is nearer to the said circle is less, because the time taken by the motion under the earth of the stars which are on the said circle is the least, and the time taken by their motion above the earth is the greatest, while, for the stars on the circles which are continually further from the said circle, the time taken by their motion above the earth is continually less, and the time taken by their motion under the earth greater; the motion above the earth takes the least time, and the motion under the earth the greatest time, in the case of the stars which are nearest the south. The stars on the circle which is the middle one of all the circles appear to take equal times to complete their motion above the earth and their motion under the earth respectively, and hence we call this circle the "equinoctial";[2] and those stars which are on circles equidistant from the equinoctial circle take equal times to describe the alternate segments; thus the segments above the earth in the northerly direction are equal to those under the earth in the southerly direction, and the segments above the earth in the southerly direction are equal to those under the earth in the northerly direction; but the sum of the times taken by the motion above the earth and by the motion under the earth continuous with it added together appears to be the same for each circle.

Further, the circle of the Milky Way and the zodiac circle, which are

[1] "Arctic circle" here refers to a circle on the celestial not the terrestrial sphere. It is the boundary between those fixed stars which never go below the horizon and those which do. Its extent therefore depends on the latitude of the observer. For one traveling north from the equator, the arctic circle grows ever larger until, when the observer is at the pole, the arctic circle embraces the entire northern celestial hemisphere. [Edd.]

[2] I.e., the equator on the celestial sphere. [Edd.]

both obliquely inclined to the parallel circles and cut one another, appear in their revolution always to show semicircles above the earth.[1]

On all the aforesaid grounds let us make it our hypothesis that the universe is spherical in shape; for if it had been in the form either of a cylinder or of a cone, the stars taken on the oblique circles bisecting the equinoctial circle would, in their revolution, have seemed to describe, not always equal semicircles, but sometimes a segment greater than a semicircle, and sometimes a segment less than a semicircle. For, *if a cone or a cylinder be cut by a plane not parallel to the base, the section arising is a section of an acute-angled cone,*[2] *which is like a shield* (an ellipse). Now it is clear that, if such a figure be cut through its centre lengthwise and breadthwise respectively, the segments respectively arising are dissimilar; it is also clear that, even if it be cut in oblique sections through the centre, the segments formed are dissimilar in that case also; but this does not appear to happen in the case of the universe. For all these reasons, the universe must be spherical in shape, and revolve uniformly about its axis, one of the poles of which is above the earth and visible, while the other is under the earth and invisible.

Let the name "horizon" be given to the plane passing through our eye which is produced to the (extremities of the) universe, and separates off the segment which we see above the earth. The horizon is a circle; for, *if a sphere be cut by a plane, the section is a circle.*

Let the name "meridian circle" be given to the circle through the poles which is at right angles to the horizon, and the name "tropics" to the circles which are touched by the circle through the middle of the signs (the zodiac) and which have the same poles as the sphere.[3]

The zodiac circle and the equinoctial circle are great circles; for they bisect one another. For the beginning of the Ram (Aries) and the beginning of the "Claws" (Libra) are diametrically opposite to one another and, both being on the equinoctial circle, the rising of the one and the setting of the other take place in conjunction, since they have between them six of the twelve signs of the zodiac, and two semicircles of the equinoctial circle, respectively, and since both beginnings, being on the equinoctial circle, take the same time to describe, the one its course above the earth, the other its course under the earth. But if a sphere rotates

---

[1] The "parallel circles" are the diurnal paths of the fixed stars. With the exception of the equator itself they are not great circles of the celestial sphere but are small circles. But the circle of the Milky Way and the zodiac, being considered as great circles, would intersect the great circle of the horizon and be bisected by it. Hence half of the Milky Way and of the zodiac would always be above the horizon. [Edd.]

[2] See p. 76, above. [Edd.]

[3] The meridian circle depends, then, on the position of the observer. This is not the case with the zodiac or the tropical circles. [Edd.]

about its own axis, all the points on the surface of the sphere describe, in the same time, similar arcs of the parallel circles on which they are carried; therefore the points in question traverse similar arcs of the equinoctial circle, on one side the arc above the earth, on the other the arc under the earth; therefore the arcs are equal; therefore both are semicircles, for the distance from rising to rising, or from setting to setting, is the whole circle; therefore the zodiac circle and the equinoctial circle bisect one another. But, if in a sphere two circles bisect one another, both of the intersecting circles are great circles; therefore the zodiac circle and the equinoctial circle are great circles.

The horizon, too, is one of the great circles. For it always bisects both the zodiac circle and the equinoctial circle; for it has always six of the twelve signs above the earth, and always a semicircle of the equinoctial circle above the earth; for the stars on the latter circle which rise and set respectively at the same time pass, in the same time, the one its course from rising to setting, the other its course from setting to rising. It is therefore manifest, from what was before proved, that there is always a semicircle of the equinoctial circle above the horizon.[1] But if, in a sphere, a circle remaining fixed bisects any of the great circles which is moving continually, the circle which cuts it is also a great circle; therefore, the horizon is one of the great circles.

The time of a revolution of the universe is the time in which each of the fixed stars passes from one rising to its next rising, or from any place whatever to the same place again.

THE THEORY OF CONCENTRIC SPHERES

The development of the theory of concentric spheres is a remarkable chapter in the history of science. It marks the first effort to account, by a single system, for all the observed motions of sun, moon, planets, and fixed stars. It was first developed by Eudoxus of Cnidus, a pupil of Plato, and Archytas, in the first half of the fourth century B.C., and improved in certain respects by Callippus of Cyzicus, a contemporary of Aristotle. Aristotle adopted the system of Callippus with modifications. The details of the system are deduced from passages in Aristotle's *Metaphysics* and a long account in Simplicius' commentary on the *De Caelo* of Aristotle (pp. 488, 493–506 [Heiberg]). The purpose of the system is to account for the observed motions and to predict future motions of the sun, moon, and planets by considering each of these bodies as a point on the surface of one of several interconnected spheres, all concentric with the earth at the center. As the spheres turn at different (but uniform) rates, around different axes, and in different directions, the resulting motion of the point is the motion of the particular heavenly body represented by that point.

For example, Eudoxus conceived a planet as a point on the equator of the innermost of four concentric spheres. The outermost sphere rotates uniformly about an axis; the poles of this axis are conceived as fixed on the surface of the next inner sphere, which, in turn, rotates uniformly, but at a different rate from the outermost, and on an axis inclined to that

[1] For an observer at either pole, however, the equator would be wholly on the horizon. [Edd.]

of the outermost; the poles of this second sphere are, in turn, fixed on the surface of the next inner sphere, which has its own rate and axis of rotation, and whose poles are, in turn, on the surface of the innermost sphere, which has its own rate and axis of rotation. It is, as we have said, on the equator of this innermost sphere that the planet is conceived as a fixed point. The Italian astronomer Giovanni Schiaparelli first worked out, on the basis of Simplicius' account, the mathematical form of the resultant of the motions of the two innermost spheres. The form is a sort of figure-of-eight on the surface of a sphere, a "spherical lemniscate." This is the curve referred to by Eudoxus as the hippopede (horse-fetter).[1] Now the motion of the two outermost spheres represents the diurnal motion and the motion in the zodiac. With these motions must be joined the motion represented by the hippopede. The resultant of all was intended to give the observed motion of the planet from day to day with its apparent changes in velocity and direction.

Eudoxus considered that the phenomena were sufficiently accounted for by assuming three spheres for the motions of the sun, three for the moon, and as we have said, four for each of the five planets, a total of 26 (not counting the sphere of the fixed stars separately). To this number Callippus added seven spheres in order to bring the theory into closer harmony with the observed motions.

Eudoxus and Callippus seem to view the spheres not as material bodies but as purely theoretical aids to understanding the motions geometrically. Furthermore, each set of concentric spheres (for sun, moon, and each of the planets) is separate from the other sets. Aristotle, on the other hand, conceives the spheres as actually existent shells and makes all the sets comprise one continuous and interconnected system. This, Aristotle held, necessitated the addition of 22 spheres to compensate for what would otherwise be the effect of the interaction of the spheres belonging to one planet upon those belonging to the planet next below.

Systems involving only concentric spheres were supplanted by systems which included eccentrics and epicycles (see p. 128). For a system of concentric spheres alone, apart from the multiplicity of spheres, which made it difficult to work with, by no means completely accounted for the phenomena. For example, the concentric spheres required that the distance of each planet from the earth be invariable. Since, however, differences in brightness, especially of Mars and Venus, and variations in the apparent size of sun and moon were observable, the theory could not hold its ground.

To the famous passage from Aristotle's *Metaphysics* is added a quotation from Sosigenes (probably the teacher of Alexander of Aphrodisias in the second century A.D.). In this quotation the variability of the apparent size of the sun or moon is confirmed by the phenomenon of the annular eclipse.

No attempt is made here to discuss completely the many technical questions involved. Reference may be made to T. L. Heath, *Aristarchus of Samos*, ch. 16–17.

Aristotle, *Metaphysics* Λ. 8, 1073*b*17–1074*a*14. Translation of W. D. Ross

Eudoxus supposed that the motion of the sun or of the moon involves, in either case, three spheres, of which the first is the sphere of the fixed stars, and the second moves in the circle which runs along the middle of the zodiac, and the third in the circle which is inclined across the breadth of the zodiac; but the circle in which the moon moves is inclined at a greater angle than that in which the sun moves. And the motion of the planets involves, in each case, four spheres, and of these also the first and

[1] See Simplicius, *op. cit.*, p. 497.3; cf. Proclus, *Commentary on Euclid Elements I*, p. 112.5, where the method of construction is reminiscent of Archytas's solution of the Delian problem.

second are the same as the first two mentioned above (for the sphere of the fixed stars is that which moves all the other spheres, and that which is placed beneath this and has its movement in the circle which bisects the zodiac is common to all), but the *poles* of the third sphere of each planet are in the circle which bisects the zodiac, and the motion of the fourth sphere is in the circle which is inclined at an angle to the equator of the third sphere; and the poles of the third sphere are different for each of the other planets, but those of Venus and Mercury are the same.

Callippus made the position of the spheres the same as Eudoxus did, but while he assigned the same number as Eudoxus did to Jupiter and to Saturn, he thought two more spheres should be added to the sun and two to the moon, if one is to explain the observed facts;[1] and one more to each of the other planets.

But it is necessary, if all the spheres combined are to explain the observed facts, that for each of the planets there should be other spheres (one fewer than those hitherto assigned), which counteract those already mentioned and bring back to the same position the outermost sphere of the star which in each case is situated below[2] the star in question; for only thus can all the forces at work produce the observed motion of the planets. Since, then, the spheres involved in the movement of the planets themselves are eight for Saturn and Jupiter and twenty-five for the others, and of these only those involved in the movement of the lowest-situated planet need not be counteracted, the spheres which counteract those of the outermost two planets will be six in number, and the spheres which counteract those of the next four planets will be sixteen; therefore the number of all the spheres—both those which move the planets and those which counteract these—will be fifty-five.[3] And if one were not to add to the moon and to the sun the movements we mentioned, the whole set of spheres will be forty-seven in number.

An Objection to the Theory of Concentric Spheres

Sosigenes in Simplicius, *Commentary on Aristotle's De Caelo*, pp. 504–505 (Heiberg).[4]
Translation of T. L. Heath, *Aristarchus of Samos*, pp. 221–223

Nevertheless the theories of Eudoxus and his followers fail to save the phenomena, and not only those which were first noticed at a later date, but

[1] According to Eudemus (in Simplicius, *Commentary on Aristotle's De Caelo*, p. 497.17) the reference is to the inequality of the seasons. [Edd.]

[2] I.e., toward the earth, the center of the universe. [Edd.]

[3] In the system of Eudoxus the sun, moon, and planets require one sphere each to account for diurnal motion. But Aristotle, for whom all the spheres are interconnected, could have accounted for the diurnal motion of all bodies with a single sphere. Had he done so, he could have reduced the total number of spheres required. [Edd.]

[4] The theory of concentric spheres required that the sun, moon, and planets be a constant distance from the earth at all times. As a matter of common observation, however,

even those which were before known and actually accepted by the authors themselves. What need is there for me to mention the generality of these, some of which, after Eudoxus had failed to account for them, Callippus tried to save—if indeed we can regard him as so far successful? I confine myself to one fact which is actually evident to the eye; this fact no one before Autolycus[1] of Pitane even tried to explain by means of hypotheses (διὰ τῶν ὑποθέσεων), and not even Autolycus was able to do so, as clearly appears from his controversy with Aristotherus. I refer to the fact that the planets appear at times to be near to us and at times to have receded. This is indeed obvious to our eyes in the case of some of them; for the star called after Aphrodite and also the star of Ares seem, in the middle of their retrogradations, to be many times as large, so much so that the star of Aphrodite actually makes bodies cast shadows on moonless nights. The moon also, even in the perception of our eye, is clearly not always at the same distance from us, because it does not always seem to be the same size under the same conditions as to medium. The same fact is moreover confirmed if we observe the moon by means of an instrument; for it is at one time a disc of eleven fingerbreadths, and again at another time a disc of twelve fingerbreadths, which when placed at the same distance from the observer hides the moon (exactly) so that his eye does not see it. In addition to this, there is evidence for the truth of what I have stated in the observed facts with regard to total eclipses of the sun; for when the centre of the sun, the centre of the moon, and our eye happen to be in a straight line, what is seen is not always alike; but at one time the cone which comprehends the moon and has its vertex at our eye comprehends the sun itself at the same time, and the sun even remains invisible to us for a certain time, while again at another time this is so far from being the case that a rim of a certain breadth on the outside edge is left visible all round it at the middle of the duration of the eclipse.[2] Hence we must conclude that the apparent

---

the planets vary from time to time in brightness. This is most obviously true of Mars and Venus. Again, the apparent size of the moon, and, to a much smaller degree, that of the sun vary. This suggests that the distance of the sun, moon, and planets from the earth is not constant. This objection to the theory of concentric spheres, an objection avoided by introducing eccentrics and epicycles, is voiced by Sosigenes, a Peripatetic philosopher, probably of the latter part of the second century A.D. The identification with the Sosigenes who aided Julius Caesar in the reform of the calendar seems to be wrong. [Edd.]

[1] Mathematician and astronomer of the latter half of the fourth century B.C. His works *On the Moving Sphere* and *On Risings and Settings* are the earliest complete mathematical treatises that have come down to us from ancient Greece, for he probably wrote before Euclid. The propositions assumed by Autolycus indicate that considerable progress had already been made in spherical geometry. [Edd.]

[2] The possibility of an annular eclipse such as that here described is cited by Sosigenes to indicate the variability of the distance of the sun from the earth. It is the first account of an annular eclipse, and, according to Heath, "shows that he [Sosigenes] had much more correct

difference in the size of the two bodies observed under the same atmospheric conditions is due to the inequality of their distances (at different times). . . . But indeed this inequality in the distances of each star at different times cannot even be said to have been unknown to the authors of the concentric theory themselves. For Polemarchus of Cyzicus[1] appears to be aware of it, but to minimize it as being imperceptible, because he preferred the theory which placed the spheres themselves about the very centre in the universe. Aristotle too, shows that he is conscious of it when, in the Physical Problems, he discusses objections to the hypotheses of astronomers arising from the fact that even the sizes of the planets do not appear to be the same always. In this respect Aristotle was not altogether satisfied with the revolving spheres, although the supposition that, being concentric with the universe, they move about its centre, attracted him. Again, it is clear from what he says in Book Λ of the Metaphysics that he thought that the facts about the movements of the planets had not been sufficiently explained by the astronomers who came before him or were contemporary with him.

THE ROTATION OF THE EARTH ON ITS AXIS

In the history of science both ancient and modern there is often considerable question of priority of discovery. This is to be expected in view of the interchange of ideas among men of science and the cumulative nature of science, since advances are made from a body of knowledge that has become common property. Certain Pythagoreans had, as we have seen, abandoned the view of the earth as central and immovable and had made the earth and all the heavenly bodies revolve about the central fire. Now the explanation of the apparent diurnal revolution of the heavens by the hypothesis of the earth at the center rotating on its axis is different from the earlier Pythagorean view. But it may represent a development of that view by the transference of the central fire to the center of the earth. In any case it had enough in common with the earlier system to cause some confusion in the later tradition. And so we find priority in the theory of the earth's axial rotation accorded in some sources to Philolaus, Hicetas, Ecphantus, and Plato, as well as to Heraclides of Pontus. Apart from the question of priority, the evidence seems to indicate that Heraclides adopted the hypothesis that the earth (whether or not at the center of the universe) rotated daily on its axis.

On the planetary theory of Heraclides the evidence is scanty and the interpretations conflicting. Some have held that he explained the variations in the brightness of Venus and Mercury by supposing that these planets revolved around the sun. On the other hand, a recent study of B. L. Van der Waerden ascribes to Heraclides the view that the earth, the sun, and the planets revolve around a common center, a view much like that of the earlier Pythagoreans. The view of Schiaparelli that Heraclides anticipated Aristarchus in holding that the earth revolved around the sun is refuted by Heath (*Aristarchus of Samos*, pp. 260–275). It is also very doubtful that Heraclides put forward the theory that the superior

---

notions on this subject than most astronomers up to Tycho Brahe."

It is to be noted that Ptolemy considered the apparent diameter of the sun approximately constant and equal to that of the moon at the latter's apogee. [Edd.]

[1] A friend of Eudoxus and probably a teacher of Callippus. [Edd.]

as well as the inferior planets revolved around the sun, a theory which, together with the notion of the sun's revolution around the earth, would foreshadow the Tychonic system.

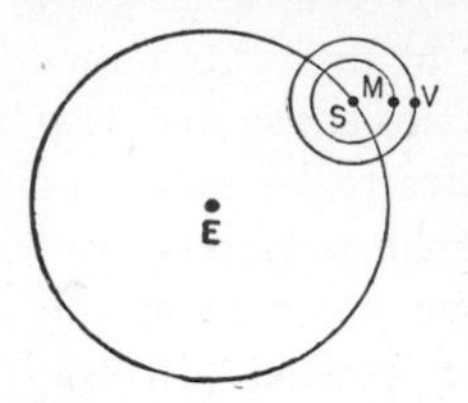

One other point may be noted. A system in which Mercury and Venus revolve about the sun, and the sun about the earth, a system which some have ascribed to Heraclides, would in effect involve epicyclic motion which played so important a role in the ultimate development of Greek astronomy; for in describing circular orbits about the sun as the sun describes a circular orbit about the earth, Venus and Mercury would trace epicyclic paths.

Cicero, *Academica* II. 39.123

Hicetas[1] of Syracuse, as Theophrastus tells us, holds that the heavens, sun, moon, and stars, all the heavenly bodies, in short, are at rest, and that nothing in the universe moves except the earth,[2] and as the earth turns and rotates about its axis at very high speed the effect is exactly the same as if the heavens were rotating and the earth at rest. And there are some who think that this is what Plato is saying in the *Timaeus*, but somewhat more obscurely.[3]

Hippolytus, *Refutation of All Heresies* I. 15

Ecphantus a Syracusan said . . . that the earth, the center of the universe, moved about its own center toward the east.

Diogenes Laertius VIII. 85

Philolaus was the first to hold that the earth rotates in a circle, though others say it was Hicetas of Syracuse.

Simplicius, *Commentary on Aristotle's De Caelo*, pp. 444.33–445.3 (Heiberg)

There have been some, among them Heraclides of Pontus[4] and Aristarchus, who thought that the phenomena could be accounted for by supposing the heaven and stars to be at rest, and the earth to be in motion

---

[1] The tradition with respect to the Pythagorean philosophers Hicetas and Ecphantus is obscure and has been variously interpreted. The agreement between their views and those of Heraclides of Pontus led P. Tannery to the hypothesis that the tradition arose from Heraclides' having expounded his views through the medium of dialogues in which Hicetas and Ecphantus appeared as characters.

[2] Perhaps Cicero is speaking merely of diurnal motion, for the system described would by no means fit the observed planetary, solar, and lunar motions.

[3] "And the earth, our foster mother, revolved (ἰλλομένην) about the axis that extends through the universe he contrived as the guardian and creator of night and day." The difficulty is with the interpretation of ἰλλομένην (or εἰλλομένην), a difficulty of which the ancients themselves were aware, as this passage of Cicero and passages of the commentators on Aristotle, *De Caelo* II. 13 (see p. 96, above) indicate.

[4] Heraclides of Pontus (*ca.* 388–*ca.* 310 B.C.) probably was a pupil of Plato and perhaps, too, of Aristotle. His literary work covered many fields, including ethics, physics, astronomy, music, literary criticism, and rhetoric.

about the poles of the equator from west [to east], making approximately one complete rotation each day. The word "approximately" is added because of the motion of the sun amounting to one degree.[1]

THE REVOLUTION OF MERCURY AND VENUS ABOUT THE SUN

Chalcidius, *Commentary on Plato's Timaeus*, 109

Finally, Heraclides of Pontus, in describing the path of Venus and of the sun, and assigning one midpoint for both, showed how Venus is sometimes above and sometimes below the sun.

Vitruvius, *On Architecture* IX. 1.6

The stars of Mercury and Venus make their retrogradations and retardations around the rays of the sun, making a crown, as it were, by their courses about the sun as center.

Martianus Capella, *On The Marriage of Philology and Mercury*, VIII. 857, 879

For though Venus and Mercury show daily risings and settings, nevertheless their circular paths do not go around the earth at all, but around the sun in freer motion. In short, they place the center of their orbits in the sun; so that they sometimes move above it and sometimes below it, i.e., nearer the earth. Venus is distant from the sun a distance of a sign and a half [45°]. But when they are above the sun, Mercury is nearer the earth, and when they are below the sun, Venus is nearer, since its circular path is broader and more extensive. . . .

I mentioned above that the circles of this star [Mercury] and Venus are epicycles. That is to say, they do not include the round earth within their own orbit, but are carried round it laterally, as it were.

THE HELIOCENTRIC HYPOTHESIS

The heliocentric hypothesis of Aristarchus had little success in antiquity. Not only did it go counter to the widely held philosophical doctrine of "natural places," with the earth as center, but also to the doctrine of a central fire which, in various forms, had become a matter of religious belief (cf. Plutarch, *De facie in orbe lunae* 922 F–923 A). The astronomers, however, generally rejected the heliocentric hypothesis on scientific grounds. If the earth revolved in an orbit about the sun, the position of the fixed stars as observed from various parts of the earth's orbit should vary. Since no such variation was observed in antiquity

[1] I.e., the sun's motion along the ecliptic, the whole circle being completed in one year and the daily motion therefore amounting to about 1°. Cf. also Aëtius III. 13.3: "Heraclides of Pontus and Ecphantus the Pythagorean suppose that the earth moves, not however with complete change in position (μεταβατικῶς), but by rotating (τρεπτικῶς) from west to east about its own center, fixed upon its axis like a wheel."

Simplicius, *op. cit.*, p. 519.9–11: "By assuming that the earth was at the center and rotated while the heaven was at rest Heraclides of Pontus thought he 'saved the phenomena.' "

(and in fact until comparatively recent times), Aristarchus was compelled to assume that the sphere of the fixed stars was incomparably greater than the sphere containing the earth's orbit. Now the geocentrists have to assume that the sphere of fixed stars is incomparably greater than the earth itself (not the sphere of its orbit). Since the assumption of the geocentrists was a somewhat easier one to make, their position was generally adopted. Again, the placing of the sun at the precise center did not succeed in saving the phenomena, e.g., the inequality of the seasons. It is probably for these reasons that Hipparchus (and later Ptolemy) adopted the geocentric hypothesis which, in connection with eccentrics, epicycles, and deferents, became the dominant astronomical system of antiquity.

Archimedes, *The Sand-Reckoner* 1.[1] Translation of T. L. Heath, *The Works of Archimedes*, p. 222

Aristarchus of Samos brought out a book consisting of some hypotheses, in which the premisses lead to the result that the universe is many times greater than that now so called. His hypotheses are that the fixed stars and the sun remain unmoved, that the earth revolves about the sun in the circumference of a circle, the sun lying in the middle of the orbit,[2] and that the sphere of the fixed stars, situated about the same centre as the sun, is so great that the circle in which he supposes the earth to revolve bears such a proportion to the distance of the fixed stars as the centre of the sphere bears to its surface. Now it is easy to see that this is impossible; for, since the centre of the sphere has no magnitude, we cannot conceive it to bear any ratio whatever to the surface of the sphere. We must however take Aristarchus to mean this: since we conceive the earth to be, as it were, the centre of the universe, the ratio which the earth bears to what we describe as the "universe" is the same as the ratio which the sphere containing the circle in which he supposes the earth to revolve bears to the sphere of the fixed stars.[3]

---

[1] The *Sand-Reckoner* or *Arenarius* of Archimedes sets forth a system of representing numbers not in the usual manner of the Greeks but by orders and periods so that any number however great, may be represented. Archimedes cites the equivalent of the number n^{n^n}, where n is 100,000,000. To show the utility of his system in the handling of large numbers Archimedes undertakes to express a number larger than the number of grains of sand necessary to fill the universe. Assuming, for his purpose, dimensions of the universe larger than those generally held at the time, he concludes the number of grains that would fill the "universe" conceived as a sphere with radius equal to the distance between the centers of the earth and sun would be less than 10^{51}, and the number required to fill the sphere of fixed stars would be less than 10^{63}. One is reminded of recent investigations into the number of elementary particles in the universe.

In the passage here quoted Archimedes refers to Aristarchus's formulation of the heliocentric hypothesis. [Edd.]

[2] The precise meaning is doubtful. O. Neugebauer (*Isis* 34 [1942] 6) translates thus: "the circumference of a circle which lies in the midst of the course [of the planets]." [Edd.]

[3] We have noted that Aristarchus tried to save the heliocentric hypothesis by attributing the absence of stellar parallax to the fact that the sphere containing (as a great circle) the orbit of the earth was incomprehensibly small in comparison with the sphere of fixed stars. Archimedes could not proceed with his problem on this basis, for the sphere of fixed stars would have been of infinite extent. And so he substitutes for Aristarchus's statement the

For he adapts the proofs of his results to a hypothesis of this kind, and in particular he appears to suppose the magnitude of the sphere in which he represents the earth as moving to be equal to what we call the "universe."[1]

THE SIZES AND DISTANCES OF THE SUN AND MOON

Aristarchus of Samos, *On the Sizes and Distances of the Sun and Moon*, Hypotheses and Proposition 7.[2] Translation of T. L. Heath (Oxford, 1913)

Hypotheses

1. That the moon receives its light from the sun.
2. That the earth is in the relation of a point and centre to the sphere in which the moon moves.[3]
3. That, when the moon appears to us halved, the great circle which divides the dark and the bright portions of the moon is in the direction of our eye.[4]

---

following proportion: as the earth is to the "universe," so is the sphere containing (as a great circle) the earth's orbit to the sphere of fixed stars. If the "universe" is the sphere with radius equal to the distance between the centers of the earth and sun, the effect of Archimedes' proportion is to make the sphere of the earth's orbit a mean proportional between the earth and the sphere of fixed stars. Such at any rate is Heath's interpretation of the controverted passage in Archimedes.

It may be noted that Copernicus had to assume that the fixed stars were so far from the earth as to make stellar parallax imperceptible. [Edd.]

[1] There are some few additional passages on the system of Aristarchus. From Plutarch, *De facie in orbe lunae* 923 A, it is clear that Aristarchus also assumed the rotation of the earth on its axis. From Plutarch, *Platonic Questions* 1006 C, and Aëtius III. 17.9 it would seem that Seleucus of Seleucia (middle of the second century B.C.) adopted the heliocentric hypothesis. He seems to have held the earth's axial rotation, if not the whole heliocentric hypothesis, as a fact, whereas for Aristarchus they remained hypotheses. [Edd.]

[2] Aristarchus of Samos (*ca.* 310–*ca.* 230 B.C.), a pupil of Strato of Lampsacus (p. 211), is best known for his anticipation of the Copernican system. This has been discussed above. Aristarchus's extant work *On the Sizes and Distances of the Sun and Moon* does not refer to the hypothesis of a heliocentric universe, perhaps because this treatise antedated his formulation of the new hypothesis. In any case, the results of Aristarchus's investigation of sizes and distances would be independent of the choice of earth or sun as center.

In this work Aristarchus on the basis of certain assumptions deduces theorems about the sizes and distances of the sun and moon. The inadequacy of his results is due not to mathematical errors but to errors in the initial assumptions. The mathematical development is most rigorous and is interesting for its handling of what are really trigonometric ratios.

The sizes and distances of the heavenly bodies were the subject of much speculation among Greek thinkers. It may be noted that in this work of Aristarchus we have the first extant work which attacks the problem on the basis of systematic mathematical deductions from certain assumptions. [Edd.]

[3] This hypothesis is probably introduced not as a fact or even as an approximation but for the purpose of simplification, to obviate the necessity of considering lunar parallax (which is by no means negligible). *A fortiori* the assumption is that the earth is in the relation of a point with reference to the spheres of the sun and of the fixed stars. [Edd.]

[4] I.e., the eye is in the same plane as the aforesaid great circle. [Edd.]

4. That, when the moon appears to us halved, its distance from the sun is then less than a quadrant by one-thirtieth of a quadrant.[1]

5. That the breadth of the [earth's] shadow is [that] of two moons.[2]

6. That the moon subtends one-fifteenth part of a sign of the zodiac.[3]

We are now in a position to prove the following propositions:

1. *The distance of the sun from the earth is greater than eighteen times, but less then twenty times, the distance of the moon (from the earth)*;[4] this follows from the hypothesis about the halved moon.

2. *The diameter of the sun has the same ratio (as aforesaid) to the diameter of the moon.*[5]

3. *The diameter of the sun has to the diameter of the earth a ratio greater than that which* 19 *has to* 3, *but less than that which* 43 *has to* 6; this follows from the ratio thus discovered between the distances, the hypothesis about the shadow, and the hypothesis that the moon subtends one-fifteenth part of a sign of the zodiac.

Proposition 7

The distance of the sun from the earth is greater than eighteen times, but less than twenty times, the distance of the moon from the earth.

For let A be the centre of the sun, B that of the earth.

Let AB be joined and produced.

Let C be the centre of the moon when halved; let a plane be carried through AB and C, and let the section made by it in the sphere on which the centre of the sun moves be the great circle ADE.

Let AC, CB be joined, and let BC be produced to D.

Then, because the point C is the centre of the moon when halved, the angle ACB will be right.

Let BE be drawn from B at right angles to BA;

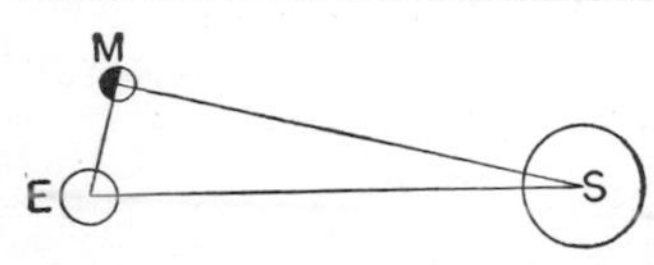

[1] I.e., $\angle MES$ in the accompanying figure is 87°. Herein lies a source of serious error. The value of the angle is actually about 89°50′. But difficulties of making the measurement, which involves the precise determination of the moment of the first or last quarter phase, were very great. [Edd.]

[2] I.e., in a lunar eclipse the diameter of the earth's shadow through which the moon must pass is twice the diameter of the moon. Hipparchus and Ptolemy, who set the ratio at $2\frac{1}{2}$–$2\frac{3}{5}$ depending on the position of the moon, were much more accurate. [Edd.]

[3] This greatly excessive assumption of the apparent angular diameter of the moon (and sun) as 2° (i.e., $\frac{1}{15}$ of 30°) is all the more surprising because Archimedes credits Aristarchus with the determination of $\frac{1}{2}$° as the value, which is a very close approximation to the actual value. Either Aristarchus's discovery was made after the publication of this treatise or else the assumption is meant to be hypothetical, and is not to be taken as referring to a fact determined by observation. [Edd.]

[4] Prop. 7, quoted below. [Edd.]

[5] Prop. 9, which follows from Proposition 7, with the additional assumption that the apparent angular diameter of the sun is the same as that of the moon. [Edd.]

then the circumference *ED* will be one-thirtieth of the circumference *EDA*; for, by hypothesis, when the moon appears to us halved, its distance from the sun is less than a quadrant by one-thirtieth of a quadrant [Hyphothesis 4].

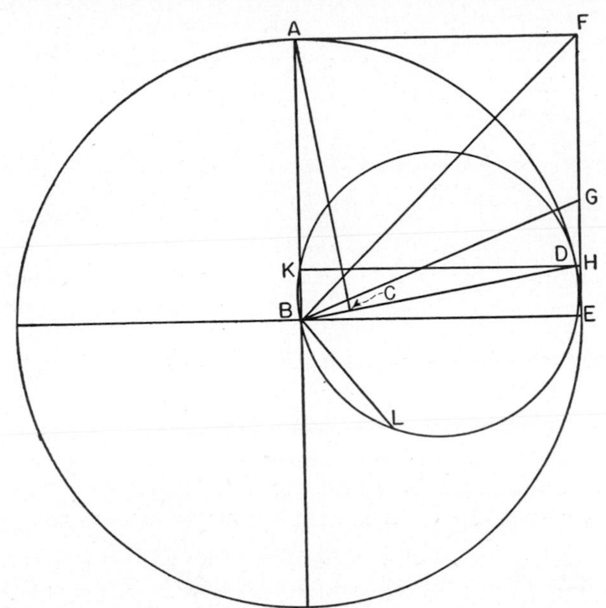

Thus the angle *EBC* is also one-thirtieth of a right angle.

Let the parallelogram *AE* be completed, and let *BF* be joined.

Then the angle *FBE* will be half a right angle.

Let the angle *FBE* be bisected by the straight line *BG*; therefore the angle *GBE* is one-fourth part of a right angle.

But the angle *DBE* is also one-thirtieth part of a right angle; therefore the ratio of the angle *GBE* to the angle *DBE* is that which 15 has to 2: for, if a right angle be regarded as divided into 60 equal parts, the angle *GBE* contains 15 of such parts, and the angle *DBE* contains 2.

Now, since *GE* has to *EH* a ratio greater than that which the angle *GBE* has to angle *DBE*,[1] therefore *GE* has to *EH* a ratio greater than that which 15 has to 2.

Next, since *BE* is equal to *EF*, and the angle *E* is right, therefore the square on *FB* is double of the square on *BE*.

But, as the square on *FB* is to the square on *BE*, so is the square on *FG* to the square on *GE*; therefore the square on *FG* is double of the square on *GE*.

---

[1] The assumption, in modern terms, is that $\tan \alpha / \tan \beta > \alpha/\beta$, where $\pi/2 > \alpha > \beta > 0$. [Edd.]

Now 49 is less than double[1] of 25,
so that the square on *FG* has to the square on *GE* a ratio greater than that which 49 has to 25;
therefore *FG* also has to *GE* a ratio greater than that which 7 has to 5.

Therefore, *componendo*, *FE* has to *EG* a ratio greater than that which 12 has to 5, that is, than that which 36 has to 15.

But it was also proved that *GE* has to *EH* a ratio greater than that which 15 has to 2;
therefore, *ex aequali*, *FE* has to *EH* a ratio greater than that which 36 has to 2, that is, than that which 18 has to 1;
therefore *FE* is greater than 18 times *EH*.

And *FE* is equal to *BE*;
therefore *BE* is also greater than 18 times *EH*;
therefore *BH* is much greater than 18 times *HE*.

But, as *BH* is to *HE*, so is *AB* to *BC*, because of the similarity of the triangles;
therefore *AB* is also greater than 18 times *BC*.

And *AB* is the distance of the sun from the earth, while *CB* is the distance of the moon from the earth; therefore the distance of the sun from the earth is greater than 18 times the distance of the moon from the earth.

Again, I say that it is also less than 20 times that distance.

For let *DK* be drawn through *D* parallel to *EB*, and about the triangle *DKB* let the circle *DKB* be described; then *DB* will be its diameter, because the angle at *K* is right.

Let *BL*, the side of a hexagon, be fitted into the circle.

Then, since the angle *DBE* is $\frac{1}{30}$th of a right angle, the angle *BDK* is also $\frac{1}{30}$th of a right angle;
therefore the circumference *BK* is $\frac{1}{60}$th of the whole circle.

But *BL* is also one-sixth part of the whole circle.

Therefore the circumference *BL* is ten times the circumference *BK*.

And the circumference *BL* has to the circumference *BK* a ratio greater than that which the straight line *BL* has to the straight line *BK*;[2]
therefore the straight line *BL* is less than ten times the straight line *BK*.

And *BD* is double of *BL*;
therefore *BD* is less than 20 times *BK*.

But, as *BD* is to *BK*, so is *AB* to *BC*;
therefore *AB* is also less than 20 times *BC*.

And *AB* is the distance of the sun from the earth,
while *BC* is the distance of the moon from the earth;

---

[1] The approximation $\frac{7}{5}$ for $\sqrt{2}$ occurs frequently (see p. 43). [Edd.]

[2] If *AB* and *CD* are unequal chords in the same circle and $AB > CD$, then $AB:CD <$ (arc AB):(arc CD). This is proved in Ptolemy's *Almagest* and is equivalent to the proposition that if α and β are acute angles and $\alpha > \beta$, $\sin\alpha/\sin\beta < \alpha/\beta$. [Edd.]

therefore the distance of the sun from the earth is less than 20 times the distance of the moon from the earth.

And it was before proved that it is greater than 18 times that distance.

Pappus, *Mathematical Collection*, pp. 554–560 (Hultsch).[1] Translation of T. L. Heath, *Greek Astronomy*

Now the first, third, and fourth of these hypotheses [of Aristarchus] practically agree with the assumptions of Hipparchus and Ptolemy. For the moon is illuminated by the sun at all times except during an eclipse, when it becomes devoid of light through passing into the shadow which results from the interception of the sun's light by the earth and which is conical in form; next the (circle) dividing the milk-white portion which owes its colour to the sun shining upon it and the portion which has the ashen colour natural to the moon itself is indistinguishable from a great circle (in the moon) when its positions in relation to the sun cause it to appear halved, at which times (a distance of) very nearly a quadrant on the circle of the zodiac is observed (to separate them); and the said dividing circle is in the direction of our eye, for this plane of the circle if produced will, in fact, pass through our eye in whatever position the moon is when for the first or second time it appears halved.

But, as regards the remaining hypotheses, the aforesaid mathematicians have taken a different view. For, according to them, the earth has the relation of a point and centre, not to the sphere in which the moon moves, but to the sphere of the fixed stars, the breadth of the earth's shadow is not (that) of two moons, nor does the moon's diameter subtend one-fifteenth part of a sign of the zodiac, that is, 2°. According to Hipparchus, on the one hand, the circle described by the moon is measured 650 times by the diameter of the moon, while the earth's shadow is measured by it 2½ times at its mean distance in the conjunctions; in Ptolemy's view, on the other hand, the moon's diameter subtends, when the moon is at its greatest distance, an arc of 0°31′20″, and, when at its least distance, of 0°35′20″, while the diameter [Pappus should have said "radius"] of the circular section of the shadow is, when the moon is at its greatest distance, 0°40′40″, and when the moon is at its least distance, 0°46′.[2]

---

[1] On Pappus, see p. 38, above. See also the notes on the Hypotheses of Aristarchus, pp. 109f. [Edd.]

[2] I.e., Hipparchus makes the angular diameter of the moon (probably mean value) 0°33′14″. The average of Ptolemy's estimates at perigee and apogee is 0°33′20″, and the actual average 0°31′9″. Ptolemy's estimate for the ratio of the diameter of the shadow to that of the moon comes to 2⅗ as against Hipparchus's estimate of 2½. The actual ratio lies between these two. For the sun Ptolemy's estimate is 0°31′20″, while the actual mean value is nearer 0°32′. Note also Archimedes' determination of the apparent diameter of the sun as between 1/200 and 1/164 of a right angle (i.e., between 27′ and 32′56″). On methods of making these estimates see pp. 139–142. [Edd.]

Hence it is that the authorities named have come to different conclusions as regards the ratios both of the distances and of the sizes of the sun and moon.

. . . As the result of the whole investigation he [Aristarchus] concludes that:

1. The sun has to the earth a greater ratio than that which 6,859 has to 27, but a less ratio than that which 79,507 has to 216;[1]
2. The diameter of the earth is to the diameter of the moon in a greater ratio than that which 108 has to 43, but in a less ratio than that which 60 has to 19;[2] and
3. The earth is to the moon in a greater ratio than that which 1,259,712 has to 79,507, but in a less ratio than that which 216,000 has to 6,859.[3]

But Ptolemy proved in the fifth book of his Syntaxis that, if the radius of the earth is taken as the unit, the greatest distance of the moon at the conjunctions is $64\frac{10}{60}$ of such units, the greatest distance of the sun is 1210, the radius of the moon $\frac{17}{60} + \frac{33}{60^2}$, the radius of the sun $5\frac{30}{60}$. Consequently, if the diameter of the moon is taken as the unit, the earth's diameter is $3\frac{2}{5}$ of such units, and the sun's diameter $18\frac{4}{5}$. That is to say, the diameter of the earth is $3\frac{2}{5}$ times the diameter of the moon, while the diameter of the sun is $18\frac{4}{5}$ times the diameter of the moon and $5\frac{1}{2}$ times the diameter of the earth.

From these figures the ratios between the solid contents are manifest, since the cube on 1 is 1 unit, the cube on $3\frac{2}{5}$ is very nearly $39\frac{1}{4}$ of the same units, and the cube on $18\frac{4}{5}$ very nearly $6{,}644\frac{1}{2}$, whence we infer that, if the solid magnitude of the moon is taken as a unit, that of the earth contains $39\frac{1}{4}$ and that of the sun $6{,}644\frac{1}{2}$ of such units; therefore the solid magnitude of the sun is very nearly 170 times greater than that of the earth.[4]

---

[1] Aristarchus, Prop. 16. [Edd.]

[2] Aristarchus, Prop. 17. [Edd.]

[3] Aristarchus, Prop. 18. [Edd.]

[4] In his treatise Aristarchus gives merely the ratios, and does not calculate the actual sizes and distances of the sun and moon. Such measurements, however, in terms of terrestrial, lunar, or solar distances are implicit in Aristarchus's results.

Thus, if l, t, and s are the diameters of the moon, earth, and sun, respectively, and S and L are the distances of the centers of the sun and moon, respectively, from that of the earth, the chief results of Aristarchus's treatise may be summed up as follows:

L/l lies between $22\frac{1}{2}$ and 30 (Prop. 11).
S/L lies between 18 and 20 (Prop. 7).
s/t lies between $6\frac{1}{3}$ and $7\frac{1}{6}$ (Prop. 15).
l/t lies between $\frac{19}{60}$ and $\frac{43}{108}$ (Prop. 17).

If $l/t = \frac{9}{25}$ (taking the mean value of $\frac{19}{60}$ and $\frac{43}{108}$),
and $L/l = 26\frac{1}{4}$ (Prop. 11, mean value),
then $L/t = 9\frac{9}{20}$, or $9\frac{1}{2}$, approximately.

Similarly the ratio $S/t = 179\frac{11}{20}$, or 180, approximately, may be obtained.

T. L. Heath's table (*Aristarchus*, p. 350), based on that of Hultsch, is appended. The

THE PRECESSION OF THE EQUINOXES

Ptolemy, *Almagest* VII. 1–2. Translation of T. L. Heath, *Greek Astronomy*

First of all we must premise, as regards the name ("fixed stars"), that, since the stars themselves always appear to keep the same figures and the same distances from each other, we may fairly call them "fixed," but, on the other hand, seeing that their whole sphere on which they are carried round as if they had grown upon it, appears itself to have an ordered movement of its own in the direct order of the signs, that is, towards the east, it would not be right to describe the sphere itself also as "fixed." We find both these facts to be as stated, judging by observations made so far as was possible in a comparatively short period. At an even earlier date Hipparchus, in consequence of the phenomena which he had recorded, became vaguely aware of the two facts,[1] but, as regards the effects over a longer

diameter of the earth, *t*, is the unit. [Edd.]

	Mean distance of Moon from Earth	Diameter of Moon	Mean distance of Sun from Earth	Diameter of Sun
According to Aristarchus......	9½	9/25 = 0.36	180	6¾
According to Hipparchus	33⅔	⅓ = 0.33	1245	12⅓
According to Posidonius......	26⅕	3/19 = 0.157	6545	39¼
According to Ptolemy........	29½	5/17 = 0.29	605	5½
In reality..................	30.2	0.27	11726	108.9

[1] By comparison of his own observations of the position of certain fixed stars with those made some 160 years earlier, Hipparchus showed that there had been a change in longitude, or, what is equivalent, a displacement of the points at which the equator and ecliptic intersect. The displacement was in a direction opposite to the diurnal motion of the universe. On the assumption that the displacement was a uniform one, Hipparchus estimated its extent as at least 36″ a year or 1° a century. This phenomenon, known as the precession of the equinoxes, is in large measure due to the fact that the earth is not a perfect sphere. The sun and moon, by their attraction of the matter at the equatorial bulge, tend to draw the plane of the equator into coincidence with that of the ecliptic. Consequently the earth's axis does not remain constantly parallel to itself, but very slowly turns about an axis passing through the center of the earth and perpendicular to the earth's orbit. That is, the celestial pole traces what is approximately a circle whose center is the pole of the ecliptic and angular radius 23½°. This circle is traced in a period of about 25,000 years. The extent of the precession is now calculated as approximately 50″ per annum.

The westward displacement of the equinox makes the sun's path along the ecliptic from one vernal equinox to the next (the tropical year) shorter than the path from a given fixed star back to the same fixed star (the sidereal year). Hipparchus gives the tropical year (Ptolemy, *Almagest* III. 1) as 365¼ less 1/300 of a day, i.e., 365d. 5h. 55m. 12s. (The modern estimate is 365d. 5h. 48m. 46s.)

The discovery of a kind of precession has sometimes been credited to the Babylonian

time, what he gave were guesses rather than facts thoroughly established, because he had come across only very few observations of the fixed stars made before his own time, and, indeed, almost the only observations he found recorded were those of Aristyllus and Timocharis,[1] and even these were neither free from doubt nor thoroughly worked out. We, for our part, have found the same result by comparing observations made to-day with those of the earlier time, but the result is now more firmly established by virtue of the fact that the inquiry has now lasted over a longer period, and the recorded data of Hipparchus about the fixed stars, with which our comparisons have mainly been made, have been handed down to us fully worked out. . . .

That the sphere of the fixed stars has a movement of its own in a sense opposite to that of the revolution of the whole universe, that is to say, in the direction which is east of the great circle described through the poles of the equator and the zodiac circle, is made clear to us especially by the fact that the same stars have not kept the same distances from the solstitial and equinoctial points in earlier times and in our time respectively, but, as time goes on, are found to be continually increasing their distance, measured in the eastward direction, from the same points beyond what it was before.

For Hipparchus, in his work "On the displacement of the solstitial and equinoctial points," comparing the eclipses of the moon, on the basis both of accurate observations made in his time, and of those made still earlier by Timocharis, concludes that the distance of Spica from the autumnal equinoctial point, measured in the inverse order of the signs, was in his own time 6°, but in Timocharis' time 8°, nearly. His words at the end are: "If then, for the sake of argument, Spica was, longitudinally with respect to the signs, at the earlier date 8° west of the autumnal-equinoctial point, but is now 6° west of it," and so on.[2] And in the case of practically all the other fixed stars the position of which he has similarly compared he shows that there has been the same amount of progression in the direct order of the signs. . . .[3]

This [i.e., an increase in longitude at the rate of about 1° in 100 years] seems to have been the idea of Hipparchus to judge by what he says, in his

astronomer Kidinnu (probably of the fourth century B.C.). O. Neugebauer, in *Journal of Cuneiform Studies* 1 (1947) 147, holds this ascription false, and indicates an alternative explanation of the supposed evidence. [Edd.]

[1] Aristyllus and Timocharis made observations at Alexandria at the beginning of the third century B.C. [Edd.]

[2] If, as seems to be the case, the variation was 2° in about 160 years, the precession would amount to 45″ per annum, a result much more accurate than 36″, given by Hipparchus as a lower limit and adopted by Ptolemy as the true value. [Edd.]

[3] Ptolemy then gives as an example the increase in longitude of the star Regulus. Comparing an observation of Hipparchus with one taken by himself, Ptolemy makes the difference in longitude 2°40′ in 265 years, or about 1° in a century. [Edd.]

work "On the length of the year": "If for this reason the solstices and the equinoxes had changed their position in the inverse order of the signs, in one year, by not less than $\frac{1}{100}°$, their displacement in 300 years should not have been less than 3°."

THE MOTION OF THE SUN AND THE PLANETS

Geminus, *Elements of Astronomy* 1.[1] Translation of T. L. Heath, *Greek Astronomy*

The summer solstice occurs when the sun comes nearest to the region where we live, describing its most northerly circle and producing the longest day of all days in the year and the shortest night. The longest day is equal to the longest night, and the shortest day to the shortest night. And the longest day contains, in the latitude of Rhodes, 14½ equinoctial hours.[2] The autumn equinox occurs when the sun in its passage from north to south is once more on the equinoctial circle and makes the day equal to the night. The winter solstice occurs when the sun is furthest away from the place where we live, and is lowest relatively to the horizon, describing its most southerly circle, and producing the longest night of all nights and the shortest day. The longest night contains, in the latitude of Rhodes, 14½ equinoctial hours.

The periods between the solstices and the equinoxes are divided as follows. From the vernal equinox to the summer solstice there are 94½ days. For in this number of days the sun traverses the Ram, the Bull, and the Twins, and, arriving at the first degree of the Crab, brings about the summer solstice. From the summer solstice to the autumnal equinox there are 92½ days, for in this number of days the sun traverses the Crab, the Lion, and the Virgin, and, arriving at the first degree of the Scales, brings about the autumnal equinox. From the autumnal equinox to the winter solstice there are 88⅛ days, for in that number of days the sun traverses the Scales, the Scorpion, and the Archer, and, arriving at the first degree of the Horned Goat, produces the winter solstice. From the winter solstice to the vernal equinox there are 90⅛ days, for in that number of days the sun traverses the remaining three signs, the Horned Goat, the Water-pourer, and the Fishes. The days forming these periods, when all added together, make up 365¼ days, which, as we saw, was the number of days in the year.

---

[1] Little is known about the life of Geminus. He may have been a pupil of Posidonius in the first century B.C. In the form in which Geminus's *Elements of Astronomy* (*Introduction to the Phenomena*) is extant, it furnishes an elementary account of astronomy as taught at the time. Of his mathematical work only fragments remain (see pp. 1, 44). [Edd.]

[2] In *Almagest* II.3 Ptolemy gives the method for ascertaining the latitude of a place from the duration of the longest day, and conversely (see p. 84, above). Rhodes is at approximately 36° north latitude. [Edd.]

At this point the question arises, why, although the four parts of the zodiac circle are equal, the sun, travelling at uniform speed all the time, yet traverses the arcs in unequal times.[1] For the hypothesis underlying the whole of astronomy is that the sun, the moon, and the five planets move at uniform speeds in circles, and in a sense contrary to that of the motion of the universe. The Pythagoreans were the first to approach such questions, and they assumed that the motions of the sun, moon, and planets are circular and uniform. For they could not brook the idea of such disorder in things divine and eternal as that they should move at one time more swiftly, at another time more slowly, and at another time stand still, which last expression refers to what are called the "stations" [or stationary points] in the case of the five planets. No one would credit such irregularity even in the case of a steady and orderly man on a journey. No doubt, the exigencies of daily life are often the cause of slowness and swiftness in men's movements; but when the stars, with their indestructible constitution, are in question, no reason can be assigned for swifter and slower motion.

As regards the remaining heavenly bodies, we shall state the cause elsewhere; here we shall show, in the case of the sun, for what reason, though moving at uniform speed, it nevertheless traverses equal arcs in unequal times.

Above all [in the celestial system] is the so-called sphere of the fixed stars, which includes the imagery of all the signs made up of fixed stars. But we must not suppose that all the stars lie on one surface, but rather that some of them are higher [i.e., more distant] and some lower [less distant]; it is only because our sight can only reach to a certain equal distance that the difference in height is imperceptible to us.

Next below the sphere of the fixed stars lies the Shining Star ["Phainon"] which goes by the name of Kronos [Saturn]. This star traverses the zodiac circle in 30 years, very nearly, and a single sign in 2 years and 6 months. Under the Shining Star, and lower than it, revolves the Bright Star ["Phaëthon"], called the star of Zeus; this traverses the zodiac circle in 12 years, and one sign in one year. Under this is ranged the Fiery Star ["Pyroëis"], that of Ares. This traverses the zodiac circle in 2 years and 6 months, and a sign in 2½ months. The next place is occupied by the sun, which traverses the zodiac circle in a year, and a sign in one month, approximately. Next lower than this lies "Phosphorus" [Lucifer], the star of Aphrodite, and this moves at approximately the same speed as the sun. Below this lies the Gleaming Star ["Stilbon"], the star of Hermes, and it also moves at equal

[1] Thales may have noted the inequality of the seasons (p. 92), but the first fairly accurate measurements that are recorded among the Greeks are those of Meton and Euctemon in the latter part of the fifth century B.C. It has been seen that Callippus improved the Eudoxan system of concentric spheres by taking account of the inequality of the seasons, which Eudoxus had neglected to do (see p. 103). [Edd.]

speed with the sun. Lower than all the rest revolves the moon, which traverses the zodiac circle in 27¼ days approximately [the "tropical" month].[1]

If, now, the sun had moved on the circle of the fixed signs, the times between the solstices and the equinoxes would have been exactly equal to one another. For, moving at uniform speed it ought, in that case, to have described equal arcs in equal times. Similarly, if the sun had moved in a circle lower than the zodiac circle, but about the same centre as that of the zodiac circle, in that case, too, the periods between the solstices and the equinoxes would have been equal. For all circles described about the same centre are similarly divided by their diameters; therefore, since the zodiac circle is divided into four equal parts by the diameters joining the solstitial and equinoctial points respectively, it would necessarily follow that the sun's circle is divided into four equal parts by the same diameters. The sun, therefore, moving at uniform speed on its own sphere [circle], would in that case have made the times corresponding to the four parts equal. But, as it is, the sun revolves at a lower level than the signs, and moves on an *eccentric* circle, as is explained below. For the sun's circle and the zodiac circle have not the same centre, but the sun's circle is displaced to one side, and, in consequence of its being so placed, the sun's course is divided into four unequal parts. The greatest of the arcs is that which lies under the quadrant of the zodiac circle which stretches from the first degree of the Ram to the 30th degree of the Twins, and the least arc is that which lies under the quadrant from the first degree of the Scales to the 30th of the Archer. . . .

The sun, then, moves at uniform speed throughout, but, because of the eccentricity of the sun's circle, it traverses the quadrants of the zodiac in unequal times.[2]

---

[1] This is the time required for the moon to make a complete revolution around the earth. The mean value is now given as 27d. 7h. 43m. 11½s. The time from new moon to new moon (the synodical period) is very accurately given by Geminus as 29½ + 1/33 days (see p. 121). Hipparchus's calculation of 29d. 12h. 44m. 3⅓s. differs by less than 1 second from the modern calculation. [Edd.]

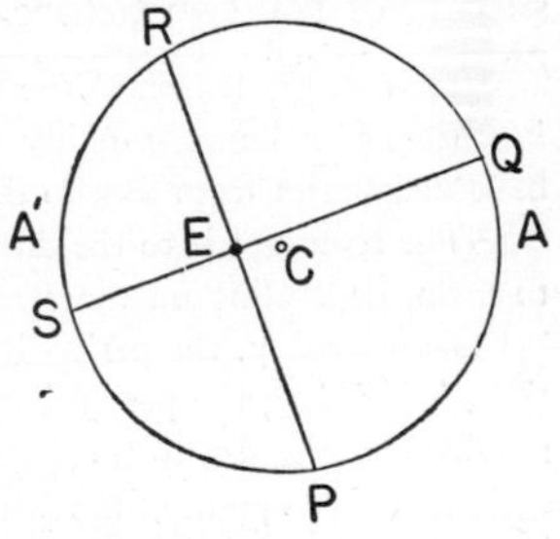

[2] Let *C* be the center of the circle in which the sun describes its path about the earth, *E*. Let *P* and *R* be the vernal and autumnal equinoxes; *Q* and *S* the summer and winter solstices. Given the length of the seasons (obtainable by observation), we can determine the position of *E* with respect to the center, i.e., the eccentricity, as well as the positions of the sun, *A*, *A′*, when farthest from and nearest to the earth, respectively (apogee and perigee). Hipparchus determined the eccentricity at 1/24 (i.e., *EC* = 1/24 the radius of the circle), and arc *PA* as 65° 30′, and assumed that these figures were constants.

This explanation of the inequality of the seasons by the assumption that the sun revolves around a center other than the center of the earth is an example of the use of eccentric circles so important in astronomical systems before Kepler (see p. 128). [Edd.]

On Day and Night

Geminus, *Elements of Astronomy* 6. Translation of T. L. Heath, *Greek Astronomy*

The word "day" is used in two senses: (1) for the time from the sun's rising to its setting; (2) for the time from the sun's rising to its rising again. The day in the second sense means the revolution of the universe plus the time taken to rise by the arc which the sun describes in its motion in a sense contrary to that of the universe during the time of the revolution of the universe.[1] Hence it is that a day and a night added together are not always equal to another day and night added together. The lengths are equal so far as our sensible perception goes, but, if exactly calculated, they show a small and imperceptible difference. The revolutions of the universe take equal times, but the times taken to rise by the arcs which the sun describes [in its own orbit] during one revolution of the universe are not equal; and it is for this reason that a day and a night added together are not always equal to another day and night added together.[2]

According to the second of the two meanings of the word "day," we say that the month has 30 days, and the year 365¼. A day and a night added together is a period of 24 equinoctial hours, and an equinoctial hour is the 24th part of a day and a night added together.

The length of the days is not the same for all countries and cities. The days are longer for those who live towards the north and shorter for those towards the south. The longest day in Rhodes has 14½ equinoctial hours, the longest in Rome 15 equinoctial hours.[3] For those farther north than the Propontis, the longest day has 16 hours, and for those still farther north, 17 and 18 hours.

Now these northern regions are thought to have been visited by Pytheas of Massilia.[4] He says, at all events, in his work "On the Ocean," that: "The barbarians showed us where the sun goes to rest. For it was found that in these regions the night was quite short, consisting in some places of two hours, in others of three, so that only a short interval elapsed from the setting of the sun before it rose again immediately." . . .

---

[1] That is, the length of the mean solar day is equal to the length of the sidereal day *plus* 1/365¼ of 24 hours, roughly 4 minutes, though, as Geminus indicates, the exact time to be added varies from day to day. [Edd.]

[2] The reference is to the difference in the length of the solar day, i.e., the time from noon to noon, depending on the season of the year.

The solar day, the period between two successive culminations of the sun, is longer than the sidereal day, the period between two successive culminations of a fixed star, by the time required for the sun to traverse its daily path in the zodiac (an average of 4 minutes). But since the sun's apparent motion in the zodiac is not uniform, the solar day, to which Geminus here refers, is not of uniform length. [Edd.]

[3] These are rough approximations. A longest day of 15 hours corresponds to a latitude of 41°9′. Rome is situated at 41°50′. [Edd.]

[4] See p. 390, below. [Edd.]

As we go further northwards, the summer-tropical circle comes to be wholly above the earth, so that at the summer solstice the day there[1] consists of 24 hours. And to those even further to the north a certain part of the zodiac circle is continually above the earth; and those for whom the space of a sign is cut off above the horizon have a day a month long; while for those with whom two signs are cut off above the earth, it is found that the longest day is of two months' duration. And, lastly, there is a place furthest of all to the north, where the pole is vertically overhead, and six signs of the zodiac are cut off above the horizon; for these people the longest day is six months long, and similarly for the night. . . .

But the increases in the lengths of days and nights are not equal in all the signs, but in the neighbourhood of the solstitial points they are quite small and imperceptible, so that the length of the days and nights remains the same for about 40 days. For, as the sun approaches and again recedes from the solstitial points, its deviations in latitude are not noticeable, so that it is reasonable that for the aforesaid number of days the sun should appear to our sensible perception to remain in the same position.[2]

The Month

Geminus, *Elements of Astronomy* 8. Translation of T. L. Heath, *Greek Astronomy*

A month is the time from one conjunction to the next conjunction, or from one full moon to the next full moon. A conjunction takes place when the sun and moon are in the same degree, that is, on the 30th day of the moon. "Full moon" means the time when the moon is diametrically opposite the sun, that is, at the "half month." A monthly period consists of $29\frac{1}{2} + \frac{1}{33}$ days.[3] In the period of a month the moon traverses the zodiac circle and in addition the arc by which the sun in the monthly period changes its place in the direct order of the signs, that is, approximately a sign. Hence, in the monthly period, the moon moves approximately through 13 signs.

---

[1] That is, at approximately $66\frac{1}{2}°$ north latitude. Cf. the passage of Strabo, p. 156, below. [Edd.]

[2] At the latitude of Rome the length of the day varies less than $\frac{1}{4}$ hour in the twenty days before and the twenty days after the solstices, whereas the variation is almost an hour in the twenty days before and again in the twenty days after the equinoxes. [Edd.]

[3] I.e., the mean lunar or synodical month, the average period between two successive conjunctions of the sun and moon (see p. 119). This is to be distinguished from the tropical month, the period between two successive passages of the equinoctial point (approximately $27\frac{1}{3}$ days). Irregularities of the motion of the moon due to the shifting of the points at which its path cuts the ecliptic (i.e., the nodes) and the shifting of the points of apogee and perigee gave rise to two other types of months, the draconitic and the anomalistic. For the lengths of all these months Hipparchus gave very accurate estimates, based largely, it seems, on Babylonian results. [Edd.]

HYPOTHESES OF PTOLEMAIC ASTRONOMY

Ptolemy, *Almagest* I. 2–3[1]

On the Order of the Theorems

2. The work which we have projected commences with a consideration of the general relation between the earth as a whole and the heavens as a whole.[2] Of the special treatments that follow, the first part[3] deals with the position of the ecliptic, the places inhabited by the human race, and the differences among the successive places, in each separate horizon, along the curvature of the earth's surface. The preliminary study of these relations makes easier the examination of the subsequent questions. The second part[4] gives an account of the motion of the sun and the moon and of the phenomena that depend on these motions. For without the previous understanding of these matters it would be impossible to set forth a complete theory of the stars. Since the theory of the stars is contained, in accordance with the general plan, in the concluding portion of this essay,[5] the investigation of the sphere of the so-called fixed stars would properly find its place there, and the material on the five so-called planets would follow. We shall try to set forth all this material using as the basic foundations for knowledge the manifest phenomena themselves and those recorded observations of the ancients and the moderns about which there is no dispute; and we shall seek to fit the propositions together by geometrical proofs.

With respect to the general portion of the treatise the following preliminary assumptions[6] are to be made: (1) that the heaven is spherical in form and rotates as a sphere; (2) that the earth, too, viewed as a complete whole, is spherical in form; (3) that it is situated in the middle of the whole heaven, like a center; (4) that by reason of its size and its distance from the sphere of fixed stars the earth bears to this sphere the relation of a point;[7] (5) that the earth does not participate in any locomotion. We shall say a few words by way of commentary on each of these propositions.

That the Heaven Rotates as a Sphere

3. It is reasonable to assume that the first ideas on these matters came to the ancients from observation such as the following. They saw the sun

[1] On Ptolemy see p. 162, below. For a discussion of his astronomy see p. 128.

[2] Book I. 3–11.

[3] Book I. 12 to the end of Book II.

[4] Books III–VI.

[5] Books VII–XIII.

[6] These are not arbitrary assumptions, for without seeking to *prove* them absolutely, the author tries to make them seem plausible (ch. 4–7).

[7] Cf. p. 108, above.

and the moon and the other stars moving from east to west in circles always parallel to each other; they saw the bodies begin to rise from below, as if from the earth itself, and gradually to rise to their highest point, and then, with a correspondingly gradual decline, to trace a downward course until they finally disappeared, apparently sinking into the earth. And then they saw these stars, once more, after remaining invisible for a time, make a fresh start and in rising and setting repeat the same periods of time and the same places of rising and setting with regularity and virtual similarity.

They were, however, led to the view of a spherical heaven chiefly by the observed circular motion described about one and the same center by those stars that are always above the horizon. For this point was, necessarily, the pole of the heavenly sphere, since the stars that are nearer this pole revolve in smaller circles, whereas those further away make larger circles, proportionately to their distance, until the distance reaches that of the stars not always visible.[1] And of these latter they observed that those stars nearer the stars that are always visible remained invisible for a shorter time while those further away remained invisible for a correspondingly longer time. And so, from these phenomena alone they first conceived the aforesaid idea, and then from the consideration of its consequences they adopted the other ideas that follow from it, since all the phenomena without qualification refuted the alternative hypotheses.

For example, if one should suppose, as some have, that the motion of the stars proceeds by a straight line without limit, how could one explain the fact that the daily motion of each star is always seen to begin from the same point? How could the stars in their unlimited motion turn back? And if they did turn back, how could this escape observation? Or how could they fail eventually to become altogether invisible, since they would appear ever smaller and smaller? In point of fact, however, they appear larger when near the region where they disappear,[2] and are only gradually occulted and, as it were, cut off by the surface of the earth.

Again, the suggestion that the stars are kindled when they rise from the earth and again are snuffed out when they return to the earth is quite contrary to reason.[3] For even if one should grant that the arrangement, size, and number of the stars, and their distances and intervals in space and time could have been the fulfillment of mere random and accidental procedure and that one part of the earth (the eastern part) had throughout it a kindling force, while the other (the western part) had an extin-

---

[1] I.e., those that are not always above the horizon.

[2] I.e., the horizon. Thus the sun and moon seem larger when rising and setting than when they are high in the heavens. On this illusion see p. 283, below. Ptolemy, in what follows, seeks to explain the phenomenon on the basis of atmospheric refraction.

[3] The view is referred to as a possibility in the Epicurean letter to Pythocles and in Lucretius. It is definitely ascribed to Epicurus by Cleomedes.

guishing force, or rather that the same part acted as a kindler from the point of view of some and as an extinguisher from the point of view of others, and that of the stars the very same ones were already kindled or extinguished, as the case might be, for some observers, but not yet for others—if, I repeat, one should grant all this, absurd as it is, what of the stars always visible, those that neither rise nor set? Why should the stars that are kindled and extinguished not rise and set everywhere? Why should those not subject to such kindling and extinguishing always be above the horizon in all latitudes? For surely the stars which for some observers are always kindled and extinguished cannot be the same as those which for other observers are never kindled and extinguished. (Yet the proponents of the hypothesis of kindling would have to assume that they are the same) for it is quite evident that the same stars rise and set for some observers (i.e., those further south) whereas they neither rise nor set for others (i.e., those further north).[1]

In a word, if one should suppose any other form of motion of the heavens save the spherical, the distances from the earth to the heavenly bodies would necessarily be unequal, however and wherever the earth itself might be supposed to lie situate. Consequently the sizes of the stars and their distances from one another would have to appear unequal to the same observers at each return, since the distances from the observers would sometimes be greater and at other times smaller. But this is not seen to be the case. For what makes the apparent size of a heavenly body greater when it is near the horizon is not its smaller distance but the vaporous moisture surrounding the earth between our eye and the heavenly body.[2] It is the same as when objects immersed in water appear larger, and in fact the more deeply immersed the larger.

The hypothesis of spherical motion finds support also in the fact that on any other hypothesis save this one alone it is impossible that the instruments for measuring hours should be correct. There is also support in the following fact. Just as the motion of the heavenly bodies is completely without hindrance and the smoothest of all motions, and the most easily moved of all shapes is the circular for plane figures and the spherical for solids, so also since the polygon with the greater number of sides is the larger of regular polygons having equal perimeters, it follows that in the case of plane figures the circle is greater than any polygon of equal perimeter, and in the case of solid figures the sphere is greater.[3] And the heaven is greater than all other bodies.

---

[1] I.e., these stars are always visible to the observers farther north.

[2] See p. 283, below.

[3] I.e., greater than any other solid of equal surface area. This proposition was not proved in antiquity but was inferred on the analogy of the proposition in the case of plane figures (see p. 82, n. 1, above).

Various physical considerations, too, lead to the same conclusion. Thus the aether consists of finer and more homogeneous parts than does any other body. Now surfaces of bodies of homogeneous parts are themselves of homogeneous parts, and the circular surface in the case of plane figures and the spherical surface in the case of solid figures are the only surfaces that consist of homogeneous parts. The aether not being a plane surface but a solid may therefore be inferred to be of spherical form. A similar inference may be made from the fact that nature has constructed all earthly and destructible bodies entirely of circular forms but forms not having homogeneous parts, while she has constructed the divine bodies in the aether of spherical form having homogeneous parts. For if these bodies were flat or quoit-shaped their form would not appear circular to all observers at the same time from different places of the earth. Hence it is reasonable to infer that the aether which encloses the heavenly bodies, being of the same nature, is of spherical form, and, because of its composition out of homogeneous parts, moves with uniform circular motion.

Ptolemy, *Almagest* I.7. Translation of T. L. Heath, *Greek Astronomy*

The Absolute Immobility of the Earth[1]

In the same way as before it can be proved that the earth cannot make any movement whatever in the aforesaid oblique direction, or ever change its position at all from its place at the centre; for the same results would, in that case, have followed as if it had happened to be placed elsewhere than at the centre. So I, for one, think it is gratuitous for any one to inquire into the causes of the motion towards the centre when once the fact that the earth occupies the middle place in the universe, and that all weights move towards it, is made so patent by the observed phenomena themselves. The ground for this conviction which is readiest to hand, seeing that the earth has been proved to be spherical and situated in the middle of the universe, is this simple fact: in all parts of the earth without exception the tendencies and the motions of bodies which have weight—I mean their own proper motions—always and everywhere operate at right angles to the (tangent) plane drawn evenly through the point of contact where the object falls.[2] That this is so makes it also clear that, if the objects were not stopped by the surface of the earth, they would absolutely reach the centre itself, since the straight line leading to the

[1] Comparison of the account given here with the passage from Aristotle (p. 143) will indicate that the basic view of the universe is the same in both the Aristotelian and the Ptolemaic or Hipparchan systems, despite the difference in the method of resolving geometrically the motions of the heavenly bodies. [Edd.]

[2] Cf. p. 146. [Edd.]

centre is always at right angles to the tangent-plane to the sphere drawn through the intersection at the point of contact.

All who think it strange that such an immense mass as that of the earth should neither move itself nor be carried somewhere seem to me to look to their own personal experience, and not to the special character of the universe, and to go wrong through regarding the two things as analogous. They would not, I fancy, think the fact in question to be strange if they could realize that the earth, great as it is, is nevertheless, when compared with the enclosing body,[1] in the relation of a point to that body. For in this way it will seem to be quite possible that a body relatively so small should be dominated and pressed upon with equal and similarly directed force on all sides by the absolutely greatest body formed of like constituents, there being no up and down in the universe any more than one would think of such things in an ordinary sphere. So far as the composite objects in the universe, and their motion on their own account and in their own nature are concerned, those objects which are light, being composed of fine particles, fly towards the outside, that is, towards the circumference, though their impulse seems to be towards what is for individuals "up," because with all of us what is over our heads, and is also called "up," points towards the bounding surface; but all things which are heavy, being composed of denser particles, are carried towards the middle, that is, to the centre, though they seem to fall "down," because, again, with all of us the place at our feet, called "down," itself points towards the centre of the earth, and they naturally settle in a position about the centre, under the action of mutual resistance and pressure which is equal and similar from all directions. Thus it is easy to conceive that the whole solid mass of the earth is of huge size in comparison with the things that are carried down to it, and that the earth remains unaffected by the impact of the quite small weights (falling on it), seeing that these fall from all sides alike, and the earth welcomes, as it were, what falls and joins it. But, of course, if as a whole it had had a common motion, one and the same with that of the weights, it would, as it was carried down, have got ahead of every other falling body, in virtue of its enormous excess of size, and the animals and all separate weights would have been left behind floating on the air, while the earth, for its part, at its great speed, would have fallen completely out of the universe itself. But indeed this sort of suggestion has only to be thought of in order to be seen to be utterly ridiculous.

Certain thinkers,[2] though they have nothing to oppose to the above arguments, have concocted a scheme which they consider more acceptable,

---

[1] I.e., the boundary of the universe. [Edd.]

[2] E.g., Heraclides of Pontus (p. 105). [Edd.]

and they think that no evidence can be brought against them if they suggest for the sake of argument that the heaven is motionless, but that the earth rotates about one and the same axis from west to east, completing one revolution approximately every day, or alternatively that both the heaven and the earth have a rotation of a certain amount, whatever it is, about the same axis, as we said, but such as to maintain their *relative* situations.

These persons forget however that, while, so far as appearances in the stellar world are concerned, there might, perhaps, be no objection to this theory in the simpler form, yet, to judge by the conditions affecting ourselves and those in the air about us, such a hypothesis must be seen to be quite ridiculous. Suppose we could concede to them such an unnatural thing as that the most rarefied and lightest things either do not move at all or do not move differently from those of the opposite character—when it is clear as day that things in the air and less rarefied have swifter motions than any bodies of more earthy character—and that (we could further concede that) the densest and heaviest things could have a movement of their own so swift and uniform—when earthy bodies admittedly sometimes do not readily respond even to motion communicated to them by other things—yet they must admit that the rotation of the earth would be more violent than any whatever of the movements which take place about it, if it made in such a short time such a colossal turn back to the same position again, that everything not actually standing on the earth must have seemed to make one and the same movement always in the contrary sense to the earth, and clouds and any of the things that fly or can be thrown could never be seen travelling towards the east, because the earth would always be anticipating them all and forestalling their motion towards the east, insomuch that everything else would seem to recede towards the west and the parts which the earth would be leaving behind it.

For, even if they should maintain that the air is carried round with the earth in the same way and at the same speed, nevertheless the solid bodies in it would always have appeared to be left behind in the motion of the earth and air together, or, even if the solid bodies themselves were, so to speak, attached to the air and carried round with it, they could no longer have appeared either to move forwards or to be left behind, but would always have seemed to stand still, and never, even when flying or being thrown, to make any excursion or change their position, although we so clearly see all these things happening, just as if no slowness or swiftness whatever accrued to them in consequence of the earth not being stationary.[1]

[1] With the apparatus available to him Ptolemy could not observe deviations due to the earth's rotation. Modern ballistics, of course, takes account of them. [Edd.]

Note on the Ptolemaic System

The *Syntaxis Mathematica* of Ptolemy (known also, through the Arabian scientists, as the *Almagest*) is the most important extant astronomical treatise of the Greeks. It was the basis of all astronomical knowledge throughout the Middle Ages and beyond.[1] For though we may think of the Copernican revolution as superseding the Ptolemaic system, it is to be noted that Copernicus, with certain modifications, used the same apparatus for the representation of motions as Ptolemy did, viz., spheres, eccentrics, deferents, and epicycles.

Ptolemy's work, in 13 books, is based in considerable measure on the work of Hipparchus, whose important astronomical treatises are no longer extant. In certain respects, especially in the theory of planetary motion, Ptolemy seems to have made significant independent contributions.

The work commences with a general introductory discussion of the fundamental hypotheses (p. 122, above). There follow the mathematical propositions leading to the construction of the Table of Chords (see p. 82, above) as well as propositions in spherical trigonometry.

The second book deals with problems of mathematical geography, e.g., the determination of the length of the longest day of the year at a place of given latitude. There is (ch. 6) a division of the inhabited latitudes into bands (*climata*) on the basis of the length of the longest day.

The third book deals with the theory of the sun and the length of the year. Here we have the Hipparchan representation of solar motion (p. 119, above) by means of an eccentric

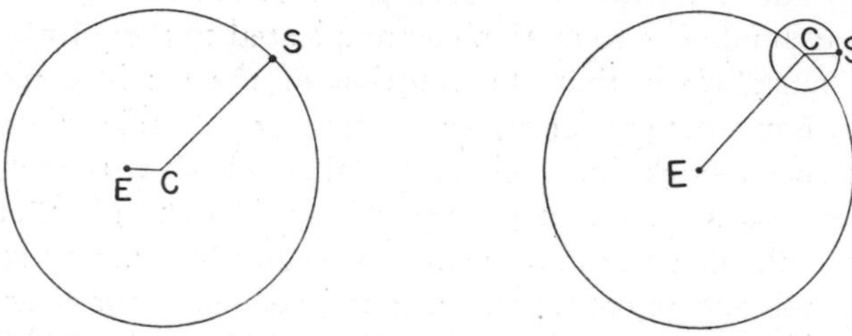

revolving about a fixed center (figure at left). Ptolemy compares this with the geometrically equivalent epicyclic hypothesis (figure at right), the center, C, of the epicycle describing a circular path, called the "deferent," about E. With the proper choice of radii, directions, and velocities, the position of the sun relative to the earth is the same in both systems.

We have seen that the equivalent of epicycles may possibly have been introduced into Greek astronomy as early as Heraclides (see p. 106). The history of the epicyclic and eccentric theory is in some respects obscure, but it had been shown, perhaps first by Apollonius of Perga, that, with the proper choice of magnitudes, for every epicyclic system an eccentric system (with fixed or movable center) could be constructed that would yield equivalent results. In any particular case the construction that was mathematically easier to work with was preferred. In the case of the solar theory Hipparchus preferred the eccentric system because it involved only one motion, whereas the epicyclic involved two. In some cases, however, combinations of epicycles and eccentrics were found necessary. These devices were resorted to because the assumption that had been made since the time of Plato—that the apparent irregularities in the motions of the heavenly bodies were to be explained as the resultant of primary *circular* motions—remained in force throughout the history of Greek astronomy.

---

[1] Another work, *Hypotheses of Planets*, in two books, one extant in Greek, the other in an Arabic version, containing in part a summary of the planetary theories of the *Almagest*, also had great influence among the Arabs.

In the theory of lunar and planetary motion, a far more difficult problem than that of solar motion, much progress seems to have been made after Hipparchus; for Ptolemy, by the use of combinations of eccentrics and epicycles and by a device that is probably his own innovation, succeeded in constructing a system that accounted for the phenomena with great accuracy.

The device referred to was the equant. The motion of the center (L) of the epicycle that carried the planet (P) was regarded as uniform not with respect to the center (C) of the deferent, or of the earth (T), but with respect to another point (E), called the "equant," i.e., $\angle LEA$ was considered as increasing uniformly. By properly locating the points E, C, and T, and determining the ratio of the diameters of the epicycle and its deferent, and choosing proper directions, velocities, and inclinations for the various circles, the apparent irregularities were accounted for. Lunar theory is discussed in Books IV–VI, along with questions of parallax, the sizes and distances of the sun and moon, and eclipses. Books IX–XIII deal with planetary theory.

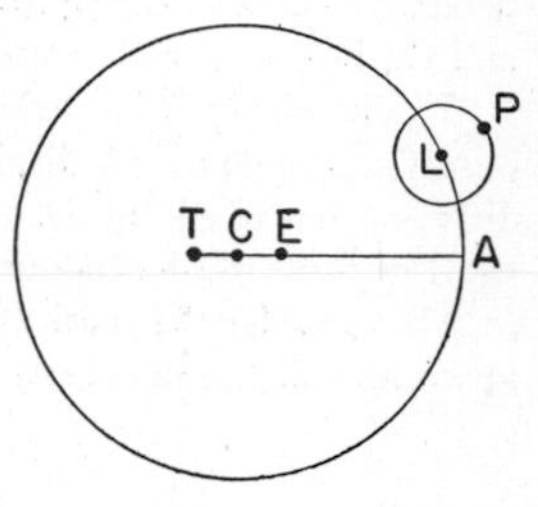

The remaining books (VII–VIII) deal with the fixed stars. The latitude, longitude, and magnitude of more than a thousand stars are given in a table arranged by constellations. There is a divergence of opinion as to how much of this catalogue represents independent observation by Ptolemy and how much is taken from predecessors, in particular Hipparchus and Menelaus, and corrected for precession.

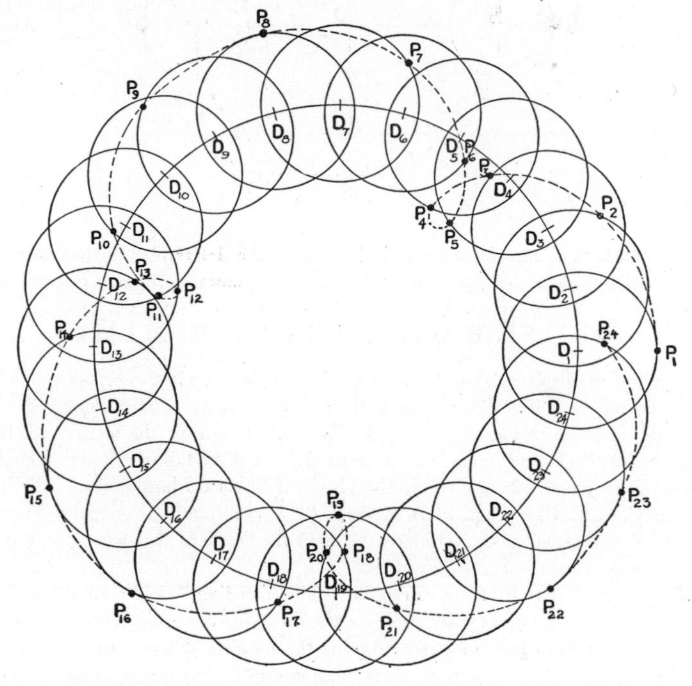

One or two additional notes may be given here with regard to the Ptolemaic planetary system. The apparatus of epicycles, deferents, eccentrics, and equants, necessitated by the apparent retrogradations and stationary points of the planets, was swept away by the system in which the orbits are elliptical (with the sun at a focus) and the velocity of the planets in their orbits varies with their position, a system introduced by Kepler and related to dynamical principles by Newton. With Copernicus the orbits are still circular, and we find Copernicus criticizing Ptolemy's use of the equant as contrary to the principle that circular motion must be uniform with respect to its own center.

We conclude this note with two diagrams taken from Brunet-Mieli, *Histoire des Sciences: Antiquité* (Paris, 1935), showing (1) how epicyclic motion when viewed from the center of the deferent produces the appearance of acceleration, retardation, station, and retrogradation, and (2) how a special case of epicyclic motion yields an elliptical orbit.

I. The planet P is carried on an epicycle whose center D describes a deferent having the observer at center. A displacement of 50° by the planet on the epicycle corresponds to a displacement of 15° by D. From the center P will appear to move rapidly or slowly, backward or forward, depending on its position on the epicycle. See p. 129, lower figure.

II. Consider the special case where the planet, moving uniformly, completes a revolution about its epicycle in exactly the same time that the center of the epicycle, moving uniformly

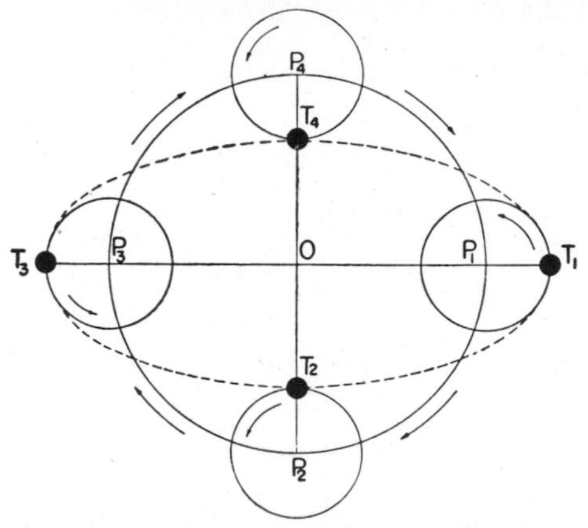

in the opposite direction, completes a revolution about the deferent. In that case, the path of the planet is an ellipse. See figure just above. The planet is T (with subscripts).

THE STAR CATALOGUE OF PTOLEMY

As an excerpt from Ptolemy's catalogue of fixed stars the stars in Cepheus, Corona Borealis, and Lyra are here given. The designation of the star by Ptolemy, and the longitude, latitude, and magnitude, as given in the *Almagest* (VII. 5) are followed by the modern identification of the star, the longitude and latitude as computed for A.D. 100 on the basis of modern star catalogues, and the magnitude as given in the Harvard Revised Photometry. All the figures are taken from C. H. F. Peters and E. B. Knobel, *Ptolemy's Catalogue of Stars* (The Carnegie Institution of Washington, 1915). References may be made to this work for a discussion of variant readings, variant identifications, etc.

It should be noted that Ptolemy's designations "preceding," and "following," and the like, with reference to a group of stars, indicate the order in which they cross the meridian in their diurnal rotation from east to west. Again, the longitudes are stated not by zodiacal signs, as in Ptolemy, but by their equivalent (e.g., not ♉ 5° but 35°). This will help comparison with the modern computations.

It will be seen that the latitudes agree very closely with the modern figures, while the longitudes are generally too small, showing a mean error, in the stars we have chosen, of

Stars	Ptolemy's figures						Modern name	Modern computation for 100 A.D.				Magnitude (Harvard Revised Photometry)
	Longitude		Latitude (all +)		Magnitude			Longitude		Latitude (all +)		
	°	′	°	′				°	′	°	′	
Constellation of Cepheus												
1. The star on the right foot.	35	0	75	40	4		1κ	37	5	75	15	4.4
2. The star on the left foot.	33	0	64	15	4		35γ	33	56	64	17	3.4
3. The star on the belt on the right side.	7	20	71	10	4		8β	9	47	71	0	3.3
4. The star touching the right shoulder from above.	346	40	69	0	3		5α	346	50	68	54	2.6
5. The star touching the right elbow from above.	339	20	72	0	4		3η	337	52	71	33	3.6
6. The star also touching the same elbow, but from below.	340	0	74	0	4		2θ	339	23	73	56	4.3
7. The star on the breast.	358	30	65	30	5		17ξ	358	11	65	45	4.4
8. The star on the left arm.	7	30	62	30	4–3		32ι	7	26	62	28	3.7
9. The southern star of the three in the tiara.	346	20	60	15	5		23ϵ	346	43	60	3	4.2
10. The middle one of the three.	347	20	61	15	4		21ζ	348	2	61	5	3.6
11. The more northern of the three.	349	0	61	20	5		22λ	350	7	61	49	5.2
Total: 11 stars, 1 of the third magnitude, 7 of the fourth, 3 of the fifth.												
Stars near Cepheus, but not in the constellation												
1. The star preceding the tiara.	343	40	64	0	5		13μ	343	50	64	9	4–5v
2. The star following the tiara.	351	20	59	30	4		27δ	351	37	59	28	3.7–4.6v
Total: 2, 1 of the fourth magnitude, 1 of the fifth.												

Stars	Ptolemy's figures					Modern name	Modern computation for A.D. 100				Magnitude (Harvard Revised Photometry)
	Longitude		Latitude (all +)		Magnitude		Longitude		Latitude (all +)		
	°	′	°	′			°	′	°	′	
Constellation of Corona Borealis											
1. The brilliant star in the crown.	194	40	44	30	2–1	5α	195	35	44	32	2.3
2. The star that precedes all the others.	191	40	46	10	4–3	3β	192	37	46	11	3.7
3. The star which follows this one and is further north.	191	50	48	0	5	4θ	192	50	48	45	4.2
4. The star which follows still later and is further north.	193	40	50	30	6	9π	195	26	50	38	5.6
5. The star that follows the brilliant star and is to the south of it.	197	10	44	45	4	8γ	198	16	44	40	3.9
6. The star that follows close behind this last one.	199	10	44	50	4	10δ	200	25	44	57	4.7
7. The star that is still further behind these.	201	20	46	10	4	13ϵ	202	31	46	16	4.2
8. The star that follows all the others in the Crown.	201	40	49	20	4	14ι	202	23	49	21	4.9
Total: 8 stars, 1 of the second magnitude, 5 of the fourth, 1 of the fifth, 1 of the sixth.											

Stars	Ptolemy's figures					Modern name	Modern computation for A.D. 100				Magnitude (Harvard Revised Photometry)
	Longitude		Latitude (all +)		Magnitude		Longitude		Latitude (all +)		
	°	′	°	′			°	′	°	′	
Constellation of the Lyre											
1. The brilliant star, called the Lyre, on the shell (Vega).	257	20	62	0	1	3α	258	45	61	51	0.14
2. The more northern of the two lying very close to this one.	260	20	62	40	4–3	$4\epsilon^1$, $5\epsilon^2$	262	20	62	33	4.7
3. The more southern of the two.	260	20	61	0	4–3	$6\zeta^1$, $7\zeta^2$	261	47	60	35	4.1
4. The star that follows these between the horns of the lyre where they emerge [from the shell].	263	40	60	0	4	$12\delta^2$	265	23	59	33	4.5
5. The more northern of the next two stars on the eastern side of the shell.	272	0	61	20	4	20η	273	50	60	54	4.5
6. The more southern of these.	272	40[1]	60	20	4–5	21θ	274	18	59	47	4.5
7. The more northern of the first two stars on the cross-bar.	261	0	56	10	3	10β	262	34	56	14	3.4–4.1v
8. The more southern of these.	260	50	55	0	4–5	$9\nu^2$	262	16	55	26	5.1
9. The more northern of the last two stars on the cross-bar.	264	10	55	20	3	14γ	265	37	55	15	3.3
10. The more southern of these.	264	0	54	45	4–5	15λ	265	50	54	41	5.1
Total: 10 stars, 1 of the first magnitude, 2 of the third, 7 of the fourth.											

[1] The manuscripts have ♉ 1°40′ (or 271°40′).

about 1°. Since the constellations chosen are in relatively high latitudes, the chance of error in the reading of longitude is greater. For the stars nearer the ecliptic the mean error in Ptolemy's longitudes is much smaller.

On the basis of the nature of the errors, Peters and Knobel held that Ptolemy's catalogue was based entirely on that of Hipparchus, with correction of 2°40′ for precession. A study of H. Vogt [*Astronomische Nachrichten* 224 (1925) 17–54] seeks to refute this view and concludes that there is no reason to doubt Ptolemy's word that he made independent observations. Systematic errors seem to be due to the nature and placing of the instruments, neglect of refraction, etc. A similar conclusion had been reached by J. L. E. Dreyer in *Monthly Notices of the Royal Astronomical Society* 77 (1917) 538; 78 (1918) 343.

A reference to Hipparchus's catalogue in Pliny (*Natural History* II. 95) may be cited here:

Now Hipparchus, who cannot be praised too highly and who, more than anyone else, exemplified man's kinship with the stars and proved that our souls are part of the heavens, detected a new star that had appeared in his own time. Because of its motion on the day on which it shone, he was led to wonder whether this happened at all frequently and whether those stars moved which we consider fixed. He went so far as to attempt —what would seem rash even for a god—to number the stars for his successors, and assign names for the various groups. For this he contrived instruments by which he might mark the location and magnitude of each star. Thus it could readily be discovered not merely whether stars perished and were born, but whether any of them moved, and whether they grew or diminished. He left the heavens as a bequest, as it were, to all who were found capable of receiving the inheritance.

SOME ASTRONOMICAL INSTRUMENTS AND THEIR USE

The instruments of the Greek astronomers and geographers were few and comparatively simple, and included sundials, dioptras, astrolabes, and water clocks. The sundials were of various types, plane and hemispherical, with markings to indicate seasonal differences. Water clocks indicating both the equinoctial hours and the civil hours that varied with the seasons were developed. The escape of water from an orifice or its constant dripping into a vessel is the principle underlying such clocks. The sand clock does not seem to have been in use among the ancient Greeks and Romans.

The dioptra, essentially a sighting tube and levelling instrument, was used for the observation of the heavenly bodies and their positions. The construction of this instrument is described in the section on Applied Mechanics (p. 336). We give here, however, two applications of it to astronomical problems—the finding of the angular distance between two stars and the determination of the angular diameter of the sun and moon.

The ring astrolabe, a kind of armillary sphere equipped with sighting apparatus, was the chief astronomical instrument of the Greeks and could be used to determine the latitude of stars. (Its construction and use are described in Ptolemy, *Almagest* V. 1 and Proclus, *Hypotyposis*, ch. 6.) This instrument is to be distinguished from the plane astrolabe, a sighting apparatus in combination with a suspensible graduated circle, useful for determining altitudes of heavenly bodies, terrestrial latitude, points of the compass, and time.

In representing the heavens the Greeks not only used star maps and catalogues but sometimes depicted the actual motions of the heavenly bodies with a type of planetarium. We have given some passages on this subject.

In connection with terrestrial maps it may be pointed out that the latitude of a place could be determined with fair accuracy by means of the dioptra or the sundial, but the absence of reliable clocks made accurate determination of longitude impossible. This will be further considered in the section on Mathematical Geography.

Sundials and Water Clocks

Vitruvius, *On Architecture* IX. 7–8. Translation of M. H. Morgan (Cambridge, Mass., 1914)

In distinction from the subjects first mentioned, we must ourselves explain the principles which govern the shortening and lengthening of the day. When the sun is at the equinoxes, that is, passing through Aries or Libra, he makes the gnomon cast a shadow equal to eight ninths of its own length, in the latitude of Rome. In Athens, the shadow is equal to three fourths of the length of the gnomon; at Rhodes to five sevenths; at Tarentum, to nine elevenths; at Alexandria, to three fifths;[1] and so at other places it is found that the shadows of equinoctial gnomons are naturally different from one another.

Hence, wherever a sundial is to be constructed, we must take the equinoctial shadow of the place. If it is found to be, as in Rome, equal to eight ninths of the gnomon, let a line be drawn on a plane surface, and

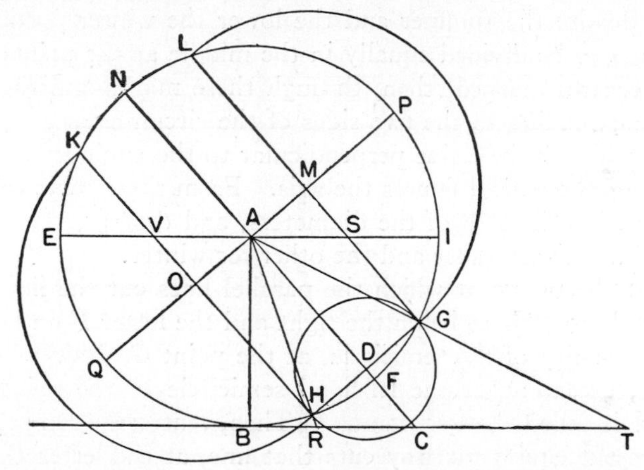

in the middle thereof erect a perpendicular, plumb to the line, which perpendicular is called the gnomon. Then, from the line in the plane, let the line of the gnomon be divided off by the compasses into nine parts, and take the point designating the ninth part as a centre, to be marked by the letter *A*. Then, opening the compasses from that centre to the line in the plane at the point *B*, describe a circle. This circle is called the meridian.

[1] The latitudes corresponding to these measurements (the earth being considered a perfect sphere) are: Rome, 41°38′, Athens 36°52′, Rhodes 35°32′, Tarentum 39°17′, Alexandria 30°58′. The latitude is the angle whose tangent is the length of the shadow (at noon of the day of the equinox) divided by the length of the gnomon. [Edd.]

Then, of the nine parts between the plane and the centre on the gnomon, take eight, and mark them off on the line in the plane to the point *C*. This will be the equinoctial shadow of the gnomon. From that point, marked by *C*, let a line be drawn through the centre at the point *A*, and this will represent a ray of the sun at the equinox. Then, extending the compasses from the centre to the line in the plane, mark off the equidistant points *E* on the left and *I* on the right, on the two sides of the circumference, and let a line be drawn through the centre, dividing the circle into two equal semicircles. This line is called by mathematicians the horizon.

Then, take a fifteenth part of the entire circumference,[1] and, placing the centre of the compasses on the circumference at the point where the equinoctial ray cuts it at the letter *F*, mark off the points *G* and *H* on the right and left. Then lines must be drawn from these (and the centre) to the line of the plane at the points *T* and *R*, and thus, one will represent the ray of the sun in winter, and the other the ray in summer. Opposite *E* will be the point *I*, where the line drawn through the centre at the point *A* cuts the circumference; opposite *G* and *H* will be the points *L* and *K*; and opposite *C*, *F*, and *A* will be the point *N*.

Then, diameters are to be drawn from *G* to *L* and from *H* to *K*. The upper will denote the summer and the lower the winter portion. These diameters are to be divided equally in the middle at the points *M* and *O*, and those centres marked; then, through these marks and the centre *A*, draw a line extending to the two sides of the circumference at the points *P* and *Q*. This will be a line perpendicular to the equinoctial ray, and it is called in mathematical figures the axis. From these same centres open the compasses to the ends of the diameters, and describe semicircles, one of which will be for summer and the other for winter.

Then, at the points at which the parallel lines cut the line called the horizon, the letter *S* is to be on the right and the letter *V* on the left, and from the extremity of the semicircle, at the point *G*, draw a line parallel to the axis, extending to the lefthand semicircle at the point *H*. This parallel line is called the Logotomus.[2] Then, centre the compasses at the point where the equinoctial ray cuts that line, at the letter *D*, and open them to the point where the summer ray cuts the circumference at the letter *H*. From the equinoctial centre, with a radius extending to the summer ray, describe the circumference of the circle of the months, which is called Menaeus. Thus we shall have the figure of the analemma.

This having been drawn and completed, the scheme of hours is next to be drawn on the baseplates from the analemma, according to the winter lines, or those of summer, or the equinoxes, or the months, and thus many

[1] Making the obliquity of the ecliptic 24°. [Edd.]

[2] For a different interpretation see O. Neugebauer, *Danske Videns. Selskab Hist.-Fil. Medd.* 26.2, p. 6 (1938). [Edd.]

different kinds of dials may be laid down and drawn by this ingenious method. But the result of all these shapes and designs is in one respect the same: namely, the days of the equinoxes and of the winter and summer solstices are always divided into twelve equal parts. . . .[1]

Methods of making water clocks have been investigated by the same writers, and first of all by Ctesibius the Alexandrian, who also discovered the natural pressure of the air and pneumatic principles. It is worth while for students to know how these discoveries came about. Ctesibius, born at Alexandria, was the son of a barber. Preëminent for natural ability and great industry, he is said to have amused himself with ingenious devices. For example, wishing to hang a mirror in his father's shop in such a way that, on being lowered and raised again, its weight should be raised by means of a concealed cord, he employed the following mechanical contrivance.

Under the roof-beam he fixed a wooden channel in which he arranged a block of pulleys. He carried the cord along the channel to the corner, where he set up some small piping. Into this a leaden ball, attached to the cord, was made to descend. As the weight fell into the narrow limits of the pipe, it naturally compressed the enclosed air, and, as its fall was rapid, it forced the mass of compressed air through the outlet into the open air, thus producing a distinct sound by the concussion.

Hence, Ctesibius, observing that sounds and tones were produced by the contact between the free air and that which was forced from the pipe, made use of this principle in the construction of the first water organs. He also devised methods of raising water, automatic contrivances, and amusing things of many kinds, including among them the construction of water clocks. He began by making an orifice in a piece of gold, or by perforating a gem, because these substances are not worn by the action of water, and do not collect dirt so as to get stopped up.

A regular flow of water through the orifice raises an inverted bowl, called by mechanicians the "cork" or "drum." To this are attached a rack and a revolving drum, both fitted with teeth at regular intervals. These teeth, acting upon one another, induce a measured revolution and movement. Other racks and other drums, similarly toothed and subject to the same motion, give rise by their revolution to various kinds of motions, by which figures are moved, cones revolve, pebbles or eggs fall, trumpets sound, and other incidental effects take place.

The hours are marked in these clocks on a column or a pilaster, and a figure emerging from the bottom points to them with a rod throughout the whole day. Their decrease or increase in length with the different days

[1] Vitruvius does not describe the markings that will achieve this result, but goes on to enumerate some dozen different forms of sundial and their inventors. [Edd.]

and months must be adjusted by inserting or withdrawing wedges. The shutoffs for regulating the water are constructed as follows: Two cones are made, one solid and the other hollow, turned on a lathe so that one will go into the other and fit it perfectly. A rod is used to loosen or to bring them together, thus causing the water to flow rapidly or slowly into the vessels. According to these rules, and by this mechanism, water clocks may be constructed for use in winter.

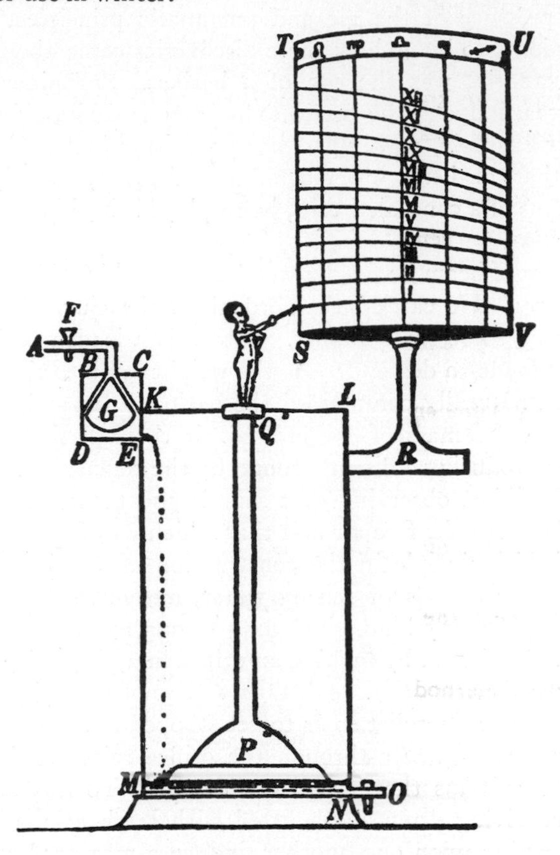

But if it proves that the shortening or lengthening of the day is not in agreement with the insertion and removal of the wedges, because the wedges may very often cause errors, the following arrangement will have to be made. Let the hours be marked off transversely on the column from the analemma, and let the lines of the months also be marked upon the column. Then let the column be made to revolve, in such a way that, as it turns continuously towards the figure and the rod with which the emerging figure points to the hours, it may make the hours short or long according to the respective months.

Astronomical Applications of the Dioptra

Hero of Alexandria, *On the Dioptra* 32

We have shown how the dioptra which we have constructed may be applied to terrestrial uses where dioptric problems are involved.[1] But the instrument also has many astronomical uses, e.g., in determining the relative distances[2] [of fixed stars or planets]. We shall now show how these distances are to be determined with the help of the dioptra.

We shall first draw a circle on the surface of the large disk of the dioptra, using the same center as that of the disk. The size of the circle will be determined by the length of the indicator attached to the sighting apparatus. We shall divide this circle into 360 parts. Now if we wish to find, in number of degrees, the [angular] distance between two stars, whether planets or fixed stars or one a fixed star and the other a planet, we proceed as follows: We remove the sighting apparatus from the disk and incline the latter until both stars in question appear at the same time in the plane of the disk. We then replace the sighting apparatus in the usual way, without disturbing the rest of the instrument, and turn the line of sight until one of the stars comes into view. Having noted the degree at which one of the indicators rests, we turn the line of sight until the other star comes into view. Again, in the same way we note the degree at which the same indicator rests. Thus we have the number of degrees between the two points sighted, and that, we can show, is the [angular] distance in degrees between the two stars.[3]

Proclus Diadochus, *Hypotyposis Astronomicarum Positionum* IV. 71–81, 87–99

Now the ancients, as Ptolemy tells us, attempted, though not on completely sound principles, to determine the apparent diameters of the sun and moon. They found, with the help of the available time measurers (water clocks or other types), the time taken for the entire diameter of each of these bodies to rise from the horizon.

But Hipparchus constructed for this purpose a dioptra, consisting of a grooved beam four cubits long with plates at right angles to it. Through these plates he sighted the diameters of these heavenly bodies and arrived at more accurate results. Ptolemy himself adopted this procedure.

Let us therefore set forth both the observations of the ancients and

---

[1] On the construction of the dioptra and its application to problems of surveying see pp. 336–342, below.

[2] I.e., the angular distances on the celestial sphere.

[3] It would be possible by combining the dioptra with a meridian instrument to obtain the declination and right ascension of each star, whence the angular distance could be computed with the help of a table of chords, but the procedure outlined in this chapter gives a rough means of reading off the distance without any intermediate steps.

the construction of the Hipparchan dioptra. First we show how it is possible to determine time by a uniform flow of water; we give the account of the engineer Hero in his work *On Water Clocks*.

A vessel is constructed with an opening in it like that in a clepsydra, through which the water can flow out uniformly as it does in such instruments.[1] The apparatus is set up beforehand and the outflow of the water is made to begin when the first ray of the sun appears above the horizon. Now the water that has been discharged during the time required for the whole disk of the sun to rise above the horizon is kept separately. And the water which flows out thereafter, uniformly and continuously for the whole period of a night and a day until the next rising of the sun, is measured in another vessel, and the ratio between the whole outflow and the outflow during the rising of the sun is found. This ratio, says Hero, will correspond to the time. That is, the ratio of the two amounts of water will be equal to the ratio of the two times.[2]

Now from this the ancients computed the ratio of the sun's diameter to the entire diurnal circle, considering that for any arc of the sun's diurnal circle could be substituted the subtending chord, that is, the sun's apparent diameter.

Other investigators made use of the other methods of measuring time, for example, the *scaphē* or some other type of sundial or else a clepsydra. They again found the time for the sun's rising and noted the interval representing the equinoctial day on the instrument.[3] Or, again, measuring the time units with a water clock they also stated that the ratio of the length of the equinoctial day to the measurement taken was equal to the ratio of the sun's entire diurnal circle to the apparent diameter of the sun.

These methods are quite worthless, says Ptolemy, for (1) it frequently happens that the hole [of the clepsydra] is stopped up; (2) the quantity of water that flows out in a night and a day is not necessarily in every case an exact multiple of the quantity taken only at the sun's rising, since in general the quantities compared are incommensurable; and (3) it is not accurate to substitute the chord for the arc which it subtends. . . .

---

[1] An overflowing reservoir maintaining a constant pressure head may have been used. For a siphon of uniform flow, see p. 244, below.

[2] The method of determining the apparent diameter by comparing the quantity of water that flows during the rising or setting of the sun or moon with that which flows during the whole day was widespread among the peoples of antiquity (see the discussion in *Papyri Osloenses* III. 31–35). Cleomedes describes the method (II. 1, 136.26 [Ziegler]) and gives the diameter of the sun as $1/750$ of the orbit (i.e., $28\frac{4}{5}'$); Martianus Capella, on the basis of a similar method, gives the diameter of the moon as $1/600$ of the orbit, or $36'$.

[3] That is, they measured the ratio of the time required for the sun's rising to the period of 12 (or 24) equinoctial hours. It would be extremely difficult to carry out the procedure described in the case of the sundial, for the light which casts the shadow comes from the whole face of the sun, not merely from its center.

For this reason Ptolemy, after rejecting all these methods, solved the problem by means of the dioptra of Hipparchus.[1] He squared off a beam at least four cubits long. Then he drew a line lengthwise down the middle of the beam, and cutting along this line he hollowed out a dovetailed groove to which he fitted at right angles a plate of suitable size. He inserted the base of this plate snugly into the cavity of the groove, so that it could move smoothly back and forth along the entire length of the beam, always remaining at right angles to the side of the beam. Similarly, he set up a second plate at the other end of the beam, also at right angles thereto. This second plate remained fixed at all times since the procedure required that sightings be taken through it. Then he bored a hole in this second plate equally distant from the sides of the plate, at a point near the base, i.e., near the beam. In the first plate, the one which I have said is movable, he next made two openings, one having the same position (relative to the base and the vertical bisector) as the opening on the fixed plate, the other at the top of the vertical bisector of the plate.

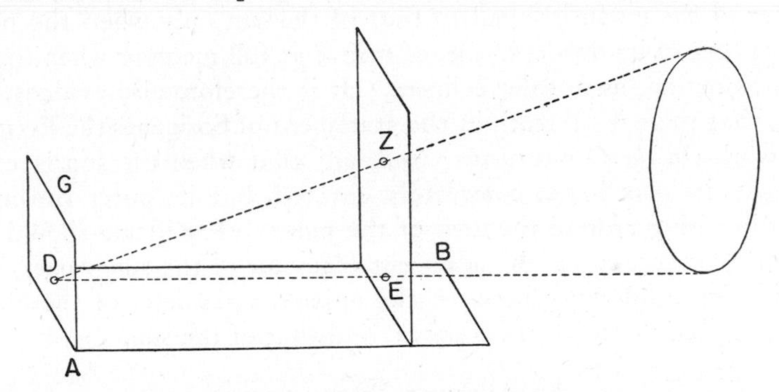

Thus let *AB* be the beam, with *A* the side from which sightings are taken and at which is fixed plate *DG*. Let *EZ* be the other plate, the one which moves along the whole length of the beam and contains the two above-mentioned openings on a perpendicular [the vertical bisector]. Let *E* be the opening at the base similarly placed to opening *D*; and *Z* the other opening near the top. The diagram of the instrument is given.

The sighting instrument may be used and set up as follows: At the time of the rising or setting of the sun, place the sighting piece in a plane parallel to the horizon, where the view of the horizon is as clear and unobstructed as possible. And let the observer stand at the immovable plate, the sun being on the side of the movable plate. The latter plate is moved back and forth until it is possible to see the lower part of the sun's disk

[1] Note, however, that a similar device seems to have been used for this purpose by Archimedes (*Sand Reckoner* I. 10), who finds the angular diameter of the sun to be between $\frac{1}{200}$ and $\frac{1}{164}$ of a right angle.

through openings *D* and *E* in the two plates, and the upper part through openings *D* and *Z*. In this way the observer obtains a view of both ends of the apparent diameter of the sun and can determine $\angle EDZ$. This angle is subtended by the apparent diameter of the sun, which is proportional to distance *EZ* on the plate. We then, says Ptolemy, marked the position of the movable plate at which the diameter of the sun can be observed. And we made similar observations in the case of the moon, and found the time at which its apparent diameter was equal to that of the sun (for the moon's apparent diameter varies according to its distance [from the earth]). The moon's apparent diameter is equal to that of the sun when the distance between the two plates on the sighting beam is the same in the case of the moon as in the case of the sun. Now the apparent diameter of the sun, as Ptolemy himself maintains, is always the same when found by the dioptra, whether the sun is at apogee or perigee. But the apparent diameter of the moon is greater or smaller according to its varying distance [from the earth]; and the apparent diameter of the moon is equal to that of the sun only when the moon is at the apogee of its own epicycle, of course at full moon or when the moon is in conjunction, as in solar eclipses. It is therefore also evident, if this is true, that there is no truth in the statement of Sosigenes the Peripatetic in his work *On the Counteracting Spheres*,[1] that when the sun is eclipsed at perigee, its disk is not completely covered, but its outer rim appears shining in a ring around the disk of the moon. For if one should admit this to be correct, either the apparent diameter of the sun varies, or else there will be a difference between the apparent diameter of the moon at apogee, as observed with the dioptra, and that of the sun.

An Ancient Planetarium

Cicero, *De Re Publica* I. 22[2]

But this type of sphere in which were exhibited the motions of the sun, moon, and the five stars called wandering or roaming stars could not be solid as was that other type of sphere.[3] All the more remarkable, then, was this contrivance of Archimedes, because he had devised a means

---

[1] See p. 103.

[2] Archimedes constructed a model of the heavens in which were represented the motions of the sun, moon, planets, and, presumably, the sphere of the fixed stars. This was famous in antiquity. It was brought to Rome by Marcellus and subsequently reproduced. See, in addition to the passages here given, the references in Proclus and Pappus, quoted on pp. 5 and 184. The obscure reference to water power in Pappus leaves the precise construction of the device still doubtful.

[3] This is a reference to another sphere on which was merely depicted a star map, and which, unlike the planetarium, could be solid. Thales, Eudoxus, and others are said to have made such star maps.

whereby a single turning would preserve the unequal and varying speeds of the different motions. When Gallus rotated this sphere, the moon followed the sun on the bronze globe by a number of revolutions equal to the number of days by which it followed it in the sky itself. Thus there occurred an eclipse of the sun on the sphere, just as in the sky, and the moon passed into the region of the earth's shadow.

Cf. *Tusculan Disputations* I. 63

For when Archimedes exhibited the motions of the moon, the sun, and the five planets on a sphere, he did what the god of Plato, who is represented in the *Timaeus* as constructing the universe, did. He made it possible by a single rotation to control motions of most diverse velocities. Now if this cannot be done without a god in the case of the universe, Archimedes surely could not have been able to imitate those same motions on his sphere without divine genius.

Cf. *On the Nature of the Gods* II. 88

Our friend Posidonius[1] recently constructed [such a sphere]. A single rotation of it produces the same effect on the sun, the moon, and the five planets as is produced in the sky itself on a single day and night.

MATHEMATICAL GEOGRAPHY

THE SHAPE AND SIZE OF THE EARTH

Aristotle, *De Caelo* II. 13–14. Translation of J. L. Stocks

13. . . . There are similar disputes about the *shape* of the earth. Some think it is spherical,[2] others that it is flat and drum-shaped.[3] For evidence they bring the fact that, as the sun rises and sets, the part concealed by the earth shows a straight and not a curved edge, whereas if the earth were spherical the line of section would have to be circular. In this they leave

[1] There is no reason to doubt that it was the philosopher Posidonius who constructed this sphere in imitation of Archimedes.

[2] There are conflicting elements in the traditional accounts of the discovery of the sphericity of the earth. There is some ground for ascribing the idea (if not a cogent proof) to Parmenides or even to the early Pythagoreans. But in recent years some have held that this discovery, as well as others in mathematics, is later than has generally been thought and belongs to the group of mathematicians of whom Archytas was a leading spirit about 400 B.C. While we cannot deal here with these controversial problems, it is important to point out the uncertainty that attends any modern reconstruction of this phase of the history of Greek science. [Edd.]

[3] The pre-Socratic philosophers, with the possible exception of the Pythagoreans and Parmenides, denied the sphericity of the earth, though they differed as to its precise shape, as the sequel indicates. [Edd.]

out of account the great distance of the sun from the earth and the great size of the circumference, which, seen from a distance on these apparently small circles, appears straight. Such an appearance ought not to make them doubt the circular shape of the earth. But they have another argument. They say that because it is at rest, the earth must necessarily have this shape. For there are many different ways in which the movement or rest of the earth has been conceived.

The difficulty must have occurred to every one. It would indeed be a complacent mind that felt no surprise that, while a little bit of earth, let loose in mid-air, moves and will not stay still, and the more there is of it the faster it moves, the whole earth, free in mid-air, should show no movement at all. Yet here is this great weight of earth and it is at rest. And again, from beneath one of these moving fragments of earth, before it falls, take away the earth, and it will continue its downward movement with nothing to stop it. The difficulty then, has naturally passed into a commonplace of philosophy; and one may well wonder that the solutions offered are not seen to involve greater absurdities than the problem itself.

By these considerations some have been led to assert that the earth below us is infinite, saying, with Xenophanes of Colophon, that it has "pushed its roots to infinity"[1]—in order to save the trouble of seeking for the cause. Hence the sharp rebuke of Empedocles, in the words "if the deeps of the earth are endless and endless the ample ether—such is the vain tale told by many a tongue, poured from the mouths of those who have seen but little of the whole." Others say that the earth rests upon water. This, indeed, is the oldest theory that has been preserved and is attributed to Thales of Miletus. It was supposed to stay still because it floated like wood and other similar substances, which are so constituted as to rest upon water but not upon air. As if the same account had not to be given of the water which carries the earth as of the earth itself! It is not the nature of water, any more than of earth, to stay in mid-air: it must have something to rest upon. Again, as air is lighter than water, so is water than earth: how then can they think that the naturally lighter substance lies below the heavier? Again, if the earth as a whole is capable of floating upon water, that must obviously be the case with any part of it. But observation shows that this is not the case. Any piece of earth goes to the bottom, the quicker the larger it is. These thinkers seem to push their inquiries some way into the problem, but not so far as they might. It is what we are all inclined to do, to direct our inquiry not by the matter itself, but by the views of our opponents: and even when interrogating oneself one pushes

[1] "We see the upper limit of the earth touching the air at our feet, but downward the earth stretches without limit." [Edd.]

the inquiry only to the point at which one can no longer offer any opposition. Hence a good inquirer will be one who is ready in bringing forward the objections proper to the genus, and that he will be when he has gained an understanding of all the differences.

Anaximenes and Anaxagoras and Democritus give the flatness of the earth as the cause of its staying still. Thus, they say, it does not cut, but covers like a lid, the air beneath it. This seems to be the way of flat-shaped bodies: for even the wind can scarcely move them because of their power of resistance. The same immobility, they say, is produced by the flatness of the surface which the earth presents to the air which underlies it; while the air, not having room enough to change its place because it is underneath the earth, stays there in a mass, like the water in the case of the water-clock.[1] And they adduce an amount of evidence to prove that air, when cut off and at rest, can bear a considerable weight. . . .

14. Let us first decide the question whether the earth moves or is at rest. For, as we said, there are some who make it one of the stars, and others who, setting it at the centre, suppose it to be "rolled" and in motion about the pole as axis.[2] That both views are untenable will be clear if we take as our starting-point the fact that the earth's motion, whether the earth be at the centre or away from it, must needs be a constrained motion. It cannot be the movement of the earth itself. If it were, any portion of it would have this movement; but in fact every part moves in a straight line to the centre. Being, then, constrained and unnatural, the movement could not be eternal. But the order of the universe is eternal. Again, everything that moves with the circular movement, except the first sphere, is observed to be passed, and to move with more than one motion. The earth, then, also, whether it move about the centre or as stationary at it, must necessarily move with two motions. But if this were so, there would have to be passings and turnings of the fixed stars. Yet no such thing is observed. The same stars always rise and set in the same parts of the earth.[3]

Further, the natural movement of the earth, part and whole alike, is to the centre of the whole—whence the fact that it is now actually situated

---

[1] The reference is not to a timekeeping device but to the type of device described on pp. 245–246, below. [Edd.]

[2] The reference is to Plato's *Timaeus*. It is generally held, however, that Plato did not teach the daily rotation of the earth and that Aristotle is himself inferring (or is alleging that others inferred) more from Plato's language than is justified (see p. 96 and n. 2.) [Edd.]

[3] Until stellar parallax was discovered, the argument that the position of rising and setting of the fixed stars from a given viewpoint was always the same constituted a powerful refutation of the earth's orbital motion. [Edd.]

at the centre—but it might be questioned, since both centres are the same, which centre it is that portions of earth and other heavy things move to. Is this their goal because it is the centre of the earth or because it is the centre of the whole? The goal, surely, must be the centre of the whole. For fire and other light things move to the extremity of the area which contains the centre. It happens, however, that the centre of the earth and of the whole is the same. Thus they do move to the centre of the earth, but accidentally, in virtue of the fact that the earth's centre lies at the centre of the whole. That the centre of the earth is the goal of their movement is indicated by the fact that heavy bodies moving towards the earth do not move parallel but so as to make equal angles,[1] and thus to a single centre, that of the earth. It is clear, then, that the earth must be at the centre and immovable, not only for the reasons already given, but also because heavy bodies forcibly thrown quite straight upward return to the point from which they started, even if they are thrown to an infinite distance. From these considerations then it is clear that the earth does not move and does not lie elsewhere than at the centre.

From what we have said the explanation of the earth's immobility is also apparent. If it is the nature of earth, as observation shows, to move from any point to the centre, as of fire contrariwise to move from the centre to the extremity, it is impossible that any portion of earth should move away from the centre except by constraint. For a single thing has a single movement, and a simple thing a simple: contrary movements cannot belong to the same thing, and movement away from the centre is the contrary of movement to it. If then no portion of earth can move away from the centre, obviously still less can the earth as a whole so move. For it is the nature of the whole to move to the point to which the part naturally moves. Since, then, it would require a force greater than itself to move it, it must needs stay at the centre. This view is further supported by the contributions of mathematicians to astronomy, since the observations made as the shapes change, by which the order of the stars is determined, are fully accounted for on the hypothesis that the earth lies at the centre. Of the position of the earth and of the manner of its rest or movement, our discussion may here end.

Its shape must necessarily be spherical. For every portion of earth has weight until it reaches the centre, and the jostling of parts greater and smaller would bring about not a waved surface, but rather compression and convergence of part and part until the centre is reached. The process

---

[1] I.e., at right angles to a tangent; if it fell otherwise than at right angles, the angles on each side of the line of fall would be unequal. . . . [It seems impossible that accurate measurements to ascertain this were carried out in antiquity. For one thing the swerve due to the earth's rotation would affect the results. Edd.]

should be conceived by supposing the earth to come into being in the way that some of the natural philosophers describe. Only they attribute the downward movement to constraint, and it is better to keep to the truth and say that the reason of this motion is that a thing which possesses weight is naturally endowed with a centripetal movement. When the mixture, then, was merely potential, the things that were separated off moved similarly from every side towards the centre. Whether the parts which came together at the centre were distributed at the extremities evenly, or in some other way, makes no difference. If, on the one hand, there were a similar movement from each quarter of the extremity to the single centre, it is obvious that the resulting mass would be similar on every side. For if an equal amount is added on every side the extremity of the mass will be everywhere equidistant from its centre, i.e., the figure will be spherical. But neither will it in any way affect the argument if there is not a similar accession of concurrent fragments from every side. For the greater quantity, finding a lesser in front of it, must necessarily drive it on, both having an impulse whose goal is the centre, and the greater weight driving the lesser forward till this goal is reached. In this we have also the solution of a possible difficulty. The earth, it might be argued, is at the centre and spherical in shape: if, then, a weight many times that of the earth were added to one hemisphere, the centre of the earth and of the whole will no longer be coincident. So that either the earth will not stay still at the centre, or if it does, it will be at rest without having its centre at the place to which it is still its nature to move. Such is the difficulty. A short consideration will give us an easy answer, if we first give precision to our postulate that any body endowed with weight, of whatever size, moves towards the centre. Clearly it will not stop when its edge touches the centre. The greater quantity must prevail until the body's centre occupies the centre. For that is the goal of its impulse. Now it makes no difference whether we apply this to a clod or common fragment of earth or to the earth as a whole. The fact indicated does not depend upon degrees of size but applies universally to everything that has the centripetal impulse. Therefore earth in motion, whether in a mass or in fragments, necessarily continues to move until it occupies the centre equally in every way, the less being forced to equalize itself by the greater owing to the forward drive of the impulse.

If the earth was generated, then, it must have been formed in this way, and so clearly its generation was spherical; and if it is ungenerated and has remained so always, its character must be that which the initial generation, if it had occurred, would have given it. But the spherical shape, necessitated by this argument, follows also from the fact that the motions of heavy bodies always make equal angles and are not parallel. This would be the natural form of movement towards what is naturally spherical. Either then the earth is spherical or it is at least naturally spherical.[1] And it is right to call anything that which nature intends it to be, and which belongs to it, rather than that which it is by constraint and contrary to nature. The evidence of the senses further corroborates this. How else would eclipses of the moon show segments shaped as we see them? As it is, the shapes which the moon itself each month shows are of every kind—straight, gibbous, and concave—but in eclipses the outline is always curved: and, since it is the interposition of the earth that makes the eclipse, the form of this line will be caused by the form of the earth's surface, which is therefore spherical.[2] Again, our observations of the stars make it evident, not only that the earth is circular, but also that it is a circle of no great size. For quite a small change of position to south or north causes a manifest alteration of the horizon. There is much change, I mean, in the stars which are overhead, and the stars seen are different, as one moves northward or southward. Indeed there are some stars seen in Egypt and in the neighbourhood of Cyprus which are not seen in the northerly regions; and stars, which in the north are never beyond the range of observation, in those regions rise and set.[3] All of which goes to show not only that the earth is circular in shape, but also that it is a sphere of no great size: for otherwise the effect of so slight a change of place would not be so quickly apparent. Hence one should not be too sure of the incredibility of the view of those who conceive that there is continuity between the parts about the pillars of Hercules and the parts about India, and that in this way the ocean is one.[4] As further evidence in favour of this they quote the case of elephants, a species occurring in each of these extreme regions, suggesting that the common characteristic of these extremes is explained by their continuity. Also, those mathematicians who try to calculate the size of the earth's circumference arrive at the figure 400,000 stades.[5] This indicates not only that the earth's mass is spherical in shape, but also that as compared with the stars it is not of great size.

---

[1] Irregularities at the surface necessitate this limitation. [Edd.]

[2] This is a strong argument for the sphericity of the earth, since the sphere is the only figure which casts a shadow such that a right section of it is always a circle. [Edd.]

[3] Though Aristotle accords more space to his *a priori* arguments for the sphericity of the earth, this and the argument from lunar eclipses are far more convincing to a modern reader. [Edd.]

[4] This view, so important in the explorations of a later period, has more point with a smaller estimate of the circumference of the earth than that to which Aristotle refers.

Cf. Strabo, *Geography* II. 3.6: "Now he [Posidonius] supposes that the length of the inhabited world, about 70,000 stades, is half of the whole circle on which it has been taken, so that, he says, by sailing straight out from the west for the same distance [70,000 stades] you would reach India." [Edd.]

[5] For methods of calculating the size of the earth, see the next selection. [Edd.]

THE MEASUREMENT OF THE CIRCUMFERENCE OF THE EARTH

Cleomedes, *On the Orbits of the Heavenly Bodies* I, 10.[1] Translation of T. L. Heath, *Greek Astronomy*

About the size of the earth the physicists, or natural philosophers, have held different views, but those of Posidonius and Eratosthenes are preferable to the rest.[2] The latter shows the size of the earth by a geometrical method; the method of Posidonius is simpler. Both lay down certain hypotheses, and, by successive inferences from the hypotheses, arrive at their demonstrations.

Posidonius says that Rhodes and Alexandria lie under the same meridian.[3] Now meridian circles are circles which are drawn through the poles of the universe and through the point which is above the head of any individual standing on the earth. The poles are the same for all these circles, but the vertical point is different for different persons. Hence we can draw an infinite number of meridian circles. Now Rhodes and Alexandria lie under the same meridian circle, and the distance between the cities is reputed to be 5,000 stades.[4] Suppose this to be the case.

All the meridian circles are among the great circles in the universe, dividing it into two equal parts and being drawn through the poles. With these hypotheses, Posidonius proceeds to divide the zodiac circle, which is equal to the meridian circles, because it also divides the universe into two equal parts, into forty-eight parts, thereby cutting each of the twelfth parts of it (i.e., signs) into four.[5] If, then, the meridian circle through

---

[1] Little is known of the life of Cleomedes. He seems to have written his elementary astronomical textbook between the time of Posidonius and Ptolemy, i.e., between the first century B.C. and the second century A.D. [Edd.]

[2] The two methods here set forth for calculating the circumference of the earth are basically the same, both depending on the difference in elevation of a given star from two observation posts on the same meridian. In the one case the star is Canopus, in the other it is the sun (its elevation measured by shadows). It is to be noted, however, though the ancients seem to have taken no account of it, that any estimate of latitude based on the length of shadows cast by a gnomon in the sun must be corrected by the addition of half the sun's angular diameter, i.e., about 16′, since the sun is not a luminous point. This is apart from the effect of atmospheric refraction, of which no account is taken in the method cited. [Edd.]

[3] Rhodes and Alexandria differ by about $1\frac{1}{2}°$ in longitude. [Edd.]

[4] Actually the distance is 380 miles. Though the matter is complicated by doubt as to which stade Posidonius used, the estimate of 5,000 stades is, in any case excessive. Eratosthenes, probably by combining his own estimate (see below) of the circumference of the earth (250,000 or 252,000 stades) with his estimate of the difference in latitude ($5\frac{2}{5}°$ or $5\frac{5}{14}°$) between the two places, gave 3,750 stades as the distance. The difference is actually 5° 13′; Posidonius's estimate of $7\frac{1}{2}°$ is quite inaccurate. [Edd.]

[5] I.e., each sign of the zodiac includes 30° and each division here indicated $7\frac{1}{2}°$. [Edd.]

Rhodes and Alexandria is divided into the same number of parts, forty-eight, as the zodiac circle, the segments of it are equal to the aforesaid segments of the zodiac. For, when equal magnitudes are divided into (the same number of) equal parts, the parts of the divided magnitudes must be respectively equal to the parts. This being so, Posidonius goes on to say that the very bright star called Canopus lies to the south, practically on the Rudder of Argo. The said star is not seen at all in Greece; hence Aratus does not even mention it in his *Phaenomena*. But, as you go from north to south, it begins to be visible at Rhodes and, when seen on the horizon there, it sets again immediately as the universe revolves.[1] But when we have sailed the 5,000 stades and are at Alexandria, this star, when it is exactly in the middle of the heaven, is found to be at a height above the horizon of one-fourth of a sign, that is, one forty-eighth part of the zodiac circle.[2] It follows, therefore, that the segment of the same meridian circle which lies above the distance between Rhodes and Alexandria is one forty-eighth part of the said circle, because the horizon of the Rhodians is distant from that of the Alexandrians by one forty-eighth of the zodiac

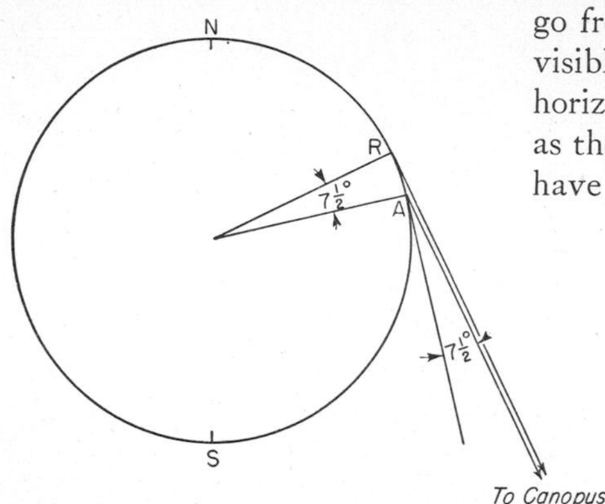

RA (Rhodes—Alexandria) = 5,000 stades. The star is assumed to be so distant that the lines to it from *R* and *A* may be considered parallel.

[1] As a matter of fact, in Posidonius's time Canopus was not only visible at Rhodes on the horizon but reached a maximum elevation of 1¾° (1⅙° if allowance be made for refraction), and remained over the horizon for about 2½ hours. [Edd.]

[2] I.e., 7½°, which is inaccurate. This error is partly offset by the excessive estimate of the distance between Rhodes and Alexandria. The resulting estimate of 240,000 stades for the circumference of the earth is fairly accurate, if the stade in question is one-tenth of a Roman mile (approximately 148.8 meters).

Another measurement of the earth's circumference, 180,000 stades, ascribed to Posidonius by Strabo (p. 157, below), has been held by some to be based on alternative measurement of the distance from Rhodes to Alexandria (3,750 instead of 5,000 stades), and by others to be based on a stade four-thirds as long as that mentioned above (i.e., two fifteenths of a Roman mile) and thus to be exactly equivalent to the first estimate of 240,000 stades. Whether the figure of 180,000 stades adopted by Ptolemy (see p. 180) was understood by him to be equivalent to 240,000 stades on the basis of the smaller stade is a matter of debate. Strabo certainly did not so interpret it (p. 157). In any case the adoption of the smaller figure by Marinus and Ptolemy seems to have affected the calculations of explorers of a later age, among them Columbus. The value of the stade (600 feet) varied with the various measurements of the foot. Lehmann-Haupt in the *Real-Encyclopädie* under "Stadion" gives the history of some seven such standards. See also an article on Posidonius and the circumference of the earth by I. E. Drabkin, *Isis* 34 (1943) 509. [Edd.]

circle. Since, then, the part of the earth under this segment is reputed to be 5,000 stades, the parts (of the earth) under the other (equal) segments (of the meridian circle) also measure 5,000 stades; and thus the great circle of the earth is found to measure 240,000 stades, assuming that from Rhodes to Alexandria is 5,000 stades; but, if not, it is in (the same) ratio to the distance. Such then is Posidonius' way of dealing with the size of the earth.

The method of Eratosthenes[1] depends on a geometrical argument and gives the impression of being slightly more difficult to follow. But his statement will be made clear if we premise the following. Let us suppose, in this case too, first, that Syene and Alexandria lie under the same meridian circle; secondly, that the distance between the two cities is 5,000 stades;[2] and thirdly, that the rays sent down from different parts of the sun on different parts of the earth are parallel; for this is the hypothesis on which geometers proceed. Fourthly, let us assume that, as proved by the geometers, straight lines falling on parallel straight lines make the alternate angles equal, and fifthly, that the arcs standing on (i.e., subtended by) equal angles are similar, that is, have the same proportion and the same ratio to their proper circles—this, too, being a fact proved by the geometers. Whenever, therefore, arcs of circles stand on equal angles, if any one of these is (say) one-tenth of its proper circle, all the other arcs will be tenth parts of their proper circles.

Any one who has grasped these facts will have no difficulty in understanding the method of Eratosthenes, which is this. Syene and Alexandria lie, he says, under the same meridian circle. Since meridian circles are great circles in the universe, the circles of the earth which lie under them are necessarily also great circles. Thus, of whatever size this method shows the circle on the earth passing through Syene and Alexandria to be, this will be the size of the great circle of the earth. Now Eratosthenes asserts, and it is the fact, that Syene lies under the summer tropic. Whenever, therefore, the sun, being in the Crab at the summer solstice, is exactly in the middle of the heaven, the gnomons (pointers) of sundials

---

[1] Eratosthenes, a contemporary of Archimedes in the third century B.C., achieved distinction in mathematics, philosophy, philology, music, astronomy, geography, and in the composition of poetry. In antiquity the nicknames Pentathlos and Beta were applied to him because of his many-sided activity and his failure to achieve complete supremacy in any one field. [Edd.]

[2] Actually the distance is 5,914 stades of 148.8 meters (see p. 150, n. 2). Again, Syene is 3° farther west than Alexandria. Eratosthenes' estimate of 250,000 stades for the circumference of the earth is equivalent to approximately 23,300 miles (on the assumption that the stade of 148.8 meters is meant). It may be noted that the change of Eratosthenes' figure from 250,000 to 252,000, whoever it was who made it, was probably occasioned by a desire to obtain a round number for 1°, viz., 700 stades. [Edd.]

necessarily throw no shadows, the position of the sun above them being exactly vertical; and it is said that this is true throughout a space three hundred stades in diameter.[1] But in Alexandria, at the same hour, the pointers of sundials throw shadows, because Alexandria lies further to the north than Syene. The two cities lying under the same meridian great circle, if we draw an arc from the extremity of the shadow to the base of the pointer of the sundial in Alexandria, the arc will be a segment of a great circle in the (hemispherical) bowl of the sundial, since the bowl of the sundial lies under the great circle (of the meridian). If now we conceive straight lines produced from each of the pointers through the earth, they will meet at the centre of the earth. Since then the sundial at Syene is vertically under the sun, if we conceive a straight line coming from the sun to the top of the pointer of the sundial, the line reaching from the sun to the centre of the earth will be one straight line. If now we conceive another straight line drawn upwards from the extremity of the shadow of the pointer of the sundial in Alexandria, through the top of the pointer to the sun, this straight line and the aforesaid straight line will be parallel, since they are straight lines coming through from different parts of the sun to different parts of the earth. On these straight lines, therefore, which are parallel, there falls the straight line drawn from the centre of the earth to the pointer at Alexandria, so that the alternate angles which it makes are equal. One of these angles is that formed at the centre of the earth, at the intersection of the straight lines which were drawn from the sundials to the centre of the earth; the other is at the point of intersection of the top of the pointer at Alexandria and the straight line drawn from the extremity of its shadow to the sun through the point (the top) where it meets the pointer.[2] Now on this latter angle stands the arc carried round from the extremity of the shadow of the pointer to its base, while on the angle at the centre of the earth stands the arc reaching from Syene to Alexandria. But the arcs are similar, since they stand on equal angles. Whatever ratio, therefore, the arc in the bowl of the sundial has to its proper circle, the arc reaching from Syene to Alexandria has that ratio to *its* proper circle. But the arc in the bowl is found to be one-fiftieth of its proper circle.[3] Therefore the distance from Syene to Alex-

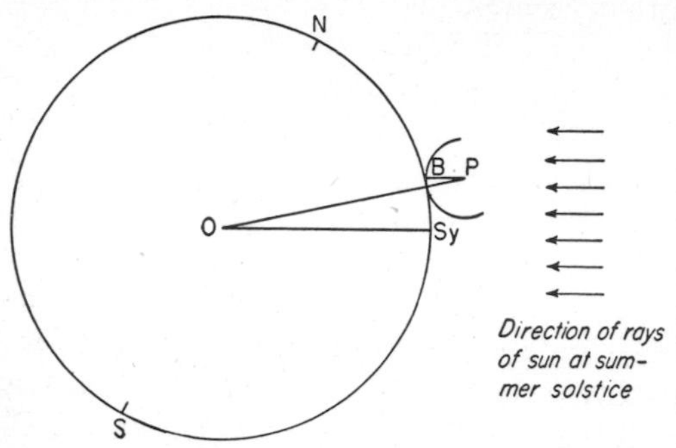

[1] I.e., the sun is not a point of light. [Edd.]

[2] The parallel lines referred to are BP and OSy, the angles those at P and O. [Edd.]

[3] I.e., $7\frac{1}{5}°$. [Edd.]

andria must necessarily be one-fiftieth part of the great circle of the earth. And the said distance is 5,000 stades; therefore the complete great circle measures 250,000 stades. Such is Eratosthenes' method.

EARLY GEOGRAPHERS

Strabo, *Geography* I.1.11. Translation of H. L. Jones (London, 1917)[1]

For the moment what I have already said is sufficient, I hope, to show that Homer was the first geographer. And, as every one knows, the successors of Homer in geography were also notable men and familiar

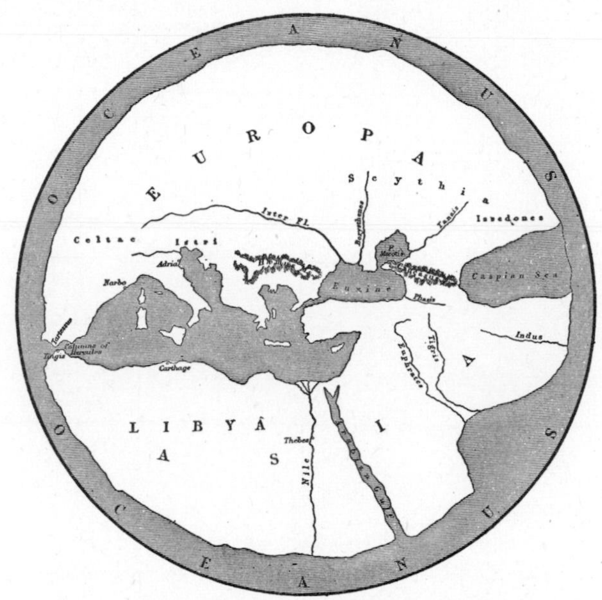

Map of the world according to Hecataeus (ca. 500 B.C.).

with philosophy. Eratosthenes declares that the first two successors of Homer were Anaximander, a pupil and fellow-citizen of Thales, and Hecataeus of Miletus; that Anaximander was the first to publish a geographical map, and that Hecataeus left behind him a work on geography, a work believed to be his by reason of its similarity to his other writings.

[1] In the development of Greek geography there are two chief currents, the mathematical and the descriptive. Mathematical geography commences with the discovery of the sphericity of the earth, perhaps by the early Pythagoreans or Parmenides, and concerns itself with questions of the size of the earth and of the inhabited portion thereof, the zones, the determination of the latitude and longitude of places, and scientific map making. The names of Eudoxus, Aristotle, Eratosthenes, Hipparchus, Marinus, and Ptolemy are outstanding in this connection. But these men, and a long list of others from Homer down, were also interested in descriptive geography. Perhaps the most important extant treatise in this field

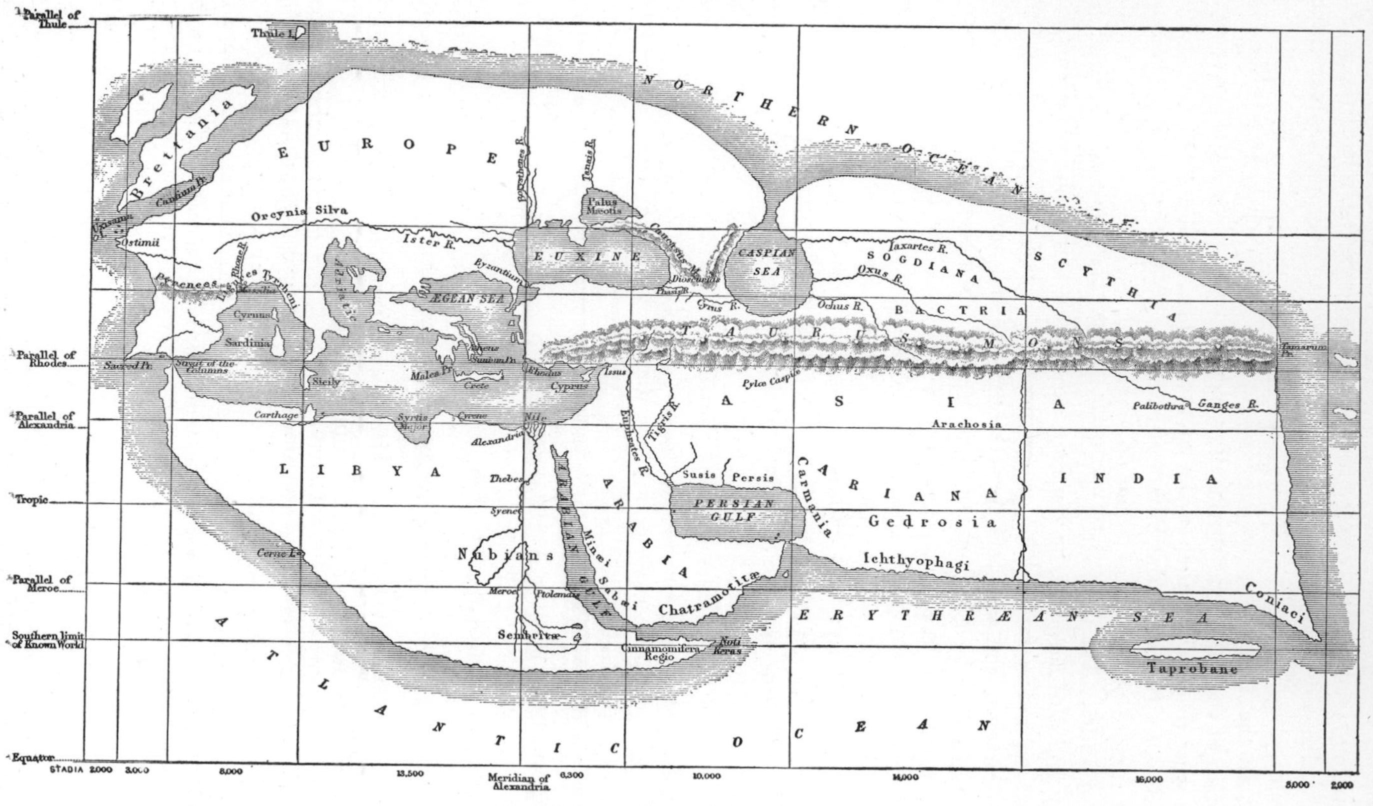

Map of the world according to Eratosthenes

THE MAP OF ERATOSTHENES

Strabo, *Geography* II. 1.1–3. Translation by H. L. Jones

1. In the Third Book of his Geography Eratosthenes, in establishing the map of the inhabited world, divides it into two parts by a line drawn from west to east, parallel to the equatorial line; and as ends of this line he takes, on the west, the Pillars of Heracles, on the east, the capes and most remote peaks of the mountain-chain that forms the northern boundary of India. He draws the line from the Pillars through the Strait of Sicily and also through the southern capes both of the Peloponnesus and of Attica, and as far as Rhodes and the Gulf of Issus. Up to this point, then, he says, the said line runs through the sea and the adjacent continents (and indeed our whole Mediterranean Sea itself extends, lengthwise, along this line as far as Cilicia); then the line is produced in an approximately straight course along the whole Taurus Range as far as India, for the Taurus stretches in a straight course with the sea that begins at the Pillars, and divides all Asia lengthwise into two parts, thus making one part of it northern, the other southern; so that in like manner both the Taurus and the Sea from the Pillars up to the Taurus lie on the parallel of Athens.

2. After Eratosthenes has said that, he thinks he must needs make a complete revision of the early geographical map; for, according to it, he says, the eastern portions of the mountains deviate considerably towards the north, and India itself is drawn up along with it and comes to occupy a more northerly position than it should. As proof of this he offers, first, an argument to this effect: the most southerly capes of India rise opposite to the regions about Meroë, as many writers agree, who judge both from the climatic conditions and from the celestial phenomena; and from the capes on to the most northerly regions of India at the Caucasus Mountains, Patrocles (the man who has particular right to our confidence, both on account of his worthiness of character and on account of his being no layman in geographical matters) says the distance is fifteen thousand stadia; but, to be sure, the distance from Meroë to the parallel of Athens is about that distance; and therefore the northerly parts of India, since they join the Caucasus Mountains, come to an end in this parallel.

3. Another proof which he offers is to this effect: the distance from the Gulf of Issus to the Pontic Sea is about three thousand stadia, if you go

---

is the *Geography* of Strabo (*ca.* 63 B.C.–after A.D. 21), a native of Pontus, who traveled widely, read extensively in the works of his predecessors, and wrote both historical and geographical works. His *Geography*, which is alone extant, consists of 17 books, of which the first two deal with general questions of geographical science, and the rest with the physical, economic, and political geography of Europe (Books III–X), Asia (XI–XVI), and Africa (XVII). There are occasional discussions of matters of geological interest. [Edd.]

towards the north and the regions round about Amisus and Sinope, a distance as great as that which is also assigned to the breadth of the mountains; and from Amisus, if you bear towards the equinoctial sunrise, you come first to Colchis; and then you come to the passage which takes you over to the Hyrcanian Sea, and to the road next in order that leads to Bactria and to the Scythians on beyond, keeping the mountains on your right; and this line, if produced through Amisus westwards, runs through the Propontis and the Hellespont; and from Meroë to the Hellespont is not more than eighteen thousand stadia, a distance as great as that from the southern side of India to the parts round about the Bactrians, if we added three thousand stadia to the fifteen thousand, some of which belonged to the breadth of the mountains, the others to that of India.

ZONES

Strabo, *Geography* II. 2.1–3; 3.1. Translation of H. L. Jones

2. Now let us see what Poseidonius has to say in his treatise on Oceanus. For in it he seems to deal mainly with geography, treating it partly from the point of view of geography properly so called, and partly from a more mathematical point of view. And so it will not be out of place for me to pass judgment upon a few of Poseidonius' statements, some of them now, and others in my discussion of the individual countries, as occasion offers, always observing a kind of standard. Now it is one of the things proper to geography to take as an hypothesis that the earth as a whole is spheroidal[1]—just as we do in the case of the universe—and accept all the conclusions that follow this hypothesis, one of which is that the earth has five zones.

Poseidonius, then, says that Parmenides was the originator of the division into five zones, but that Parmenides represents the torrid zone as almost double its real breadth, inasmuch as it falls beyond both the tropics and extends into the two temperate zones, while Aristotle calls "torrid" the region between the tropics, and "temperate" the regions between the tropics and the "arctic circles." But Poseidonius censures both systems, and with justice, for by "torrid," he says, is meant only the region that is uninhabitable on account of heat; and, of the zone between the tropics, more than half is uninhabitable if we may base a conjecture upon the Ethiopians who live south of Egypt—if it be true, first, that each division of the torrid zone made by the equator is half the whole breadth of that zone and,[2] secondly, that, of this half, the part that reaches to

---

[1] The meaning here is "spherical." [Edd.]

[2] Strabo proceeds to give a definite estimate of the inhabited and uninhabited portions of the torrid zone north of the equator. But, for the division of the zone south of the equator,

Meroë from Syene (which is a point on the boundary line of the summer tropic)[1] is five thousand stadia in breadth, and the part from Meroë to the parallel of the Cinnamon-producing Country, on which parallel the torrid zone begins, is three thousand stadia in breadth. Now the whole of these two parts can be measured, for they are traversed both by water and by land; but the rest of the distance, up to the equator, is shown by calculation based upon the measurement which Eratosthenes made of the earth[2] to be eight thousand eight hundred stadia. Accordingly, as is the ratio of the sixteen thousand eight hundred stadia[3] to the eight thousand eight hundred stadia, so would be the ratio of the distance between the two tropics to the breadth of the torrid zone.[4] And if, of the more recent measurements of the earth, the one which makes the earth smallest in circumference be introduced—I mean that of Poseidonius, who estimates its circumference at about one hundred and eighty thousand stadia[5]—this measurement, I say, renders the breadth of the torrid zone somewhere about half the space between the tropics, or slightly more than half, but in no wise equal to, or the same as, that space. And again, Poseidonius asks how one could determine the limits of the temperate zones, which are non-variable, by means of the "arctic circles," which are neither visible among all men nor the same everywhere. Now the fact that the "arctic circles" are not visible to all could be of no aid to his refutation of Aristotle, because the "arctic circles" must be visible to all who live in the temperate zone, with reference to whom alone the term "temperate" is in fact used. But his point that the "arctic circles" are not everywhere visible in the same way, but are subject to variations, has been well taken.[6]

When Poseidonius himself divides the earth into the zones,[7] he says that five of them are useful with reference to the celestial phenomena; of these five, two—those that lie beneath the poles and extend to the regions that have the tropics as arctic circles—are "periscian";[8] and the two that come next and extend to the people who live beneath the tropics are

he can only assume that a similar estimate applies. By so assuming he reaches a conclusion for the whole zone, in the form of a ratio.

[1] The north and south temperate zones had also the name of summer and winter zones; and hence the summer tropic is the northern tropic.

[2] 250,000 or 252,000 stadia (see pp. 151ff., above). [Edd.]

[3] The distance between the northern tropic and the equator.

[4] That is, 16,800:8,800 = 33,600:17,600. The ratio is 21:11, and the breadth of the torrid zone 17,600 stadia. . . .

[5] See p. 150. [Edd.]

[6] Since the "arctic circle," in the sense here used, varies with the geographical latitude of the observer (see p. 99, n. 1), it is unsuitable as a boundary of the temperate zone. [Edd.]

[7] Seven.

[8] That is, the frigid zones, where the shadows describe an oval in the summer time.

"heteroscian";[1] and the zone between the tropics, "amphiscian."[2] But for purposes of human interest there are, in addition to these five zones, two other narrow ones that lie beneath the tropics and are divided into two parts by the tropics; these have the sun directly overhead for about half a month each year. These two zones, he says, have a certain peculiarity, in that they are parched in the literal sense of the word, are sandy, and produce nothing except silphium and some pungent fruits that are withered by the heat; for those regions have in their neighbourhood no mountains against which the clouds may break and produce rain, nor indeed are they coursed by rivers; and for this reason they produce creatures with woolly hair, crumpled horns, protruding lips, and flat noses (for their extremities are contorted by the heat); and the "fish-eaters" also live in these zones. Poseidonius says it is clear that these things are peculiar to those zones from the fact that the people who live farther south than they do have a more temperate atmosphere, and also a more fruitful, and a better-watered, country.

3. Polybius[3] makes six zones: two that fall beneath the arctic circles, two between the arctic circles and the tropics, and two between the tropics and the equator. However, the division into five zones seems to me to be in harmony with physics as well as geography; with physics, in relation both to the celestial phenomena and to the temperature of the atmosphere; in relation to the celestial phenomena, because, by means of the "periscian" and the "heteroscian" and the "amphiscian" regions (the best way to determine the zones), the appearance of the constellations to our sight is at the same time determined; for thus, by a kind of rough-outline division, the constellations receive their proper variations; and in relation to the temperature of the atmosphere, because the temperature of the atmosphere, being judged with reference to the sun, is subject to three very broad differences—namely, excess of heat, lack of heat, and moderate heat, which have a strong bearing on the organisations of animals and plants, and the semi-organisations[4] of everything else beneath the air or in the air itself. And the temperature of the atmosphere receives its

---

[1] That is, the temperate zones, where the shadows are thrown in opposite directions at noon; the shadow in the northern zone falling north and in the southern falling south.

[2] That is, the torrid zone, where the shadow for any point at noon is north part of the year and south part of the year.

[3] Polybius of Megalopolis in Arcadia (*ca.* 204–122 B.C.), an important figure in the transmission of Greek culture to Rome, wrote a universal history. Of the 40 books the first five and fragments of others are extant. His contributions to geography are in the descriptive and physical rather than in the mathematical aspects of that study. Polybius also exerted considerable personal influence on Roman culture during his stay in Rome as hostage (166–150), when he was a member of the Scipionic circle. [Edd.]

[4] Seeds, for example. [The text and interpretation are uncertain. Edd.]

proper determination by this division of the earth into five zones: for the two frigid zones imply the absence of heat, agreeing in the possession of one characteristic temperature; and in like manner the two temperate zones agree in one temperature, that of moderate heat; while the one remaining is consistent in having the remaining characteristic, in that it is one and torrid in temperature. And it is clear that this division is in harmony with geography. For geography seeks to define by boundaries that section of the earth which we inhabit by means of the one of the two temperate zones. Now on the west and on the east it is the sea that fixes its limits, but on the south and the north the nature of the air; for the air that is between these limits is well-tempered both for plants and for animals, while the air on both sides of these limits is harsh-tempered, because of excess of heat or lack of heat. It was necessary to divide the earth into five zones corresponding to these three differences of temperature; indeed, the cutting of the sphere of the earth by the equator into two hemispheres, the northern hemisphere in which we live and the southern hemisphere, suggested the three differences of temperature. For the regions on the equator and in the torrid zone are uninhabitable because of the heat, and those near the pole are uninhabitable because of the cold; but it is the intermediate regions that are well-tempered and inhabitable. But when he adds the two zones beneath the tropics, Poseidonius does not follow the analogy of the five zones, nor yet does he employ a like criterion; but he was apparently representing zones by the ethnical criteria also, for he calls one of them the "Ethiopic zone," another the "Scythico-Celtic zone," and a third the "intermediate zone."

THE EXISTENCE OF ANTIPODES

Pliny, *Natural History* II. 161–166.[1] Translation of John Bostock and H. T. Riley (London, 1855)

On this point there is a great contest between the learned and the vulgar. We maintain, that there are men dispersed over every part of the earth, that they stand with their feet turned towards each other, that the vault of the heavens appears alike to all of them, and that they, all of them, appear to tread equally on the middle of the earth. If any one should ask, why those situated opposite to us do not fall, we directly ask in return, whether those on the opposite side do not wonder that we do not fall. But I may make a remark, that will appear plausible even to the most unlearned, that if the earth were of the figure of an unequal globe, like the seed of a pine, still it may be inhabited in every part.

[1] On Pliny see p. 389. Even in Pliny's time there seems to have been considerable skepticism among the Romans as to the existence of antipodes, quite apart from the question of the sphericity of the earth. A few centuries later the denial of antipodes was quite general. [Edd.]

But of how little moment is this, when we have another miracle rising up to our notice! The earth itself is pendent and does not fall with us; it is doubtful whether this be from the force of the spirit which is contained in the universe, or whether it would fall, did not nature resist, by allowing of no place where it might fall. For as the seat of fire is nowhere but in fire, nor of water except in water, nor of air except in air, so there is no situation for the earth except in itself, everything else repelling it. It is indeed wonderful that it should form a globe, when there is so much flat surface of the sea and of the plains. And this was the opinion of Dicaearchus,[1] a peculiarly learned man, who measured the heights of mountains, under the direction of the kings, and estimated Pelion, which was the highest, at 1,250 paces perpendicular, and considered this as not affecting the round figure of the globe.[2] But this appears to me to be doubtful, as I well know that the summits of some of the Alps rise up by a long space of not less than 50,000 paces. But what the vulgar most strenuously contend against is, to be compelled to believe that the water is forced into a rounded figure; yet there is nothing more obvious to the sight among the phaenomena of nature. For we see everywhere, that drops, when they hang down, assume the form of small globes, and when they are covered with dust, or have the down of leaves spread over them, they are observed to be completely round; and when a cup is filled, the liquid swells up in the middle. But on account of the subtile nature of the fluid and its inherent softness, the fact is more easily ascertained by our reason than by our sight. And it is even more wonderful, that if a very little fluid only be added to a cup when it is full, the superfluous quantity runs over, whereas the contrary happens if we add a solid body, even as much as would weigh 20 denarii. The reason of this is, that what is dropt in raises up the fluid

[1] Dicaearchus of Messina, a pupil of Aristotle, worked in the fields of geography, political science, cultural history, philology, and philosophy. His estimate of the height of Pelion (1,250 paces = 6,250 Roman feet) is about 20 per cent too large. The estimate of 50 miles as the height of some of the Alps may be a textual error or may refer not to perpendicular height but to the distance to be traversed in ascending from base to summit. [Edd.]

[2] Cf. Simplicius, *Commentary on Aristotle's De Caelo*, pp. 549.32–550.4 (Heiberg): "In comparison with the great size of the earth the protrusion of mountains is not sufficient to deprive it of its spherical shape or to invalidate measurements based on its spherical shape. For Eratosthenes shows that the perpendicular distance from the highest mountain tops to the lowest regions is ten stades. This he shows with the help of dioptras which measure magnitudes at a distance."

Cf. also Theon of Smyrna, p. 124.19 (Hiller): "The perpendicular distance from the highest mountains to the lowest lying lands of the earth is ten stades, as Eratosthenes and Dicaearchus are said to have discovered. These distances are found with the help of dioptras, the instruments which measure magnitudes at a distance." Theon goes on to show how little effect even the highest mountains have on the generally spherical shape of the earth, since 10 stades are, as he points out, only 1/8000 of the earth's diameter. [Edd.]

at the top, while what is poured on it slides off from the projecting surface. It is from the same cause that the land is not visible from the body of a ship when it may be seen from the mast; and that when a vessel is receding, if any bright object be fixed to the mast, it seems gradually to descend and finally to become invisible. And the ocean, which we admit to be without limits, if it had any other figure, could it cohere and exist without falling, there being no external margin to contain it? And the same wonder still recurs, how is it that the extreme parts of the sea, although it be in the form of a globe, do not fall down? In opposition to which doctrine, the Greeks to their great joy and glory, were the first to teach us, by their subtile geometry, that this could not happen, even if the seas were flat, and of the figure which they appear to be. For since water always runs from a higher to a lower level, and this is admitted to be essential to it, no one ever doubted that the water would accumulate on any shore, as much as its slope would allow it. It is also certain that the lower anything is, so much the nearer is it to the centre, and that all the lines which are drawn from this point to the water which is the nearest to it, are shorter than those which reach from the beginning of the sea to its extreme parts. Hence it follows, that all the water, from every part, tends towards the centre, and, because it has this tendency, does not fall.

Aristotle, *Meteorologica* II. 5, 362*a*32–*b*30. Translation of E. W. Webster

There are two inhabitable sections of the earth: one near our upper, or northern pole, the other near the other or southern pole; and their shape is like that of a tambourine. If you draw lines from the centre of the earth they cut out a drum-shaped figure. The lines form two cones; the base of the one is the tropic, of the other the ever visible circle,[1] their vertex is at the centre of the earth. Two other cones towards the south pole give corresponding segments of the earth. These sections alone are habitable. Beyond the tropics no one can live: for there the shade would not fall[2] to the north, whereas the earth is known to be uninhabitable before the sun is

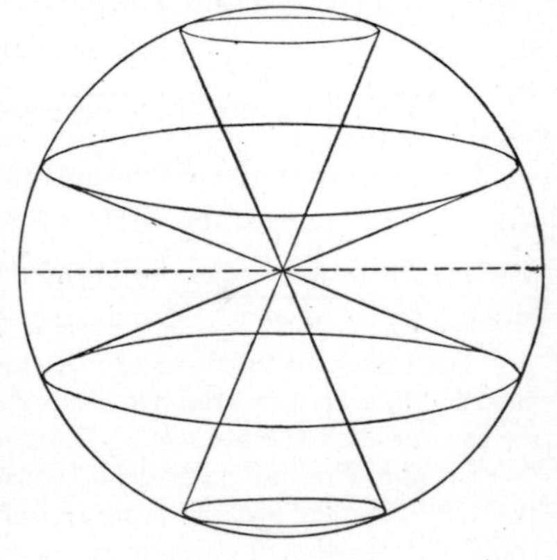

[1] This term, originally used for the boundary circle of circumpolar stars, hence dependent on the latitude of the observer, may here be used of a definite latitude of observation, e. g., that of Athens, or possibly to correspond to what we call the arctic circle, where the sun is, at the summer solstice, visible for 24 hours. [Edd.]

[2] I.e., would not always fall. [Edd.]

in the zenith or the shade is thrown to the south: and the regions below the Bear[1] are uninhabitable because of the cold.

[The Crown, too, moves over this region: for it is in the zenith when it is on our meridian.]

So we see that the way in which they now describe the geography of the earth is ridiculous. They depict the inhabited earth as round, but both ascertained facts and general considerations show this to be impossible. If we reflect we see that the inhabited region is limited in breadth, while the climate admits of its extending all round the earth. For we meet with no excessive heat or cold in the direction of its length but only in that of its breadth; so that there is nothing to prevent our travelling round the earth unless the extent of the sea presents an obstacle anywhere. The records of journeys by sea and land bear this out. They make the length far greater than the breadth. If we compute these voyages and journeys the distance from the Pillars of Heracles to India exceeds that from Aethiopia to Maeotis and the northernmost Scythians by a ratio of more than 5 to 3, as far as such matters admit of accurate statement. Yet we know the whole breadth of the region we dwell in up to the uninhabited parts: in one direction no one lives because of the cold, in the other because of the heat.

But it is the sea which divides as it seems the parts beyond India from those beyond the Pillars of Heracles and prevents the earth from being inhabited all round.[2]

THE ELEMENTS OF GEOGRAPHY

Ptolemy, *Geography* I. 1–5 (Müller)[3]

1. Wherein geography differs from chorography

Geography is the representation, by a map, of the portion of the earth known to us, together with its general features. Geography differs from chorography in that chorography concerns itself exclusively with partic-

[1] I.e., where the Bear is in the zenith when it is on the meridian.

[2] The passage is interesting for its rational conclusion that, if habitability depends on latitude, there are two habitable bands stretching completely around the earth, one north of the equator and the other south. [Edd.]

[3] Claudius Ptolemy made great contributions to mathematical astronomy (pp. 82, 128), optics (p. 271), and musical theory (p. 302). Of the highest importance is his work in mathematical geography, in which he carries on the tradition of Eudoxus, Aristotle, Eratosthenes, Hipparchus, and Marinus of Tyre, and in his turn influences all geographical thought for the next fifteen hundred years. The *Geography* commences with a discussion (Book I) of the general principles of the science, the methods of determining the latitude and longitude of places, and the principles of cartography, in particular the methods of representing a spherical surface on a plane. The rest of the work (Books II–VIII) is almost entirely taken up (to the exclusion of descriptive geography) with cataloguing the latitude and longitude of some 8,000 different places—cities, islands, mountains, river mouths, etc.—in the various parts

ular regions and describes each separately, representing practically everything of the lands in question, even the smallest details, such as harbors, villages, districts, streams branching from the principal rivers, and the like. It is the task of geography, on the other hand, to present the known world as one and continuous, to describe its nature and position, and to include only those things that would be contained in more comprehensive and general descriptions, such as gulfs, large cities and nations, the more important rivers, in short the more significant instances of each type.

The purpose of chorography is the description of the individual parts, as if one were to draw merely an ear or an eye; but the purpose of geography is to gain a view of the whole, as, for example, when one draws the whole head.

Now in the case of the drawings just referred to, the principal parts are of necessity fitted in first, and must be properly proportioned, when seen from a suitable distance, to the spaces on the surface used for the drawing (whether the drawing be of an entire object or of only a part), in order that the representation may be perceived as a whole. It is consequently both reasonable and proper that chorography should represent the smaller particular details and geography the regions themselves with the general features that belong to them. For the principal features of the inhabited earth and those most readily fitted in on the proper scale are the various regions themselves in their proper location; while the principal features of these regions are the peculiar details of the land in question.

Again, chorography deals, for the most part, with the nature rather than with the size of the lands. It has regard everywhere for securing a likeness but not, to the same extent, for determining relative positions. Geography, on the other hand, is concerned with quantitative rather than with qualitative matters, since it has regard in every case for the correct proportion of distances, but only in the case of the more general features does it concern itself with securing a likeness, and then only with respect to configuration.

---

of the earth. There are thus present the materials for maps, maps which, in all probability, Ptolemy himself made or had others make, of the regions of the earth—ten maps for the parts of Europe, four for those of Africa, and twelve for those of Asia. Despite abundant inaccuracies in the determination of latitude and longitude (not merely in the case of less known regions but in the Mediterranean region itself), Ptolemy's work represents a considerable advance over that of his predecessors. The substantial reduction of Marinus's excessive estimates of the extent of the known world in terms of angular latitude and longitude is a case in point. Ptolemy's conclusions depend on his adoption of the estimate of 180,000 stadia for the circumference of the earth. On this see p. 150, n. 2, above.

The passages chosen set forth general principles of geographical science and map making, and a brief summary of the map of the world.

Therefore chorography has need of topography, and no one can be a chorographer unless he is also skilled in drawing. But geography has no such absolute need of topography, for by using mere lines and annotations it shows positions and general outlines. For this reason, while chorography does not require the mathematical method, in geography this method plays the chief part.

For geography must first consider the form of the whole earth as well as its size and its position with reference to the heavens, so that it may be able to tell the size and nature of the known portion of the earth and under what parallel circles of the celestial sphere each place lies. From this it will be possible to learn the length of nights and days, the fixed stars that are overhead and those that at all times are above or below the horizon, as well as all other information that we include in an account of habitable regions.

These are studies that form a part of the most sublime and beautiful theoretic science, for it is with the aid of mathematics that they reveal to man's understanding the heaven itself in its true nature (for we can observe it as it rotates about us); while they represent the earth through a model, since the real earth, which is so great and does not surround us, cannot be traversed in its entirety or in its individual parts by the same men.

2. The foundations of geography

The above account should suffice as a general outline of the purpose of the geographer and wherein he differs from the chorographer.

Since we have now undertaken to map our inhabited world in such a way as to approximate the proportions of the real world as closely as possible, we think that we should explain beforehand that in this procedure the descriptions given by travelers are of the highest importance. For we gain our greatest knowledge from the reports of those travelers through the various regions who have theoretic understanding of geography. Again, their observations and reports may be based either on geometrical or astronomical measurement. Now geometry indicates the relative positions of places by simple measurement of distances, while astronomy uses observational data obtained with astrolabes and sundials. The astronomical method is self sufficient and more accurate, but the geometrical method is rougher and requires the astronomical to supplement it.

Now since in either case it is necessary to determine in what direction the line joining two given places lies (for it is necessary to know not merely the distance between the places but also the direction, whether it be, e.g., to the north or the east or some line intermediate between them), it is impossible to perform this investigation accurately without observation by means of the aforesaid instruments. For with their help the position of the meridian may be obtained at any time and place, and, as a consequence, the direction of distances traversed.

But even if this is found, the measurement in stades does not give us an accurate knowledge of the true distance, since our course rarely coincides with straight lines but, on the contrary, shows many deviations both on land and on sea. Now in the case of journeys on land it is necessary, in order to find the length in a straight line, to subtract from the whole number of stades traversed the excess due to the nature and number of the deviations, as estimated. In the case of sea voyages allowance must be made for variations due to the winds, which do not in general preserve a constant intensity. But even if the distance between two places is accurately measured, this does not give us the ratio of that distance to the whole circumference of the earth or its position with respect to the equator and the poles.

But measurement based on celestial observation gives each of these things accurately. It shows how great are the arcs mutually intercepted by the parallel circles and the meridians (i.e., the arcs of the meridians falling between the parallel circles and the equator, and the arcs of the equator and the parallels falling between the meridians); it also shows how great an arc the two places in question intercept on a great circle of the earth drawn through them. And this method does not require us to compute the number of stades in order to obtain the ratio of the various parts of the earth to one another and for the general procedure of drawing our map. For it is sufficient to take the circumference of the earth, divided into as many parts as desired, and to show how many such parts of a great circle of the earth there are in each of the several distances examined.[1]

But naturally this method does not suffice for the division of the whole circumference or parts of it into actual distances familiar to us by our own measurements. And for this reason alone it has become necessary to compare some straight distance on the earth with the similar arc of a great circle on the celestial sphere, and, obtaining by observation the proportion of this arc to the whole circle, and, by measurement based on some given part, the number of stades in the terrestrial distance under this arc, to find the number of stades in the whole circumference.[2]

---

[1] I.e., distances may be represented merely in degrees of arc independently of the number of stades to a degree.

[2] A given terrestrial distance, lying theoretically on the arc of a great circle, is measured in stades; the ratio of this arc to the whole circumference is then determined by astronomical observation, whence the measurement of the whole circumference in stades may be found. The measurements by Eratosthenes and Posidonius are examples of this method (see pp. 149–153, above).

For we assume on the basis of mathematics[1] that the surface of the earth and sea taken together is substantially spherical, having as center the center of the celestial sphere, so that every plane passed through the center cuts the aforesaid surfaces[2] in great circles, and the angles subtended at the center intercept similar arcs on these circles.[3] Now, though the number of stades in terrestrial distances along a straight line can be obtained by measurement, the ratio of this distance to the whole circumference of the earth can not be obtained by measurement because of the impossibility of a comparison.[4] This ratio can, however, be obtained from the fact that the arc subtended on the celestial sphere is similar.[5] For it is possible to obtain the ratio of this arc to the whole celestial circumference. And the ratio of a similar arc on the earth to the great circle of the earth is the same.

3. A method of finding the number of stades in the circumference of the earth, given any distance in stades along a straight line, not necessarily on the same meridian; and conversely

Now our predecessors sought not only a rectilinear terrestrial distance (so that it would constitute an arc of a great circle)[6] but one lying in the plane of a single meridian. Observing by sun dials the points overhead[7] at the two places between which the distances lay, they immediately obtained an arc on the meridian joining these places equal to the arc over which a journey would be made, because the places lay, as we said, on the same plane.[8] Moreover, straight lines drawn through the places in question to the points directly overhead met each other when produced, and the common center of the circles was the point of intersection of these lines. Now they took the ratio of the arc between the points overhead to the whole circle passing through the celestial poles as equal to the ratio of the terrestrial distance in question to the circumference of the whole earth.

But even if the great circle on the earth's surface that includes the measured distance does not pass through the poles, but is any great circle,

---

[1] See the proofs of Aristotle and Archimedes, p. 236.

[2] I.e., the surfaces of the terrestrial and celestial spheres.

[3] Since any central angle intercepts similar arcs on concentric circles, i.e., arcs equal in circular, but not in linear, measure.

[4] I.e., a comparison between the terrestrial distance and the terrestrial circumference.

[5] I.e., equal in degrees to the arc represented by the distance in question on the terrestrial sphere.

[6] The "rectilinear" or shortest distance between two points on the surface of a sphere is the arc of the great circle joining the points.

[7] I.e., the angular distance from the sun (e.g., at noon of the equinox) to the zenith.

[8] I.e., on the same plane passing through the earth's diameter, or, in other words, on the same meridian. The arc thus measures the difference in latitude between the two places.

we have demonstrated that our problem may still be solved by similar observations of the elevation of the pole at the terminal points and the position which the terrestrial distance has with reference to each meridian. This we have done by constructing an instrument[1] for viewing the heavenly bodies. With this instrument we easily obtain many very useful facts, but in particular (1) on any day or night the elevation of the north pole at the point from which the observation is made;[2] (2) at every hour not only the position of the meridian but also the position with reference to it of any land route, i.e., the angle made by the great circle drawn through the route and the meridian passing through the zenith.[3]

Using these angles we may alike determine both the required arc, with the sole help of the instrument, and also the arc cut off by two meridians on parallels other than the equator. And so, with this method, by measuring only one straight distance on the earth we can find the number of stades in the whole circumference, and furthermore, knowing this we can then find without actual measurement the number of stades in any other distance. And we can do this even when the distances are not entirely straight[4] and do not lie on the same meridian or parallel of latitude. It is necessary, however, that the precise direction of the distance be carefully observed, as well as the latitude of the termini, for it is from the ratio that the arc representing the distance in question bears to the great circle that the number of stades in such distance can be easily computed, given the circumference of the whole earth.

---

[1] The reference is to the ring astrolabe described in the *Almagest* (see p. 134, above).

[2] This would be the latitude of the observer.

[3] This angle together with the readily obtained latitude of the two places would, theoretically, determine the distance, in circular measure, of the route in question. The practical difficulties of measuring longitude in antiquity would be reflected in the application of the method.

[4] Perhaps the reference is to places separated by mountains. In any case, however, Ptolemy would theoretically be able, either by trigonometry or by using globe and compass, to find the distance, along an arc of a great circle, between any two places whose latitude and longitude have been determined. But the chief difficulty lay in the determination of the longitude of a place before instantaneous communication became possible. The best available method was that of noting the time when a lunar eclipse was seen in two different cities. Since the eclipse is seen at about the same time in both places, the difference in the *local* time (i.e., the number of hours after sunrise or sunset) will indicate the difference in longitude, at the rate of 15° for each hour. Because of the unreliability of timekeeping instruments in antiquity the method described furnished only a rough approximation, and reports of distances as given by travelers came to be the chief basis of determining longitude. There are also references to the use of royal "pacers" in Ptolemaic Egypt for the purpose of estimating distances. The use of the hodometer (p. 342) in practice is doubtful. Of course, for short distances surveying was done with such instruments as the dioptra and the groma.

The method of determining the distance between Rome and Alexandria is taken up by Hero of Alexandria (*Dioptra*, 35); see O. Neugebauer, *op. cit.* (p. 136, n. 2, above).

4. Actual observations of phenomena are to be preferred to reports of travelers

This, then, being the case, if those who have traveled over the several countries had made use of this type of observation, it would have been possible to make a completely accurate map of the inhabited earth. But Hipparchus alone gave us the elevation of the north pole [i.e., the latitude] for a few cities of the multitude that must be included in a description of the earth, and indicated places that lie on the same parallels, while some of his successors recorded some of the cities that lie opposite each other (not those equally distant from the equator, but merely those lying on the same meridian), on the basis of sailings made between these places with the aid of north or south winds. Again, most distances, especially those to the east or west, were given quite roughly, not by reason of the carelessness of those who made these investigations but probably because of the fact that the more strictly mathematical method of observation had not yet been adopted, and it had not been considered important to record sufficiently numerous observations of lunar eclipses on the same occasion in different places. I refer to an eclipse like that observed at Arbela at the fifth hour and at Carthage at the second, from which the distance of the two places from each other eastward or westward could have been determined in equinoctial hours.[1] For these reasons it would be proper for one who is to make a map of the earth in accordance with these principles to use the data obtained by more accurate observations as the foundation for the map, and to fit in with these data those obtained from other sources, until the positions indicated in the latter data both in relation to one another and in relation to the fundamental data are in accord, in so far as is possible, with the more accurate traditions.

5. Attention must be paid to more recent accounts because of changes in the earth with passage of time

The construction of the maps, therefore, would be best accomplished on the basis of the principles described. But in the case of regions which are not completely known either because of their great size or because

[1] As a matter of fact, however, the case illustrates the difficulty of accurately determining longitude by lunar eclipses or any other means in the absence of dependable clocks. The three hours' difference corresponds to a difference in longitude of 45°, more than 11° greater than the actual difference between Carthage and Arbela. The eclipse, which is referred to by many ancient authors, is probably that of Sept. 20, 331 B.C. Cf. Pliny, *Natural History* II. 180: "At Arbela, upon the victory of Alexander the Great, the moon is said to have been eclipsed at the second hour of the night, whereas the eclipse took place in Sicily as the moon was rising." The reference to the hour of occurrence is fairly accurate in Pliny's account but not in Ptolemy's, according to modern computations. (See H. v. Mžik, *Des Klaudios Ptolemaios Einführung in die darstellende Erdkunde*, pt. I [Vienna, 1938], p. 21, n. 3.) But the eclipse seems to have preceded the battle by 11 days.

they do not always present the same aspect, later reports give us a more accurate account each time, and this is true in connection with geography, too. For the various historical accounts themselves agree that many divisions of the inhabited portion of our earth have not yet come to be known because of the difficulties occasioned by their size. Again, other divisions have not been accurately described because of carelessness on the part of those who received reports, and some parts are themselves different from what they were formerly because of destruction or changes that have taken place in them. For these reasons it is necessary that here, too, we pay heed in general to the last reports that come to us, distinguishing that which is worthy of belief and that which is not, both in our use of the new data and in our judgment of what had previously been reported.

PRINCIPLES OF CARTOGRAPHY

Ptolemy, *Geography* I. 21–24 (Müller)

21. Precautions to be taken in the case of a map drawn on a plane

For the aforesaid reasons it would be well to keep the lines which represent meridians straight, and to make those which represent parallels of latitude arcs of circles drawn about one and the same center. Now from this center, taken at the north pole, the straight meridian lines will have to be drawn in such a way that above all the resemblance, both as to form and appearance, to a spherical surface may be preserved. For the meridian lines cut the parallels of latitude here, too,[1] at right angles and at the same time meet at the same common pole.

Now since it is impossible to preserve the spherical proportions through all the parallels of latitude, it would be sufficient to do so for the parallel through Thule and for the equator. In this way the boundaries which encompass our latitudes may be accurately proportioned, and the parallel through Rhodes on which the most numerous measurements of longitudinal distance have been made may be divided in accordance with the ratio it bears to the meridian circle, that is, in accordance with the ratio of approximately 4 to 5 measured along equal arcs.[2] This is Marinus's method. Thus the length of the better known part of the inhabited earth would be in proper relation to[3] its breadth. The method of doing this we shall make clear after we have set forth how a map may be made on a sphere.

[1] I.e., as on the spherical surface itself. Admittedly a strained interpretation of τε πάλιν, the reading of the manuscripts. But Müller's conjecture, τὸ πολύ, is hardly an improvement.

[2] I.e., an arc representing 5° of longitude on the parallel through Rhodes is equal in length to an arc representing 4° of latitude, or, what amounts to the same thing, the parallel circle through Rhodes is four-fifths as long as the equator or any great circle on the earth.

[3] Perhaps what is meant is that the parallel through Rhodes over the entire length of the inhabited world (180°, equal to 144° on a great circle) would be about twice as long as the meridian distance between the parallels through Thule and Cattigara (see pp. 171 f.).

22. How the inhabited earth may be represented on a sphere

The decision as to the size of the sphere will depend on the number of details the mapmaker wishes to include. And this, in turn, will depend on his skill and ambition, for as the sphere is increased the amount of detail and the accuracy of the map will likewise be increased. But whatever the size of the sphere is to be, we determine its poles and then carefully attach to it through the poles a semicircle raised just enough from the spherical surface to keep it from rubbing against the sphere as it rotates. This semicircle should be narrow so that it may not cover up many places. One of its edges should extend exactly between the points which mark the poles so that we may draw the meridian lines with that edge. We shall divide the latter into 180 parts and mark the numbers beginning with the midpoint of the semicircle where it will intersect the equator. Now similarly we shall draw the equator, and divide one of the semicircles comprising it into 180 parts, again setting the numbers opposite these parts, beginning with the boundary through which we are to draw the westernmost meridian line.

Now we shall make our map on the basis of the tables of degrees of longitude and latitude for each of the places to be represented, using the divisions on the semicircles, viz., the equator and the movable meridian. We turn the latter to the degree of longitude indicated, that is, to the division of the equator corresponding to that number, and we measure the latitudinal distance from the equator according to the divisions on the meridian. We place a mark corresponding to the indicated number of degrees just as we make a star map on a solid sphere.

Similarly, it will be possible to draw meridians at intervals of as many degrees of longitude as we wish, using as a ruler the aforesaid divided edge of the semicircular ring. It will also be possible to draw parallels of latitude at as great intervals as we wish, by placing the marker next to that number on the edge which indicates the latitude desired, and turning the marker and the semicircular ring as far as the meridians that indicate the limits of the known portion of the earth.

23. Table of meridians and parallels drawn on the map

Now these outermost meridians encompass an interval of twelve hours, as we have shown above.[1] And the parallel which marks the southern boundary will be drawn just as far south of the equator[2] as the parallel through Meroe is north of the equator. Now we have thought it con-

[1] In a previous section Ptolemy indicates the total longitude of the inhabited world as approximately 180° (12 hours). Marinus had made this 225° (15 hours). Actually the longitude from the Fortunate Isles to Sera is about 127°.

[2] I.e., 16°25′ south latitude. This figure is an approximation to the latitude where the longest day of the year is 13 hours. Meroe itself is about 16°56′ north latitude.

venient to draw meridians at intervals of a third part of an equinoctial hour, i.e., at every five of the divisions on the equator,[1] and to draw parallels north of the equator, so that

1. The first differs by a quarter of an hour from the equator[2] and is 4°15′ distant from it along the meridian, an approximation based on geometrical demonstrations;
2. The second differs by half an hour, and is 8°25′ distant;
3. The third differs by three-quarters of an hour and is 12°30′ distant;
4. The fourth differs by one hour and is 16°25′ distant, and is drawn through Meroe;
5. The fifth differs by 1¼ hours and is 20°15′ distant;
6. The sixth, which is under the summer tropic, differs by 1½ hours and is 23°50′ distant, and is drawn through Syene;
7. The seventh differs by 1¾ hours and is 27°10′ distant;
8. The eighth[3] differs by 2 hours and is 30°20′ distant;
9. The ninth differs by 2¼ hours and is 33°20′ distant;
10. The tenth differs by 2½ hours and is 36° distant, and is drawn through Rhodes;
11. The eleventh differs by 2¾ hours and is 38°35′ distant;
12. The twelfth[4] differs by 3 hours and is 40°55′ distant;
13. The thirteenth differs by 3¼ hours and is 43°5′ distant;
14. The fourteenth[5] differs by 3½ hours and is 45° distant;
15. The fifteenth[6] differs by 4 hours and is 48°30′ distant;
16. The sixteenth differs by 4½ hours and is 51°30′ distant;
17. The seventeenth differs by 5 hours and is 54° distant;
18. The eighteenth differs by 5½ hours and is 56°10′ distant;
19. The nineteenth differs by 6 hours and is 58° distant;
20. The twentieth differs by 7 hours and is 61° distant;
21. The twenty-first differs by 8 hours and is 63° distant, and is drawn through Thule.[7]

---

[1] I.e., at intervals of 5°.

[2] I.e., the longest day of the year at this latitude is 12¼ hours. In the *Almagest* Ptolemy gives the procedure for determining the length of the longest day of the year at a given place from the latitude of the place (see p. 84, above; see also H. v. Mžik, *op. cit.*, pp. 90–92). The division of the map by parallels corresponding to the length of the longest day was used by Eratosthenes and, in greater detail, by Hipparchus and Ptolemy. Cf. the system of "climata" in Strabo II. 5.34–43.

[3] Some manuscripts add "drawn through Alexandria," though this is probably a later interpolation (see Müller *ad loc.*).

[4] Some manuscripts add "through the Hellespont."

[5] Some manuscripts add "through the middle of Pontus."

[6] Some manuscripts add "through the Borysthenes."

[7] In the *Almagest* (II. 6) there are, including the equator, 33 parallels of latitude from 0° to the arctic circle (66°8′40″). The two works show small divergences in the computation of latitudes.

Another parallel will be drawn south of the equator and distant one-half hour from it. This parallel, through Cape Rhaptus and Cattigara, will be approximately the same distance in degrees from the equator as is the parallel on the other side, viz., 8°25′.

24. A method for representing the inhabited earth on a plane to correspond to its position on a sphere

I

In drawing a map on a plane surface our method of obtaining the proper proportions of the outermost parallels will be as follows. We

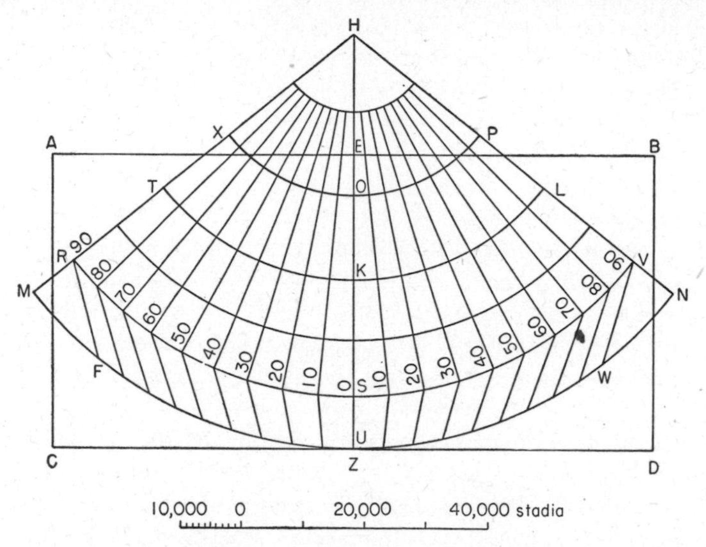

shall construct a rectangular table *ABDC* with side *AB* approximately double side *AC*. Let us assume that *AB* in the upper portion will represent the northern part of the map. Then, bisecting *AB* with line *EZ* at right angles to it, we shall apply congruent to *AB* a straight ruler in such a way that the [perpendicular] line passing through the midpoint of its length forms a single straight line with *EZ*, that is line *EH*.[1] Let *EH* contain 34 units and *HZ* 131⁵⁄₁₂. With *H* as a center and a radius of 79 units along *HZ* we shall describe an arc *TKL* representing the parallel through Rhodes.

[1] In the projection here described the habitable part of the world is represented on a plane made from the surface of a right cone tangent to the original sphere at the parallel through Rhodes (36° north latitude). The vertex is the center from which the parallels of latitude are drawn, while the meridians are straight lines intersecting at the vertex. For a mathematical treatment of the projection and an explanation of how the numbers are arrived at, see H. v. Mžik, *op. cit.*, pp. 93–99.

To obtain the limits of longitude, a distance of six hours on either side of *K*, we shall take a distance of four units on *HZ*, the line representing the central meridian, as equal to five units on the parallel through Rhodes, because of the ratio of five to four between the great circle and the parallel through Rhodes.[1] Taking 18 such units on either side of *K* along arc *TKL*, we shall have the points through which it will be necessary to draw from *H* the meridians representing intervals of one-third of an hour, including the two boundary meridians *HTM* and *HLN*.

The parallel through Thule will accordingly be drawn 52 units distant from *H* along *HZ*, viz., *XOP*, and the equator, similarly, 115 units distant from *H*, viz., *RSV*. The southernmost parallel *MUN*, opposite that through Meroe, will be 131$\frac{5}{12}$ units distant from *H*.

Therefore the ratio between *RSV* and *XOP* will be 115 to 52, corresponding to the ratio between these parallels on a sphere, since *HS* is to *HO* as 115 to 52, and arc *RSV* is to arc *XOP* as *HS* is to *HO*.

The meridian distance *OK*, i.e., the distance from the parallel through Thule to that through Rhodes, will be found to be 27 units; the distance *KS*, from the parallel through Rhodes to the equator, 36 units; the distance *SU*, from the equator to the parallel lying opposite that which passes through Meroe, 16$\frac{5}{12}$ units. Again, the distance *OU*, the breadth of the known portion of the earth, is 79$\frac{5}{12}$ or in whole numbers 80 units, and the parallel *TKL*, the mean longitudinal distance, will contain 144 of these units, which accords with the basic assumptions underlying our demonstrations. For the ratio of the breadth [of our inhabited portion of the world] to its length along the parallel through Rhodes, 40,000 stades to 72,000, is substantially the same.[2] Now we shall draw the other parallel circles, if we desire, with *H* again as center and with radii equal to the distances from *S* by which these parallels are separated from the equator, as has been set forth.

Now it will be possible for us to draw the lines which represent meridians not all the way to the parallel circle *MUN* but only as far as the equator *RSV*, and then, dividing *MUN* into 180 parts equal in size and number to those on the parallel through Meroe, to draw to the divisions on *MUN* from the divisions on the equator intermediate straight lines representing the meridians, so that their position on the other side of the equator and sloping toward the south may be indicated by the change in direction, e.g., lines *RF* and *VW*.

Furthermore, in order that we may more easily indicate the position of the places to be represented, we shall make a narrow ruler equal in

---

[1] The length of the parallel through Rhodes is four-fifths that of the equator (144:180).

[2] I.e., 80:144 = 40,000:72,000. The measurement in stades is based on the estimate of 180,000 stades as the circumference of the earth (see p. 150, n. 2, above).

length either to *HZ* or merely to *HS*. Fixing this ruler at *H* as a pole, so that as it turns over the whole length of the map we may carefully apply one of its sides to the straight lines representing the meridians (the side is so cut out as to pass directly through the pole), we shall divide this side into 131⅚₁₂ units extending over *HZ* or 115 extending merely over *HS*, and shall mark the numbers beginning with the mark at the equator. And from these numbers it will be possible also to indicate the parallels, so that it will not be necessary on the map itself to divide the meridian *OU* into all its parts and to mark the numbers along the divisions. For to do this would make indistinct the names of the places near that meridian.

And so, after dividing the equator, too, into the 180 parts representing the twelve hours, and after indicating the numbers beginning with the meridian furthest to the west, we shall bring the side of the ruler to the required degree of longitude. Then, using the divisions on the ruler we shall arrive at the required position with respect to latitude and make the proper mark in each case, as was pointed out in the case of places on the sphere.

II

But we can make a still truer and better proportioned map of the inhabited earth on a plane surface if we make the meridian lines appear as they do on the surface of a sphere when the axis of vision[1] passes through the center of the sphere and that point of the sphere at which the meridian bisecting the length of the known portion of the earth and the parallel bisecting its breadth intersect. In this way the opposite boundaries are symmetrical with respect to the vision and are so perceived.[2]

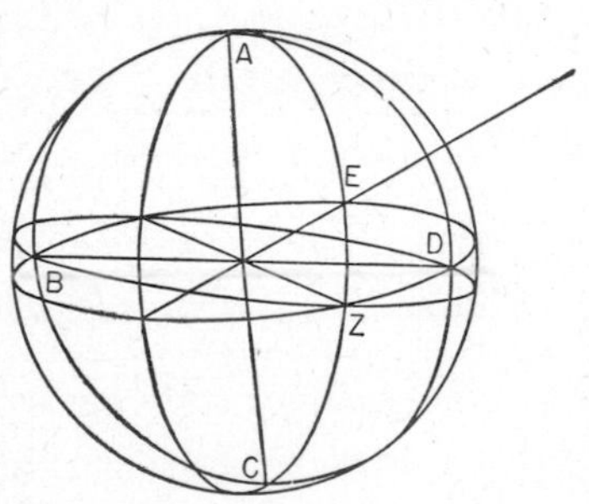

Now, in the first place, in order to find the amount of inclination of the parallel circles to the plane that passes through the aforesaid intersection and the center of the sphere and is perpendicular to the central meridian of longitude, let the great circle which forms the boundary of the visible hemisphere be *ABCD*. And let the semicircle of the meridian which bisects this hemisphere be *AEC*, and the point of inter-

[1] I.e., the axis of the cone containing the visual rays.

[2] The projection here described is, as Ptolemy points out, more accurate but more difficult than the preceding one. Only one of the meridians—that bisecting the length of the map—is a straight line. For mathematical details see H. v. Mžik, *op. cit.*, pp. 100–104. The projections described in this section were not used by Ptolemy in the 26 special maps of the various parts of the oecumene; in these maps the meridians of longitude were drawn parallel to one another and perpendicular to the parallels of latitude.

section of this semicircle and the parallel that bisects the breadth of the known portion of the earth be the point *E*, directly in the axis of vision. Now let a semicircle *BED* of a great circle be drawn through *E* at right angles to *AEC*. The plane of this semicircle will clearly include the axis of vision.[1] Cutting off arc *EZ* of 23°50′—for this is the distance from the equator to the parallel through Syene, which parallel approximately bisects the breadth of the inhabited portion of the earth—draw through *Z* a semicircle, *BZD*, of the equator. Therefore the plane of the equator and the planes of the other parallels will appear inclined to the plane [*BED*] which includes the axis of vision by an amount equal to arc *EZ*, that is, by 23°50′.

Now let the straight lines *AEZC* and *BED* be considered as representing arcs, *BE* bearing to *EZ* the ratio of 90 to $23\frac{5}{6}$. With *CA* produced, let the center about which the arc of circle *BZD* will be drawn fall at *H*. And let it be required to find the ratio of *HZ* to *EB*. Draw line *ZB*, bisect it at *T*, and draw *TH* which is, of course, perpendicular to *BZ*.

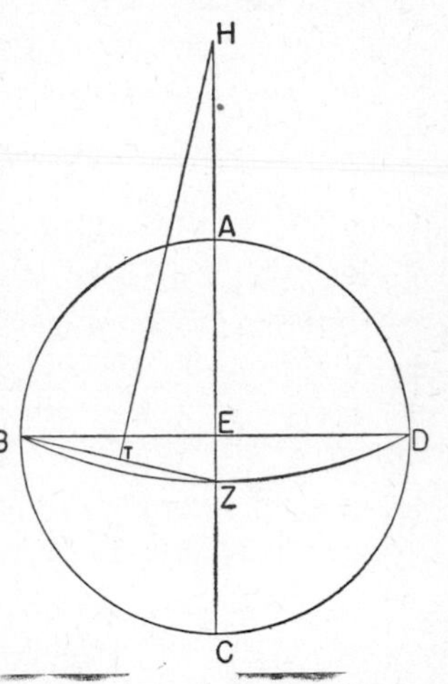

Since, then, *EZ* has been assumed to contain $23\frac{5}{6}$ units of the kind of which *BE* contains 90, the hypotenuse *BZ* will contain $93\frac{1}{10}$. Again, angle *BZE* will contain $150\frac{1}{3}$ units of the kind of which two right angles contain 360, and the remaining angle *THZ* will contain $29\frac{2}{3}$ such units. Hence the ratio of *HZ* to *ZT* is $181\frac{5}{6}$ to $46\frac{11}{20}$. And if line *TZ* contains $46\frac{11}{20}$ units, *BE* will contain 90 such units. Therefore line *HZ* will contain $181\frac{5}{6}$ units of the kind of which *BE* contains 90 and *ZE* $23\frac{5}{6}$. And *H* will be the point from which all the parallels will be drawn on the map made on a plane.

Having found these data, let *ABDC* [p. 176] be the surface of the map, with *AB* double *AC*, *AE* equal to *EB*, and *EZ* perpendicular to *AE* and *EB*. Let a line equal to *EZ* be divided into the 90 parts that represent a quadrant of a circle. If we lay off *ZH* containing $16\frac{5}{12}$ such parts, *HT* $23\frac{5}{6}$, and *HK* 63, and consider *H* as on the equator, then *T* will be on the point through which will be drawn the parallel through Syene, approximately bisecting the breadth of the known portion of the earth, *Z* the point through which will be drawn the parallel forming its southern boundary and lying opposite the parallel through Meroe, and *K* the point through which will be drawn the parallel forming its northern boundary and passing through the island of Thule.

[1] Since it includes two points of this line, *E* and the center of the sphere.

Now prolonging line *ZE* and taking *HL* on that line equal to 181⅚ units, or merely 180 (for the difference on the map will be negligible), with *L* as center and radii *LZ*, *LT*, and *LK*, respectively, we shall draw arcs *PKR*, *XTO*, and *MZN*. Thus the proper degree of inclination of the parallel circles to the plane which includes the axis of vision[1] will be preserved, since in this case too the axis of vision must be directed toward *T* and must be perpendicular to the plane of the map so that here too the opposite boundaries of the map may be symmetrically envisioned.

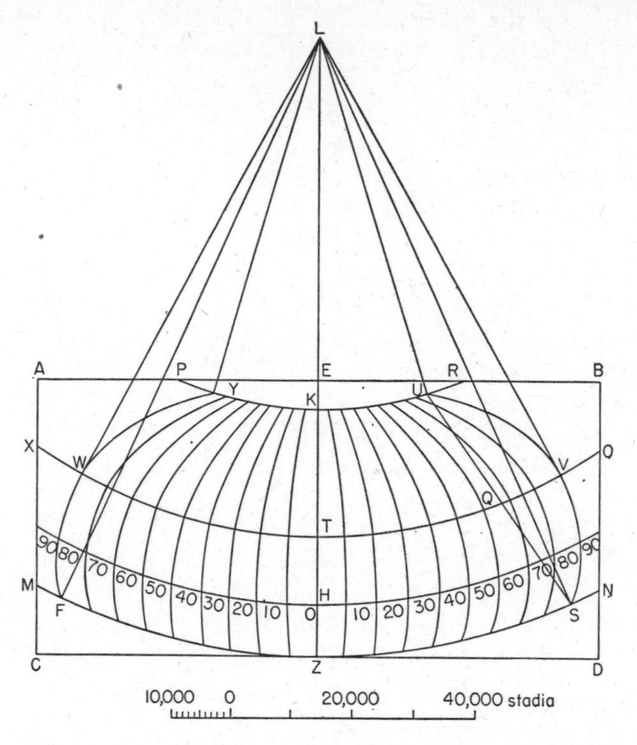

Now in order that the length too may properly correspond to the breadth of the inhabited portion of the earth (for on the assumption that the length of the great circle of the sphere is 5 units, the parallel circle through Thule is 2¼ units, that through Syene 4 $\frac{7}{12}$ units, and that through Meroe 4⅚ units), we must place on either side of the straight line representing meridian *ZK* eighteen meridians at intervals of one-third of an equinoctial hour to complete the total number of meridian semicircles included within the whole extent of longitude. Thus we shall make divisions on each of the three aforesaid parallels at intervals of five degrees, the number of degrees in one-third

[1] I.e., the plane perpendicular to the surface of the map and determined by the eye and the central meridian.

of an equinoctial hour.[1] In the case of these parallels our divisions from *K* will be at intervals of 2¼ units, from *T* at intervals of 4⁷⁄₁₂ units, and from *Z* at intervals of 4⅚ units, line *EZ* being assumed to contain 90 such units.

Then having drawn arcs in each case through the three corresponding points, which will be sufficient to represent all the meridians, among them arcs *SVU* and *FWY* which bound the entire extent of longitude, we shall also add arcs to represent the other parallels using as center *L* and as radii the intervals along *ZK* corresponding to the distances of the respective parallels from the equator.

It is immediately clear that such a map as we have just described corresponds more closely to the form of a map on a sphere than does the map previously described. For in the case of a sphere too, when it remains unmoved and is not rotated (a necessary condition in the case of a map on a plane), and the axis of vision is directed to the center of the map, only one meridian, the central one, falling in the plane which includes the axis of vision, can appear as a straight line. All the other meridians on either side of this one appear bent concavely towards it, the more so the further they are removed from it. Now this curvature will be preserved to the proper degree in the case of the map on the plane as it was just described. So also will the proper proportion of the parallels of latitude to one another be preserved, not alone in the case of the equator and the parallel through Thule, as in the map previously described, but also, as nearly as possible, in the case of the other parallels, as will be evident upon trial. Again, the proper proportion of the entire latitude to the entire longitude will be preserved, not merely, as in the map previously described, on the parallel drawn through Rhodes, but with substantial accuracy on all the parallels.

For if in this case too, as in the former map, we draw *SQU* as a straight line, arc *TQ* will, clearly, bear a smaller ratio to arcs *ZS* and *KU* than is the proper ratio on this map, i.e., the ratio that was taken along the whole arc *TV*, determined according to the equator. And if we make this arc [*TQ*] the proper length in relation to *KZ*, the extent of latitude of the known portion of the earth, arcs *SZ* and *KU* will be longer than they should properly be in relation to *ZK*, as indeed arc *TV* is. If, again, we keep arcs *ZS* and *KU* of the proper length to correspond to arc *ZK*, arc *TQ* will be smaller than it should be in relation to arc *KZ*, just as it is smaller than arc *TV*.

In these respects, therefore, this method has the advantage over that previously described, but is inferior in point of convenience of drawing. For in the case of the map previously described it was possible, by applying and pivoting the ruler, after only one of the parallels had been drawn and divided, to fix the position of every place. In the map just described, however, this easy procedure is not available because of the curvature of

[1] Each of the 36 meridians drawn will be determined by three corresponding points on the parallels.

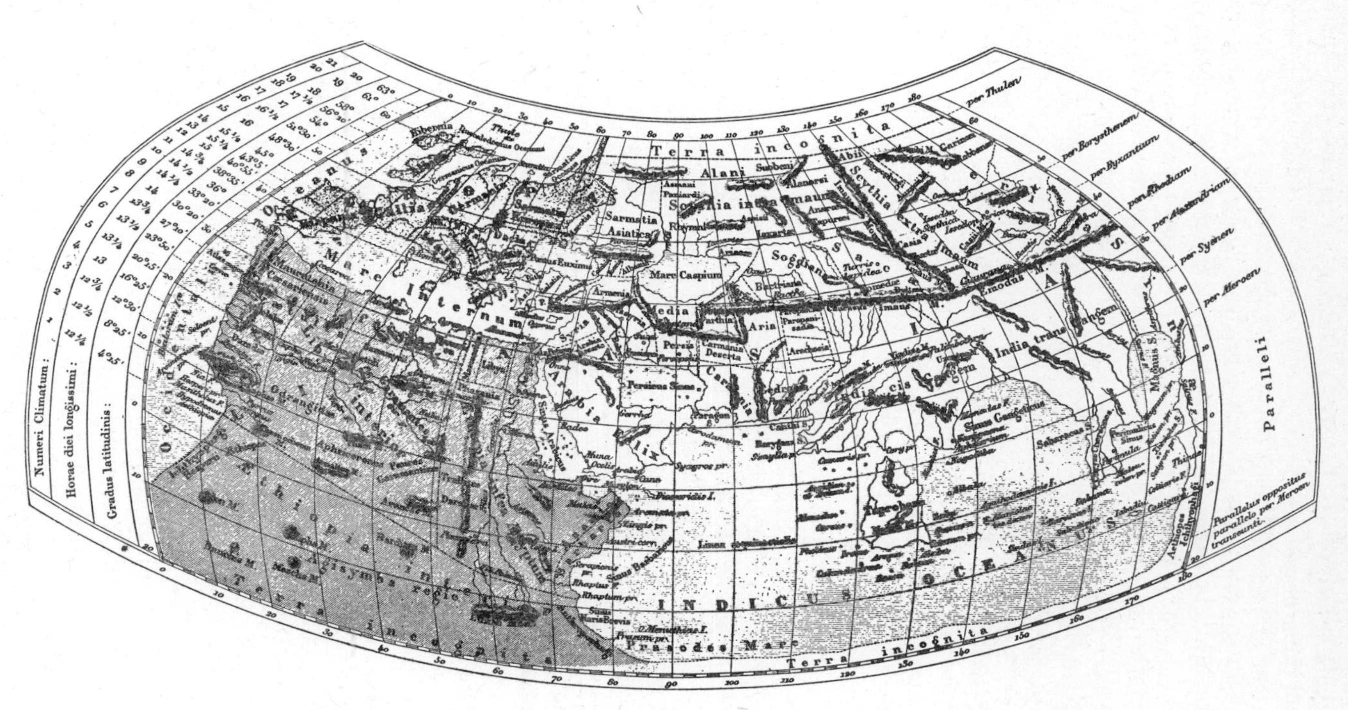

Map of the world based on Ptolemy's description. Note the latitude and length of the longest day at the various "climata."

the meridian lines with respect to the central one. It is necessary, therefore, to draw all the arcs and, in the case of positions falling inside the boxed off areas[1] to get at them by a calculation of their relation to all four sides enclosing the area in question, a calculation based on the latitude and longitude assigned each such position. This being the case, though both here and everywhere I think that the better and more difficult method is to be more highly regarded than the inferior and easier method, both methods that have been described should still be kept available, for the sake of those who by reason of their unwillingness to exert themselves will have recourse to the easier method.

A GENERAL DESCRIPTION OF THE MAP OF THE INHABITED PORTION OF THE EARTH

Ptolemy, *Geography* VII. 5 (Nobbe)

The portion of the earth that we inhabit[2] was divided into three continents by the ancients, who gave accurate accounts about each of them and have left, as the fruits of their investigation, written descriptions dealing with each. Now we, too, on the basis in part of our own observations and in part of careful selection from their work, have endeavored to make a representation of the entire inhabited portion of the earth in the form of a map, so that every useful detail may be available to lovers of knowledge, useful, that is, for equipping their minds with information, and capable of arousing the keen interest which is natural to them.

The portion of the earth that we inhabit is bounded on the east by an unknown land that adjoins the eastern peoples of Asia Major, the Sinae and the inhabitants of Serica; on the south too by unknown lands, which enclose the Indian Sea and bound the country in the south of Libya called Ethiopia Agisymba; on the west by the unknown land that embraces the Ethiopian gulf of Libya, and thereafter by the western Ocean which lies along the westernmost parts of Libya and Europe; and on the north by the Ocean continuous [with the western Ocean] which surrounds the Britannic Islands and the northernmost parts of Europe and is called the Duecaledonian and the Sarmatic Seas, and also by the unknown land that borders on the northernmost countries of Asia Major, Sarmatia, Scythia, and Serica.

Of the seas contained within the inhabited portion of the earth, our sea,[3] together with the smaller seas that are a part of it—the Adriatic, the Aegean, the Propontis and the Pontus, and the Sea of Maeotis—opens

[1] The reference is to the areas cut off between two successive meridian lines and two successive parallel circles represented on the map.

[2] The use of "we" (καθ' ἡμᾶς) in this technical phrase seems to recognize the possibility of the existence of another inhabited region situated in the southern hemisphere (see p. 159), but Ptolemy is not consistent in his usage.

[3] I.e., the Mediterranean.

into the Ocean only at the straits of Heracles, like a peninsula having as its isthmus, as it were, these straits of Heracles. But the Hyrcanian or Caspian Sea is surrounded on all sides by land, like an island with land and sea reversed.

Similarly, the whole Indian Sea, along with the gulfs that are connected with it, the Arabian, Persian, Gangetic, and that properly called the Great Gulf, is entirely surrounded by land.

And so, of the three continents Asia is joined to Libya by the isthmus of Arabia, which also separates our sea from the Arabian Gulf, and also by the unknown land that surrounds the Indian Sea. Again Asia is joined to Europe by the isthmus between the Sea of Maeotis and the Sarmatic Ocean at the passage of the Tanais River. Libya is separated from Europe only by the strait,[1] and while not directly joined to Europe, is yet indirectly connected with it through Asia. For Asia is continuous with both Europe and Libya, being contiguous with them in the east.[2]

Now the largest of the three continents is Asia, second largest Libya, and third Europe. Of the seas that were said to be entirely surrounded by land, the largest is the Indian Sea, second largest our own, third the Hyrcanian or Caspian. Again, the more important gulfs, in order of size, are as follows: first the Gangetic, second the Persian, third the Great Gulf, fourth the Arabian,[3] fifth the Ethiopian, sixth Pontus, seventh the Aegean, eighth the Sea of Maeotis, ninth the Adriatic, and tenth the Propontis.

Of the more important islands or peninsulas the largest is Taprobane,[4] second largest Albion of the Britannic Islands, third the Golden Chersonese,[5] fourth Hibernia of the Britannic Islands, fifth the Peloponnesus, sixth Sicily, seventh Sardinia, eighth Cyrnus,[6] ninth Crete, tenth Cyprus.

On the assumption that a great circle contains 360°, the southern limit of the known portion of the earth is indicated by the parallel of latitude 16°25′ south of the equator, the parallel through Meroe being the same distance north of the equator. The northern limit is indicated by the parallel 63° north of the equator, drawn through the island of Thule. Thus the entire latitude of the known earth is 79°25′ or, approximately, 80°, and in stades approximately 40,000. That is, one degree contains 500 stades, as has been ascertained by the more exact measurements. The circumference of the whole earth is 180,000 stades.[7]

---

[1] I.e., of Heracles.

[2] I.e., *east* from the point of view of Europe and Libya.

[3] Not the modern Arabian Sea but the Red Sea.

[4] Ceylon.

[5] The Malay Peninsula.

[6] Corsica.

[7] See p. 150, n. 2.

Again, the eastern limit of the known portion of the earth is indicated by the meridian drawn through the chief city of the Sinae, 119°30′ east of the meridian through Alexandria as measured on the equator, i.e., approximately eight equinoctial hours. And the western limit is indicated by the meridian drawn through the Fortunate Isles, 60°30′, or four equinoctial hours, west of the meridian through Alexandria. The westernmost meridian is 180°, i.e., a semicircle, or twelve equinoctial hours, distant from the easternmost.

Thus the entire longitude of the known portion of the earth, measured along the equator, contains 90,000 stades; measured along its southernmost parallel approximately 86,330⅓; measured along the northernmost parallel 40,000 stades; measured along the parallel through Rhodes, 36° distant from the equator, the parallel on which measurements have generally been made, approximately 72,000;[1] measured along the parallel through Syene, 23°50′ distant from the equator and approximately bisecting the entire breadth, 82,336. This corresponds to the ratio between the aforesaid parallels and the equator.

Thus the longitude of the known portion of the earth is greater than its latitude, in the climata furthest north, by approximately 1/60 of the latitude; in those near the parallel through Rhodes by approximately 5/6; in those near the parallel through Syene by the amount of the latitude plus approximately 1/18 thereof; in the southernmost parts by the amount of the latitude plus approximately 1/6 thereof; in the climata near the equator by the amount of the latitude plus ¼ thereof.

And the length of the longest day or night on the southernmost of the aforesaid parallels is 13 equinoctial hours (just as on the parallel through Meroe), on the equator 12 hours, on the parallel through Syene 13½ hours, on that through Rhodes 14½, on the northernmost parallel, that through Thule, 20. The difference over the whole latitude is nine equinoctial hours.[2]

POST-PTOLEMAIC GEOGRAPHY

After Ptolemy there was no advance in mathematical geography in antiquity. In fact, despite the interest in the traditional knowledge that we find in such authors as Macrobius, Capella, and some of the Church Fathers, the general tendency was to discard this knowledge. With the literal interpretation of the Bible the denial of the existence of antipodes became quite general, and there was also a widespread reversion to the conception of a flat earth. The decay is strikingly seen in a comparison of the maps of Ptolemy with those of a few centuries later and with such conceptions of the world as that of Cosmas Indicopleustes in the sixth century.

---

[1] A variant, 72,812, accords better with the fraction 5/6, below. But see p. 173.

[2] At the summer solstice the length of the day on the parallel through Thule is 20 hours, whereas on the southernmost parallel it is 11 hours.

PHYSICS

The ancient Greek interest in nature which manifested itself in their philosophy was not purely speculative. The earliest theories of the Ionian and Italian schools were closely associated with keen observation of nature and with attempts to explain the phenomena on the basis of laws, in some cases laws of a mathematical form. The mathematical relation between the pitch and length of a vibrating string was probably known to the early Pythagoreans. The explicit formulation of the law of the lever had been made by the time of Aristotle.

From such beginnings there followed a considerable development in the Alexandrian period. Archimedes treats the theoretical principles of mechanics geometrically, and writers on mechanics show various practical applications of the theoretical treatment. In his work *On Floating Bodies* Archimedes lays the foundations of the science of hydrostatics. The development of geometrical optics can be traced in the extant treatises of Euclid, Hero, and Ptolemy. From the fundamental discovery of the corporeality of air in the sixth or fifth century B.C. there arises a widespread interest in the principles and applications of pneumatics, culminating in the work of Ctesibius of Alexandria. The nature of this work is best illustrated in the extant treatises of Philo of Byzantium and Hero of Alexandria.

In the field of general dynamics unsuccessful attempts were made by Aristotle and others to formulate mathematical laws. These failures prevented a more thorough integration of the various branches of physics and constituted a barrier that was not completely removed until the seventeenth century.

We have included passages that set forth the theoretical aspects of the various branches of physics and also passages that describe practical applications. It is to be remembered, however, that our best sources for the technological applications are not literary but archaeological.

PHYSICS AND MATHEMATICS

Aristotle, *Physics* II.2 (193*b*22–194*a*12). Translation of R. P. Hardie and R. K. Gaye (Oxford, 1930)

The next point to consider is how the mathematician differs from the physicist. Obviously physical bodies contain surfaces and volumes, lines and points, and these are the subject-matter of mathematics.

Further, is astronomy different from physics or a department of it? It seems absurd that the physicist should be supposed to know the nature of sun or moon, but not to know any of their essential attributes, particularly as the writers on physics obviously do discuss their shape also and whether the earth and the world are spherical or not.

Now the mathematician, though he too treats of these things,[1] nevertheless does not treat of them as the limits of a physical body; nor does he

[1] Surfaces, etc.

consider the attributes indicated as the attributes of such bodies. That is why he separates[1] them; for in thought they are separable from motion, and it makes no difference, nor does any falsity result, if they are separated. The holders of the theory of Forms do the same, though they are not aware of it; for they separate the objects of physics, which are less separable than those of mathematics. This becomes plain if one tries to state in each of the two cases the definitions of the things and of their attributes. "Odd" and "even," "straight" and "curved," and likewise "number," "line," and "figure," do not involve motion; not so "flesh" and "bone" and "man"[2]—*these* are defined like "snub nose,"[3] not like "curved."

Similar evidence is supplied by the more physical of the branches of mathematics, such as optics, harmonics, and astronomy. These are in a way the converse of geometry. While geometry investigates physical lines but not *qua* physical, optics investigates mathematical lines, but *qua* physical, not *qua* mathematical.

MECHANICS

The Study of Mechanics

Pappus, *Mathematical Collection* VIII. 1–5

The science of mechanics, my dear Hermodorus, is not merely useful for many important practical undertakings, but is justly esteemed by philosophers and is diligently pursued by all who are interested in mathematics, since it is fundamentally concerned with the doctrine of nature with special reference to the material composition of the elements in the cosmos. For it examines bodies at rest, their natural tendency,[4] and their locomotion in general, not only assigning causes of natural motion, but devising means of impelling bodies to change their position, contrary to their natures, in a direction away from their natural places. In this the science of mechanics uses theorems suggested to it by a consideration of matter itself.

Now the mechanicians of Hero's[5] school tell us that the science of mechanics consists of a theoretical and a practical part. The theoretical

---

[1] I.e., abstracts. [Edd.]

[2] Aristotle's point is that while mathematicians are justified in studying lines, surfaces, etc., in complete abstraction from physical bodies, the Platonists are not justified in introducing such abstraction into physics, too. For physical entities cannot, like the objects of mathematics, be defined apart from motion. [Edd.]

[3] Aristotle uses this more than once as an example of a concept requiring reference to a material object for its definition. [Edd.]

[4] The natural "tendency" is that of a body toward its natural place, as in Aristotelian physics.

[5] On Hero of Alexandria see p. 197.

part includes geometry, arithmetic, astronomy, and physics, while the practical part consists of metal-working, architecture, carpentry, painting, and the manual activities connected with these arts. One who has had instruction from boyhood in the aforesaid theoretical branches, and has attained skill in the practical arts mentioned, and possesses a quick intelligence, will be, they say, the ablest inventor of mechanical devices and the most competent master-builder. But since it is not generally possible for a person to master so many mathematical branches and at the same time to learn all the aforesaid arts, they advise a person who is desirous of engaging in mechanical work to make use of those special arts which he has mastered for the particular ends for which they are useful.

The most important of the mechanical arts from the point of view of practical utility are the following.[1] (1) The art of the *manganarii*,[2] known also, among the ancients, as mechanicians. With their machines they need only a small force to overcome the natural tendency of large weights and lift them to a height. (2) The art of the makers of engines of war, who are also called mechanicians. They design catapults to fling missiles of stone and iron and the like a considerable distance. (3) The art of the contrivers of machines, properly so-called. For example, they build water-lifting machines by which water is more easily raised from a great depth. (4) The art of those who contrive marvelous devices. They too are called mechanicians by the ancients. Sometimes they employ air pressure, as does Hero in his *Pneumatica*; sometimes ropes and cables to simulate the motions of living things, e.g., Hero in his works on *Automata* and *Balances*; and sometimes they use objects floating on water, e.g., Archimedes in his work *On Floating Bodies*, or water clocks, e.g., Hero in his treatise on that subject, which is evidently connected with the theory of the sun dial.[3] (5) The art of the sphere makers, who are also considered mechanicians. They construct a model of the heavens [and operate it] with the help of the uniform circular motion of water.[4]

Now some say that Archimedes of Syracuse mastered the principles and the theory of all these branches. For he is the only man down to

---

[1] Cf. the account of Geminus, p. 2.

[2] Strictly the *manganon* refers only to the pulley, but it is used in a wider sense to include all machines operating on the same general principle. So the *manganarius* is, in general, the machinist.

[3] The work on *Automata* gives directions for the construction of a complete puppet theater and figures that enact dramatic scenes. Nothing remains of Hero's work on *Balances* and there is but one fragment of that *On Water Clocks* (see p. 140). The connection between the water clock and the sundial is not merely that they are both instruments of horology but that the latter is useful in determining adjustments in the former when it is desired to vary the length of the hour according to the season of the year.

[4] The meaning seems to be that the motions of the heavenly bodies in the ancient "planetarium" were effected by a form of water power. On the Archimedean Sphere see p. 142.

our time who brought a versatile genius and understanding to them all, as Geminus the mathematician tells us in his discussion of the relationship of the branches of mathematics. But Carpus of Antioch says somewhere that Archimedes of Syracuse wrote only one book on a mechanical subject, that on sphere-construction, but did not consider any of the other mechanical branches worthy of literary treatment. Now this wonderful man, a man so richly endowed that his name will be celebrated forever by all mankind, is extolled by most people for his achievement in mechanics. But his chief concern was the composition of works dealing with the principal matters of geometric and arithmetic theory, even those parts often held to be least important. Evidently he was so devoted to these branches that he did not permit himself to add to them anything extraneous.

But Carpus and others have made use of geometry as a basis for various arts, and properly so. For in aiding numerous arts geometry is in no wise harmed by the association with them. Since geometry is, so to speak, the mother of these arts, it is not harmed by aiding in the construction of engines or in the work of the master-builder, or by association with geodesy, horology, mechanics, and scene-painting. On the contrary, geometry obviously promotes these arts and is justly honored and glorified by them.

Such, then, is the nature of mechanics, which is both a science and an art; and such are the parts into which it is divided. Now I consider it well to set forth more concisely, clearly, and rigorously than my predecessors have done, the most important theorems proved geometrically by the old writers on the subject of the motion of heavy bodies, as well as the theorems which I succeeded in discovering for myself. I cite as examples:

1. If a given weight is drawn by a given force on a horizontal plane, to find the force by which the weight will be drawn up a plane inclined to the horizontal at a given angle.[1] This proposition is useful to those mechanicians who construct machines for lifting weights, for by adding a force of men to the force found to be theoretically required they may be confident that the weight will be drawn up;

2. Given two unequal straight lines to find two mean proportionals in continued proportion. By this theorem every solid figure may be augmented or decreased in any given ratio;[2]

3. Given a wheel with a known number of cogs or teeth, to find the diameter of a second wheel to be engaged with the first and having a given number of teeth. This proposition is generally useful and in particular for machine makers in connection with the fitting of cogged wheels.

---

[1] Pappus's treatment of the problem of the inclined plane is given below, p. 194.

[2] I.e., a solid may be constructed similar to a given solid, the volume of one being in a given ratio to that of the other. This is a generalization of the problem of doubling the cube (see p. 62).

Each of these propositions will be elucidated in its proper place along with other propositions useful to the master-builder and the mechanician. But first let us discuss those things which have to do with the matter of centers of gravity.

We do not have to discuss at this time what is meant by the heavy and the light, what is the cause of the upward or downward tendency of bodies, and in fact what significance attaches to the terms up and down and by what limits each is bounded. These matters have been treated by Ptolemy in his *Mathematica*. But we should consider just what we mean by the center of weight of a given body, for that is the fundamental element in the whole subject of centers of gravity on which depend all the other parts of mechanical theory. For the other theorems in this field can be clear, in my opinion, if this fundamental concept is clear. Now we define the center of gravity of any given body as a point within the body such that, if we imagine the body to be suspended from that point, the body will be at rest, maintaining its original position without any tendency to turn.

Statics

THE LAW OF THE LEVER AND SOME APPLICATIONS

Archimedes, *On the Equilibrium of Planes* I, Postulates and Propositions 1–7.[1]
Translation and paraphrase of T. L. Heath, *The Works of Archimedes*, pp. 189–194

"I postulate the following:

"1. Equal weights at equal distances are in equilibrium, and equal weights at unequal distances are not in equilibrium but incline towards the weight which is at the greater distance.

"2. If, when weights at certain distances are in equilibrium, something be added to one of the weights, they are not in equilibrium, but incline towards that weight to which the addition was made.

"3. Similarly, if anything be taken away from one of the weights, they are not in equilibrium but incline towards the weight from which nothing was taken.

---

[1] Archimedes' treatise *On the Equilibrium of Planes*, in two books, develops the principle of the lever and deals with the determination of the center of gravity of various figures. The treatment is geometrical throughout. In the case of the theory of the lever a comparison of Archimedes' treatment with that of the Aristotelian *Mechanics* given below is very instructive. Archimedes bases his deductions on assumptions with respect to equilibrium and centers of gravity; the Aristotelian treatment, on the other hand, seeks to embrace the problem under the general laws of motion. There has been considerable discussion of the advantages and limitations of the two methods. The question of the extent to which the postulates of Archimedes are dependent on experience and the extent to which the law of the lever is itself presupposed in the development of the proof have been widely discussed, e.g., in works of Mach, Duhem, and Vailati.

In the part quoted, we omit the proofs of Propositions 1–5. The portion within quotation marks is a translation, the rest a paraphrase. [Edd.]

"4. When equal and similar plane figures coincide if applied to one another, their centres of gravity similarly coincide.

"5. In figures which are unequal but similar the centres of gravity will be similarly situated. By points similarly situated in relation to similar figures I mean points such that, if straight lines be drawn from them to the equal angles, they make equal angles with the corresponding sides.

"6. If magnitudes at certain distances be in equilibrium, (other) magnitudes equal to them will also be in equilibrium at the same distances.

"7. In any figure whose perimeter is concave in (one and) the same direction the centre of gravity must be within the figure."

Proposition 1. Weights which balance at equal distances are equal. . . .

Proposition 2. Unequal weights at equal distances will not balance but will incline towards the greater weight. . . .

Proposition 3. Unequal weights will balance at unequal distances, the greater weight being at the lesser distance. . . .

Proposition 4. If two equal weights have not the same centre of gravity, the centre of gravity of both taken together is at the middle point of the line joining their centres of gravity. . . .

Proposition 5. If three equal magnitudes have their centres of gravity on a straight line at equal distances, the centre of gravity of the system will coincide with that of the middle magnitude. . . .

Corollary 1. The same is true of any odd number of magnitudes if those which are at equal distances from the middle one are equal, while the distances between their centres of gravity are equal.

Corollary 2. If there be an even number of magnitudes with their centres of gravity situated at equal distances on one straight line, and if the two middle ones be equal, while those which are equidistant from them (on each side) are equal respectively, the centre of gravity of the system is the middle point of the line joining the centres of gravity of the two middle ones.

Propositions 6, 7

Two magnitudes, whether commensurable [Proposition 6] or incommensurable [Proposition 7], balance at distances reciprocally proportional to the magnitudes.

I. Suppose the magnitudes A, B to be commensurable, and the points A, B to be their centres of gravity. Let DE be a straight line so divided at C that

$$A:B = DC:CE.$$

We have then to prove that, if A be placed at E and B at D, C is the centre of gravity of the two taken together.

Since A, B are commensurable, so are DC, CE. Let N be a common measure of DC, CE. Make DH, DK each equal to CE, and EL (on CE produced) equal to CD. Then $EH = CD$, since $DH = CE$. Therefore LH is bisected at E, as HK is bisected at D.

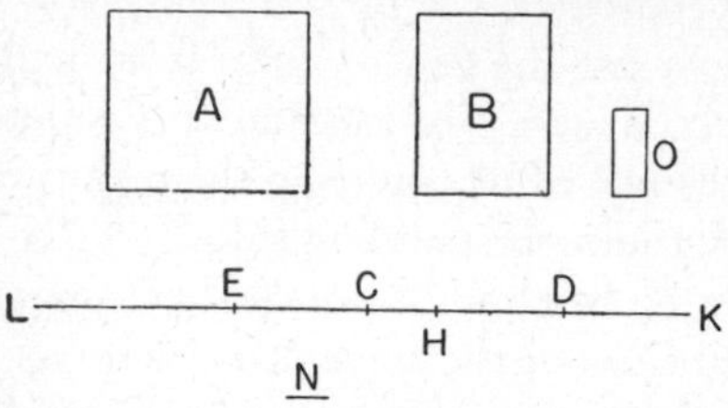

A B O

L E C D K H N

Thus LH, HK must each contain N an even number of times.

Take a magnitude O such that O is contained as many times in A as N is contained in LH, whence

$$A:O = LH:N.$$

But

$$B:A = CE:DC$$
$$= HK:LH.$$

Hence, *ex aequali*, $B:O = HK:N$, or O is contained in B as many times as N is contained in HK.

Thus O is a common measure of A, B.

Divide LH, HK into parts each equal to N, and A, B into parts each equal to O. The parts of A will therefore be equal in number to those of LH, and the parts of B equal in number to those of HK. Place one of the parts of A at the middle point of each of the parts N of LH, and one of the parts of B at the middle point of each of the parts N of HK.

Then the center of gravity of the parts of A placed at equal distances on LH will be at E, the middle point of LH [Proposition 5, Corollary 2], and the centre of gravity of the parts of B placed at equal distances along HK will be at D, the middle point of HK.

Thus we may suppose A itself applied at E, and B itself applied at D.

But the system formed by the parts O of A and B together is a system of equal magnitudes even in number and placed at equal distances along LK. And, since $LE = CD$, and $EC = DK$, $LC = CK$, so that C is the middle point of LK. Therefore C is the centre of gravity of the system ranged along LK.

Therefore A acting at E and B acting at D balance about the point C.

II. Suppose the magnitudes to be incommensurable, and let them be $(A + a)$ and B respectively. Let DE be a line divided at C so that

$$(A + a):B = DC:CE.$$

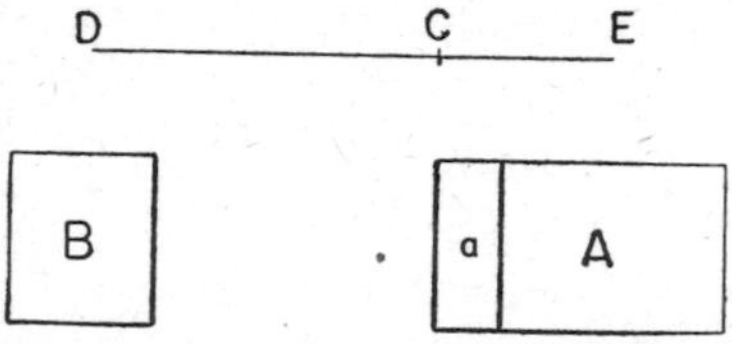

D C E

B a A

Then, if $(A + a)$ placed at E and B placed at D do not balance about C, $(A+a)$ is either too great to balance B, or not great enough.

Suppose, if possible, that $(A + a)$ is too great to balance B. Take from $(A+a)$ a magnitude a smaller than the deduction which would make the remainder balance B, but such that the remainder A and the magnitude B are commensurable.

Then, since A, B are commensurable, and

$$A:B < DC:CE,$$

A and B will not balance [Proposition 6], but D will be depressed.

But this is impossible, since the deduction a was an insufficient deduction from $(A + a)$ to produce equilibrium, so that E was still depressed.

Therefore $(A + a)$ is not too great to balance B; and similarly it may be proved that B is not too great to balance $(A + a)$.

Hence $(A + a)$, B taken together have their centre of gravity at C.

[Aristotle], *Mechanics* 1–4, 21. Translation of E. S. Forster (Oxford, 1913)[1]

1. First, then, a question arises as to what takes place in the case of the balance. Why are larger balances more accurate than smaller? And the fundamental principle of this is, why is it that the radius which extends further from the centre is displaced quicker than the smaller radius, when the near radius is moved by the same force? Now we use the word "quicker" in two senses; if an object traverses an equal distance in less time, we call it quicker, and also if it traverses a greater distance in equal time. Now the greater radius describes a greater circle in equal time; for the outer circumference is greater than the inner.

The reason of this is that the radius undergoes two displacements. Now if the two displacements of a body are in any fixed proportion, the resulting displacement must necessarily be a straight line, and this line is the diagonal of the figure made by the lines drawn in this proportion. . . .[2]

And the converse is also true. It is plain that, if a point be moved along the diagonal by two displacements, it is necessarily moved according to the proportion of the sides of the parallelogram; for otherwise it will

---

[1] The *Mechanics* consists of a theoretical discussion of the lever followed by a series of discussions of problems involving, in the main, applications of the theory of the lever and related machines, as well as more general physical questions. The work, though probably not by Aristotle, reflects the viewpoint of the Peripatetic School and is generally considered to have been written not long after Aristotle's time. The author seeks an "explanation" of the law of the lever in the increased velocity with which a weight turns the farther it is from the fulcrum. Particularly noteworthy is this connection of the problem of the lever with the equations of forced motion in Aristotle's system, in which velocity is directly proportional to the force applied and inversely proportional to the weight moved (p. 203). Contrast the discussion of Archimedes above, which is based solely on considerations of equilibrium and is not connected with the laws of motion. (English letters are substituted throughout for the Greek letters used by the translator.) [Edd.]

[2] There follows a proof of the parallelogram (or rectangle) of velocities similar to that given on p. 223, below. In Aristotle's system, force is proportional to the distance through which a given weight is moved in a given time and the direction of the motion is the direction of the force. Hence the deduction of the parallelogram of forces from the parallelogram of velocities is an easy step. [Edd.]

not be moved along the diagonal. If it be moved in two displacements in no fixed ratio for any time, its displacement cannot be in a straight line. For let it be a straight line. This then being drawn as a diagonal, and the sides of the parallelogram filled in, the point must necessarily be moved according to the proportion of the sides; for this has already been proved. Therefore, if the same proportion be not maintained during any interval of time, the point will not describe a straight line; for, if the proportion were maintained during any interval, the point must necessarily describe a straight line, by the reasoning above. So that, if the two displacements do not maintain any proportion during any interval, a curve is produced.

Now that the radius of a circle has two simultaneous displacements is plain from these considerations, and because the point from being vertically above the centre comes back to the perpendicular, so as to be again perpendicularly above the centre.

Let *ABC* be a circle, and let the point *B* at the summit be displaced to *D* by one force, and come eventually to *C* by the other force. If then it were moved in the proportion of *BD* to *DC*, it would move along the diagonal *BC*. But in the present case, as it is moved in no such proportion, it moves along the curve *BEC*. And, if one of two displacements caused by the same forces is more interfered with and the other less, it is reasonable to suppose that the motion more interfered with will be slower than the motion less interfered with; which seems to happen in the case of the greater and less of the radii of circles. For on account of the extremity of the lesser radius being nearer the stationary centre than that of the greater, being as it were pulled in a contrary direction, towards the middle, the extremity of the lesser moves more slowly. This is the case with every radius, and it moves in a curve, naturally along the tangent, and unnaturally towards the centre.[1] And the lesser radius is always moved more in respect of its unnatural motion; for being nearer to the retarding centre it is more constrained. And that the less of two radii having the same centre is moved more than the greater in respect of the unnatural motion is plain from what follows.

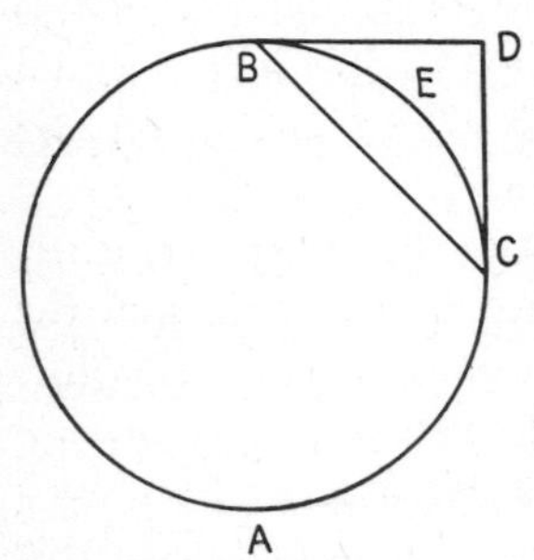

Let *BCED* be a circle, and *XNMV* another smaller circle within it, both having the same centre *A*, and let the diameters be drawn, *CD* and *BE* in the large circle, and *MX* and *NV* in the small; and let the rectangle

[1] The view of circular motion as the resultant of two components, one tangential and the other centripetal, is noteworthy. This view was not applied by Aristotle to the motion of the heavenly bodies. [Edd.]

DYPC be completed. If the radius *AB* comes back to the same position from which it started, i.e., to *AB*, it is plain that it moved towards itself; and likewise *AX* will come to *AX*. But *AX* moves more slowly than *AB*, as has been stated, because the interference is greater and *AX* is more retarded.

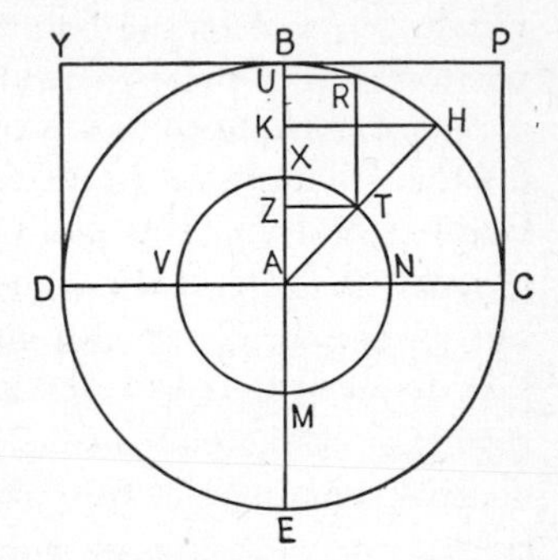

Now let *ATH* be drawn, and from *T* a perpendicular upon *AB* within the circle, *TZ*; and, further, from *T* let *TR* be drawn parallel to *AB*, and *RU* and *HK* perpendiculars on *AB*; then *RU* and *TZ* are equal. Therefore *BU* is less than *XZ*; for in unequal circles equal straight lines drawn perpendicular to the diameter cut off smaller portions of the diameter in the greater circles; *RU* and *TZ* being equal.[1]

Now the radius *AT* describes the arc *XT* in the same time as the extremity of the radius *BA* has described an arc greater than *BR* in the greater circle; for the natural displacement is equal and the unnatural less, *BU* being less than *XZ*. Whereas they ought to be in proportion, the two natural motions in the same ratio to each other as the two unnatural motions.

Now the radius *AB* has described an arc *BH* greater than *BR*. It must necessarily have described *BH* in the time in which *X* describes *XT*; for that will be its position when in the two circles the proportion between the unnatural and natural movements holds good. If, then, the natural movement is greater in the greater circle, the unnatural movement, too, would agree in being proportionately greater in that case only, where *B* is moved along *BH* while *X* is moved along *XT*. For in that case the point *B* comes by its natural movement to *H*, and by its unnatural movement to *K*, *HK* being perpendicular from *H*. And as *HK* to *BK*, so is *TZ* to *XZ*. Which will be plain, if *B* and *X* be joined to *H* and *T*.[2] But if the arc described by *B* be less or greater than *HB*, the result will not be the same, nor will the natural movement be proportional to the unnatural in the two circles.

So that the reason why the point further from the centre is moved quicker by the same force, and the greater radius describes the greater circle, is plain from what has been said; and hence the reason is also clear why larger balances are more accurate than smaller. For the cord by

[1] According to the parallelogram of distances, the result ought to be: *BU*: *UR* :: *XZ*: *TZ*, but it is proved that *UR* and *TZ* are equal, but *BU* and *XZ* unequal, so that the theory of the parallelogram fails. Why is this? The answer is that the same force moves longer radii quicker than shorter.

[2] For the triangles *BKH* and *XZT* are similar.

which a balance is suspended acts as the centre, for it is at rest, and the parts of the balance on either side form the radii. Therefore by the same weight the end of the balance must necessarily be moved quicker in proportion as it is more distant from the cord, and some weight must be imperceptible to the senses in small balances, but perceptible in large balances; for there is nothing to prevent the movement being so small as to be invisible to the eye. Whereas in the large balance the same load makes the movement visible. In some cases the effect is clearly seen in both balances, but much more in the larger on account of the amplitude of the displacement caused by the same load being much greater in the larger balance. And thus dealers in purple, in weighing it, use contrivances with intent to deceive, putting the cord out of centre and pouring lead into one arm of the balance, or using the wood towards the root of a tree for the end towards which they want it to incline, or a knot, if there be one in the wood; for the part of the wood where the root is is heavier, and a knot is a kind of root.

2. How is it that if the cord is attached to the upper surface of the beam of a balance, if one takes away the weight when the balance is depressed on one side, the beam rises again; whereas, if the cord is attached to the lower surface of the beam, it does not rise but remains in the same position? Is it because, when the cord is attached above, there is more of the beam on one side of the perpendicular than on the other, the cord being the perpendicular? In that case the side on which the greater part of the beam is must necessarily sink until the line which divides the beam into two equal parts reaches the actual perpendicular, since the weight now presses on the side of the beam which is elevated.

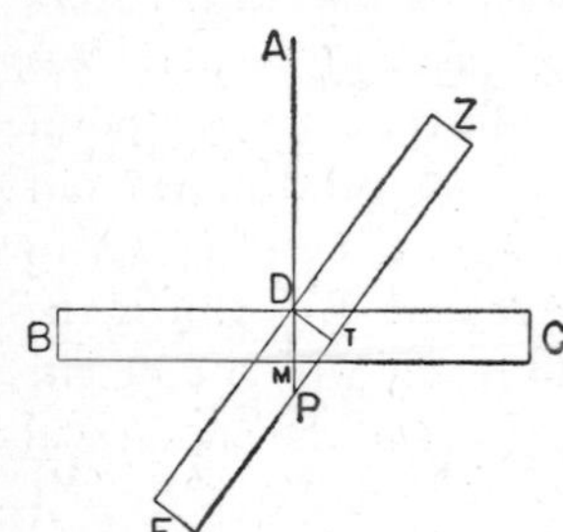

Let *BC* be a straight beam, and *AD* a cord. If *AD* be produced it will form the perpendicular *ADM*. If the portion of the beam towards *B* be depressed, *B* will be displaced to *E* and *C* to *Z*; and so the line dividing the beam into two halves, which was originally *DM*, part of the perpendicular, will become *DT* when the beam is depressed; so that the part of the beam *EZ* which is outside the perpendicular *AM* will be greater by *TP* than half the beam. If therefore the weight at *E* be taken away, *Z* must sink, because the side towards *E* is shorter. It has been proved then that when the cord is attached above, if the weight be removed the beam rises again.

But if the support be from below, the contrary takes place. For then the part which is depressed is more than half of the beam, or in other words, more than the part marked off by the original perpendicular; it does not therefore rise, when the weight is removed, for the part that is elevated is lighter. Let *NV* be the beam when horizontal, and *KLM* the perpendicular dividing *NV* into two halves. When the weight is placed at *N*, *N* will be displaced to *O* and *V* to *P*, and *KL* to *LT*, so that *KO* is greater than *LP* by *TLK*.[1] If the weight, therefore, is removed the beam must necessarily remain in the same position; for the excess of the part in which *OK* is over half the beam acts as a weight and remains depressed.

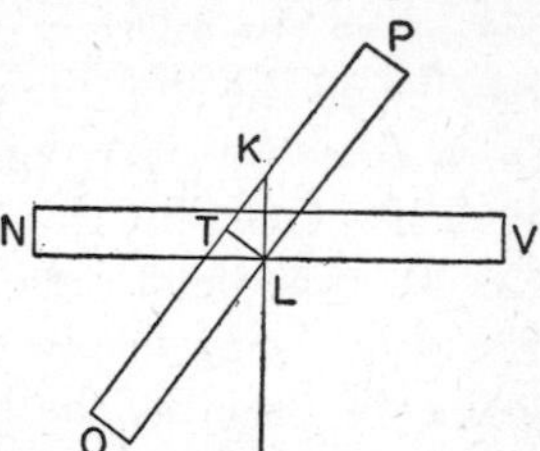

3. Why is it that, as has been remarked at the beginning of this treatise, the exercise of little force raises great weights with the help of a lever, in spite of the added weight of the lever; whereas the less heavy a weight is, the easier it is to move, and the weight is less without the lever? Does the reason lie in the fact that the lever acts like the beam of a balance with the cord attached below and is divided into two unequal parts? The fulcrum, then, takes the place of the cord, for both remain at rest and act as the centre. Now since a longer radius moves more quickly than a shorter one under pressure of an equal weight; and since the lever requires three elements, viz., the fulcrum—corresponding to the cord of a balance and forming the centre—and two weights, that exerted by the person using the lever and the weight which is to be moved; this being so, as the weight moved is to the weight moving it, so, inversely, is the length of the arm bearing the weight to the length of the arm nearer to the power.[2] The further one is from the fulcrum, the more easily will one raise the weight; the reason being that which has already been stated, namely, that a longer radius describes a larger circle. So with the exertion of the same force the motive weight will change its position more than the weight which it moves, because it is further from the fulcrum.[3]

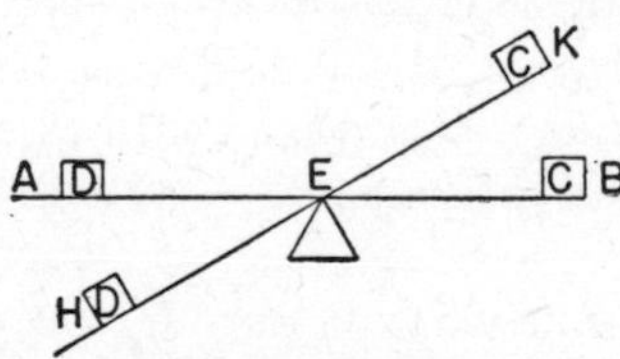

Let *AB* be a lever, *C* the weight to be lifted, *D* the motive weight, and *E* the fulcrum; the position of *D* after it has raised the weight will be *H*, and that of *C*, the weight raised, will be *K*.

4. Why is it that those rowers who are amidships move the ship most? Is it because the oar acts as a lever? The fulcrum then is the thole-pin (for it remains in the same place); and the weight is the sea which the oar displaces; and the power that moves the lever is the rower. The further

[1] More precisely, by twice *TLK*. [Edd.]

[2] We have here the quantitative formulation of the principle of the lever. [Edd.]

[3] E.g., in a case where it is required to lift a large weight with a comparatively small force, as in the case of the crowbar. [Edd.]

he who moves a weight is from the fulcrum, the greater is the weight which he moves; for then the radius becomes greater, and the thole-pin acting as the fulcrum is the centre. Now amidships there is more of the oar inside the ship than elsewhere; for there the ship is widest, so that on both sides a longer portion of the oar can be inside the two walls of the vessel. The ship then moves because, as the blade presses against the sea, the handle of the oar, which is inside the ship, advances forward, and the ship, being firmly attached to the thole-pin, advances with it in the same direction as the handle of the oar. For where the blade displaces most water, there necessarily must the ship be propelled most; and it displaces most water where the handle is furthest from the thole-pin. This is why the rowers who are amidships move the ship most; for it is in the middle of the ship that the length of the oar from the thole-pin inside the ship is greatest.

21. How is it that dentists extract teeth more easily by applying the additional weight of a tooth-extractor than with the bare hand only? Is it because the tooth is more inclined to slip in the fingers than from the tooth-extractor, or does not the iron slip more than the hand and fail to grasp the tooth all round, since the flesh of the fingers being soft both adheres to and fits round the tooth better? The truth is that the tooth-extractor consists of two levers opposed to one another, with the same fulcrum at the point where the pincers join; so they use the instrument to draw teeth, in order to move them more easily.

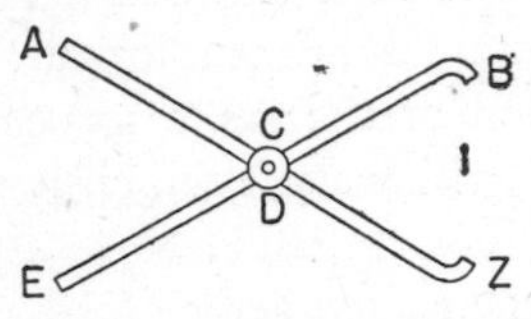

Let *A* be one extremity of the tooth-extractor and *B* the other extremity which draws the tooth, and *ADZ* one lever and *BCE* the other, and *CD* the fulcrum, and let the tooth, which is the weight to be lifted, be at the point *I*, where the two levers meet. The dentist holds and moves the tooth at the same time with *B* and *Z*; and when he has moved it, he can take it out more easily with his fingers than with the instrument.

FORCES, WEIGHTS, AND THE INCLINED PLANE

Pappus, *Mathematical Collection*, VIII, Prop. 9 (pp. 1054.4–1058.26 [Hultsch])

A given force is needed to draw a given weight along a horizontal plane. It is required to find the force needed to draw the weight up another plane inclined at a given angle to the horizontal plane.

Let the horizontal plane pass through *MN*, and let the plane inclined to the horizontal at the given angle, $\angle KMN$, pass through *MK*. Let *A* be the weight and *C* the force required to move it over the horizontal plane. Consider a sphere with center *E* and weight equal to that of *A*. Place this sphere on the inclined plane passing through M and K. The sphere will be tangent to the plane at L, as is shown in the third theorem of the *Spherics*.[1] EL will therefore be perpendicular to the plane (for this is also shown in the *Spherics*, Theorem IV), and also to KM. Pass a plane through KM and EL cutting the sphere in circle LHX. Draw ET through center E parallel to MN, and draw LZ, from L, perpendicular to ET. Now since $\angle ETL$ is given (for it is equal to the given $\angle KMN$), $\angle ELZ$ is also given, for $\angle ELZ = \angle ETL$ (since triangles ETL and ELZ are similar). Therefore $\triangle ELZ$ is given in form. Hence the ratio $EL{:}EZ$, that is, $EH{:}EZ$, is known, as is also $(EH - EZ){:}EZ$, that is $ZH{:}EZ$.

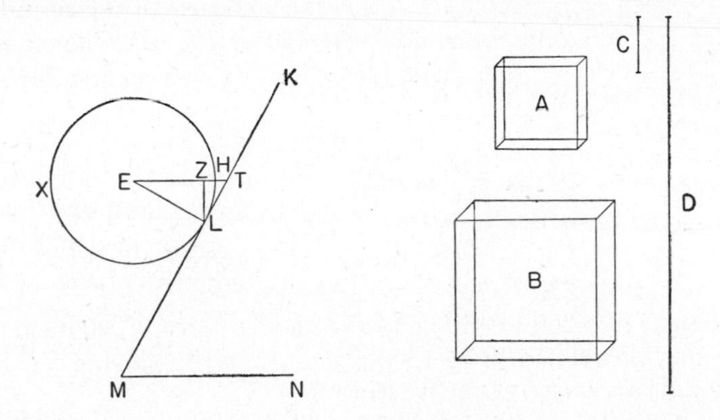

Let weight A be to weight B and force C to force D, as HZ is to ZE. Now C is the force required to move A. Therefore the force required to move B on the same plane will be D.[2] Since weight A : weight B = $HZ{:}ZE$, it follows that if E and H are the centers of gravity of weights A and B, respectively, the weights will be in equilibrium if balanced at point Z.[3] But weight A has its center of gravity at E (for the sphere represents A). Therefore, if weight B is placed so that its center is at H, it will so balance the sphere that the latter will not move down because of the slope of the plane, but will remain unmoved, as if it were on the horizontal plane. But weight A required force C to move it in the horizontal plane. Therefore, to be moved up the inclined plane it will require a force which is the sum of forces C and D, where D is the force required

---

[1] The reference here and in the next sentence is to a work of Theodosius of Bithynia, who lived at the end of the second or beginning of the first century B.C. and wrote mathematical and astronomical works. The best known is the *Spherics*, on the geometry of the sphere.

[2] The assumption is that the force required to move a weight on a horizontal plane is proportional to the weight. Though this may seem strange to a modern reader, it is to be noted that some types of frictional resistance may, as a first approximation, be treated as proportional to the weight of the load.

[3] By an application of the principle of the lever.

to move the weight B in the horizontal plane. Force D, moreover, is given.[1]

The geometrical solution of the problem has been indicated. However, to set forth the method and proof in a typical case, let weight A be, say, 200 talents, and let the moving force, C, required to draw the weight in a horizontal plane be equal to the force of 40 men. Let $\angle KMN\ (= \angle ETL)$ be ⅔ of a right angle. $\angle ZLT$ is therefore ⅓ of a right angle and, since $\angle ELT$ is a right angle, $\angle ELZ$ is also ⅔ of a right angle. Of the 360 equal parts into which four right angles are divided, $\angle ELZ$ contains 60. Therefore, if a circle be circumscribed about right $\triangle EZL$, the arc subtended by chord EZ will contain 120 of the 360 parts of the circumference, and chord EZ will itself be almost 104/120 of EL, the diameter of that circle. This is clear from the table of chords in the first book of the *Mathematica* of Ptolemy.[2]

Therefore

$$\frac{EL}{EZ} = \frac{EH}{EZ} = \frac{120}{104},$$

and

$$\frac{HZ}{ZE} = \frac{16}{104} = \frac{\text{weight } A}{\text{weight } B} = \frac{\text{force } C}{\text{force } D}.$$

But weight A is 200 talents, and the moving force, C, 40 men.

Therefore weight B will be 1300 talents, and the moving force, D, 260 men (for $16:104 = 200:1300 = 40:260$).

Hence, if 40 men are required to move a weight, A, of 200 talents on a plane parallel to the horizon, it follows that the sum of 40 and 260, that is, 300 men, will be required to move the same weight up a plane inclined to the horizon at $\angle KMN$, ⅔ of a right angle.[3]

---

[1] For the first three terms of the proportion, $HZ:ZE$ = force C: force D, are given.

[2] For this table see p. 84, above. The table gives us the ratio between EZ (chord of an arc of 120°) and EL (diameter) $103^p55'23'' : 120^p$.

[3] If F is the force just required to maintain weight A in motion up a plane inclined at angle α to the horizontal, and C, as in Pappus, the force just required to keep A in motion on a horizontal plane, Pappus's formula of $F = C + D$ reduces to $F = C + C \sin \alpha/(1 - \sin \alpha)$ [for $D = BC/A$ and $B/A = \sin \alpha/(1 - \sin \alpha)$]. That is, $F = C/(1 - \sin \alpha)$. The unsatisfactory character of Pappus's analysis is immediately clear from this formula, which requires, as α approaches 90°, that force F increase without limit rather than merely approach the force necessary to lift weight A vertically.

The problem of the inclined plane is discussed not only in this passage but in that of Hero of Alexandria cited below (p. 197). In Hero's analysis the resistance due to friction is neglected, whereas in Pappus's analysis this resistance is, by the very terms of the problem, made a part of the problem. In this there is certainly no fallacy, as some have supposed, but merely an additional complication of the problem. Both Hero and Pappus fail because of their incorrect notion of what force is necessary to achieve equilibrium, that is, because of their inability to resolve the force exerted by the weight into a component perpendicular to the plane and a component parallel to the plane.

For the problem given at the outset of this note the modern solution, neglecting friction,

Hero of Alexandria, *Mechanics* I.20–23[1]

20. Many people have the erroneous belief that weights placed on the ground may be moved only by forces equivalent to these weights. Let us demonstrate that weights placed as described may [theoretically] be moved by a force less than any given force, and let us explain the reason why this is not the case in practice.[2] Suppose that a weight, symmetrical, smooth, and quite solid, rests on a plane surface, and that this plane is capable of inclining toward both sides, that is, toward the right and the

---

is $F = A \sin \alpha$. (Hero's formulation for a cylinder is $F = [2A(\alpha + \sin \alpha \cos \alpha)]/\pi$.) If friction be taken into account, and be considered, as a first approximation, proportional to the component of the weight perpendicular to the plane over which the motion takes place, the modern formulation would be (to keep the notation used above) $F = A \sin \alpha + C \cos \alpha$.

[1] In Hero of Alexandria we have the culmination of the tradition of Ctesibius of Alexandria (p. 137) and Philo of Byzantium (p. 255). The superior importance of Hero lies in the fact that a large number of his works are extant. These works, which are concerned chiefly with the applications of science and only secondarily with its theoretical aspects, are our chief source of information on numerous developments in Greek science.

The date of Hero, though widely discussed, remains uncertain, the estimates ranging from the second century B.C. to the third A.D.

Of his works the following survive either in the original or in translation: *Pneumatics*, devoted chiefly to the description of devices operating by air pressure (see p. 248); *Mechanics*, concerned with the theory and applications of machines, as well as with general matters of statics and dynamics; *On the Dioptra*, dealing mainly with the construction and use of this instrument of levelling and angular measurement; *Belopoeica* and *Chirobalistra*, on the construction and operation of instruments of artillery; *On the Construction of Automata*, a work describing the construction of automatic figures that give a dramatic performance and apparatus for securing various stage effects; *Catoptrics*, a work dealing with the theory and applications of mirrors (attributed in the extant Latin version to Ptolemy, but probably by Hero—see p. 261); *Metrica*, dealing, in the main, with the application of formulas (including approximations) for areas and volumes of various figures.

There are numerous other geometrical works attributed to Hero but probably often recast through the centuries and overlaid with non-Heronian material, e.g., *Definitions*, *Geometria*, *Geodaesia*, *Stereometrica*, *Mensurae*, and *Liber Geoponicus*. In addition there are a few works of which only fragments survive, e.g., a work on water clocks and a commentary on Euclid's *Elements* (though considerable Heronian material is present in al-Nairīzī's ninth- or tenth-century Arabic commentary).

The *Mechanics* is extant in an Arabic version in three books. Of the original Greek only fragments are extant, the longest of which is a passage, which Pappus quotes (p. 224, below), on the five simple machines. The first book deals with various problems in statics, dynamics, and kinematics, the second discusses the simple machines and includes a collection of problems similar to that in the Aristotelian *Mechanics*, while the third deals with the construction and operation of machines, especially those for lifting large weights. For the selections given here the German translation of L. Nix has been used as a basis. Professor A. S. Halkin has helped interpret the Arabic text.

[2] The general point of this and the following section (I.21) is clear. The passage constitutes an important step in the development of the principle of inertia. Particularly to be noted is the treatment of friction and the means of reducing it. There is some difficulty in the precise interpretation of certain points; see the following note.

left. Suppose it inclines first toward the right. In that case we see that the given weight moves down toward the right, since it is in the nature of weights to move downward unless something supports them and hinders their motion. If, now, the side sloping downward is again lifted to the horizontal plane and restored to equilibrium, the weight will remain fixed in this position.

Again, if the plane is inclined toward the other side, that is, toward the left, the weight, too, will tend toward the lowered side, even if the slope is extremely small. The weight, in this case, does not require a force to set it in motion but rather a force to keep it from moving. Now when the weight again returns to equilibrium and does not tend in either direction, it remains in position without any force to support it. It continues to be at rest until the plane is made to slope towards either side, in which case the weight, too, tends in that direction. Thus it follows that the weight, which is prone to move in any desired direction, requires, for its motion, only a very small force equal to the force which inclines it.[1] Therefore the weight will be moved by any small force.

21. Pools of water that lie on non-sloping planes do not flow but remain still, not tending toward either side. But if the slightest inclination is imparted to them they flow completely toward that side, until not the least particle of water remains in its original position (unless there are declivities in the plane in the recesses of which small parts of water remain, as sometimes happens in the case of vessels).

Now this is the case with water because its parts are not strongly cohesive but are easily separable. Since, however, bodies that cohere strongly do not, naturally, have smooth surfaces and are not easily smoothed down, the result is that because of their roughness they support one another. That is, they are engaged like cogged wheels in a machine, and are consequently prevented [from rolling].[2]

For when the parts are numerous and closely bound to one another by reason of mutual cohesion, a large coordinated force is required [to produce motion of one body made up of such parts over another]. Experience

---

[1] The idea seems to be that the force required to set the weight in motion on the horizontal plane is equal to the resistance that enables the weight to continue at rest while the plane is inclined until the critical angle is reached beyond which the weight moves down the plane. If a plane be inclined at this critical angle, α, we should say that the force required to draw weight W up the plane is $W \sin \alpha + R$, where R is the component due to friction. Hero's point seems to be, but it is not put very clearly, that the same force would be necessary to set the weight in motion on the horizontal plane. Of course, where R approached zero, α would approach zero, as would also the force necessary to set the weight in motion on the horizontal plane.

[2] I.e., even in the case where the surface is somewhat inclined. The text is not clear but the reference may be to the interaction of a body and the surface on which it rests.

has taught men to lay logs with cylindrical surfaces under tortoises,[1] so that these logs touch only a small part of the plane, whence only the smallest amount of friction results. Logs are thus used to move weights easily, but the weight of the moving apparatus must exceed that of the load to be moved. Others plane down boards to render them smooth, fasten them together on the ground, and coat them with grease, so that whatever roughness there is may be smoothed out. Thus they move the load with little force. Columns [cylinders], even if they are heavy, may be moved easily if they lie upon the ground in such a way that only one line is in contact with the ground. This is true also of the sphere,[2] which we have already discussed.

22. Now if it is desired to raise a weight to a higher place, a force equal to the weight is needed. Consider a rotating pulley suspended perpendicular to the plane and turning about an axis at its midpoint. Let a cord be passed around the pulley and let one end be fastened to the weight and the other be operated by the moving force. I say that this weight may be moved by a force equal to it. For suppose that, instead of a force, there is, at the other end of the cord, a second weight. It will be seen that if the two weights are equal the pulley will not turn toward either side. The first weight is not strong enough to overbalance the second, and the second is not strong enough to overbalance the first, since both are equal. But if a slight addition is made to one weight, the other will be drawn up. Therefore, if the force that is to move the load is greater than the load, it will be strong enough to move the latter, unless friction in the turning of the pulley or the stiffness of the cords interferes with the motion.

23. Weights on an inclined plane have a tendency to move downward, as is the case with all bodies. If such movement does not take place we must invoke the explanation given above.[3] Suppose we wish to draw a weight up an inclined plane the surface of which is smooth and even, as is also the surface of that part of the weight which rests on the plane. For our purpose we must have a force or weight operating on the other side and just balancing the given weight, that is, conserving the equilibrium so that any addition of force will be sufficient to move the weight up the plane.

To prove our contention, let us demonstrate it in the case of a given cylinder. The cylinder has a natural tendency to roll downward because no large part of it touches the surface of the plane. Consider a plane perpendicular to the inclined plane and passing through the line of tangency between the cylinder and the inclined plane.

[1] The reference is to sheds moved by rollers and used in military operations.

[2] Here the tangency approaches a single point.

[3] I.e., the effect of friction.

Clearly, the new plane will pass through the axis of the cylinder and divide the cylinder into two halves. For, given a circle and a tangent, a line drawn from the point of contact at right angles to the tangent will pass through the center of the circle. Now pass a second plane through the same line (i.e., the line at which the cylinder touches the inclined plane) perpendicular to the horizon. This plane will not coincide with the plane previously constructed, but will divide the cylinder into two unequal parts, of which the smaller lies above and the larger below. The larger part, because it is larger, will outweigh the smaller, and the cylinder will roll down. If, now, we suppose that from the larger [of the two parts into which this plane perpendicular to the horizon divides the cylinder] that amount be removed by which the larger exceeds the smaller portion, the two parts will then be in equilibrium and their joint weight will remain unmoved on the line of tangency to the inclined plane, tending neither upward nor downward. We need, therefore, a force equivalent to this difference to preserve equilibrium.[1] But if the slightest addition be made to this force, it will overbalance the weight.

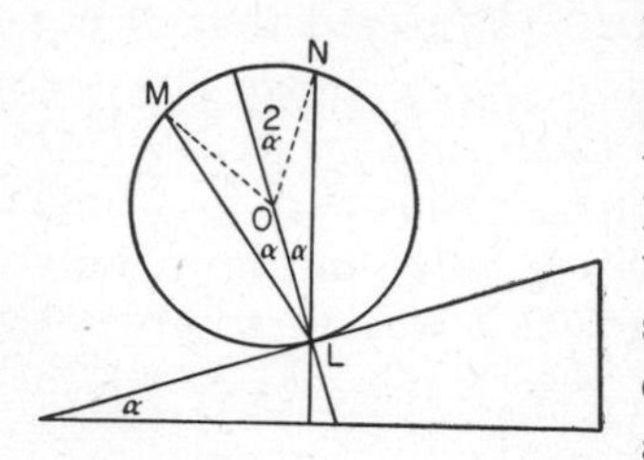

DYNAMICS

NATURAL (FREE) AND UNNATURAL (FORCED) MOTION ACCORDING TO ARISTOTLE

In the following passage we have Aristotle's doctrine of natural and forced motion and the inference of the existence of a fifth element, the aether, different from the terrestrial elements, earth, air, fire, and water. Each of the latter has its own "natural" place and "natural" motion toward that place. The natural place of earth is in the center of the universe, and its natural motion is downwards (i.e., toward the center); the natural place of fire is at the outermost part of the sublunary sphere, and its natural motion is upward (i.e., away from the center); the natural place of water is the region directly outside the central region, and its natural motion downward except in the central region; the natural place of air is the region between those of water and fire, and its natural motion upward except in the region of fire (and air itself). The natural motion of compounds is that of the predominating element therein. Motion of an element or a compound other than its natural motion is forced or unnatural motion.

---

[1] The method of Hero amounts to holding that (in the case of a cylinder of radius r, height h, and density d, rolling on a plane inclined to the horizontal at angle α) the force (F) required to draw the weight (W) up the plane is $F = W[2(\alpha + \cos\alpha \sin\alpha)/\pi] = 2r^2hd\ (\alpha + \cos\alpha \sin\alpha)$. The modern formulation is $F = W \sin\alpha$. Hero's formula is obtained by noting that he takes the part of the cylinder standing over LMN as the part to be balanced by force F, so that F/W = area $LMN/(\pi r^2)$, and area LMN = sector $MON + 2\ \triangle NOL = r^2\ (2\alpha + \sin 2\alpha)$. Compare the problem of the inclined plane as formulated and treated by Pappus (p. 194). The modern solution was not discovered until the days of Stevinus and Galileo.

The natural motion of all the terrestrial elements is rectilinear. The natural motion of the heavens, on the other hand, is circular (it does not seem possible to characterize such unchanging and apparently eternal motion as forced). From this Aristotle infers that the substance composing the heavenly bodies and the heavenly spheres (see p. 43) is a new element, aether.

To those accustomed to the ideas of modern physics, the distinction between natural and unnatural motion seems highly "artificial" and inept. But if we think of the natural motion of a body as that which it would follow if not constrained by some visible external force, the Aristotelian distinction seems a reasonable one, in the sense that it so readily suggests itself that it requires some reason to reject it. But there are, in fact, good reasons for rejecting the Aristotelian view. (1) Fire and air do not naturally or freely go upward; they are forced to do so by the descent of colder air. (2) Circular motion may be conceived as the resultant of two rectilinear components (making it unnecessary to posit a new element). It is curious that this latter analysis, though it was made in the Peripatetic treatise *Mechanics* for the rotary motion of terrestrial matter, was not systematically applied to celestial motion before Newton. Newton's achievement, however, presupposes Galileo's conception of inertia, that (in the absence of friction) the continued application of force is manifested not in uniform motion but in accelerated motion.

The doctrine of the aether as a fifth element was of fundamental importance in the later Peripatetic and medieval theory of matter. See also p. 352.

Aristotle, *De Caelo* I. 2–3 (268*b*14–270*a*12). Translation of J. L. Stocks

All natural bodies and magnitudes we hold to be, as such, capable of locomotion; for nature, we say, is their principle of movement. But all movement that is in place, all locomotion, as we term it, is either straight or circular or a combination of these two, which are the only simple movements. And the reason of this is that these two, the straight and the circular line, are the only simple magnitudes. Now revolution about the centre is circular motion, while the upward and downward movements are in a straight line, "upward" meaning motion away from the centre, and "downward" motion towards it. All simple motion, then, must be motion either away from or towards or about the centre. This seems to be in exact accord with what we said above: as body found its completion in three dimensions, so its movement completes itself in three forms. . . .

We may take it that all movement is either natural or unnatural, and that the movement which is unnatural to one body is natural to another—as, for instance, is the case with the upward and downward movements, which are natural and unnatural to fire and earth respectively. It necessarily follows that circular movement, being unnatural to these bodies, is the natural movement of some other. Further, if, on the one hand, circular movement is *natural* to something, it must surely be some simple and primary body which is ordained to move with a natural circular motion, as fire is ordained to fly up and earth down. If, on the other hand, the movement of the rotating bodies about the centre is *unnatural*, it would be remarkable and indeed quite inconceivable that this movement alone should be continuous and eternal, being nevertheless contrary to nature.

At any rate the evidence of all other cases goes to show that it is the unnatural which quickest passes away. And so, if, as some say, the body so moved is fire, this movement is just as unnatural to it as downward movement; for any one can see that fire moves in a straight line away from the centre. On all these grounds, therefore, we may infer with confidence that there is something beyond the bodies that are about us on this earth, different and separate from them; and that the superior glory of its nature is proportionate to its distance from this world of ours.

In consequence of what has been said, in part by way of assumption and in part by way of proof, it is clear that not every body either possesses lightness or heaviness. As a preliminary we must explain in what sense we are using the words "heavy" and "light," sufficiently, at least, for our present purpose: we can examine the terms more closely later, when we come to consider their essential nature. Let us then apply the term "heavy" to that which naturally moves towards the centre, and "light" to that which moves naturally away from the centre. The heaviest thing will be that which sinks to the bottom of all things that move downward, and the lightest that which rises to the surface of everything that moves upward. Now, necessarily, everything which moves either up or down possesses lightness or heaviness or both—but not both relatively to the same thing: for things are heavy and light relatively to one another; air, for instance, is light relatively to water, and water light relatively to earth. The body, then, which moves in a circle cannot possibly possess either heaviness or lightness. For neither naturally nor unnaturally can it move either towards or away from the centre. Movement in a straight line certainly does not belong to it *naturally*, since one sort of movement is, as we saw, appropriate to each simple body, and so we should be compelled to identify it with one of the bodies which move in this way. Suppose, then, that the movement is *unnatural*. In that case, if it is the downward movement which is unnatural, the upward movement will be natural; and if it is the upward which is unnatural, the downward will be natural. For we decided that of contrary movements, if the one is unnatural to anything, the other will be natural to it. But since the natural movement of the whole and of its part—of earth, for instance, as a whole and of a small clod—have one and the same direction, it results, in the first place, that this body can possess no lightness or heaviness at all (for that would mean that it could move by its own nature either from or towards the centre, which, as we know, is impossible); and, secondly, that it cannot possibly move in the way of locomotion by being forced violently aside in an upward or downward direction. For neither naturally nor unnaturally can it move with any other motion but its own, either itself or any part of it, since the reasoning which applies to the whole applies also to the part.

THE EQUATIONS OF ARISTOTELIAN DYNAMICS

Aristotle, *Physics* VII. 5.[1] Translation of R. P. Hardie and R. K. Gaye (Oxford, 1930)

Now since wherever there is a movent, its motion always acts upon something, is always in something, and always extends to something (by "is always in something" I mean that it occupies a time: and by "extends to something" I mean that it involves the traversing of a certain amount of distance: for at any moment when a thing is causing motion, it also has caused motion, so that there must always be a certain amount of distance that has been traversed and a certain amount of time that has been occupied). If, then, *A* the movent have moved *B* a distance *C* in a time *D*, then in the same time the same force *A* will move ½*B* twice the distance *C*, and in ½*D* it will move ½*B* the whole distance *C*: for thus the rules of proportion will be observed. Again if a given force move a given weight a certain distance in a certain time and half the distance in half the time, half the motive power will move half the weight the same distance in the same time. Let *E* represent half the motive power *A* and *Z* half the weight *B*: then the ratio between the motive power and the weight in the one case is similar and proportionate to the ratio in the other, so that each force will cause the same distance to be traversed in the same time.[2]

But if *E* move *Z* a distance *C* in a time *D*, it does not necessarily follow that *E* can move twice *Z* half the distance *C* in the same time.[3] If, then,

[1] This passage reveals not only a noteworthy achievement but also the failure of Aristotle's dynamics. The achievement consists in applying mathematics to physical phenomena, and in making certain abstractions which such treatment requires, e.g., in neglecting, as irrelevant, differences in the bodies moved other than weight (and, by implication, shape), in considering the medium perfectly homogeneous, which it never is in nature, and in defining force quantitatively in terms of the effect produced. But in reaching the result that, in "forced" motion, the distance would vary directly as the product of force and time, and inversely as the weight, Aristotle fails to carry abstraction and analysis far enough. The effect of friction and the resistance of the medium are included in the factor of weight. The basic case of a body moving without friction in a medium devoid of resistance, under the influence of a single constant force, is not considered. [Edd.]

[2] In considering these formulations it is to be noted that for Aristotle there can be no motion without the continuous application of force throughout the duration of the motion. Even when a projectile is hurled, Aristotle holds that the projecting force (albeit in diminishing amount) is still acting, through the motion of the medium, upon the missile. The relation of Aristotle's formulations to the notions of "work" and "power" in Newtonian physics should be noted. [Edd.]

[3] According to the Aristotelian equation of "forced" motion, in which distance varies directly as the product of force and time and inversely as the weight of the object moved, the application of a force no matter how small should always result in some motion of translation. Since observed facts do not bear this out, Aristotle is forced to make an exception and to limit his equation only to cases where the force is large enough to produce any motion. The necessity for this limitation might have suggested the inadequacy of the formulation in which distance (and velocity) is proportional to force. The error lies again in the insufficient

A move *B* a distance *C* in a time *D*, it does not follow that *E*, being half of *A*, will in the time *D* or in any fraction of it cause *B* to traverse a part of *C* the ratio between which and the whole of *C* is proportionate to that between *A* and *E* (whatever fraction of *A E* may be): in fact it might well be that it will cause no motion at all; for it does not follow that, if a given motive power causes a certain amount of motion, half that power will cause motion either of any particular amount or in any length of time: otherwise one man might move a ship, since both the motive power of the ship-haulers and the distance that they all cause the ship to traverse are divisible into as many parts as there are men. Hence Zeno's reasoning is false when he argues that there is no part of the millet that does not make a sound: for there is no reason why any such part should not in any length of time fail to move the air that the whole bushel moves in falling. In fact it does not of itself move even such a quantity of the air as it would move if this part were by itself: for no part even exists otherwise than potentially.[1]

If on the other hand we have two forces each of which separately moves one of two weights a given distance in a given time, then the forces in combination will move the combined weights an equal distance in an equal time: for in this case the rules of proportion apply.

THE NECESSITY OF A MEDIUM: EFFECT OF THE DENSITY OF THE MEDIUM ON VELOCITY

Aristotle, *Physics* IV.8.[2] Translation of R. P. Hardie and R. K. Gaye

Further, in point of fact things that are thrown move though that which gave them their impulse is not touching them, either by reason of mutual replacement, as some maintain, or because the air that has been

---

analysis of the various forces involved in an observed motion. The resistance of friction and medium is included in the factor of weight. In Newtonian physics, however, this resistance is treated as a term to be added to the product of mass and acceleration in computing the force required to produce a given acceleration in a given body (which includes the case of uniform motion, where acceleration is zero). [Edd.]

[1] Simplicius tells us that when Protagoras said that a single grain of millet or a thousandth part thereof did not in falling produce a sound, Zeno of Elea, the propounder of the famous paradoxes on motion, asked why a single grain or any fraction thereof should not produce a sound bearing the same ratio to that produced by a medimnus (bushel) as the said single grain or fraction thereof bore to the medimnus. [Edd.]

[2] The question of the existence of the void was much discussed in antiquity, the atomists affirming and the other schools denying its existence. Modern physical problems of continuity and discontinuity, action at a distance, and the existence of the ether, are, in a sense, lineal descendants of the ancient problem. Aristotle argues at great length against the existence of a void. We have included only the part that deals with kinetics, viz., (1) the theory that a medium is necessary to sustain the motion of projectiles, and (2) the theory that the velocity of "natural" (i.e., free) motion is inversely proportional to the density of the medium. For views opposing those of Aristotle see pp. 217, 221, below. [Edd.]

pushed pushes them with a movement quicker than the natural locomotion of the projectile wherewith it moves to its proper place.[1] But in a void none of these things can take place, nor can anything be moved save as that which is carried is moved.

Further, no one could say why a thing once set in motion should stop anywhere; for why should it stop *here* rather than *here*? So that a thing will either be at rest or must be moved *ad infinitum*, unless something more powerful get in its way.[2]

Further, things are now thought to move into the void because it yields; but in a void this quality is present equally everywhere, so that things should move in all directions.

Further, the truth of what we assert is plain from the following considerations. We see the same weight or body moving faster than another for two reasons, either because there is a difference in what it moves through, as between water, air, and earth, or because, other things being equal, the moving body differs from the other owing to excess of weight or of lightness.

Now the medium causes a difference because it impedes the moving thing, most of all if it is moving in the opposite direction, but in a secondary degree even if it is at rest; and especially a medium that is not easily divided, i.e., a medium that is somewhat dense.

A, then, will move through *B* in time *C*, and through *D*, which is thinner, in time *E* (if the length of *B* is equal to *D*), in proportion to the density of the hindering body.[3] For let *B* be water and *D* air; then by so much as air is thinner and more incorporeal than water, *A* will move through *D* faster than through *B*. Let the speed have the same ratio to the speed, then, that air has to water. Then if air is twice as thin, the body will traverse *B* in twice the time that it does *D*, and the time *C* will be twice the time *E*. And always, by so much as the medium is more incorporeal and less resistant and more easily divided, the faster will be the movement.

Now there is no ratio in which the void is exceeded by body, as there is no ratio of 0 to a number. . . . Similarly the void can bear no ratio to the full, and therefore neither can movement through the one to movement

---

[1] The reference is to the doctrine according to which the medium, e.g., air, is viewed as a necessary condition for the motion of projectiles rather than as a factor tending to impede such motion (see p. 217). [Edd.]

[2] Aristotle here, as elsewhere, is on the point of stating the principle of inertia but is prevented by his exclusive concern with existential data rather than with hypothetical or ideal conditions and limiting cases. [Edd.]

[3] Fundamental for Aristotle's argument, and fallacious according to the modern notion of density, is the assumption that speed of motion is inversely proportional to the density of the medium. (For the effect of weight on the speed of "natural" motion, see pp. 207–212.) Since the void is of no density, motion through it would have to be of infinite speed. Such motion being impossible in the finite universe of Aristotle, it follows that the void does not exist in nature. [Edd.]

through the other, but if a thing moves through the thinnest medium such and such a distance in such and such a time, it moves through the void with a speed beyond any ratio. For let *Z* be void, equal in magnitude to *B* and to *D*. Then if *A* is to traverse and move through it in a certain time, *H*, a time less than *E*, however, the void will bear this ratio to the full. But in a time equal to *H*, *A* will traverse the part *T* of *D*. And it will surely also traverse in that time any substance *Z* which exceeds air in thickness in the ratio which the time *E* bears to the time *H*. For if the body *Z* be as much thinner than *D* as *E* exceeds *H*, *A*, if it moves through *Z*, will traverse it in a time inverse to the speed of the movement, i.e., in a time equal to *H*. If, then, there is *no* body in *Z*, *A* will traverse *Z* still more quickly. But we supposed that its traverse of *Z* when *Z* was void occupied the time *H*. So that it will traverse *Z* in an equal time whether *Z* be full or void. But this is impossible. It is plain, then, that if there is a time in which it will move through any part of the void, this impossible result will follow: it will be found to traverse a certain distance, whether this be full or void, in an equal time; for there will be some *body* which is in the same ratio to the other body as the time is to the time.

To sum the matter up, the cause of this result is obvious, viz., that between any two movements there is a ratio (for they occupy time, and there is a ratio between any two times, so long as both are finite), but there is no ratio of void to full.

These are the consequences that result from a difference in the media; the following depend upon an excess of one moving body over another. We see that bodies which have a greater impulse either of weight or of lightness, if they are alike in other respects,[1] move faster over an equal space, and in the ratio which their magnitudes bear to each other. Therefore they will also move through the void with this ratio of speed. But that is impossible; for why should one move faster? (In moving through *plena* it must be so; for the greater divides them faster by its force. For a moving thing cleaves the medium either by its shape, or by the impulse which the body that is carried along or is projected possesses.) Therefore all will possess equal velocity.[2] But this is impossible.

It is evident from what has been said, then, that, if there is a void, a result follows which is the very opposite of the reason for which those who believe in a void set it up.

---

[1] E.g., in shape. [Edd.]

[2] Note that the modern view that all bodies falling from the same point through a vacuum fall with finite speed would not (apart from the question of relative weights) contradict Aristotle, who would insist that what we call "vacuum" is not a void or absolutely empty space but space suffused with some material body. But the experimental fact that as a medium is rarefied the speed of motion increases not indefinitely but toward a finite limit renders Aristotle's idea of density unfruitful for science. [Edd.]

ARISTOTLE'S LAWS FOR THE MOTION OF FREELY FALLING AND FREELY RISING BODIES

Aristotle's treatment of the subject of freely falling bodies is often seized upon by detractors of Greek science. Indeed this often constitutes their sole acquaintance with ancient science and is remembered merely in connection with the dramatic experiment, popularly attributed to Galileo, in which balls of different weight were dropped from the tower of Pisa. It should be noted, however, that if this experiment ever took place it was not in a vacuum but in the air, and the resistance or retarding force of the air is not the same for bodies of different weights any more than for bodies of different shapes. In the air heavier bodies *do* fall more rapidly than lighter ones of the same volume and shape, but the ratio of the velocities is not, as Aristotle assumed, equal to that of the weights, but is a far more complicated ratio.

Aristotle's theory of "natural" motion is in harmony with his theory of "forced" motion (see p. 200). In the latter case, the average velocity of a body over a given distance is proportional to the force exerted (continuously) upon it; in the former case, weight, which measures the internal tendency of a heavy body to move toward the center of the universe, is substituted for force. The result is the strange doctrine that the velocity of a freely falling body is proportional to weight. Some of the problems in interpreting Aristotle's doctrine are discussed by I. E. Drabkin in *American Journal of Philology* 59 (1938) 60–84.

The failure of Aristotle here is due (just as in the case of his discussion of "forced" motion) to his not having analyzed a phenomenon of great complexity far enough into its component factors. He arrived, indeed, at the conclusion that all bodies falling freely in a void would fall at an equal velocity, but his sole use of this conclusion was to deny the possibility of a void in nature, since the velocity in question would have to be indefinitely large (see p. 205).

This brings us to the other difficulty with Aristotle's analysis, the notion that the velocity of a freely falling body is inversely proportional to the density of the medium (see the preceding selection). Density is left as a qualitative term undefined in terms of weight per unit of volume. But if some such quantitative notion of density as this were assumed, clearly the proper approach would be to consider that the velocity of fall through a medium is equal to the velocity of fall in a void *minus* a term dependent (*inter alia*) on the density of the medium, not *divided by* such a term.

Now there was a school in antiquity that made this very approach. The selection from Philoponus (p. 217) is typical of this school. But it is to be noted that Philoponus's views are vitiated by the insistence that in the (hypothetical) void bodies fall at velocities proportional to their weights.

Certain of the atomists, again, hold that the velocity of fall in the void is equal (and finite) for all bodies, but no quantitative theory is developed that would lead to or flow from a fruitful principle of inertia.

Acceleration in connection with freely falling bodies is discussed below (p. 208). Throughout our discussion of freely falling bodies in Aristotle's theory, analogous propositions with respect to freely rising bodies are to be supplied.

Translation of Aristotle's *Physics* by R. P. Hardie and R. K. Gaye and of *De Caelo* by J. L. Stocks

De Caelo 273*b*30–274*a*2: A given weight moves a given distance in a given time; a weight which is as great and more moves the same distance in a less time, the times being in inverse proportion to the weights. For instance, if one weight is twice another, it will take half as long over a given movement.

De Caelo 290*a*1–2: Whenever bodies are moving with their proper motion, the larger moves quicker.

Physics 216*a*13–16: We see that bodies which have a greater impulse either of weight or of lightness, if they are alike in other respects,[1] move faster over an equal space, and in the ratio which their magnitudes bear to each other.

De Caelo 308*a*29–33: By absolutely light, then, we mean that which moves upward or to the extremity, and by absolutely heavy that which moves downward or to the centre. By lighter or relatively light we mean that one, of two bodies endowed with weight and equal in bulk, which is exceeded by the other in the speed of its natural downward movement.[2]

De Caelo 277*b*4–5: The greater the mass of fire or earth the quicker always is its movement towards its own place.

De Caelo 309*b*11–15: For any two portions of fire, small or great, will exhibit the same ratio of solid to void; but the upward movement of the greater is quicker than that of the less, just as the downward movement of a mass of gold or lead, or of any other body endowed with weight, is quicker in proportion to its size.[3]

De Caelo 294*a*13–15: A little bit of earth, let loose in mid-air, moves and will not stay still, and the more there is of it the faster it moves. . . .

Cf. Hero of Alexandria, *Mechanics* II. 34*d*: Why do heavier bodies fall to the ground in shorter time than lighter bodies? The reason is that, just as heavy bodies move more readily the larger is the *external* force by which they are set in motion, so they move more swiftly the larger is the *internal* force within themselves. And in *natural* motion this internal force and downward tendency are greater in the case of heavier bodies than in the case of lighter.

ACCELERATION OF "NATURAL" (FREE) MOTION IN ARISTOTLE AND OTHERS

Aristotle, *De Caelo* I. 8, 277*a*27–33. Translation of J. L. Stocks

This conclusion that local movement is not continued to infinity is corroborated by the fact that earth[4] moves more quickly the nearer it is

[1] E.g., in shape. [Edd.]

[2] The proposition is not specifically stated, but a combination of this passage with *De Caelo* 309*b*11 and 273*b*30, cited here, implies the doctrine that the ratio of the velocities of free fall of two bodies of the same volume (and shape) and unequal weights is equal to the ratio of their weights. [Edd.]

[3] I.e., the ratio of the velocities of the free fall of two bodies of the same substance (and shape) and unequal volumes (and, therefore, weight) is equal to the ratio of their weights. [Edd.]

[4] I.e., freely falling earth. [Edd.]

to the center, and fire the nearer it is to the upper place. But if movement were infinite, speed would be infinite also; and if speed then weight and lightness. For as superior speed in downward movement implies superior weight, so infinite increase of weight necessitates infinite increase of speed.

Aristotle, *Physics* V.6, 230*b*24–28. Translation of R. P. Hardie and R. K. Gaye

But whereas the velocity of that which comes to a standstill seems always to increase, the velocity of that which is carried violently seems always to decrease. . . . Moreover, "coming to a standstill" is generally recognized to be identical or at least concomitant with the locomotion of a thing to its proper place.

Simplicius, *Commentary on Aristotle's De Caelo* 264.20–267.6 (Heiberg)

Now in this discussion there is general agreement that bodies move more swiftly as they approach their natural places, but various explanations are adduced. Aristotle holds that as bodies approach the whole mass of their own element they acquire a greater force therefrom and recover their form more perfectly; that thus it is by reason of an increase of weight that earth moves more swiftly when it is near the center.

Hipparchus,[1] on the other hand, in his work entitled *On Bodies Carried Down by Their Weight* declares that in the case of earth thrown upward it is the projecting force that is the cause of the upward motion, so long as the projecting force overpowers the downward tendency of the projectile, and that to the extent that this projecting force predominates, the object moves more swiftly upwards; then, as this force is diminished (1) the upward motion proceeds but no longer at the same rate, (2) the body moves downward under the influence of its own internal impulse, even though the original projecting force lingers in some measure, and (3) as this force continues to diminish the object moves downward more swiftly, and most swiftly when this force is entirely lost.

Now Hipparchus asserts that the same cause operates in the case of bodies let fall from above. For, he says, the force which held them back remains with them up to a certain point, and this is the restraining factor which accounts for the slower movement at the start of the fall. Alexander[2] replies: "This may be true in the case of bodies moved by force or kept by force in the place opposite their natural place, but the argument no longer applies to bodies which on coming into being move in accordance with their own nature to their proper place."

---

[1] Hipparchus of Nicaea (second century B.C.) is generally considered the greatest astronomer of antiquity. We have already referred to some of Hipparchus's scientific achievements (pp. 82, 115, 128, 134, 139).

[2] See p. 90.

On the subject of weight, too, Hipparchus contradicts Aristotle for he [Hipparchus] holds that bodies are heavier the further removed they are from their natural places.[1] This, too, fails to convince Alexander. "For," writes Alexander, "it is far more reasonable to suppose that when bodies change their natures, as when light bodies become heavy, they still retain something of their former nature when they are still at the very beginning of their downward fall and are just changing to that form by virtue of which they are carried downward, and that they become heavier as they go along, than to suppose that they still keep the force imparted to them by that which originally kept them up and prevented them from moving downward. Furthermore, if it is in the nature of the heavy to be below (for this is why its natural motion is toward that place), objects would be heaviest and would have assumed their proper form in this regard whenever they were below; and, since they have their perfection in downward movement, it would be reasonable to suppose that they receive an additional weight the nearer they come to that place. For if these bodies move downward more swiftly in proportion to their distance from above, it would be unreasonable to suppose that they exhibit this property in proportion as they are less heavy. For to hold such a view is to deny that these bodies move downward because of weight. . . ."

Now there are not a few who assert that bodies move downward more swiftly as they draw nearer their goal because objects higher up are supported by a greater quantity of air, objects lower down by a lesser quantity, and that heavier objects fall more swiftly because they divide the underlying air more easily. For just as in the case of bodies which sink in water the lighter they are the more does the water seem to hold them up and resist the downward motion, so it is fair to suppose that the same thing happens in air, and that the greater the amount of underlying air, the more do lighter objects seem buoyed up. Similarly, a greater amount of fire moves upward more swiftly since it divides the air above it more easily, and in proportion as the quantity of air above it is greater, that which moves upward through the air moves more slowly. And though air may not be similar by nature to water, air does, since it is corporeal, impede the motion of objects passing through it. If this is the case, acceleration is due not to addition of weight, but to diminution of the resistant medium. . . .

"The reason given by Aristotle for acceleration in natural motion, namely, an addition of weight or lightness, is," says Alexander, "a sounder reason and more in accordance with nature. Aristotle would hold that acceleration is due to the fact that as the body approaches its natural place it attains its form in a purer degree, that is, if it is a heavy body, it becomes heavier, and if light, lighter."

[1] See p. 200.

Now in the first place I think it worth while to investigate how the acceleration of bodies as they approach their natural places (a fact said to be universally acknowledged) is to be explained. Again, if it is a case of addition of weight or lightness, it follows that a body weighed in air at the surface of the earth should appear heavier than if weighed in air from a high tower, or tree, or sheer precipice (the weigher stretching himself out over the edge). Now this seems impossible, unless, indeed, one were to say that in this case the difference in weight is imperceptible.

Simplicius, *Commentary on Aristotle's Physics*, p. 916.4–30 (Diels)

It is universally asserted as self-evident that bodies moving naturally to their natural places undergo acceleration. As to the cause of such acceleration some say, and reasonably, that the bodies are endowed with greater force as they approach more closely to their proper wholeness, that is, as they achieve greater perfection of form. Others, again, say that it is the quantity of intervening air that impedes bodies moving upward or downward, until these bodies approach their natural places, when only a small amount of the medium is left to be traversed. But few adduce any proof of the fact itself, that when bodies moving naturally are near their natural places they move more swiftly. It may therefore not be out of place to set forth the indications [of acceleration] given by Strato, the Physicist.[1] For in his treatise On Motion, after asserting that a body so moving completes the last stage of its trajectory in the shortest time, he adds: "In the case of bodies moving through the air under the influence of their weight this is clearly what happens. For if one observes water pouring down from a roof and falling from a considerable height, the flow at the top is seen to be continuous, but the water at the bottom falls to the ground in discontinuous parts. This would never happen unless the water traversed each successive space more swiftly." By "this" Strato means the breaking up of the continuity of the object as it approaches the ground.

Strato also adduces another argument, as follows: "If one drops a stone or any other weight from a height of about an inch, the impact made on the ground will not be perceptible, but if one drops the object from a height of a hundred feet or more, the impact on the ground will be a powerful one.

[1] Strato of Lampsacus, known as the Physicist, was the successor of Theophrastus in the leadership of the Peripatetic School. He was the author of numerous works on logic, ethics, physics, metaphysics, psychology, and physiology. None of these can with certainty be identified with any that have come down to us, but it seems probable that we have in the Introduction to the *Pneumatics* of Hero of Alexandria (see p. 248) a considerable fragment which, if not actually from a work of Strato, represents his thought quite closely. Strato seems to have combined both atomistic and Aristotelian theories in his physical system. He tries to prove, with the aid of experiments, not merely that air is corporeal but that it contains fine vacua amid its particles of matter, that all matter, in fact (except possibly the diamond) contains such "discontinuous" vacua, and that a "continuous" vacuum does not exist naturally but can be produced artificially.

Now there is no other cause for this powerful impact. For the weight of the object is not greater, the object itself has not become greater, it does not strike a greater space of ground, nor is it impelled by a greater [external force]. It is merely a case of acceleration. And it is because of this acceleration that this phenomenon and many others take place."

The proof given, I think, indicates that an object raised but slightly from the earth is slow to move, for it is still, as it were, on the earth, but when an object moves toward its natural place from a distance considerably above the earth its power always keeps increasing as it approaches.

Cf. Plutarch, *Platonic Questions* 7.5: Weights thrown [downward] cleave the air as they fall upon it with impact and scatter it. The air, moving to the rear, follows the falling body, for it is the nature of air to rush in and fill an empty place. This action accelerates the motion of the falling body.

THE KINETICS OF ATOMS IN THE SYSTEM OF EPICURUS

Lucretius, *On the Nature of Things* II. 80–164, 216–332.[1] Translation of Cyril Bailey (Oxford, 1921)

If you think that the first-beginnings of things can stay still, and by staying still beget new movements in things, you stray very far away from true reasoning. For since they wander through the void, it must needs be that all the first-beginnings of things move on either by their own weight[2] or sometimes by the blow of another. For when quickly, again and again,

[1] Lucretius (*ca.* 98–*ca.* 55 B.C.), Roman poet-philosopher, takes as his theme the philosophical system of Epicurus. With a fervor that is rarely relaxed he preaches a materialism that will enable men to free themselves of the chief cause of unhappiness, the fear of the gods and the fear of death. Though neglected in the Middle Ages, Lucretius has been studied since the Renaissance with increased attention.

The first two books of the *De Rerum Natura* contain the doctrine of matter and motion, matter as consisting of atoms (of various sizes and shapes) and their combinations, motion as taking place in the void; the third deals with the mind and the soul as impermanent combinations of atoms; the fourth with the sensations and their atomic basis; the fifth with the formation of our world and the evolution of life and civilization on the earth; the sixth takes up various meteorological and geological phenomena, suggesting, in the Epicurean manner, various hypotheses that might account for the phenomena.

In their view of the universe the Epicureans adopted, with important modifications, the atomic physics of Leucippus and Democritus. [Edd.]

[2] Herein lies an important difference between Democritean and Epicurean atomism. In the former system weight is neither a primary quality of the atom nor a cause of motion. Weight first enters into the system when a world is in the process of being formed, and then only as a measure of the degree of resistance to the attendant whirl, wherein central and outer parts are differentiated. Before that stage, the motion of the atom is not due to weight but is a specific characteristic of atomicity and there is no differentiation of "up" and "down." The system of Epicurus differs, as our text indicates. [Edd.]

they have met and clashed together, it comes to pass that they leap asunder at once this way and that; for indeed it is not strange, since they are most hard with solid heavy bodies, and nothing bars them from behind. And the more you perceive all the bodies of matter tossing about, bring it to mind that there is no lowest point in the whole universe, nor have the first-bodies any place where they may come to rest, since I have shown in many words, and it has been proved by true reasoning, that space spreads out without bound or limit, immeasurable towards every quarter everywhere. And since that is certain, no rest, we may be sure, is allowed to the first-bodies moving through the deep void, but rather plied with unceasing, diverse motion, some when they have dashed together leap back at great space apart, others too are thrust but a short way from the blow. And all those which are driven together in more close-packed union and leap back but a little space apart, entangled by their own close-locking shapes, these make the strong roots of rock and the brute bulk of iron and all other things of their kind. Of the rest which wander through the great void, a few leap far apart, and recoil afar with great spaces between; these supply for us thin air and the bright light of the sun. Many, moreover, wander on through the great void, which have been cast back from the unions of things, nor have they anywhere else availed to be taken into them and link their movements. And of this truth, as I am telling it, a likeness and image is ever passing presently before our eyes. For look closely, whenever rays are let in and pour the sun's light through the dark places in houses: for you will see many tiny bodies mingle in many ways all through the empty space right in the light of the rays, and as though in some everlasting strife wage war and battle, struggling troop against troop, nor ever crying a halt, harried with constant meetings and partings; so that you may guess from this what it means that the first-beginnings of things are for ever tossing in the great void. So far as may be, a little thing can give a picture of great things and afford traces of a concept. And for this reason it is the more right for you to give heed to these bodies, which you see jostling in the sun's rays, because such jostlings hint that there are movements of matter too beneath them, secret and unseen. For you will see many particles there stirred by unseen blows change their course and turn back, driven backwards on their path, now this way, now that, in every direction everywhere. You may know that this shifting movement comes to them all from the first-beginnings. For first the first-beginnings of things move of themselves; then those bodies which are formed of a tiny union, and are, as it were, nearest to the powers of the first-beginnings, are smitten and stirred by their unseen blows, and they in their turn, rouse up bodies a little larger. And so the movement passes upwards from the first-beginnings, and little by little comes forth to our senses, so that those bodies move too, which we can descry in the sun's light; yet it is not clearly seen by what blows they do it.[1]

Next, what speed of movement is given to the first-bodies of matter, you may learn, Memmius, in a few words from this. First, when dawn strews the land with new light, and the diverse birds flitting through the distant woods across the soft air fill the place with their clear cries, we see that it is plain and evident for all to behold how suddenly the sun is wont at such a time to rise and clothe all things, bathing them in his light. And yet that heat which the sun sends out, and that calm light of his, is not passing through empty space; therefore, it is constrained to go more slowly, while it dashes asunder, as it were, the waves of air. Nor again do the several particles of heat move on one by one, but entangled one with another, and joined in a mass; therefore they are at once dragged back each by the other, and impeded from without, so that they are constrained to go more slowly. But the first-beginnings, which are of solid singleness, when they pass through the empty void, and nothing checks them without, and they themselves, single wholes with all their parts, are borne, as they press on, towards the one spot which they first began to seek, must needs, we may be sure, surpass in speed of motion, and be carried far more quickly than the light of the sun, and rush through many times the distance of space in the same time in which the flashing light of the sun crowds the sky.

Herein I would fain that you should learn this, too, that when first-bodies are being carried downwards straight through the void by their own weight, at times quite undetermined, and at undetermined spots they push a little from their path:[2] yet only just so much as you could call a change of trend. But if they were not used to swerve, all things would fall downwards through the deep void like drops of rain, nor could collision

---

[1] Just as free atoms move with an incomprehensibly great velocity, so the atoms within compounds retain this velocity as they are jostled about by the other atoms within the compound. The extent of the trajectories depends on the closeness of the union. The compound, of course, whose component atoms are thus jostled in all directions, cannot itself be said to move with atomic velocity. Instead, as the atoms unite to form larger and larger bodies, the motion of the latter is retarded more and more until finally a body is produced large enough to be perceived and with motion slow enough to be perceived. Such is the dust particle in the sunbeam—a favorite example of the ancient atomists. Note throughout the emphasis on analogies. [Edd.]

[2] Having assumed that all free atoms fall in the same direction (downward) in the void, and at equal speed, Epicurus, in order to account for the initial impacts of atom upon atom, further assumes that at unpredictable times and places an atom may swerve ever so slightly from the perpendicular. Again, this swerve without external cause releases the universe from the inexorable chain of causality, introduces an indeterminism not present in the Democritean system, and accounts for the free will manifested by living beings. Analogies have sometimes been drawn between certain features of modern physical theory and the abandonment of strict causality in the system of Epicurus. [Edd.]

come to be, nor a blow brought to pass for the first-beginnings: so nature would never have brought aught to being.

But if perchance any one believes that heavier bodies, because they are carried more quickly straight through the void, can fall from above on the lighter, and so bring about the blows which can give creative motions, he wanders far away from true reason. For all things that fall through the water and thin air, these things must needs quicken their fall in proportion to their weights, just because the body of water and the thin nature of air cannot check each thing equally, but give place more quickly when overcome by heavier bodies. But, on the other hand, the empty void cannot on any side, at any time, support anything, but rather, as its own nature desires, it continues to give place; *wherefore all things must needs be borne on through the calm void, moving at equal rate with unequal weights.*[1] The heavier will not then ever be able to fall on the lighter from above, nor of themselves bring about the blows, which make diverse the movements, by which nature carries things on. Wherefore, again and again, it must needs be that the first-bodies swerve a little; yet not more than the very least, lest we seem to be imagining a sideways movement, and the truth refute it. For this we see plain and evident, that bodies, as far as in them lies, cannot travel sideways, since they fall headlong from above, as far as you can descry. But that nothing at all swerves from the straight direction of its path, what sense is there which can descry?

Once again, if every motion is always linked on, and the new always arises from the old in order determined, nor by swerving do the first-beginnings make a certain start of movement to break through the decrees of fate, so that cause may not follow cause from infinite time; whence comes this free will for living things all over the earth, whence, I ask, is it wrested from fate, this will whereby we move forward, where pleasure leads each one of us, and swerve likewise in our motions neither at determined times nor in a determined direction of place, but just where our mind has carried us? For without doubt it is his own will which gives to each one a start for this movement, and from the will the motions pass flooding through the limbs. Do you not see too how, when the barriers are flung open, yet for an instant of time the eager might of the horses cannot burst out so suddenly as their mind itself desires? For the whole store of matter throughout the whole body must be roused to movement, that then aroused through every limb it may strain and follow the eager longing of the mind; so that you see a start of movement is brought to pass from the heart, and comes forth first of all from the will of the mind, and then afterwards is spread through all the body and limbs. Nor is it the same as when we move forward

---

[1] This is the fundamental proposition about freely falling bodies in our physics. It was to avoid the consequences of its application to freely falling atoms that Epicurus introduced the "swerve" (see also p. 206). [Edd.]

impelled by a blow from the strong might and strong constraint of another. For then it is clear to see that all the matter of the body moves and is hurried on against our will, until the will has reined it back throughout the limbs. Do you not then now see that, albeit a force outside pushes many men and constrains them often to go forward against their will and to be hurried away headlong, yet there is something in our breast, which can fight against it and withstand it? And at its bidding too the store of matter is constrained now and then to turn throughout the limbs and members, and, when pushed forward, is reined back and comes to rest again. Wherefore in the seeds too you must needs allow likewise that there is another cause of motion besides blows and weights, whence comes this power born in us, since we see that nothing can come to pass from nothing. For weight prevents all things coming to pass by blows, as by some force without. But that the very mind feels not some necessity within in doing all things, and is not constrained like a conquered thing to bear and suffer, this is brought about by the tiny swerve of the first-beginnings in no determined direction of place and at no determined time.

Nor was the store of matter ever more closely packed nor again set at larger distances apart. For neither does anything come to increase it nor pass away from it. Wherefore the bodies of the first-beginnings in the ages past moved with the same motion as now, and hereafter will be borne on for ever in the same way; such things as have been wont to come to being will be brought to birth under the same law, will exist and grow and be strong and lusty, inasmuch as is granted to each by the ordinances of nature. Nor can any force change the sum of things; for neither is there anything outside, into which any kind of matter may escape from the universe, nor whence new forces can arise and burst into the universe and change the whole nature of things and alter its motions.[1]

Herein we need not wonder why it is that, when all the first-beginnings of things are in motion, yet the whole seems to stand wholly at rest, except when anything starts moving with its entire body. For all the nature of the first-bodies lies far away from our senses, below their purview; wherefore, since you cannot reach to look upon them, they must needs steal away their motions from you too; above all, since such things as we can look upon, yet often hide their motions, when withdrawn from us on some distant spot. For often the fleecy flocks cropping the glad pasture on a hill creep on whither each is called and tempted by the grass bejewelled with fresh dew, and the lambs fed full gambol and butt playfully; yet all this seems blurred to us from afar, and to lie like a white mass on a green hill. Moreover, when mighty legions fill the spaces of the plains with their

[1] In this paragraph we have the doctrine of conservation of matter and momentum in its simplest form. [Edd.]

chargings, awaking a mimic warfare, a sheen rises there to heaven and all the earth around gleams with bronze, and beneath a noise is roused by the mighty mass of men as they march, and the hills smitten by their shouts turn back the cries to the stars of the firmament, and the cavalry wheel round and suddenly shake the middle of the plains with their forceful onset, as they scour across them. And yet there is a certain spot on the high hills, whence all seems to be at rest and to lie like a glimmering mass upon the plains.

ANTI-ARISTOTELIAN VIEWS ON THE LAWS OF FALLING BODIES

Ioannes Philoponus, *Commentary on Aristotle's Physics*, pp. 678.24–684.10 (Vitelli) [1]

Weight, then, is the efficient cause of downward motion, as Aristotle himself asserts. This being so, given a distance to be traversed, I mean through a void where there is nothing to impede motion, and given that the efficient cause of the motion differs,[2] the resultant motions will inevitably be at different speeds, even through a void. . . . Clearly, then, it is the natural weights of bodies, one having a greater and another a lesser downward tendency, that cause differences in motion. For that which has a greater downward tendency divides a medium better. Now air is more

[1] Philoponus, who lived at the end of the fifth and beginning of the sixth century A.D., wrote on philosophy, grammar, rhetoric, theology, mathematics, and science. He composed commentaries on no less than eleven works of Aristotle. After conversion to Christianity he published *De Aeternitate Mundi*, an attack on paganism and Neoplatonism, and *De Opificio Mundi*, a defense of biblical cosmogony. He is the author of a treatise on the plane astrolabe (see p. 134) and of a commentary on the *Arithmetic* of Nicomachus.

Of the greatest interest are some of his views on basic questions of physics, in which he combats Aristotle and approaches the modern position more closely than did Aristotle. This is particularly true of his argument against *antiperistasis* (see p. 221), and his insistence, against Aristotle, that, so far from being essential to the motion of a projectile, the medium acts merely to resist such motion. The motion of the projectile is due entirely, according to Philoponus, to a force impressed on it by the projector. In connection with freely falling bodies, he argues that in the void bodies would fall with finite velocities and that the time required for a body to fall a given distance through a medium is equal to that required for the body to fall the same distance through a void *plus* an amount proportional to the density of the medium. Aristotle would make the total time required for the fall proportional to the density of the medium. It is to be noted, however, that for Philoponus bodies falling in the void fall not with equal velocities but with finite velocities proportional to their weights, while for Aristotle, though the velocities in the (hypothetical) void would be equal, they would not be finite. It is difficult to say to what extent Philoponus is original in these views and to what extent he is recording an anti-Aristotelian tradition.

Philoponus is often known by the name John the Grammarian (Ioannes Grammaticus). Whether he is to be identified with John of Alexandria, to whom are ascribed certain medical works, has been questioned. "Philoponus" (lover of labor) seems to be an honorific surname. That our author merited it is clear from the fact that his extant Aristotelian commentaries—only one part of his work—contain some 3,000 large pages of closely printed Greek.

[2] I.e., that there are differences in weight.

effectively divided by a heavier body. To what other cause shall we ascribe this fact than that that which has greater weight has, by its own nature, a greater downward tendency, even if the motion is not through a plenum? . . .

And so, if a body cuts through a medium better by reason of its greater downward tendency, then, even if there is nothing to be cut, the body will none the less retain its greater downward tendency. . . . And if bodies possess a greater or a lesser downward tendency in and of themselves, clearly they will possess this difference in themselves even if they move through a void. The same space will consequently be traversed by the heavier body in shorter time and by the lighter body in longer time, even though the space be void. The result will be due not to greater or lesser interference with the motion but to the greater or lesser downward tendency, in proportion to the natural weight of the bodies in question. . . .

Sufficient proof has been adduced to show that if motion took place through a void, it would not follow that all bodies would move therein with equal speed. We have also shown that Aristotle's attempt to prove that they would so move does not carry conviction. Now if our reasoning up to this point has been sound it follows that our earlier proposition is also true, namely, that it is possible for motion to take place through a void in finite time. . . .

Thus, if a certain time is required for each weight, in and of itself, to accomplish a given motion, it will never be possible for one and the same body to traverse a given distance, on one occasion through a plenum and on another through a void, in the same time.

For if a body moves the distance of a stade through air, and the body is not at the beginning and at the end of the stade at one and the same instant, a definite time will be required, dependent on the particular nature of the body in question, for it to travel from the beginning of the course to the end (for, as I have indicated, the body is not at both extremities at the same instant), and this would be true even if the space traversed were a void. But a certain *additional time* is required because of the interference of the medium. For the pressure of the medium and the necessity of cutting through it make motion through it more difficult.

Consequently, the thinner we conceive the air to be through which a motion takes place, the less will be the *additional time* consumed in dividing the air. And if we continue indefinitely to make this medium thinner, the additional time will also be reduced indefinitely, since time is indefinitely divisible. But even if the medium be thinned out indefinitely in this way, the total time consumed will never be reduced to the time which the body consumes in moving the distance of a stade through a void. I shall make my point clearer by examples.

If a stone move the distance of a stade through a void, there will necessarily be a time, let us say an hour, which the body will consume in moving the given distance. But if we suppose this distance of a stade filled with water, no longer will the motion be accomplished in one hour, but a certain additional time will be necessary because of the resistance of the medium. Suppose that for the division of the water another hour is required, so that the same weight covers the distance through a void in one hour and through water in two. Now if you thin out the water, changing it into air, and if air is half as dense as water, the time which the body had consumed in dividing the water will be proportionately reduced. In the case of water the additional time was an hour. Therefore the body will move the same distance through air in an hour and a half.[1] If, again, you make the air half as dense, the motion will be accomplished in an hour and a quarter. And if you continue indefinitely to rarefy the medium, you will decrease indefinitely the time required for the division of the medium, for example, the additional hour required in the case of water. But you will never completely eliminate this additional time, for time is indefinitely divisible.

If, then, by rarefying the medium you will never eliminate this additional time, and if in the case of motion through a plenum there is always some portion of the second hour to be added, in proportion to the density of the medium, clearly the stade will never be traversed by a body through a void in the same time as through a plenum. . . .

But it is completely false and contrary to the evidence of experience to argue as follows:[2] "If a stade is traversed through a plenum in two hours, and through a void in one hour, then if I take a medium half as dense as the first, the same distance will be traversed through this rarer medium in half the time, that is, in one hour: hence the same distance will be traversed through a plenum in the same time as through a void." *For Aristotle wrongly assumes that the ratio of the times required for motion through various media is equal to the ratio of the densities of the media. . . .*

Now this argument of Aristotle's seems convincing and the fallacy is not easy to detect because it is impossible to find the ratio which air bears to water, in its composition, that is, to find how much denser water is than air, or one specimen of air than another.[3] But from a consideration of the moving bodies themselves we are able to refute Aristotle's contention. For if, in the case of one and the same body moving through two different

---

[1] For the *additional* time in the case of air will be only half an hour.

[2] This is one of Aristotle's arguments in denying the existence of the void. Vitelli's conjectural restoration of a lacuna here has been adopted.

[3] It is not merely that instruments were lacking. The ancients in general lacked the quantitative notion of density as mass per unit of volume, though the work of Archimedes had prepared the way, and Menelaus may also have investigated the subject.

media, the ratio of the times required for the motions were equal to the ratio of the densities of the respective media, then, since differences of velocity are determined not only by the media but also by the moving bodies themselves, the following proposition would be a fair conclusion: "in the case of bodies differing in weight and moving through one and the same medium, the ratio of the times required for the motions is equal to the inverse ratio of the weights." For example, if the weight were doubled, the time would be halved. That is, if a weight of two pounds moved the distance of a stade through air in one-half hour, a weight of one pound would move the same distance in one hour.[1] Conversely, the ratio of the weights of the bodies would have to be equal to the inverse ratio of the times required for the motions.

But this is completely erroneous, and our view may be corroborated by actual observation more effectively than by any sort of verbal argument. *For if you let fall from the same height two weights*[2] *of which one is many times as heavy as the other, you will see that the ratio of the times required for the motion does not depend on the ratio of the weights, but that the difference in time is a very small one.* And so, if the difference in the weights is not considerable, that is, if one is, let us say, double the other, there will be no difference, or else an imperceptible difference, in time, though the difference in weight is by no means negligible, with one body weighing twice as much as the other.

Now if, in the case of different weights in motion through the same medium, the ratio of the times required for the motions is not equal to the inverse ratio of the weights, and, conversely, the ratio of the weights is not equal to the inverse ratio of the times, the following proposition would surely be reasonable: "If identical bodies move through different media, like air and water, the ratio of the times required for the motions through the air and water, respectively, is not equal to the ratio of the densities of air and water, and conversely."

Now if the ratio of the times is not determined by the ratio of the densities of the media, it follows that a medium half as dense will not be traversed in half the time, but in longer than half. Furthermore, as I have indicated above, in proportion as the medium is rarefied, the shorter is the *additional* time required for the division of the medium. But this additional time is never completely eliminated; it is merely decreased in proportion to the degree of rarefaction of the medium, as has been indicated. . . . And so, if the *total* time required is not reduced in proportion to the degree of

---

[1] The text has *one hour . . . one half hour*, either through a copyist's carelessness or possibly through a confusion of natural and forced motion.

[2] It is not to be assumed that Philoponus originated this experiment any more than that Galileo did.

rarefaction of the medium, and if the time added for the division of the medium is diminished in proportion to the rarefaction of the medium, but never entirely eliminated, it follows that a body will never traverse the same distance through a plenum in the same time as through a void.

ANTI-ARISTOTELIAN VIEWS ON "FORCED" MOTION

Ioannes Philoponus, *Commentary on Aristotle's Physics*, pp. 639.3–642.9 (Vitelli) [1]

Such, then, is Aristotle's account in which he seeks to show that forced motion and motion contrary to nature[2] could not take place if there were a void. But to me this argument does not seem to carry conviction. For in the first place really nothing has been adduced, sufficiently cogent to satisfy our minds, to the effect that motion contrary to nature or forced motion is caused in one of the ways enumerated by Aristotle. . . .

For in the case of *antiperistasis*[3] there are two possibilities; (1) the air that has been pushed forward by the projected arrow or stone moves back to the rear and takes the place of the arrow or stone, and being thus behind it pushes it on, the process continuing until the impetus of the missile is exhausted, or, (2) it is not the air pushed ahead but the air from the sides that takes the place of the missile. . . .

Let us suppose that *antiperistasis* takes place according to the first method indicated above, namely, that the air pushed forward by the arrow gets to the rear of the arrow and thus pushes it from behind. On that assumption, one would be hard put to it to say what it is (since there seems to be no counter force) that causes the air, once it has been pushed forward, to move back, that is along the sides of the arrow, and, after it reaches the rear of the arrow, to turn around once more and push the arrow forward. For, on this theory, the air in question must perform three distinct motions: it must be pushed forward by the arrow, then move back, and finally turn and proceed forward once more. Yet air is easily moved, and once set in motion travels a considerable distance. How, then, can the air, pushed by the arrow, fail to move in the direction of the impressed impulse, but instead, turning about, as by some command, retrace its course? Furthermore, how can this air, in so turning about, avoid being

[1] On Philoponus, see p. 217. The passage here quoted is of interest because of its refutation of the Aristotelian view that the action of the medium (e.g., air) sustains the motion of a projectile. Philoponus argues that such motion is sustained by the force impressed by the projector in the projectile, while the resistance of the medium tends to oppose the motion.

[2] See p. 201. "Forced" motion is a more general term than motion "contrary to nature." The former refers to any but "natural" motion, the latter to motion in a direction exactly opposite to that of "natural" motion. The terms are not always used in this strict sense, however.

[3] The term is used in general of the process whereby P_1 pushes P_2 into P_3's place, P_2 pushes P_3 into P_4's place, . . . , P_{n-1} pushes P_n into P_1's place.

scattered into space, but instead impinge precisely on the notched end of the arrow and again push the arrow on and adhere to it? Such a view is quite incredible and borders rather on the fantastic.

Again, the air in front that has been pushed forward by the arrow is, clearly, subjected to some motion, and the arrow, too, moves continuously. How, then, can this air, pushed by the arrow, take the place of the arrow, that is, come into the place which the arrow has left? For before this air moves back, the air from the sides of the arrow and from behind it will come together and, because of the suction caused by the vacuum, will instantaneously fill up the place left by the arrow, particularly so the air moving along with the arrow from behind it. Now one might say that the air pushed foward by the arrow moves back and pushes, in its turn, the air that has taken the place of the arrow, and thus getting behind the arrow pushes it into the place vacated by the very air pushed forward (by the arrow) in the first instance. But in that case the motion of the arrow would have to be discontinuous. For before the air from the sides, which has taken the arrow's place, is itself pushed, the arrow is not moved. For this air does not move it. But if, indeed, it does, what need is there for the air in front to turn about and move back? And in any case, how or by what force could the air that had been pushed forward receive an impetus for motion in the opposite direction? . . .

So much, then, for the argument which holds that forced motion is produced when air takes the place of the missile (*antiperistasis*). Now there is a second argument which holds that the air which is pushed in the first instance [i.e., when the arrow is first discharged] receives an impetus to motion, and moves with a more rapid motion than the natural [downward] motion of the missile, thus pushing the missile[1] on while remaining always in contact with it until the motive force originally impressed on this portion of air is dissipated. This explanation, though apparently more plausible, is really no different from the first explanation by *antiperistasis*, and the following refutation will apply also to the explanation by *antiperistasis*.

In the first place we must address the following question to those who hold the views indicated: "When one projects a stone by force, is it by pushing the air behind the stone that one compels the latter to move in a direction contrary to its natural direction? Or does the thrower impart a motive force to the stone, too?" Now if he does not impart any such force to the stone, but moves the stone merely by pushing the air, and if the bowstring moves the arrow in the same way, of what advantage is it for the stone to be in contact with the hand, or for the bowstring to be in contact with the notched end of the arrow?

---

[1] Reading *αὐτό* for *αὐτόν* (of Vitelli).

For it would be possible, without such contact, to place the arrow at the top of a stick, as it were on a thin line,[1] and to place the stone in a similar way, and then, with countless machines, to set a large quantity of air in motion behind these bodies. Now it is evident that the greater the amount of air moved and the greater the force with which it is moved the more should this air push the arrow or stone, and the further should it hurl them. But the fact is that even if you place the arrow or stone upon a line or point quite devoid of thickness and set in motion all the air behind the projectile with all possible force, the projectile will not be moved the distance of a single cubit.

If, then, the air, though moved with a greater force,[2] could not impart motion to the projectile, it is evident that, in the case of the hurling of missiles or the shooting of arrows, it is not the air set in motion by the hand or bowstring that produces the motion of the missile or arrow. For why would such a result be any more likely when the projector is in contact with the projectile than when he is not? And, again, if the arrow is in direct contact with the bowstring and the stone with the hand, and there is nothing between, what air behind the projectile could be moved? If it is the air from the sides that is moved, what has that to do with the projectile? For that air falls outside the [trajectory of the] projectile.

From these considerations and from many others we may see how impossible it is for forced motion to be caused in the way indicated.[3] *Rather is it necessary to assume that some incorporeal motive force is imparted by the projector to the projectile*, and that the air set in motion contributes either nothing at all or else very little to this motion of the projectile. If, then, forced motion is produced as I have suggested, it is quite evident that if one imparts motion "contrary to nature" or forced motion to an arrow or a stone the same degree of motion will be produced much more readily in a void than in a plenum. And there will be no need of any agency external to the projector. . . .

THE PARALLELOGRAM OF VELOCITIES

Hero of Alexandria, *Mechanics* I. 8[4]

We shall now prove that a point moved by two motions, each of uniform velocity, may traverse unequal distances [in a given time]. Let *ABDC* represent a rectangle with diagonal *AD*. Let point *A* move with constant velocity along line *AB*, and let line *AB* [at the same time] move with constant velocity along lines *AC* and *BD*. Let the time which point

[1] To reduce friction to a minimum.

[2] I.e., greater than the force that is used by one who hurls a projectile.

[3] I.e., by the motion of air.

[4] See p. 197, n. 1, above.

A takes to reach B be equal to the time which line AB takes to reach CD. I say that point A in a given time moves along two unequal lines.

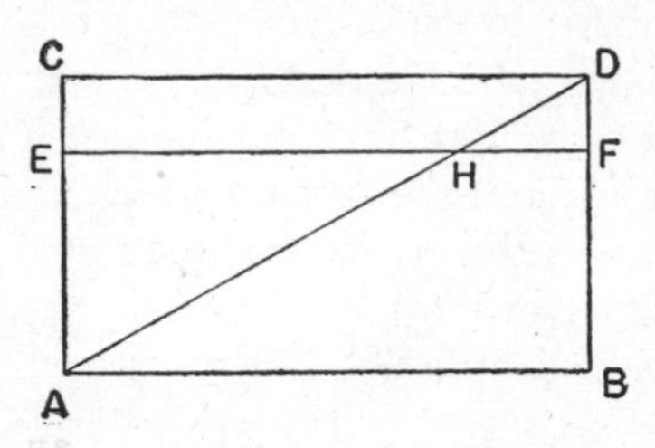

Proof: when line AB has moved for a given time, and has reached the position EF, point A, which moves along line AB, will, at the given time, also be on line EF. And there is a constant proportion. The ratio of line AC to line AB (i.e., to line CD) is equal to the ratio of line AE to the line extending from point E to the point moving on it. But $AC:CD = AE:EH$. Therefore the point moving along line AB will, at H, be on line AD, the diagonal. Similarly it can be shown that the point moving on line AB is always moving along line AD, and traverses, in a given time, both lines AB and AD. But AD and AB are unequal. Therefore a point moving with a constant velocity will, in a given time, traverse two unequal lines. But the motion of the point on line AB is, as we have pointed out, a simple motion, whereas its motion on diagonal AD is composed of (1) the motion of line AB on the two lines AC and BD and (2) the motion of A along line AB. Therefore point A will in a given time and with constant velocity traverse two unequal lines. Q.E.D.[1]

The Dynamics of Machinery

THE FIVE SIMPLE MACHINES

Hero of Alexandria, *Mechanics* II. 1–6, in Pappus, *Mathematical Collection*, pp. 1114.22–1124.4 1128.3–1130.3 (Hultsch)

1. We shall now give, from the works of Hero, an abridged account of the five aforesaid machines. We shall also set forth, for the benefit of students, the essential facts with regard to cranes of one, two, three, and four masts on the chance that one who seeks the books in which these subjects have been treated may not find them available. For we have met with many copies which have been considerably mutilated, lacking both beginning and end.[2]

There are five machines by the use of which a given weight is moved by a given force, and it will be our task to give the forms, the applications, and the names of these machines. Now both Hero and Philo[3] have shown that these machines, though they differ considerably from one another in their external form, are reducible to a single principle. The names are as follows: wheel and axle, lever, system of pulleys, wedge, and, finally, the so-called endless screw.

[1] There is a similar treatment of the parallelogram of velocities in *Mechanics* 1 of the Aristotelian Corpus (see p. 198, above).

[2] The reference seems to be to copies of Hero's works.

[3] I.e., Philo of Byzantium (see p. 255).

The wheel and axle is constructed in the following way. One must take a strong log squared off like a beam, make its ends round by planing,[1] and fit bronze end-pivots to the axle, so that when inserted in round openings in the immovable framework, they turn easily. For these openings also have a bronze lining upon which the end-pivots rest. Now this beam is called the axle. At the middle of this axle is placed a wheel having a square opening to fit the axle so that the axle and wheel turn together.

Having indicated the construction we must now speak of the use of the machine. If we wish to move a large weight with a smaller force, we attach a cable to the weight and fasten this cable around the curved portion of the axle. We then insert spokes in holes bored in the wheel and turn the wheel by pressing on the spokes. In this way the weight will readily be lifted by a smaller force, as the cable is wound around the axle, or else bunched to keep the whole cable from covering the axle.[2] The size of this machine must be adapted to the weight which one intends to move; and the ratio of the parts of the machine[3] must be adapted to the ratio of the weight moved to the moving force, as will be proved below.[4]

2. The second machine is the lever, which may well have been man's first discovery in connection with the moving of large weights. For, wishing to move large weights, men found it necessary first to lift them from the ground. But they had no means of grasping the weight because all parts of its base rested on the ground. They therefore dug a small groove under the weight, inserted here the end of a long wooden pole, and placed a stone, called hypomochlion (fulcrum), under the pole near the weight. They then pressed down on the other end. When they saw that the motion was quite easy they realized that large weights could be moved in this way. Now the pole is called a lever and is either squared or rounded. The nearer the fulcrum is placed to the weight, the more easily is the weight moved, as will be shown below.

[1] The Greek is obscure. As the sequel shows, the entire part about which the cable turned was also to be rounded.

[2] This last clause, though bracketed by Hultsch, is present in the Arabic version.

[3] I.e., the ratio of the diameter of the wheel to that of the axle.

[4] Hero, *Mechanics* II.22, not quoted by Pappus but extant in the Arabic version (p. 232, below).

3. The third machine is the system of pulleys. If we wish to lift a given weight we may attach a rope to it and pull with a force equal to the weight in question. If, however, we untie[1] the rope from the weight, attach one end to an immovable beam, pass the other end around a pulley-wheel made fast to the weight, and pull this free end we shall move the weight more easily. Again, if to the fixed beam we attach another pulley-wheel, pass the free end of the rope over it, and then pull, we shall move the weight still more easily. And still further, if we attach another pulley-wheel to the weight, pass the free end of the rope over it and pull, we shall move the weight still more easily. Thus, as we attach more pulley-wheels both to the fixed beam and to the weight, and pass the free end of the rope successively around the wheels, we shall move the weight more and more easily. The greater the number of parts into which the rope is divided by the successive turns, the more easily will the weight be moved. But that end of the rope which is not pulled must be attached to the fixed beam.[2] However, to avoid the necessity of attaching each separate pulley-wheel to the fixed beam or to the weight, those pulleys which were described as attached to the fixed beam are instead inserted into a single wooden frame, called a pulley-block (manganon), and permitted to rotate about their axes. This pulley-block is itself attached by another cable to the fixed beam. Again, those pulley-wheels connected with the weight are inserted into a second pulley-block which is like the first and is, in turn, connected only with the weight. The pulley-wheels must be so arranged in the blocks that the cord lengths do not become entangled with one another and hard

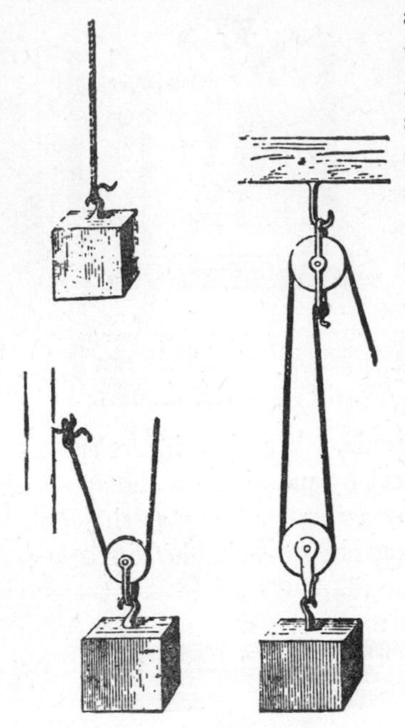

[1] Reading ἐκλύσαντες instead of ἑλκύσαντες (Hultsch).

[2] I.e., directly or indirectly (see the figures). The last two sentences, though bracketed by Hultsch, are present in the Arabic version.

to move. We shall show below[1] why an increase in the number of turns of the rope makes the lifting of the weight easier, and why one end of the rope is connected with the fixed beam.

4. The next machine, the wedge, also has important applications, for example in pressing out unguents, or in making possible perfect joints in carpentry. But the most important of all the uses of the wedge is in the removal of stones from quarries, when the stones must be separated from the solid bed-rock beneath. None of the other machines can accomplish

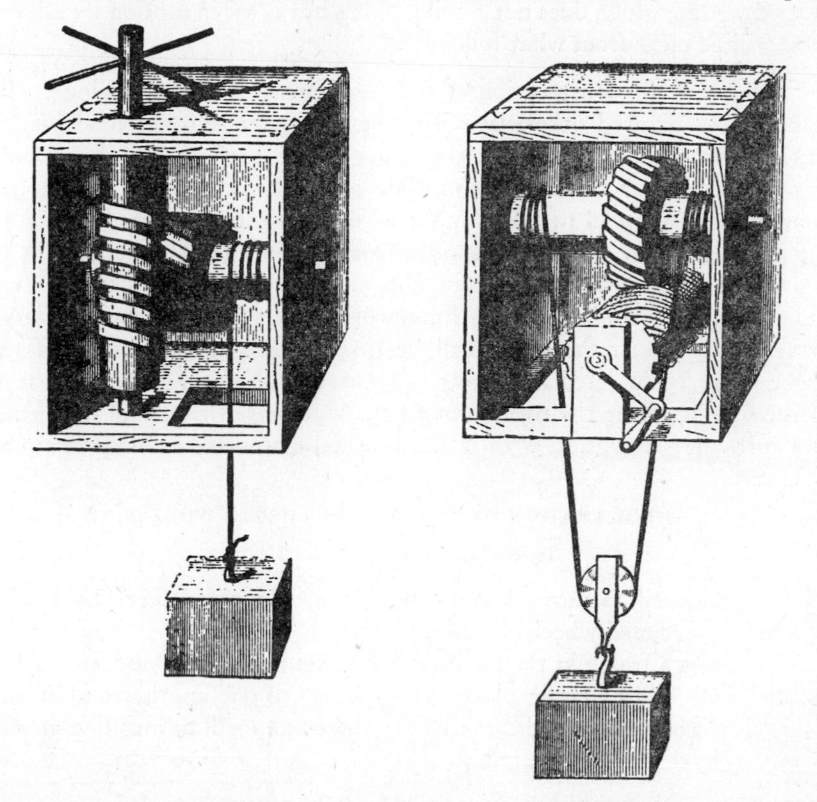

this, even if they are all joined together. Only the wedge functions in this case, with the help of the force that happens to be available. There is, furthermore, no slackening of the action of the wedge in the intervals between the successive blows of the workmen, but the tension continues high. This is clear from the fact that, on occasion, even when the wedge is not being struck, sounds are heard and breaking takes place under the action of the wedge. The smaller the angle of the wedge the easier is its

[1] Hero, *Mechanics* II.24, not quoted by Pappus but extant in the Arabic version (p. 233, below).

operation, that is, the less powerful is the required blow, as we shall show below.

5. The construction of each of the machines which we have described is clear and such as to make the machine complete in itself, as is in many cases obvious from the applications. The screw, however, both in construction and in application involves some complexity. Sometimes, to be sure, it operates alone, but at other times is used in conjunction with another machine. It is to be noted, however, that the screw is merely a spiral-shaped wedge, which does not receive blows but is set in motion by a lever.[1] This will be clear from what follows. . . .[2]

6. Such is the construction of the screw when it operates alone. But it may also be used otherwise. For we may take another machine, the so-called wheel and axle, with the drum about the axle toothed, and a screw placed next to this drum. The screw may be placed either perpendicular or parallel to the ground, with its spiral thread fitting into the teeth of the drum and its extremities rotating in circular openings in the framework, as we have indicated above. Now one end of the screw is permitted to protrude from the framework. At this end either a crank is attached by which the screw will be turned, or holes are bored so that spokes may be inserted and the screw turned. Now if we attach ropes to the weight, place the ropes around the axle on both sides of the drum,[3] and turn the screw, and with it the toothed drum, we shall draw up the weight.

THE OPERATION OF A SYSTEM OF COGGED WHEELS

Hero of Alexandria, *Dioptra* 37[4]

It is required to move a given weight with a given force by the use of a series of cogged wheels.

Construct a frame in the form of a rectangular box. Between the long parallel walls of this frame place axles parallel to one another and at such intervals that cogged wheels attached to these axles will fit into one another in the manner to be indicated.

---

[1] Hultsch adds "and by turning" but the addition is not found in the Arabic version.

[2] There follows (pp. 1124.4–1128.2) a description of the method of constructing the screw.

[3] As in the second figure, p. 227.

[4] The description of the system of cogged wheels was, in all probability, interpolated in the *Dioptra* from another work of Hero. It has nothing to do with the essential subject matter of the *Dioptra*. A very similar Greek text (but involving different numbers) is quoted by Pappus in his *Mathematical Collection* (pp. 1060–1068) as from the Βαρουλκός (*Weight-lifter*) of Hero. The same material also appears in the extant Arabic version of Hero's *Mechanics* (see p. 197), though at the very beginning, where it also seems out of place.

The basic principle of simple machines is assumed throughout the discussion, viz., that the product of the weight moved by the distance through which it is moved is equal to the product of the motive force by the distance through which it acts.

Let *ABCD* represent the aforesaid box, and *EZ* an axle set therein in the manner described and able to be turned freely. To this axle attach a cogged wheel, *HG*, having a diameter five times, let us say, that of *EZ*.

To illustrate the arrangement by an example, let us suppose the weight to be lifted is 1,000 talents, and the moving force available, that is, the force which a man or boy can exert by himself without the aid of a machine, is five talents.[1] Now if a rope attached to the weight is passed through a hole in base *AB* and wound around axle *EZ*, this rope will, as it is wound up, lift the weight. But in order that wheel *HG* be moved, a force greater than 200 talents must be exerted upon it, since the diameter of the wheel is, as we have assumed, five times as great as that of the axle. For this was shown in the discussion of the five simple machines. But we have available a force not of 200 talents but only of five.

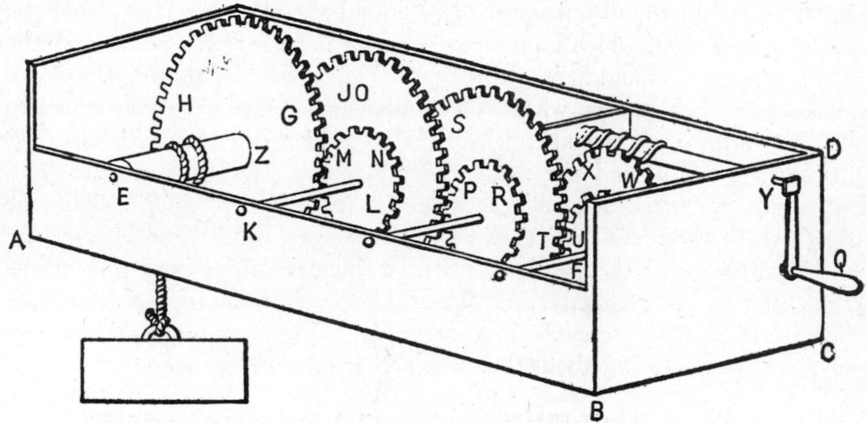

Therefore, place a second axle *KL*, with a cogged wheel *MN* attached to it, parallel to *EZ*. Let wheel *HG* be so geared that its teeth fit into those of wheel *MN*. Now attach to axle *KL* a wheel *JO* with diameter, in turn, five times that of *MN*. A force, then, of 40 talents will be needed by one who wishes to move the weight by the application of force to wheel *JO*. For 40 talents is one-fifth of 200 talents.

Therefore, again place another cogged wheel *PR*, in such a way as to mesh with cogged wheel *JO*, and attach to *PR*[2] still another such wheel *ST* having a diameter five times that of *PR*. The force required at wheel *ST* to balance the weight will be, according to the proportion, eight talents. But the force at our disposal was given as five talents.

Consequently, place still another cogged wheel *UF* so as to mesh with cogged wheel *ST*. To the axle of wheel *UF* attach a cogged wheel *XW*,

[1] A force of five talents seems, in the context, to mean the force just necessary to keep lifting five talents vertically.

[2] That *PR* is attached to an axle parallel to *EZ* and *KL* and that *ST* is attached to the same axle is implied here but expressed in Pappus's quotation.

of which the diameter bears to the diameter of wheel *UF* the same ratio as eight talents to five talents, the given available force.

Now if we suppose the box *ABCD* raised and the wheels and axles arranged as indicated, and if we suspend the weight from axle *EZ* and apply the lifting force at wheel *XW*, then neither the weight nor the force will prevail (provided that the axles turn readily and the geared wheels mesh properly), but the force will just support the weight, as on a balance.

Again, if we increase either of the two by a small additional weight, the side to which the weight has been added will overbalance the other. For example, if a weight of but one mina be added to the force of five talents, it will overbalance and draw up the other weight.

Now instead of adding a weight, place a screw next to wheel *XW*, the thread fitting the teeth of the wheel and the screw turning easily on pegs inserted in round openings. Let one of the pegs extend outside the frame at wall *CD*, which is at right angles to the screw. Square off the protruding portion and fit it with a crank *YQ*, so that by turning the crank one may turn the screw, wheel *XW*, and wheel *UF*, which is attached to *XW*. In this way the next wheel *ST* will be turned, as will also *PR* (attached to *ST*), and the next wheel *JO*, and *MN* (attached to *JO*), and the next wheel *HG*, and the axle *EZ* (attached to *HG*). Thus the rope, wound about *EZ* and attached to the weight, will lift the weight.

That it will lift the weight is obvious since the force operating at the crank is even more effective, for the crank describes a circle greater than the perimeter of the screw.[1] For the superior power of larger circles over smaller circles rotating about the same center has been demonstrated.[2]

THE RELATION OF FORCE AND VELOCITY IN SIMPLE MACHINES

Hero of Alexandria, *Mechanics* II. 21–26

21. We assert that of all figures the circle possesses the greatest and easiest mobility, whether it revolves about its center or on a plane to which it is perpendicular. The same holds true of the figures related to the circle, that is, spheres and cylinders, for their motion is circular, as we have shown in the preceding book.

Suppose we wish, in the first place, to move a large weight with a small force by the use of the wheel and axle, without the difficulty [discussed in the preceding chapter].[3] Let the weight that we desire to move be 1,000

---

[1] That is, the available force of five talents applied to the crank would produce at wheel *XW* a force much greater than the five talents theoretically required there to balance the weight. The precise reading is doubtful, though the general intent is clear.

[2] The reference is to a discussion in the *Mechanics* similar to that of the Aristotelian *Mechanics* (pp. 189 ff, above).

[3] It had been indicated that if a single wheel and axle is used, and if it is required, for example, to lift a 1,000-talent weight with a force of five talents, the diameter of the wheel must be more than 200 times as large as that of the axle.

talents and the force with which we wish to move it be five talents. Now we must, in the first place, bring the force and the weight into equilibrium, since, after equilibrium is attained, we can make the force predominate over the weight by the addition of a slight force to the machine. We place at *A* the axle on which turns the cable attached to the weight, and we place at *B* the wheel connected with that axle. To facilitate the construction of the apparatus we make the diameter of the wheel five times that of the axle. In that case the force required to move wheel *B*, which balances the weight of 1,000 talents, will be 200 talents. But the force we have available, is, as we assumed, only five talents. With this force, therefore, we cannot move the given weight merely with wheel *B*. Hence

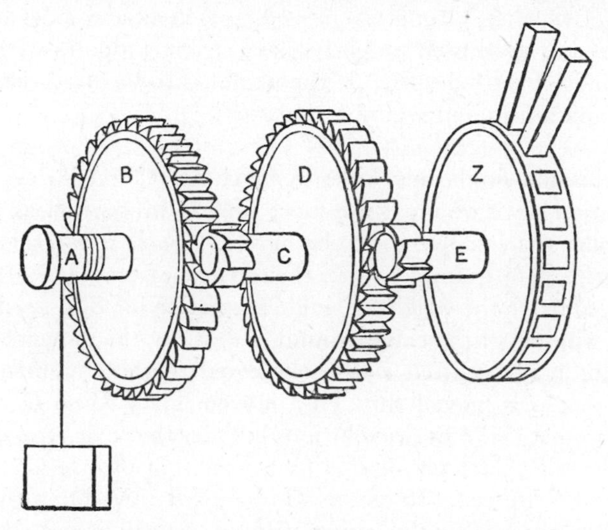

we construct a geared axle, *C*, which fits the teeth of wheel *B* in such a way that if axle *C* moves, wheel *B* is also set in motion together with the first axle, *A*, with the result that the weight moves if axle *C* turns. This axle is set in motion by the force which moves wheel *B*, for we have shown that all objects revolving about fixed centers may be moved by a small force. Hence it makes no difference whether the weight is moved by wheel *B* or by axle *C*. Now let a wheel, *D*, be attached to axle *C* and let the diameter of *D* be, say, five times that of axle *C*. In that case a force of 40 talents is required at wheel *D* to hold the weight in equilibrium. Again, if we take another axle, *E*, fitting into wheel *D*, the moving force at *E* will likewise be 40 talents. Now let there be still another wheel, *Z*, attached to axle *E*, of diameter eight times that of axle *E*, since the force of 40 talents is eight times the force of five talents. The force required at *Z* to hold the weight in equilibrium will then be five talents, which was the given available force. In order, however, that the force overcome the weight, we must either make wheel *Z* a little larger or axle *E* a little smaller.

If we do this the force will overcome the weight. If we wish to use several wheels and axles in this process we must employ the same ratio, since the combination of all the ratios must correspond to the weight if we wish to bring force and weight into equilibrium.[1] But if we want the force to overbalance the weight we must apply an amount of force just in excess of that required, according to the combined ratios, to preserve the equilibrium of the weight.[2]

Thus a given weight may be moved by an axle passing through a wheel. If, however, we do not wish to have geared wheels, we may wind cables around the wheels and axles. In that case the same work may be performed, since the first axle, that one which draws the load, will be moved by the wheel which is ultimately moved.[3] Wheels and axles of this type must, if they are to be used properly, have strong supports with holes into which the ends of the axles fit. If the weight is to be lifted these supports must be set on a firm and secure foundation.

22. In this and similar machines in which great force is developed there is a retardation since we must use more time in proportion as the moving force is smaller than the weight to be moved. *Force is to force as time is to time, inversely.* For example, when the force at wheel *B* is 200 talents and is sufficient to move the weight, a single revolution [of *B*] is needed to effect a sufficient winding of the cable around *A* that, by the movement of wheel *B* the weight may be lifted a distance equal to the circumference of *A*. But if the weight is moved through a movement of wheel *D*, the circumference of *C* must make five revolutions in order that axle *A* make one revolution, since the diameter of *B* is five times that of axle *C*. Hence five revolutions of *C* are equal to one revolution of *B* if we bring the axles and wheels back to their original position. If not, we will have a corresponding ratio.[4] Wheel *D* moves according as *B* moves, and five revolutions of *D* require five times as much time as a single revolution of *B*. Now 200 talents are five times 40 talents. Thus the ratio of moving force to moving force is equal inversely to the ratio of time to time. This is also the case with several axles and wheels. The proof is the same.

---

[1] Though it is not put very clearly, the idea seems to be that no matter how many or how few wheels and axles are used, the same rules for computing what we call the "mechanical advantage" obtain and determine whether our force is sufficient to balance the weight.

[2] Note that the efficiency of the machine is not discussed. Any force just larger than the amount *theoretically* required to keep the weight in equilibrium is presumed to be sufficient to lift the weight.

[3] We might expect the designations of "first" and "ultimate" to be interchanged.

[4] The meaning seems to be that not merely in the case of complete revolutions but in general the arc traveled by a point on the circumference of *C* is, in circular measure, five times that traveled by a point on the circumference of *B*.

23. Let it now be required to move the same weight with the same force by the machine known as the block and tackle [system of pulleys]. Let *A* represent the weight, *B* the point from which it is to be lifted, *C* the point directly above, that is, the fixed point of support to which we are to lift the weight. Suppose the block and tackle has, let us say, five pulleys, and that the pulley by which the weight is drawn originally is at point *D*. The force at *D* must then, in order to balance the 1,000 talents, be 200 talents. But the force available to us is only five talents. Therefore we draw a cord from pulley *D* to a block and tackle at point *E*. Let

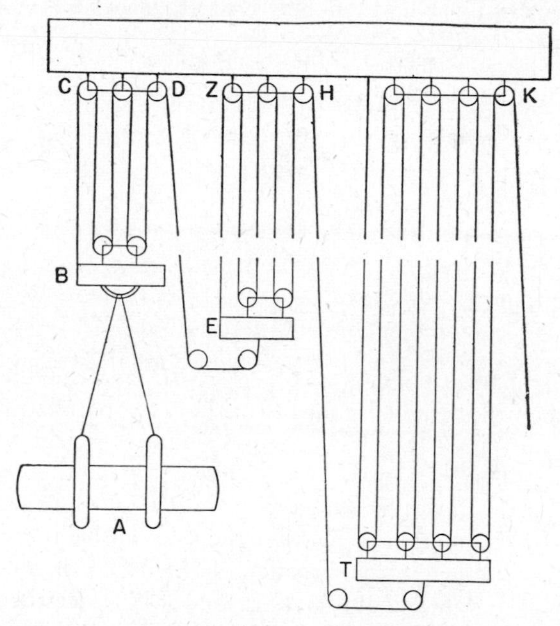

there be a fixed point of support at *Z* directly above *E*. Let there be at this fixed point of support and at *E*, in its vicinity, five pulleys, with the pulley at which the force is applied at *H*. In that case the force at *H* must be a force of 40 talents. Now if, again, we draw the end of the cord at *H* to still another block and tackle at *T*, with fixed point of support at *K*, at which the pull is exerted, it follows that since 40 talents are eight times five talents, the block and tackle must have eight pulleys for a force of five talents at *K* to balance the 1,000 talents. However, for the force at *K* to over-balance the weight, the number of pulleys must be more than eight. In that case the force will overbalance the weight.

24. A similar retardation obviously takes place in the case of this machine, too, since the same proportion is maintained [as in the case of the wheel and axle]. For when the force at *D* which amounts to 200 talents

lifts the weight from *B* to *C*, it winds [each of] the five cords around the five pulleys to the extent of the distance between *B* and *C*, while the force at *H* winds [each of] the five cords there five times as much. Now if we make *BC* equal to *EZ*, then for every unit length of rope passing through the pulleys at *BC* five units will pass through those at *EZ*, since if the weight moves the distance *BC*, five ropes each of length *BC* must pass through the pulleys. Hence time is to time as the moving force to the moving force, inversely. That the multiplication of cords should not be too great, distance *EZ* should be five times distance *BC*, and *TK* eight times *EZ*. In this case the sets of block and tackle will rise together.[1]

25. Again, the same weight [1,000 talents] may be moved by the same force [five talents] by means of the lever, according to a similar procedure.

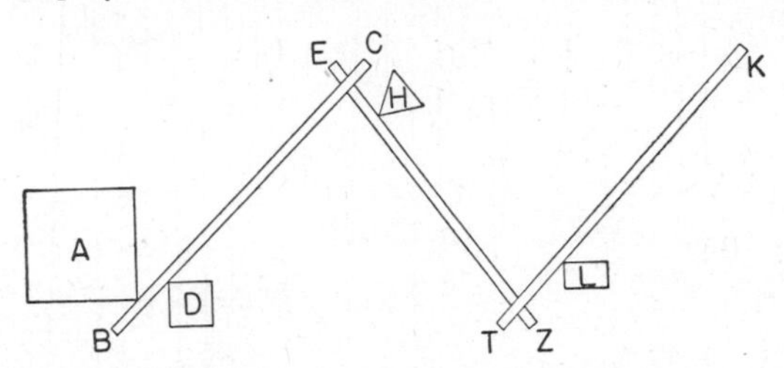

Let the weight be at point *A*, the lever at *BC*, and the fulcrum at point *D*. Let us move the weight by means of the lever parallel to the ground, making *CD* five times *BD*. Thus the force at *C* sufficient to balance the 1,000 talents will be 200 talents. Now let there be another lever, *EZ*, and let *E*, the end of this lever, be joined with point *C* so that the latter will be set in motion by the motion of *E*. Let the fulcrum be at point *H* and let arm *E* of the lever move in the direction of *D*. If *ZH* be made five times *HE*, the force at *Z* will amount to 40 talents. Now let there be still another lever, *TK*, end *T* of which is connected with *Z* and moves in a direction opposite to that in which *E* moves. If the fulcrum of this lever be at point *L*, and if this lever move in a direction opposite to that in which *E* moves, and *KL* be eight times *LT*, the force at *K* will be five talents and will balance the weight. If, however, we wish the force to overbalance the weight we must make *KL* larger than eight times *LT*. Or, if *KL* is eight times *LT*, *ZH* five times *HE*, and *CD* more than five times *DB*, the force will lift the weight.

26. Also in this case there is a retardation in the same ratio [as in the other machines], since there is no difference between these levers and the combinations of wheel and axle moving about a center. For the levers, like axles, move about points *D*, *H*, and *L*, the fulcra round which the

[1] I.e., will reach the upper support at the same time.

levers turn. Corresponding to the circles described by axles are the circles which points B, E, and T describe; corresponding to the circles described by wheels are those which points C, Z, and K describe. In the same way as we proved in the case of axles that the ratio of force to force is equal, inversely, to the ratio of time to time, so we may prove it also in the case of levers.

Hydrostatics

THE PRINCIPLES OF HYDROSTATICS

Archimedes, *On Floating Bodies* I, Postulate 1 and Propositions 1–7.[1] Translation and paraphrase of T. L. Heath, *The Works of Archimedes*, pp. 253–258

Postulate 1

"Let it be supposed that a fluid is of such a character that, its parts lying evenly and being continuous, that part which is thrust the less is driven along by that which is thrust the more; and that each of its parts is thrust by the fluid which is above it in a perpendicular direction if the fluid be sunk in anything and compressed by anything else."[2]

Proposition 1

If a surface be cut by a plane always passing through a certain point, and if the section be always a circumference [*of a circle*] *whose centre is the aforesaid point, the surface is that of a sphere.*[3]

For, if not, there will be some two lines drawn from the point to the surface which are not equal.

Suppose O to be the fixed point, and A, B to be two points on the surface such that OA, OB are unequal. Let the surface be cut by a plane

[1] In the work *On Floating Bodies*, in two books, Archimedes lays the foundations of the science of hydrostatics. In the first book the propositions lead to the so-called "Principle of Archimedes," that a body weighed in a fluid is lighter by the weight of the fluid displaced. It is this principle, with the notion of specific gravity that is involved in it, that enabled Archimedes to solve the problem of Hiero's crown. Book II investigates the conditions of stability of a right segment of a paraboloid of revolution of varying proportions and specific gravities floating in a given fluid.

The material within quotation marks is a translation, the rest a paraphrase, based on William of Moerbeke's Latin version, dated 1269. The Constantinople manuscript discovered in 1906 (see p. 69) contains a substantial portion of the original Greek text, but since Heath's version is largely a paraphrase, few changes are required by the now available Greek text. [Edd.]

[2] This postulate involves the modern notion of fluids as characterized by the complete absence of elasticity of form.

According to the Greek, the last clause of the postulate may mean: "unless the fluid is enclosed in something or is under pressure from something else." [Edd.]

[3] This proposition is needed in the proof that the surface of a fluid at rest is spherical (with the same center as that of the earth). [Edd.]

passing through OA, OB. Then the section is, by hypothesis, a circle whose centre is O.

Thus $OA = OB$; which is contrary to the assumption. Therefore the surface cannot but be a sphere.

Proposition 2

The surface of any fluid at rest is the surface of a sphere whose centre is the same as that of the earth.

Suppose the surface of the fluid cut by a plane through O, the centre of the earth, in the curve $ABCD$.

$ABCD$ shall be the circumference of a circle.

For, if not, some of the lines drawn from O to the curve will be unequal. Take one of them, OB, such that OB is greater than some of the lines from O to the curve and less than others. Draw a circle with OB as radius. Let it be EBF, which will therefore fall partly within and partly without the surface of the fluid.

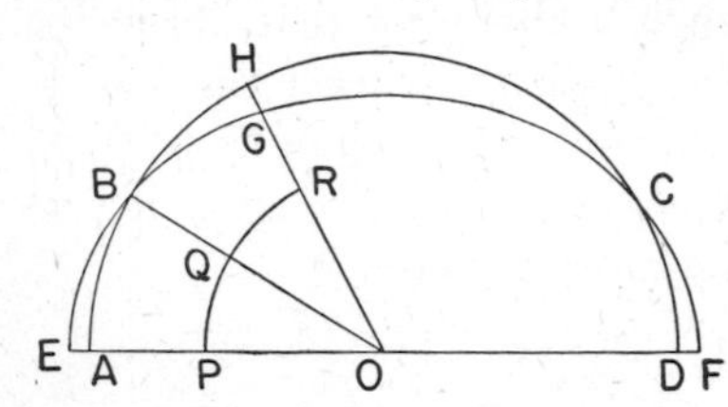

Draw OGH making with OB an angle equal to the angle EOB, and meeting the surface in H and the circle in G. Draw also in the plane an arc of a circle PQR with centre O and within the fluid.

Then the parts of the fluid along PQR are uniform and continuous, and the part PQ is compressed by the part between it and AB, while the part QR is compressed by the part between QR and BH. Therefore the parts along PQ, QR will be unequally compressed, and the part which is compressed the less will be set in motion by that which is compressed the more.

Therefore there will not be rest; which is contrary to the hypothesis.

Hence, the section of the surface will be the circumference of a circle whose centre is O; and so will all other sections by planes through O.

Therefore the surface is that of a sphere with centre O.[1]

---

[1] Compare Aristotle's proof, *De Caelo* II.4 (translation of J. L. Stocks):

"The surface of water is seen to be spherical if we take as our starting-point the fact that water naturally tends to collect in a hollow place—'hollow' meaning 'nearer the center.' Draw from the center the lines AB, AC, and let their extremities be joined by the straight line BC. The line AD, drawn to the base of the triangle, will be shorter than either of the radii. Therefore the place in which it terminates will be a hollow place. The water then will collect there until equality is established, that is, until the line AE is equal to the two radii. Thus water forces its way to the ends of the radii, and there only will it rest: but the line which connects the extremities of the radii is circular: therefore the surface of the water BEC is spherical." [Edd.]

Proposition 3

Of solids those which, size for size, are of equal weight with a fluid will, if let down into the fluid, be immersed so that they do not project above the surface but do not sink lower. . . .[1]

Proposition 4

A solid lighter than a fluid will, if immersed in it, not be completely submerged, but part of it will project above the surface. . . .

Proposition 5

Any solid lighter than a fluid will, if placed in the fluid, be so far immersed that the weight of the solid will be equal to the weight of the fluid displaced. . . .

Proposition 6

If a solid lighter than a fluid be forcibly immersed in it, the solid will be driven upwards by a force equal to the difference between its weight and the weight of the fluid displaced. . . .

Proposition 7

A solid heavier than a fluid will, if placed in it, descend to the bottom of the fluid, and the solid will, when weighed in the fluid, be lighter than its true weight by the weight of the fluid displaced.

1. The first part of the proposition is obvious, since the part of the fluid under the solid will be under greater pressure, and therefore the other parts will give way until the solid reaches the bottom.

2. Let A be a solid heavier than the same volume of the fluid, and let $(G + H)$ represent its weight, while G represents the weight of the same volume of the fluid.

Take a solid B lighter than the same volume[2] of the fluid, and such that the weight of B is G, while the weight of the same volume of the fluid is $(G + H)$.

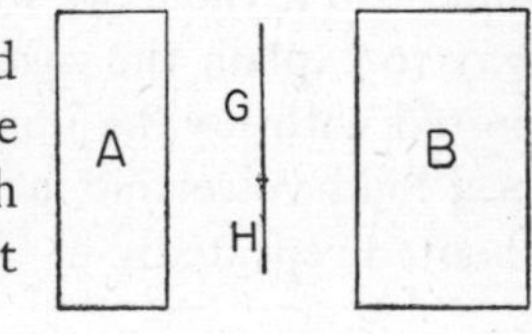

Let A and B be now combined into one solid and immersed. Then, since $(A + B)$ will be of the same weight as the same volume of fluid, both weights being equal to $(G + H) + G$, it follows that $(A + B)$ will remain stationary in the fluid.

Therefore the force which causes A by itself to sink must be equal to the upward force exerted by the fluid on B by itself. This latter is equal

[1] We have stated Propositions 3–6 without the accompanying proofs. [Edd.]

[2] I.e., the same volume as the volume of B. [Edd.]

to the difference between $(G + H)$ and G [Proposition 6]. Hence A is depressed by a force equal to H, i.e., its weight in the fluid is H, or the difference between $(G + H)$ and G.[1]

Cf. Vitruvius, *On Architecture* VII. 8.3

When it [quicksilver] is poured into a vessel and a hundred-pound weight of stone is placed over it, the stone floats at the surface, being unable by its weight to press down, squeeze out, and divide the fluid. But when the hundred-pound stone weight is removed and a scruple[2] of gold is placed on the quicksilver, the gold will not float but will, of itself, sink to the bottom. Thus it is not to be denied that the gravity[3] of bodies does not depend on their total weight but on their specific nature.

The Problem of the Crown

Vitruvius, *On Architecture* IX, Introduction 9–12. Translation of M. H. Morgan

In the case of Archimedes, although he made many wonderful discoveries of diverse kinds, yet of them all, the following, which I shall relate, seems to have been the result of a boundless ingenuity. Hiero, after gaining the royal power in Syracuse, resolved, as a consequence of his successful exploits, to place in a certain temple a golden crown which he had vowed to the immortal gods. He contracted for its making at a fixed price and weighed out a precise amount of gold to the contractor. At the appointed time the latter delivered to the king's satisfaction an exquisitely finished piece of handiwork, and it appeared that in weight the crown corresponded precisely to what the gold had weighed.

But afterwards a charge was made that gold had been abstracted and an equivalent weight of silver had been added in the manufacture of the crown. Hiero, thinking it an outrage that he had been tricked, and yet not knowing how to detect the theft, requested Archimedes to consider the matter. The latter, while the case was still on his mind, happened to go to the bath, and on getting into a tub observed that the more his body sank into it the more water ran out over the tub. As this pointed out the way to explain the case in question, without a moment's delay and transported with joy, he jumped out of the tub and rushed home naked, crying in a loud voice that he had found what he was seeking; for as he ran he shouted repeatedly in Greek, "Εὕρηκα, εὕρηκα."[4]

---

[1] Archimedes' principle of hydrostatics is utilized in the determination of the volume of irregular solids (Hero, *Metrica* II, *ad fin.*). [Edd.]

[2] About a gram.

[3] What we call "specific gravity" is really meant here.

[4] The difference between the process of discovery and of deductive proof is well illustrated by the present passage in contrast with that from the work *On Floating Bodies*. [Edd.]

Taking this as the beginning of his discovery, it is said that he made two masses of the same weight as the crown, one of gold and the other of silver. After making them, he filled a large vessel with water to the very brim and dropped the mass of silver into it. As much water ran out as was equal in bulk to that of the silver sunk in the vessel. Then, taking out the mass, he poured back the lost quantity of water, using a pint measure, until it was level with the brim as it had been before. Thus he found the weight of silver corresponding to a definite quantity of water.

After this experiment, he likewise dropped the mass of gold into the full vessel and, on taking it out and measuring as before, found that not so much water was lost, but a smaller quantity: namely, as much less as a mass of gold lacks in bulk compared to a mass of silver of the same weight. Finally, filling the vessel again and dropping the crown itself into the same quantity of water, he found that more water ran over for the crown than for the mass of gold of the same weight. Hence, reasoning from the fact that more water was lost in the case of the crown than in that of the mass, he detected the mixing of silver with the gold and made the theft of the contractor perfectly clear.[1]

WATERS OF DIFFERENT DENSITIES

[Aristotle], *Problemata* XXIII. 13. Translation of E. S. Forster (Oxford, 1927)

Why is it easier to swim in the sea than in a river? Is it because the swimmer always leans on the water as he swims, and we receive more support from that which is of a more corporeal nature, and sea water is more corporeal than river water, for it is thicker and able to offer more resistance to pressure?

Plutarch, *Quaestiones Conviviales* I. 9.2

And Theon replied: "Aristotle long ago answered the question[2] you have just propounded to us. He based his answer on the earthy matter in sea water. This rough, earthy matter being mingled with sea water

[1] The mere fact that the crown displaced a volume of water greater than that displaced by an equal weight of gold would indicate (to one familiar with the principle of Archimedes) the presence of an alloy.

To find the proportion of gold and silver in the crown, Archimedes may have proceeded, as Vitruvius says, to compare the volumes of water displaced first by the crown itself, then by a weight of gold equal to the weight of the crown, and finally by a weight of silver equal to the weight of the crown. If the volumes of water displaced are in the three cases V, V_1, and V_2, respectively, the proportion of gold to silver, by weight, will be $(V_2 - V)/(V - V_1)$.

The same result would be obtained if, instead of measuring the amount of water displaced in each case, Archimedes weighed the three bodies in water and noted the apparent loss of weight in each case. This may have been the method actually employed by Archimedes. [Edd.]

[2] The question was why fresh water washes clothes better than salt water.

makes it salt. And on this account sea water buoys up and supports swimmers and other weights more than does fresh water, since the latter is thin and weak, being unmixed and pure. . . ."

The Hydrometer

Synesius, Letter 15, to Hypatia.[1] Hercher, *Epistolographi Graeci*, p. 649

I am so unfortunate as to need a hydroscope. Have a bronze one constructed and assembled for me. It consists of a cylindrical tube the size and shape of a flute. On this tube in a vertical line are the notches by which we tell the weight of waters we are testing. A cone fits over one end of the tube so evenly and closely that cone and tube have a common base. This cover is the baryllium. Now whenever you place the tube in water it will remain erect. You will then be able to read the scale and this will indicate the weight of the water.

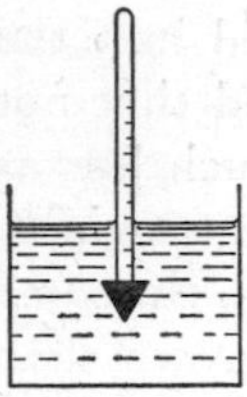

Hydrodynamics

There was no possibility of real progress in hydrodynamics before the discovery of the laws of falling bodies and the discovery that the velocity of the efflux of a liquid from an orifice varies as the square root of the "head," that is, the height of the liquid above the orifice. These discoveries were made in the seventeenth century. There are, however, passages in ancient authors indicating an appreciation of the dependence of velocity on "head" and also of the effect of the length of a pipe in reducing the volume of water delivered. In addition to passages from Frontinus we have cited a passage from Hero of Alexandria, which shows an appreciation of the dependence of the volume of water delivered not only on the cross section of the stream but on the velocity of flow. The problem of securing a uniform flow in the water clock was treated by Hero (cf. p. 244 for uniform flow from a siphon).

Sextus Julius Frontinus, *On the Water Supply of the City of Rome* 35, 113.[2] Translation of C. E. Bennett (London, 1925)

35. Let us remember that every stream of water, whenever it comes from a higher point and flows into a reservoir after a short run, not only comes up to its measure, but actually yields a surplus; but whenever it comes from a lower point, that is, under less pressure, and is conducted a longer distance, it shrinks in volume, owing to the resistance of its conduit;

---

[1] Synesius of Cyrene (*ca.* 370–415), friend and pupil of Hypatia, became Bishop of Ptolemais about 410. The letter refers to a hydrometer needed, it would seem, for testing of waters during an illness of the writer.

[2] Sextus Julius Frontinus (*ca.* 35–103), praetor in 70, governor of Britain in 75, and superintendent of aqueducts in 97, is the author of two extant treatises, *De Aquis Urbis Romae* in two books, and *Strategematica* in three (the authenticity of a fourth is doubtful). Fragments of a work on surveying ascribed to Frontinus also exist. The *De Aquis* contains valuable historical and descriptive material on the Roman water supply. [Edd.]

and that, therefore, on this principle it needs either a check or a help in its discharge.[1]

113. In setting ajutages also, care must be taken to set them on the level, and not place the one higher and the other lower down. The lower one will take in more; the higher one will suck in less, because the current of water is drawn in by the lower one . . . .

Hero of Alexandria, *Dioptra* 31

Given a spring, to determine its flow, that is, the quantity of water which it delivers.

One must, however, note that the flow does not always remain the same. Thus, when there are rains the flow is increased, for the water on the hills being in excess is more violently squeezed out. But in times of dryness the flow subsides because no additional supply of water comes to the spring. In the case of the best springs, however, the amount of flow does not contract very much.

Now it is necessary to block in all the water of the spring so that none of it runs off at any point, and to construct a lead pipe of rectangular cross section. Care should be taken to make the dimensions of the pipe considerably greater than those of the stream of water. The pipe should then be inserted at a place such that the water in the spring will flow out through it. That is, the pipe should be placed at a point below the spring so that it will receive the entire flow of water. Such a place below the spring will be determined by means of the dioptra. Now the water that flows through the pipe will cover a portion of the cross-section of the pipe at its mouth. Let this portion be, for example, 2 digits [in height]. Now suppose that the width of the opening of the pipe is 6 digits. $6 \times 2 = 12$. Thus the flow of the spring is 12 [square] digits.

It is to be noted that in order to know how much water the spring supplies it does not suffice to find the area of the cross section of the flow which in this case we say is 12 square digits. It is necessary also to find the speed of flow, for the swifter is the flow, the more water the spring supplies, and the slower it is, the less. One should therefore dig a reservoir under the stream and note with the help of a sundial how much water flows into the reservoir in a given time, and thus calculate how much will flow in a day. It is therefore unnecessary to measure the area of the cross section of the stream.[2] For the amount of water delivered will be clear from the measure of the time.

---

[1] I.e., to make the pipe discharge the normal quantity allotted to a pipe of that size.

[2] I.e., by this much more accurate method of measuring the flow, as opposed to that described above.

THE SIPHON

Hero of Alexandria, *Pneumatics* I. 1, 2, 4.[1] Translation of J. G. Greenwood (*The Pneumatics of Hero of Alexandria*, ed. B. Woodcroft, London, 1851)

1. Let *ABC* be a bent siphon, or tube, of which the leg *AB* is plunged into a vessel *DE* containing water. If the surface of the water is in *FG*, the leg of the siphon, *AB*, will be filled with water as high as the surface, that is, up to *H*, the portion *HBC* remaining full of air. If, then, we draw off the air by suction through the aperture *C*, the liquid also will follow from the impossibility, explained above, of a continuous vacuum.[2] And, if the aperture *C* be level with the surface of the water, the siphon, though full, will not discharge the water, but will remain full: so that, although it is contrary to nature for water to rise, it has risen so as to fill the tube *ABC*; and the water will remain in equilibrium, like the beams of a balance, the portion *HB* being raised on high, and the portion *BC* suspended. But if the outer mouth of the siphon be lower than the surface *FG*, as at *K*, the water flows out; for the liquid in *KB*, being heavier,[3] overpowers and draws toward it the liquid in *BH*. The discharge, however, continues only until the surface of the water is on a level with the mouth *K*, when, for the same reason as before, the efflux ceases. But if the outer mouth of the tube be lower than *K*, as at *L*, the discharge continues until the surface of the water reaches the mouth *A*. If then we wish all the water in the vessel to be drawn out, we must depress the siphon so far that the mouth *A* may reach the bottom of the vessel, leaving only a passage for the water.

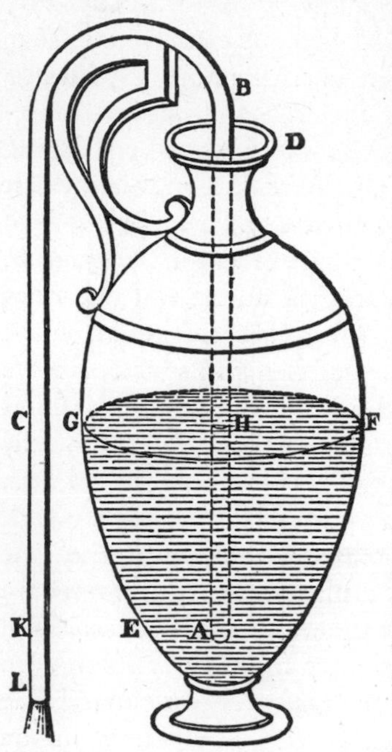

2. Now some writers have given the above explanation of the action of the siphon, saying that the longer leg, holding more, attracts the shorter. But that such an explanation is incorrect, and that he who believes so would be greatly mistaken if he were to attempt to raise water from a lower level,

[1] On the *Pneumatics* of Hero, see p. 248, below. The part here quoted follows the Introduction. [Edd.]

[2] See pp. 249 ff. below. [Edd.]

[3] This explanation is rejected in the next paragraph insofar as it considers merely the weight of the water and not also the lengths of the respective arms (i.e., the weight of water for each unit of area at the ends of the tube). The reading βαθύτερον (deeper) for βαρύτερον (heavier) may, however, be correct. [Edd.]

we may prove as follows. Let there be a siphon with its inner leg longer and narrow, and the outer much less in length but broader so as to contain more water than the longer leg. Then, having first filled the siphon with water, plunge the longer leg into a vessel of water or a well. Now, if we allow the water to flow, the outer leg, containing more than the inner, should draw the water out of the longer leg, which will at the same time draw up the water in the well; and the discharge having begun will exhaust all the water or continue forever, since the liquid without is more than that within. But this is not found to be the case; and therefore the alleged cause is not the true one. Let us then examine into the natural cause. The surface of every liquid body, when at rest, is spherical and concentric with that of the earth; and, if the liquid be not at rest, it moves until it attains such a surface.[1] If then we take two vessels and pour water into each, and, after filling the siphon and closing its extremities with the fingers, insert one leg into one vessel plunging it beneath the water, and the other into the other, all the water will be continuous, for each of the liquids in the vessels communicates with that in the siphon. If, then, the surfaces of the liquids in the vessels were at the same level before, they will both remain at rest when the siphon is plunged in. But if they were not, as soon as the water is continuous it must inevitably flow into the lower vessel through the channel of communication, until either all the water in both vessels stands at the same height, or one of the vessels is emptied. Suppose that the liquids stand at the same height; they will of course be at rest, so that the liquid in the siphon will also be at rest. If, then, the siphon be conceived to be intersected by a plane in the surface of the liquids in the vessels, even now the liquid in the siphon will be at rest, and, if raised without being inclined to either side, it will again be at rest, and that, whether the siphon is of equal breadth throughout or one leg is much larger than the other. For the reason why the liquid remained at rest did not lie in this, but in the fact that the apertures of the siphon were at the same level. The question now arises why, when the siphon is raised, the water is not borne down by its own weight, having beneath it air which is lighter than itself. The answer is that a continuous void cannot exist;[2] so that, if the water is to descend, we must first fill the upper part of the siphon, into which no air can possibly force its way. But if we pierce a hole in the upper part of the siphon, the water will immediately be rent in sunder, the air having found a passage. Before the hole is bored, the liquid in the siphon, resting on the air beneath, tends to drive it away, but the air having no means of escape does not allow the water to pass out: when however the air has obtained a passage through

---

[1] See p. 236, above. [Edd.]

[2] Such would be the case in the top of the siphon if the water began to flow out of the arms. [Edd.]

the hole, being unable to sustain the pressure of the water, it escapes. It is from the same cause that, by means of a siphon, we can suck wine upwards, though this is contrary to the nature of a liquid; for, when we have received into the body the air which was in the siphon, we become fuller than before, and a pressure is exerted on the air contiguous to us, and this in turn presses on the atmosphere at large, until a void has been produced at the surface of the wine, and then the wine undergoing pressure itself will pass into the exhausted space of the siphon;[1] for there is no other place into which it can escape from the pressure. It is from this cause that its unnatural upward movement arises. . . .

[A Siphon of Uniform Flow]

4. It is evident from what has been proved above that as long as the siphon is stationary the stream through it will be of irregular velocity, for the result is the same as in a discharge through a hole pierced in the bottom of a vessel, where the stream is irregular from the pressure of a greater weight on the discharge at its commencement, and, of a less, as the contents of the vessel are reduced. In like manner, in proportion as the excess of the outer leg of the siphon is greater, the velocity of the stream is greater; for a greater pressure is exerted on the discharge than when the projection of the outer leg below the surface of the water in the vessel is less. Therefore we have said that the discharge through the siphon is always of variable velocity. But we must contrive a siphon in which the velocity of the discharge shall be uniform.

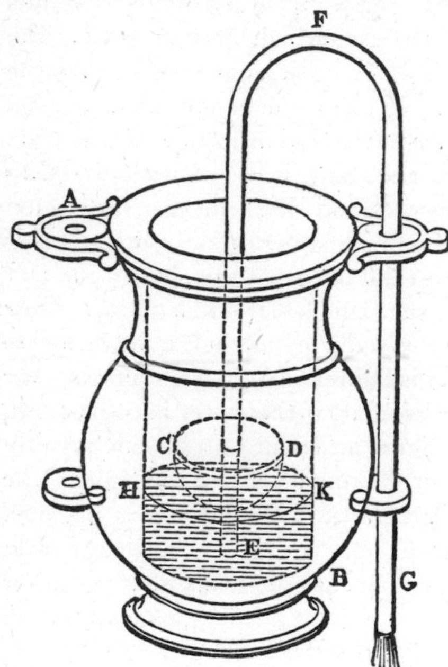

Let there be a vessel, *AB*, containing water, on which a small basin, *CD*, floats, having its mouth covered with the lid *CD*. Through this lid and the bottom of the basin insert one leg of the siphon soldering it into the holes with tin. Let the other leg be outside the vessel *AB*, having its mouth lower than the surface of the water in *AB*. If we draw the air in the siphon through the outer extremity, the water will at once

[1] This part of the explanation is correct, but the setting forth of the preliminary steps shows that the action of the siphon is not really understood. [Edd.]

follow because of the impossibility of a continuous vacuum in the siphon; and the siphon, having begun to flow, flows on until it has exhausted all the water in the vessel: but the discharge will be uniform, since the projection of the outer leg below the surface of the water does not vary; for, as the vessel becomes empty, the basin sinks with the siphon. The greater the excess of the outer leg the greater will be the velocity of the discharge, yet still uniform.[1] In the figure, *EFG* is the siphon described, and the surface of the water is in the line *HK*.

Pneumatics

From the commonplace phenomena of the bellows and siphon and the inflation of wine-skins, Greek thinkers at least as early as the fifth century B.C. concluded that air was corporeal. At the beginning of the Alexandrian age the inventive genius of Ctesibius seems to have given the impetus for the construction of more complicated devices using air pressure and suction. These were generally contrived for amusement and magical effect, as we see from the *Pneumatics* of Philo of Byzantium and Hero of Alexandria. But there were also more significant inventions based on the same principle, e.g., the hydraulic organ, artillery pieces operated by air pressure, and water pumps of various types. On the theoretical side, philosophers and naturalists sought to explain the constitution of air, the apparent "horror vacui" which seemed to be at the basis of the contrivances, the action of heat in increasing air pressure, and the action of air in supporting combustion. There were also some attempts to weigh air, but these were unsuccessful because of the inadequacy of the available instruments and procedures. The notion of air pressure as a physical force is somewhat confused among the Greeks, and the absence of a sound theory of dynamics undoubtedly prevented progress in incorporating pneumatics into the general framework of dynamics.

We have given some passages dealing with the theory of pneumatics and also some practical applications. Reference should also be made to the discussion of the siphon, which we have given under the heading of Hydrodynamics (p. 242), though it is also a part of ancient pneumatics. Devices in which the technological interest seems to outweigh the theoretical have been given under the heading of Applied Mechanics (p. 326). Mention may also be made of the importance of pneumatic theory in Greek medicine (e.g., p. 486).

THE CORPOREALITY OF AIR

[Aristotle], *Problemata* XVI. 8.[2] Translation of E. S. Forster

Of the phenomena which occur in the water-clock[3] the cause seems to be in the general that ascribed by Anaxagoras; for the air which is cut off within it is the cause of the water not entering when the tube has been

---

[1] Theoretically, the velocity of flow depends on the difference in height between the surface of water and the end of the tube from which the water flows. This difference remaining constant, the discharge is, theoretically, uniform. Hero goes on to describe (ch. 5) a siphon in which the rate of uniform flow can be adjusted. On this see A. G. Drachmann, *Journal of Hellenic Studies* 52 (1932) 116–118.

[2] On the *Problemata* see p. 300, n. 3. [Edd.]

[3] The water clock (clepsydra) here referred to is not the instrument for timekeeping but apparently a device for transferring liquids. It consists of a globular vessel with small holes at the bottom and a pipe handle open at the top (see figure, p. 327). When the clepsydra

closed. The air, however, by itself is not the cause; for if one plunges the water-clock obliquely into the water, having first blocked up the tube, the water will enter. So Anaxagoras does not adequately explain how the air is the cause; though, as has been said, it certainly is the cause. Now air, whether impelled along or travelling of itself without any compelling force, naturally travels in a straight line like the other elements. When therefore the water-clock is plunged obliquely into the water, the air preserving its straight course is driven out by the water through the holes opposite to those which are in the water, and, as it goes out, the water flows in. But if the water-clock is plunged upright into the water, the air not being able to pass straight up, because the upper parts are closed, remains round the first holes; for it cannot contract into itself. The fact that the air can keep out the water by its immobility can be illustrated by an experiment with the water-clock itself. For if you fill the bulb itself of the water-clock with water, having stopped up the tube, and invert it with the tube downwards, the water does not flow along the tube to the outlet. And when the outlet is opened, it does not immediately flow out along the tube but only after a moment's interval, since it is not already at the outlet of the tube but passes along it afterwards, when it is opened. But when the water-clock is full and in an upright position, the water passes through the strainer[1] as soon as ever the tube is opened, because it is in contact with the strainer, whereas it is not in contact with the extremities of the tube. The water does not, therefore, flow into the water-clock, for the reason already mentioned, but flows out when the tube is opened because the air in it being set in motion up and down causes considerable movement in the water inside the water-clock. The water then, being thrust downwards and having itself also a tendency in that direction, naturally flows out, forcing its way through the air outside the water-clock, which is set in motion and is equal in force to the air which impels it but weaker than it in its power of resistance, because the interior air, since it passes through

---

is placed in a vessel containing liquid, this liquid enters the clepsydra through the holes in the bottom. Then, if the hole at the top is covered with the thumb, the clepsydra may be removed without the loss of the liquid in it. When the thumb is taken from the opening, the liquid flows out of the holes at the bottom. This device seems also to have served as a sort of toy (cf. the well-known fragment of Empedocles, in which respiration is likened to the action of the clepsydra, [Aristotle], *On Respiration* 473*b*–474*a*).

The water clock used for timekeeping had nothing to do with the principle of air pressure involved in the device here described but had the same name (clepsydra) from the fact that in its earliest form it measured the lapse of time by the amount of water that flowed through an opening at its base. In the later form of the water clock the water flowed into the vessel and the hour was marked by a mechanical device that depended on the varying level of the water in the vessel. The pneumatic device in the clock sometimes ascribed to Plato (p. 334) may have been used merely to mark an hour audibly. [Edd.]

[1] I.e., the small holes at the bottom of the vessel. [Edd.]

the tube, which is narrow, flows more quickly and violently and forces the water on. The reason why the water does not flow when the tube is closed is that the water on entering into the water-clock drives the air forcibly out of it. (That this is so is shown by the breath and noise engendered in it as the water enters.)

Cf. [Aristotle], *Problemata* XXV. 1. Translation of E. S. Forster

Why is it that pain is caused if the limbs are enclosed in inflated skins? Is it due to the pressure of the air? For just as the air does not yield to pressure applied to the skin from outside but repels it, so the air also presses upon the limbs enclosed within. Or is it because the air is held within by force and is compressed, and so, having naturally an outward impetus in every direction, it presses against the body enclosed within?

THE WEIGHT OF AIR IN AIR

Aristotle, *De Caelo* 311*b*6–10

Therefore earth and everything which consists mainly of earth must have weight everywhere, water has weight everywhere except in earth, and air has weight except in water and earth. Each of these bodies has weight in its own place except fire. Air is no exception, as is proved by the fact that an inflated bladder weighs more than an empty one.[1]

Simplicius, *Commentary on Aristotle's De Caelo*, pp. 710.14–711.12 (Heiberg)

Ptolemy, the mathematician, in his work *On Weights* holds a view contrary to that of Aristotle and seeks to show that neither air nor water has weight in its own place. That water has no weight in its own place he shows from the fact that divers, even those who go down to considerable depth, do not feel the weight of the water above them. One might say in reply that it is the cohesion of the water forming a framework not only above the diver but below him and on either side that prevents him from feeling the weight, as with animals in the holes of walls. For though they may be touching the wall on all sides yet they are not weighed down by it

[1] Until the science of hydrostatics was founded by Archimedes, very little progress was made by the Greeks toward quantitative notions of weights per unit of volume and of specific weight. Even thereafter, in fact, the progress was slight.

Air has no weight in air (of the same composition and at the same temperature, pressure, etc.), in the sense that theoretically no portion of a homogeneous body is heavier, volume for volume, than any other portion. But this could not very well be determined by the experiment of the inflated skin referred to in these passages. For one thing, the air within, even if at the same temperature as the external air, would presumably be in a higher state of compression and would, volume for volume, weigh more than the external air. In any case, the measurements required are too delicate to have been performed satisfactorily in antiquity, as the contradictory results show. The appeal to experiment to confirm physical theory is, however, worthy of note.

because it supports itself on every side. In the case of the water, if a separate mass of it rested upon one, its weight would, no doubt, be felt.

Now Ptolemy seeks to prove the proposition that air has no weight in its own medium by the same experiment of the inflated skin. Not only does he contradict Aristotle's view that the skin when inflated is heavier than when uninflated, but he maintains that the inflated skin is actually lighter.

I performed the experiment with the greatest possible care and found that the weight of the skin when inflated and uninflated was the same. One of my predecessors who tried the experiment wrote that he found the weights to be the same, or rather that the skin was a trifle heavier before inflation, a result which agrees with that of Ptolemy.

Now if the result of my experiment is correct, it follows, clearly, that in their respective natural places the elements are without weight, having neither heaviness nor lightness. Ptolemy agrees with this so far as water is concerned. And this is reasonable, for if weight is a striving toward the natural place, things which are at their natural places should not show any striving or tendency in that direction, since they already are there. That which is sated does not reach for food.

But if, as Ptolemy holds, the skin when inflated is lighter than when uninflated, not only does air possess lightness in its own natural place but it would seem to follow by similar reasoning that water possesses heaviness in its own natural place.

Cf. [Aristotle], *Problemata* XXV. 13. Translation of E. S. Forster

[Why is it that an inflated skin floats?] Is it because the air in it is carried upwards? For when the skin is empty it sinks; but when it is inflated, it remains on the surface, because the air supports it. But if the air makes it lighter and prevents it from sinking, why does a skin become heavier when it is inflated? And how is it that when it is heavier it floats, and when it is lighter it sinks?

THE PRINCIPLES OF PNEUMATICS AND SOME DEMONSTRATIONS

Hero of Alexandria, *Pneumatics* I, Introduction; II. 11.[1] Translation of J. G. Greenwood

Introduction

The investigation of the properties of Atmospheric Air having been deemed worthy of close attention by the ancient philosophers and mech-

[1] In the *Pneumatics* Hero adds his own discoveries to those of his predecessors, chiefly Ctesibius and Philo. The work, in two books, commences with a theoretical discussion of the void, which has been thought to be based, in large part, on the work of Strato of Lampsacus. The bulk of the treatise deals with some 80 devices which achieve amusing and striking effects by air pressure, often in combination with mechanical contrivances. Whether the devices

anists, the former deducing them theoretically, the latter from the action of sensible bodies, we also have thought proper to arrange in order what has been handed down by former writers, and to add thereto our own discoveries: a task from which much advantage will result to those who shall hereafter devote themselves to the study of mathematics. We are further led to write this work from the consideration that it is fitting that the treatment of this subject should correspond with the method given by us in our treatise, in four books, on water-clocks. For, by the union of air, earth, fire, and water, and the concurrence of three, or four, elementary principles, various combinations are effected, some of which supply the most pressing wants of human life, while others produce amazement and alarm.

But, before proceeding to our proper subject, we must treat of the vacuum.[1] Some [2] assert that there is absolutely no vacuum; others that, while no continuous vacuum is exhibited in nature, it is to be found distributed in minute portions through air, water, fire, and all other substances: and this latter opinion, which we will presently demonstrate to be true from sensible phenomena, we adopt. Vessels which seem to most men empty are not empty, as they suppose, but full of air. Now the air, as those who have treated of physics are agreed, is composed of particles minute and light, and for the most part invisible. If, then, we pour water into an apparently empty vessel, air will leave the vessel proportioned in quantity to the water which enters it. This may be seen from the following experiment. Let the vessel which seems to be empty be inverted, and, being carefully kept upright, pressed down into water; the water will not enter it even though it be entirely immersed: so that it is manifest that the air, being matter, and having itself filled all the space in the vessel, does not allow the water to enter. Now, if we bore the bottom of the vessel, the water will enter through the mouth, but the air will escape through the hole. Again, if, before perforating the bottom, we raise the vessel vertically, and turn it up, we shall find the inner surface of the vessel entirely free from moisture, exactly as it was before immersion. Hence it must be assumed that the air is matter. The air when set in motion becomes wind (for wind is nothing else but air in motion), and if, when the bottom

---

were in all cases actually constructed is doubtful, for the difficulties of securing the type of vacuum often required must have been insurmountable in antiquity. A few demonstrations of pneumatic principles are given here; pneumatic devices are described in the section on Applied Mechanics, pp. 326–336, below.

The *Pneumatics* of Hero is relatively more complete than that of Philo of Byzantium and is available in the original Greek, whereas we have only Latin and Arabic versions of the work of Philo. [Edd.]

[1] See p. 211, n. 1, above. [Edd.]

[2] E.g., Aristotle (see pp. 204 ff.). [Edd.]

of the vessel has been pierced and the water is entering, we place the hand over the hole, we shall feel the wind escaping from the vessel; and this is nothing else but the air which is being driven out by the water. It is not then to be supposed that there exists in nature a distinct and continuous vacuum, but that it is distributed in small measures through air and liquid and all other bodies.[1] Adamant[2] alone might be thought not to partake of this quality, as it does not admit of fusion or fracture, and, when beaten against anvils or hammers, buries itself in them entire. This peculiarity however is due to its excessive density: for the particles of fire, being coarser than the void spaces in the stone, do not pass through them, but only touch the outer surface; consequently, as they do not penetrate into this, as into other substances, no heat results. The particles of the air are in contact with each other, yet they do not fit closely in every part, but void spaces are left between them, as in the sands on the sea shore: the grains of sand must be imagined to correspond to the particles of air, and the air between the grains of sand to the void spaces between the particles of air. Hence, when any force is applied to it, the air is compressed, and, contrary to its nature, falls into the vacant spaces from the pressure exerted on its particles: but when the force is withdrawn, the air returns again to its former position from the elasticity of its particles, as is the case with horn shavings and sponge, which, when compressed and set free again, return to the same position and exhibit the same bulk. Similarly, if from the application of force the particles of air be divided and a vacuum be produced larger than is natural, the particles unite again afterwards; for bodies will have a rapid motion through a vacuum, where there is nothing to obstruct or repel them, until they are in contact. Thus, if a light vessel with a narrow mouth be taken and applied to the lips, and the air be sucked out and discharged, the vessel will be suspended from the lips, the vacuum drawing the flesh towards it that the exhausted space may be filled. It is manifest from this that there was a continuous vacuum in the vessel. The same may be shown by means of the egg-shaped cups used by physicians, which are of glass, and have narrow mouths. When they wish to fill these with liquid, after sucking out the contained air, they place the finger on the vessel's mouth and invert them into the liquid; then, the finger being withdrawn, the water is drawn up into the exhausted space, though the upward motion is against its nature. Very similar is the operation of cupping-glasses, which, when applied to the body, not only do not fall though of considerable weight, but even draw the contiguous matter toward them through the apertures of the body. The

[1] Hero (or Strato) holds here that there is a void between any two atoms of a material body, but that a complete void between two bodies of gross matter cannot exist naturally, though it can, as the sequel shows, be produced artificially. [Edd.]

[2] The reference here is to the diamond. [Edd.]

explanation is that the fire placed in them consumes and rarefies the air they contain, just as other substances, water, air or earth, are consumed and pass over into more subtle substances. . . .[1]

When, therefore, the air in the cupping-glasses, being in like manner consumed and rarefied by fire, issues through the pores in the sides of the glass, the space within is exhausted and draws towards it the matter adjacent, of whatever kind it may be. But, if the cupping-glass be slightly raised, the air will enter the exhausted space and no more matter will be drawn up.

They, then, who assert that there is absolutely no vacuum may invent many arguments on this subject, and perhaps seem to discourse most plausibly though they offer no tangible proof. If, however, it be shewn by an appeal to sensible phenomena that there is such a thing as a continuous vacuum, but artificially produced; that a vacuum exists also naturally, but scattered in minute portions; and that by compression bodies fill up these scattered vacua, those who bring forward such plausible arguments in this matter will no longer be able to make good their ground.

Provide a spherical vessel, of the thickness of metal plate so as not to be easily crushed, containing about 8 cotylae [2 quarts].[2] When this has been tightly closed on every side, pierce a hole in it, and insert a siphon, or slender tube, of bronze, so as not to touch the part diametrically opposite to the point of perforation, that a passage may be left for water. The other end of the siphon must project about 3 fingers' breadth [2 inches][2] above the globe, and the circumference of the aperture through which the siphon is inserted must be closed with tin applied both to the siphon and to the outer surface of the globe, so that when it is desired to breathe through the siphon no air may possibly escape from the vessel. Let us watch the result. The globe, like other vessels commonly said to be empty, contains air, and as this air fills all the space within it and presses uniformly against the inner surface of the vessel, if there is no vacuum, as some suppose, we can neither introduce water nor more air, unless the air contained before make way for it; and if by the application of force we make the attempt, the vessel, being full, will burst sooner than admit it. For the particles of air cannot be condensed, as there must in that case be interstices between them, by compression into which their bulk may become less; but this is not credible if there is no vacuum: nor again, as the particles press against one another throughout their whole surface and likewise against the sides of the vessel, can they be pushed away so as to make room if there is no vacuum. Thus in no way can anything from without be introduced into the globe unless some portion of the previously

---

[1] There follow, in the part here omitted, instances of transformation of the elements through intermixture. [Edd.]

[2] The figure in brackets is only approximate. [Edd.]

contained air escape; if, that is to say, the whole space is closely and uniformly filled, as the objectors suppose. And yet, if any one, inserting the siphon in his mouth, shall blow into the globe, he will introduce much wind without any of the previously contained air giving way. And, this being the uniform result, it is clearly shown that a condensation takes place of the particles contained in the globe into the interspersed vacua. The condensation however is effected artificially by the forcible introduction of air. Now if, after blowing into the vessel, we bring the hand close to the mouth, and quickly cover the siphon with the finger, the air remains the whole time pent up in the globe; and on the removal of the finger the introduced air will rush out again with a loud noise, being thrust out, as we stated, by the expansion of the original air which takes place from its elasticity. Again, if we draw out the air in the globe by suction through the siphon, it will follow abundantly, though no other substance take its place in the vessel, as has been said in the case of the egg. By this experiment it is completely proved that an accumulation of vacuum goes on in the globe; for the particles of air left behind cannot grow larger in the interval so as to occupy the space left by the particles driven out. For if they increase in magnitude when no foreign substance can be added, it must be supposed that this increase arises from expansion, which is equivalent to a re-arrangement of the particles through the production of a vacuum. But it is maintained that there is no vacuum; the particles therefore will not become larger, for it is not possible to imagine for them any other mode of increase. It is clear, then, from what has been said that certain void spaces are interspersed between the particles of the air, into which, when force is applied, they fall contrary to their natural action.

The air contained in the vessel in water does not undergo much compression, for the compressing force is not considerable, seeing that water, in its own nature, possesses neither weight nor power of excessive pressure.[1] Whence it is that, though divers to the bottom of the sea support an immense weight of water on their backs, respiration is not compelled by the water, though the air contained in their nostrils is extremely little. It is worth while here to examine what reason is given why those who dive deep, supporting on their backs an immense weight of water, are not crushed. Some say that it is because water is of uniform weight: but these give no reason why divers are not crushed by the water above. The true reason may be shown as follows: Let us imagine the column of liquid which is directly over the surface of the object under pressure (in immediate contact with which the water is) to be a body of the same weight and form as the superincumbent liquid, and that this is so placed in the water that its under surface coincides with the surface of the body

[1] The Greek text may mean that water naturally has no weight in water (cf. Aristotle's view, p. 247), though the translator does not so interpret it. [Edd.]

pressed, resting upon it in the same manner as the previously superincumbent liquid, with which it exactly corresponds. It is clear, then, that this body does not project above the liquid in which it is immersed, and will not sink beneath its surface. For Archimedes has shewn, in his work *On Floating Bodies*, that bodies of equal weight with any liquid, when immersed in it, will neither project above nor sink beneath its surface:[1] therefore they will not exert pressure on objects beneath.[2] Again, such a body, if all objects which exert pressure from above be removed, remains in the same place; how then can a body which has no tendency downward exert pressure? Similarly, the liquid displaced by the body will not exert pressure on objects beneath; for, as regards rest and motion, the body in question does not[3] differ from the liquid which occupies the same space.

Again, that void spaces exist may be seen from the following considerations: for, if there were not such spaces, neither light, nor heat, nor any other material force could penetrate through water, or air, or any body whatever. How could the rays of the sun, for example, penetrate through water to the bottom of the vessel?[4] If there were no pores in the fluid, and the rays thrust the water aside by force, the consequence would be that full vessels would overflow, which however does not take place. Again, if the rays thrust the water aside by force, it would not be found that some were reflected while others penetrated below; but now all those rays that impinge upon the particles of the water are driven back, as it were, and reflected, while those that come in contact with the void spaces, meeting with but few particles, penetrate to the bottom of the vessel. It is clear, too, that void spaces exist in water from this, that, when wine is poured into water, it is seen to spread itself through every part of the water, which it would not do if there were no vacua in the water. Again, one light traverses another; for, when several lamps are lighted, all objects are brilliantly illuminated, the rays passing in every direction through each other. And indeed it is possible to penetrate through bronze, iron, and all other bodies, as is seen in the instance of the marine torpedo.[5]

---

[1] See p. 237. [Edd.]

[2] But this is true only because the water exerts buoyant force from below. In other words, this explanation is not essentially different from the other if that be understood to mean that the pressure at any point below the surface is equal in all directions (see pp. 247 f.). [Edd.]

[3] The translator supplies this word, as necessary to the sense, though it is not in the manuscripts. The text of W. Schmidt (Leipzig, 1899) may be rendered thus: "for only with regard to rest and motion does the body in question differ from the liquid which occupies the same place." [Edd.]

[4] The assumption is that light is corporeal, an assumption that Aristotle sought to refute (see p. 285). [Edd.]

[5] The reference is to the conduction by metals of the electric shock of the torpedo (see p. 430). [Edd.]

That a continuous vacuum can be artificially produced has been shewn by the application of a light vessel to the mouth and by the egg of physicians.[1] With regard, then, to the nature of the vacuum, though other proofs exist, we deem those that have been given, and which are founded on sensible phenomena, to be sufficient. It may, therefore, be affirmed in this matter that every body is composed of minute particles, between which are empty spaces less than the particles of the body[2] (so that we erroneously say that there is no vacuum except by the application of force, and that every place is full either of air, or water, or some other substance) and, in proportion as any one of these particles recedes, some other follows it and fills the vacant space: that there is no continuous vacuum except by the application of some force: and again, that the absolute vacuum is never found, but is produced artificially.

These things having been clearly explained, let us treat of the theorems resulting from the combination of these principles; for, by means of them, many curious and astonishing kinds of motion may be discovered.[3]

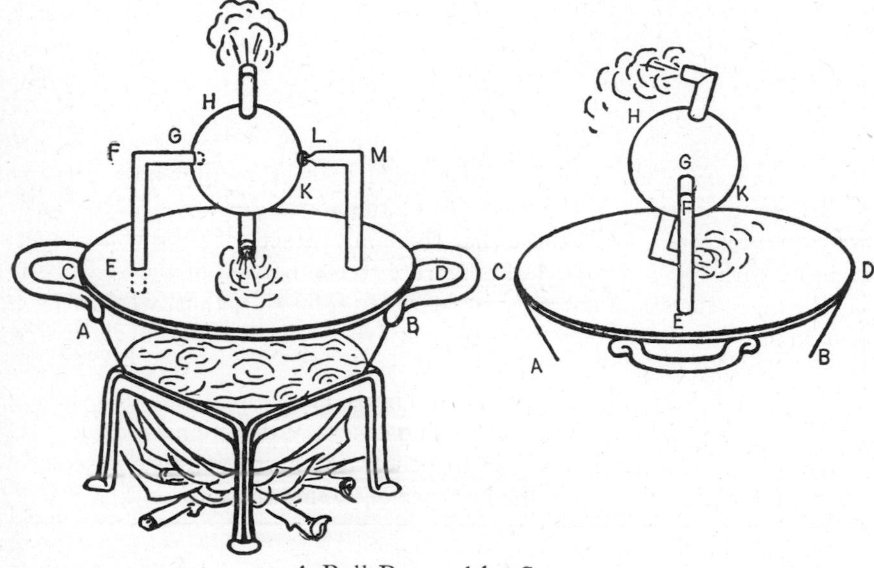

A Ball Rotated by Steam

11. Place a cauldron over a fire: a ball shall revolve on a pivot. A fire is lighted under a cauldron, *AB* [see figure], containing water, and covered at the mouth by the lid *CD*: with this the bent tube *EFG* communicates, the

[1] I.e., the cupping glass. [Edd.]

[2] It would seem to follow that within a body of gross matter the atoms occupy more space than the void. But wide variations among substances in their weight per unit of volume would be excluded, on this theory, if the atoms differed only in size and shape. [Edd.]

[3] There follows the passage on the siphon (see pp. 242 ff.). [Edd.]

extremity of the tube being fitted into a hollow ball, *HK*. Opposite to the extremity *G* place a pivot, *LM*, resting on the lid *CD*; and let the ball contain two bent pipes, communicating with it at the opposite extremities of a diameter, and bent in opposite directions, the bends being at right angles and across the lines *FG*, *LM*. As the cauldron gets hot it will be found that the steam, entering the ball through *EFG*, passes out through the bent tubes towards the lid, and causes the ball to revolve, as in the case of the dancing figures.[1]

Philo of Byzantium, *Pneumatics* 7, 8 (Schmidt)[2]

A Thermoscope

7. Fire, too, by its nature is closely connected with air, and for that reason air is drawn along with it.[3] This will be proved by what follows.

Take a sphere of lead, hollowed out, so that there is room within, and of moderate size. It should not be too thin, lest it be easily broken; nor should it be too heavy. For our purposes it should be quite dry.[4] Pierce the sphere on top and insert a bent tube reaching almost to the bottom. Place the other end of the same tube in another vessel filled with water. Let this end also, as in the other case, reach almost to the

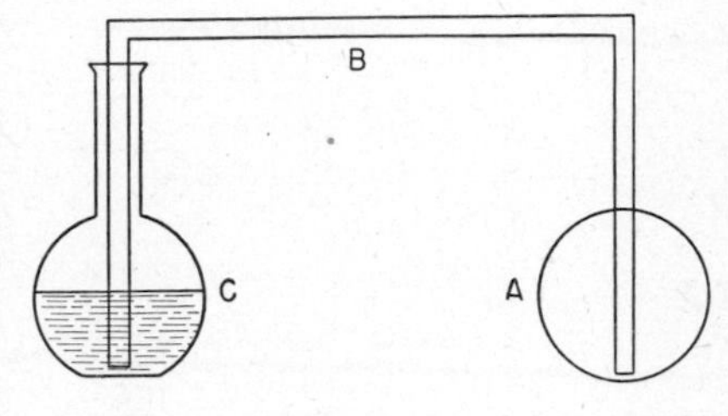

[1] The passage of the steam is accompanied by a reaction that causes the sphere to rotate in the opposite direction. In a similar device operated by warmed air, not by steam, as here, a disk upon which are placed small figures is made to rotate. It is to this that the "dancing figures" refer. Cf. the action of the modern rotating lawn sprinkler. It may be noted that the translator entitles this chapter "The Steam-Engine." [Edd.]

[2] Little is known of the life of Philo of Byzantium. He may have been a pupil of Ctesibius and in any case carried on the type of work of which Ctesibius had laid the foundations. He probably lived in the third century B.C and wrote a comprehensive treatise on various branches of mechanics, particularly on mechanics in warfare, e.g., in the conduct of siege operations, the defense of cities, and the construction of artillery. He wrote also about applications of the lever, pneumatics, the construction of automata, and the building of harbors. Of his work the portion dealing with artillery construction, *Belopoeica* (p. 318), and some parts dealing, in the main, with siege and defense operations are extant in the original. The part dealing with pneumatics is extant, in part, in a medieval Latin translation of a no longer extant Arabic version, and also in a much more complete Arabic version. Many passages, e.g., on the corporeality of air, on siphons, clepsydras, etc., are essentially the same as those found in the later work of Hero of Alexandria. The text used here is the Latin translation referred to above. Cf. Hero of Alexandria, *Pneumatics* II. 8.

[3] The precise meaning is doubtful. The next selection would seem to indicate that the meaning is that the air is attracted by the fire and destroyed, but here it is the expansion of air that is involved.

[4] I.e., "water-proof," as the extant Arabic version indicates.

bottom, so that the flow of water may be facilitated.[1] Call the sphere *A*, the tube *B*, and the vessel *C*.

I say, then, that if you expose the sphere to the sun, part of the air enclosed in the tube will pass out when the sphere becomes hot. This will be evident because the air will descend from the tube into the water, agitating it and producing a succession of bubbles.

Now if the sphere is put back in the shade, that is, where the sun rays do not reach it, the water will rise and pass through the tube until it descends into the sphere. If you then put the sphere back in the sun the water will return to the vessel; but it will flow back to the sphere once more if you place the sphere in the shade. No matter how many times you repeat the operation the same thing will always happen.

In fact, if you heat the sphere with fire, or even if you pour hot water over it, the result will be the same. And if the sphere is then cooled, water passes from the vessel to the sphere.

Combustion and Air

8. . . . Hence we shall prove that a place cannot be empty of air and of all other bodies as well. For example, pour water into a vessel, *A*. In the center of *A* let a sort of candle-holder, *B*, be set up protruding over the water, and let a lighted candle, *C*, be placed at the top of *B*. Over *C* invert vessel *D* in such a way that its mouth is near the water[2] and the candle is in the center of *D*. A short while after this is done you will see water rise from the lower to the upper vessel. Now this will not happen except for the reason we have indicated, namely, that the air enclosed in vessel *D* is destroyed by the fire, because air cannot remain in proximity to fire. After the air has been destroyed by the action of the fire, the latter will raise the water in proportion to the quantity of air which is lost. This is similar to what takes place in the case of the tube described above.[3] Thus the air in this vessel (*D*) placed over the candle is destroyed because it is, so to speak, dissolved by the fire. For this reason the water is raised and entering fills the place left by the air, since that place was empty. The figure is appended.[4]

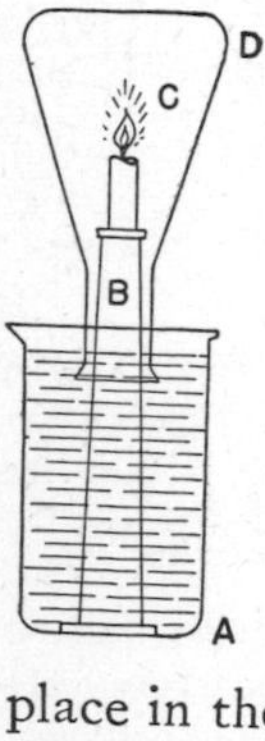

[1] For this purpose the end must always be *immersed* in water.

[2] So both the Latin and the extant Arabic versions. But for the experiment to succeed the mouth must be *under* the water.

[3] *Pneumatics*, ch. 7.

[4] Though the idea of destruction of air in combustion is here expressed, there is, of course, no notion of oxidation.

Cf. Galen IV. 487–488 (Kühn): "For clearly we see these [flames], just as living things, swiftly extinguished when they are deprived of air. If a physician's cupping instrument or any narrow or concave vessel be put over the flames so as to cut off the access of air they are soon snuffed out. Now if we could discover why flames are in these cases extinguished, we should perhaps discover what advantage the heat in animals derives through respiration."

OPTICS

Various phases of optics were the object of speculation and investigation by the Greeks. Philosophers developed theories about the nature of light, color, and vision. Physiologists sought to explain the mechanism of seeing. Mathematicians and scientists studied perspective and mirrors with the help of the concept of visual rays; they arrived at the fundamental laws of reflection. Refraction was investigated empirically as well as mathematically, and its importance in connection with astronomical observation was not overlooked. There were practical applications of optical theory in the arts, e.g., in scene painting for the theater, and in the construction of devices employing mirrors. As for the literary remains, apart from the wealth of material in the writings of philosophers and physicians, there are treatises specially devoted to optics, extant either in the original Greek or in translation, by Euclid, Hero of Alexandria, Ptolemy (?), and Damianus or Heliodorus of Larissa. We have given selections representing the various branches. Note also Geminus's classification of these branches (p. 4, above). On theories of vision see pp. 543–546.

Introduction to the Theory of Perspective

Euclid, *Optics*, Definitions and Propositions I–VIII, XLV, XLVIII (Heiberg) [1]

Definitions

Let it be assumed

1. That the rectilinear rays proceeding from the eye diverge indefinitely;[2]

2. That the figure contained by a set of visual rays is a cone of which the vertex is at the eye and the base at the surface of the objects seen;

3. That those things are seen upon which visual rays fall and those things are not seen upon which visual rays do not fall;

---

[1] On Euclid, see p. 37. The *Optics* is extant in two versions, of which the earlier form is thought to be Euclid's own arrangement and the later that of Theon of Alexandria (latter part of the fourth century A.D.) The work consists of definitions (or rather assumptions) followed by 58 theorems geometrically demonstrated and constituting a treatise on perspective. It is the earlier version, as edited by Heiberg, from which the present translations have been made. On the *Catoptrics* attributed to Euclid, see p. 261. For a criticism of the *Optics* of Euclid from the point of view of modern optical theory see G. Ovio, *L'Ottica di Euclide* (Milan, 1918), and the introduction to Paul Ver Eecke's translation of Euclid's *Optics* and *Catoptrics* (Paris, 1938).

Tradition ascribed works on perspective to Democritus and Anaxagoras. See Vitruvius VII. Preface 11.

[2] The adoption of the theory of vision in which the visual rays proceed from the eye to the object, rather than from the object to the eye, does not affect the geometric development of the theory of perspective—the object of Euclid's work.

The precise meaning of the definition is doubtful and the version given conjectural, though according to a scholion the meaning would be substantially the same as that of Theon's revision, which reads: "Let it be assumed that the rays from the eye move in straight lines diverging from one another" (note the immediate application to Prop. I). Contrast the treatment in Ptolemy (?), *Optics* pp. 24f. (Govi).

4. That things seen under[1] a larger angle appear larger, those under a smaller angle appear smaller, and those under equal angles appear equal;

5. That things seen by higher visual rays appear higher, and things seen by lower visual rays appear lower;

6. That, similarly, things seen by rays further to the right appear further to the right, and things seen by rays further to the left appear further to the left;

7. That things seen under more angles are seen more clearly.[2]

Proposition I

No visible object is seen completely at one time.

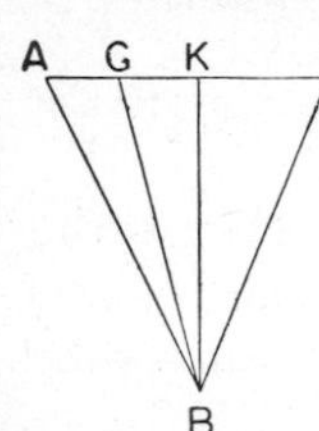

Let *AD* be a visible object, *B* the eye, *BA*, *BG*, *BK*, and *BD* visual rays from *B* to the object. Then, since the incident rays move at an interval from one another, they cannot fall continuously over *AD*.[3] Hence there are intervals along *AD* upon which the rays will not fall. The whole of *AD* will, therefore, not be seen at one time. We think that we see the whole of *AD* at one time because the rays move along the object very quickly.

Proposition II

Of equal magnitudes situated at a distance ihose that are nearer are seen more clearly.

Let *B* be the eye, and *GD* and *KL* the visible objects, which we are to consider as equal and parallel, *GD* being nearer the eye. Let *BG*, *BD*, *BK*, and *BL* be incident visual rays. The visual rays to *KL* will not pass through points *G* and *D*. For if they did, in the resulting triangle, *BDLKGB*, *KL* would be larger than *GD*. But they were assumed to be equal. Therefore, *GD* will be seen by more visual rays[4] than will *KL*. *GD* will, consequently, be seen more clearly than will *KL*. For objects seen under a larger number of angles are seen more clearly.

---

[1] The angle referred to is that at the vertex of the cone (Definition 2).

[2] The meaning is brought out in Prop. II and is essentially contained in Definition 4, the angles being those between each pair of successive rays.

[3] The assumption of discontinuity (see Definition 1 and note) seems to run counter to the geometrical continuity of the *Elements*. Euclid is here seeking to explain geometrically facts that are really due to limitations in the sensitivity of the retina. He does not, however, fully explain the meaning of the discontinuity he assumes.

[4] The discontinuity of the rays is also assumed here and leads to an application of Definition 7.

Proposition III

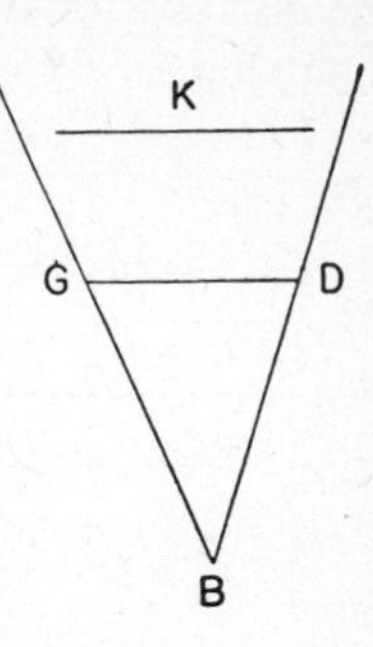

For every object there is a distance at which it is no longer seen.

Let B be the eye and GD the visible object. I say that at a certain distance GD will no longer be seen. For suppose that GD is situated in an interval, K, between visual rays.[1] Hence none of the visual rays from B will fall upon K. But an object upon which visual rays do not fall is not seen. Therefore, for each object there is a distance at which it is no longer seen.

Proposition IV

Of equal intervals on the same straight line those seen from a greater distance appear smaller.

Let AB, BG, and GD be equal intervals on the same straight line. Draw AE perpendicular to this line; let the eye be at E. I say that AB will appear larger than BG, and BG larger than GD.

Let EB, EG, and ED be incident visual rays. Draw BZ through B parallel to GE. $AZ = ZE$, for since BZ was drawn parallel to one side, GE, of $\triangle AEG$, it follows that $EZ:ZA = GB:BA$.

Hence, as we have said, $AZ = ZE$.
But $BZ > ZA$.
Therefore $BZ > ZE$.
and $\angle ZEB > \angle ZBE$.
But $\angle ZBE = \angle BEG$.
Therefore $\angle ZEB > \angle BEG$.

Consequently AB will appear larger than BG.

Similarly, if a parallel to DE be drawn through G, it may be shown that BG will appear larger than GD.

Proposition V

Equal magnitudes situated at different distances from the eye appear unequal, and the nearer always appears larger.[2] . . .

Proposition VI

Parallel lines when seen from a distance appear to be an unequal distance apart.

---

[1] I.e., between proximate visual rays, discontinuity being assumed.

[2] The proof is similar to that of Prop. II.

Let *AB* and *GD* be two parallels, and *E* be the eye. I hold that *AB* and *GD* seem to be an unequal distance apart, and that the interval between them at a point nearer the eye seems greater than at a point more remote from the eye.

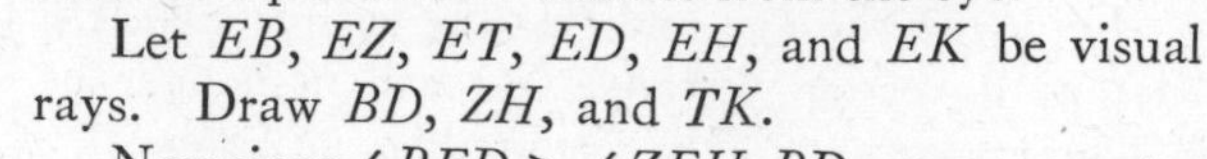

Let *EB*, *EZ*, *ET*, *ED*, *EH*, and *EK* be visual rays. Draw *BD*, *ZH*, and *TK*.

Now since $\angle BED > \angle ZEH$, *BD* appears greater than *ZH*.

Again, since $\angle ZEH > \angle TEK$, *ZH* appears greater than *TK*.

That is, $BD > ZH > TK$ *in appearance.*

The intervals, then, between parallels will not appear equal but unequal.[1] . . .

Proposition VII

Equal but non-contiguous intercepts on the same straight line if unequally distant from the eye appear unequal. . . .

Proposition VIII

Equal and parallel magnitudes unequally distant from the eye do not appear [*inversely*] *proportional to their distances from the eye.*

Let *AB* and *GD* be two such magnitudes unequally distant from the eye, *E*. I say that it is not the case that the apparent size of *GD* is to the apparent size of *AB* as *BE* is to *ED*, as might seem plausible.

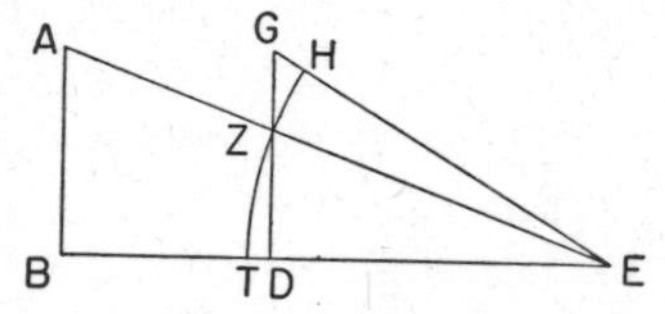

Let *AE* and *EG* be visual rays. With *E* as center and *EZ* as radius describe arc *HZT*.

Since $\triangle EZG > \text{sector } EZH$,
and $\triangle EZD < \text{sector } EZT$,
it follows that $\triangle EZG/\text{sector } EZH > \triangle EZD/\text{sector } EZT$,
and, by alternation, $\triangle EZG/\triangle EZD > \text{sector } EZH/\text{sector } EZT$.
Whence, by composition, $\triangle EGD/\triangle EZD > \text{sector } EHT/\text{sector } EZT$.
But $GD/DZ = \triangle EDG/\triangle EZD$,
and $GD = AB$.
Again, $BE/ED = AB/DZ$.
Therefore $BE/ED > \text{sector } EHT/\text{sector } EZT$.
But $\text{sector } EHT/\text{sector } EZT = \angle HET/\angle ZET$.
Therefore $BE/ED > \angle HET/\angle ZET$.
Now *GD* is seen under $\angle HET$, and *AB* under $\angle ZET$.

---

[1] The theorem of convergence is fundamental in the theory of perspective. There follows a proof of convergence for the case where the eye is not in the same plane as the parallels.

The equal magnitudes do not, therefore, appear in [inverse] proportion to their distances from the eye.[1]

Proposition XLV

There is a common point from which unequal magnitudes appear equal.

Let *BG* be greater than *GD*. About *BG* describe a segment of a circle greater than a semicircle, and about *GD* describe a segment of a circle similar to that about *BG*, i.e., a segment containing an angle equal to that contained in segment *BZG*. The segments, then, will intersect, let us say at *Z*. Draw *ZB*, *ZG*, and *ZD*.

Since angles inscribed in similar segments are equal, the angles in segments *BZG* and *GZD* are equal. But things seen under equal angles appear equal. Therefore, if the eye is placed at point *Z*, *BG* will appear equal to *GD*. But $BG > GD$. There is, then, a common point from which unequal magnitudes appear equal.

Proposition XLVIII

To find points from which a given magnitude will appear half as large or a fourth as large, or, in general, in any fraction in which the angle may be divided.

Let magnitude *AZ* be equal to *BC*. Describe a semicircle about line *AZ* and inscribe right angle *K* therein.

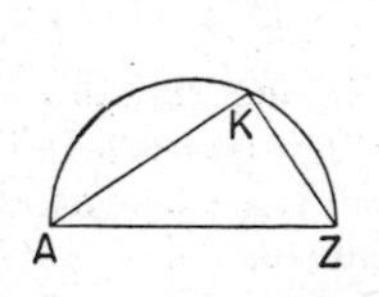

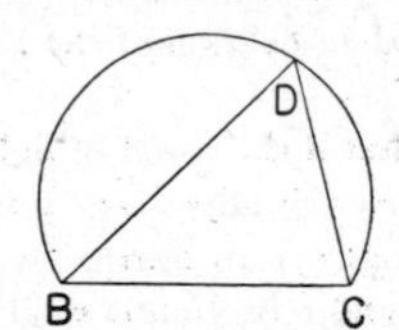

Let line *BC* be equal to *AZ* and around *BC* describe a segment of a circle such that an angle inscribed therein will be half of angle *K*. Then angle *K* is double angle *D*, and *AZ* will, therefore, appear twice as large as *BC* when the eye is on circumferences *AKZ* and *BCD*, respectively.[2]

Reflection

THE THEORY OF MIRRORS

Hero, *Catoptrics* 1–6, 7, 10, 15, 18 (Schmidt)[3]

1. . . . The science of vision is divided into three parts: optics, diop-

[1] In this proposition Euclid in effect proves $\tan a/\tan b < a/b$ (where a and b are acute angles and $a < b$).

[2] The process may then be repeated indefinitely, the magnitude appearing ¼, ⅛, . . . as large.

[3] The subject of mirrors, catoptrics, was a branch of the scientific study of optics in antiquity (see p. 4, above). Archimedes wrote a treatise on the subject, but it is not extant. The *Catoptrics* ascribed to Euclid is probably a compilation by Theon of Alexandria at the

trics,[1] and catoptrics. Now optics has been adequately treated by our predecessors and particularly by Aristotle,[2] and dioptrics we have ourselves treated elsewhere as fully as seemed necessary. But catoptrics, too, is clearly a science worthy of study and at the same time produces spectacles which excite wonder in the observer. For with the aid of this science mirrors are constructed which show the right side as the right side, and, similarly the left side as the left side, whereas ordinary mirrors by their nature have the contrary property and show the opposite sides. It is also possible with the aid of mirrors to see our own backs,[3] and to see ourselves inverted, standing on our heads, with three eyes, and two noses, and features distorted, as if in intense grief. The study of catoptrics, however, is useful not merely in affording diverting spectacles but also for necessary purposes. For who will not deem it very useful that we should be able to observe, on occasion, while remaining inside our own house, how many people there are on the street and what they are doing?[4] And will anyone not consider it remarkable to be able to tell the hour, night or day, with the aid of figures appearing in a mirror? For as many figures appear as there are hours of the day or of the night, and if a [given] part of the day has passed a

end of the fourth century A.D. The *Catoptrics* of Hero of Alexandria is therefore our earliest extant work on the subject. The original Greek is lost but we have a Latin version thought to have been made by William of Moerbeke in the thirteenth century. This text, called *De Speculis*, was generally ascribed to Ptolemy until Ptolemy's (?) *Optics*, which contains in its third book a treatment of catoptrics (pp. 268, 271), became known in the Latin translation made in the twelfth century from an Arabic version. Since there is independent evidence for the ascription of the *De Speculis* to Hero, the identification of the *De Speculis* with Hero's *Catoptrics* seems plausible.

Hero bases his treatment on the proposition that if the speed of light is incomprehensibly great (that is probably the sense in which he uses the adjective "infinite"), it travels by a straight line and is so reflected that the path from eye to mirror to object is a minimum. From this the equality of the angles of incidence and reflection is easily deduced.

Quite apart from the method of proof, it is impossible to say when this fundamental principle of catoptrics was first formulated. That the formulation is pre-Aristotelian is shown by C. B. Boyer in *Isis* 36 (1946), 94–95. The reflection of sound at equal angles is referred to, e.g., in [Aristotle], *Problemata* XI. 23.

We may mention here an extant fragment of a work on burning mirrors by Anthemius (beginning of sixth century) in which ellipsoidal and paraboloidal reflectors as well as combinations of plane mirrors are discussed. There is a tradition (in Lucian, Galen, and others), probably without foundation, that Archimedes set fire to the fleet of Marcellus by using burning mirrors.

[1] The word "dioptrics" is probably not here used in its modern sense, in which it refers to the study of refraction. The reference in the following sentence seems to be to Hero's work *On the Dioptra*, an instrument for taking sightings (see pp. 139, 336).

[2] E.g., in the *De Anima* or in the *De Sensu*, where the theory of vision is treated. Ancient lists of Aristotle's works also mention one on optics.

[3] The meaning might possibly be "to see those who are behind us," but see ch. 15.

[4] This interpretation (different from that of Schmidt) seems to be supported by ch. 16 of the work.

[given] figure will appear. Again, who will not be astonished when he sees, in a mirror, neither himself nor another, but whatever we desire that he see? Such, then, being the scope of the science, I think it necessary and proper to describe the views held by my predecessors, that my account may not be incomplete.

2. Practically all who have written of dioptrics and of optics have been in doubt as to why rays proceeding from our eyes are reflected by mirrors and why the reflections are at equal angles. Now the proposition that our sight is directed in straight lines proceeding from the organ of vision may be substantiated as follows. For whatever moves with unchanging velocity moves in a straight line.[1] The arrows we see shot from bows may serve as an example. For because of the impelling force the object in motion strives to move over the shortest possible distance, since it has not the time for slower motion, that is, for motion over a longer trajectory. The impelling force does not permit such retardation. And so, by reason of its speed, the object tends to move over the shortest path.[2] But the shortest of all lines having the same end points is the straight line.

That the rays proceeding from our eyes move with infinite velocity may be gathered from the following consideration. For when, after our eyes have been closed, we open them and look up at the sky, no interval of time is required for the visual rays to reach the sky. Indeed, we see the stars as soon as we look up, though the distance is, as we may say, infinite. Again, if this distance were greater the result would be the same, so that, clearly, the rays are emitted with infinite velocity. Therefore they will suffer neither interruption, nor curvature, nor breaking,[3] but will move along the shortest path, a straight line.

3. That our vision is directed along a straight line has, then, been sufficiently indicated. We shall now show that rays incident on mirrors and also on water and on all plane surfaces are reflected. Now the essential characteristic of polished bodies is that their surfaces are compact. Thus, before they are polished, mirrors have some porosities upon which the rays fall and so cannot be reflected. But these mirrors are polished by rubbing until the porosities are filled by a fine substance; then the rays incident upon the compact surface are reflected. For just as a stone violently hurled against a compact body, such as a board or wall, rebounds, whereas

[1] The meaning seems to be that as long as an object moves very swiftly it retains its rectilinear motion. A similar notion with respect to projectile motion prevailed until the time of Galileo.

[2] The point seems to be that if visual rays were not propagated in straight lines they would not traverse a given distance as swiftly as possible.

[3] The reference is not to reflection or refraction but to a change in the continuous rectilinear character of the motion.

a stone hurled against a soft body, such as wool or the like, does not (for the projecting force[1] accompanies the stone and then, in the case of the hard obstacle, gives way, not being able to accompany the stone any further or move it forward, while in the case of the soft obstacle, the force merely slackens and is separated from the stone), so the rays that are emitted by us with great velocity, as we have shown, also rebound when they impinge on a body of compact surface. Now in the case of water and glass not all such rays are reflected since both these substances have irregularities, composed as they are of units having minute parts, and of solid particles.[2] For in looking through glass and water we see our own reflection and also what lies beyond the surface of the glass or water. That is, in the case of standing water, we see what is at the bottom, and in the case of glass, what lies beyond its surface. For those rays which fall upon solid bodies are themselves turned back and reflected, while those which penetrate through porous bodies enable us to see that which lies beyond. Hence images reflected from such bodies are imperfectly seen because not all the visual rays are reflected to the objects, but some of them, as we have indicated, are lost through the pores.

4. That rays incident upon polished bodies are reflected has, then, in our opinion, been adequately proved. Now by the same reasoning, that is, by a consideration of the speed of the incidence and the reflection, we shall prove that these rays are reflected at equal angles[3] in the case of plane and spherical mirrors. For our proof must again make use of minimum lines.[4] I say, therefore, that of all incident rays [from a given point] reflected to a given point by plane and spherical mirrors the shortest are those that are reflected at equal angles; and if this is the case the reflection at equal angles is in conformity with reason.

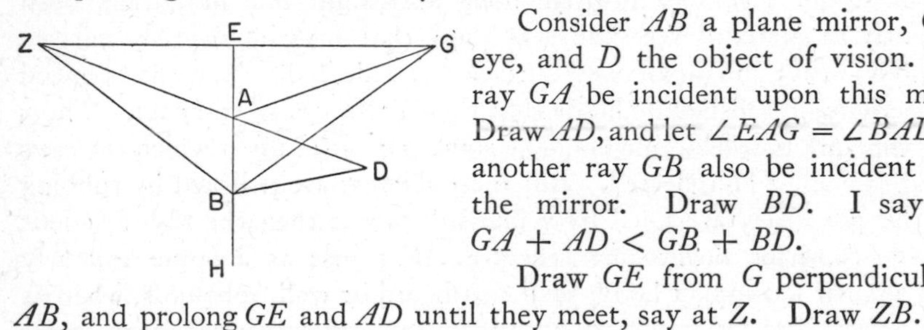

Consider AB a plane mirror, G the eye, and D the object of vision. Let a ray GA be incident upon this mirror. Draw AD, and let $\angle EAG = \angle BAD$. Let another ray GB also be incident upon the mirror. Draw BD. I say that $GA + AD < GB + BD$.

Draw GE from G perpendicular to AB, and prolong GE and AD until they meet, say at Z. Draw ZB.

---

[1] Reading *emittens* at p. 322.21 (see p. 410 of the edition of Schmidt). Some have seen here an early instance of the anti-Aristotelian doctrine of *vis impressa* (p. 223, above).

[2] A type of atomism, not very clearly defined, is invoked here to aid the explanation, as in the *Pneumatics* (see pp. 249 ff.)

[3] I.e., the angle of incidence will equal the angle of reflection.

[4] I.e., the sum of the incident ray from eye to mirror and the reflected ray from mirror to object must be a minimum.

Now $\angle BAD = \angle EAG$,
and $\angle ZAE = \angle BAD$ (as vertical angles).
Therefore $\angle ZAE = \angle EAG$.
And since the angles at E are right angles,
$$ZA = AG$$
and $$ZB = BG.$$
But $$ZD < ZB + BD$$
and $$ZA = AG, ZB = BG.$$
Therefore $GA + AD < GB + BD$.
Now $\angle EAG = \angle BAD$,
and $\angle EBG < \angle EAG$,
and $\angle HBD > \angle BAD$.
Therefore $\angle HBD$ is, *a fortiori*, greater than $\angle EBG$.[1]

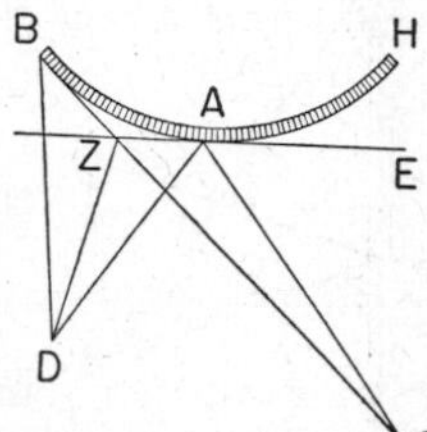

5. Let AB be the surface of a spherical mirror, G the eye, and D the object seen. Let GA and AD make equal angles with the mirror, while GB and BD make unequal angles. I say that $GA + AD < GB + BD$.

Draw EAZ tangent at A.
Then $\angle HAE = \angle BAZ$, and the remainder[2] $\angle EAG = \angle ZAD$.
If ZD be drawn, $GA + AD < GZ + ZD$, as was proved above.
But $GZ + ZD < GB + BD$.
Therefore $GA + AD < GB + BD$.

In general, then, in the case of mirrors [both plane and spherical], one must consider whether there is or is not a point from which incident rays may be reflected at equal angles in such a way that the ray incident from the organ of vision and the ray reflected to the object of vision, when added together, make a sum less than that of all other pairs of rays similarly incident and reflected.[3]

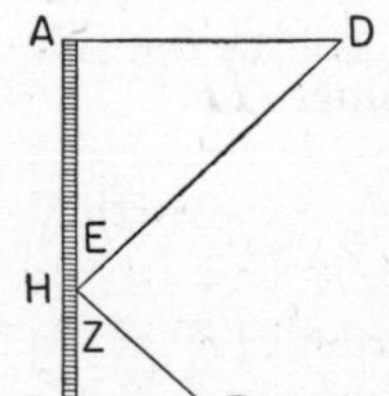

6. In the case of plane mirrors there is a place at the covering of which an image will no longer be seen.

Let AG be a plane mirror . . . , B the eye, and D the visible object. Draw AD and BG perpendicular to the mirror, and divide AG at H in such a way that $AD{:}BG = AH{:}HG$. I say, then, that if H is covered the image of D is no longer seen.

[1] That is, not only is the path of the ray shortest when the angles of incidence and reflection are equal, but there is only one incident ray which can be reflected at equal angles.

[2] I.e., the remainder of the angle between the visual ray and the mirror. The equality was proved in ch. 4, for the tangent plane may be considered a plane mirror.

[3] But not reflected at an angle equal to the angle of incidence. There are probably lacunae in the text as we have it, but the reference seems to be to a test whether the image of a given object will be seen in a mirror (plane or spherical) from a given fixed point.

For if BH and HD are drawn, the triangles, because of the proportionality of their sides, will be similar. Hence $\angle E = \angle Z$, and D will be visible through point H. Therefore, if this point is covered with wax or some other material, D will no longer be visible.

If, however, the covering at H is removed from the mirror, the image will again appear in the mirror. For all rays incident upon a mirror will be reflected at equal angles.

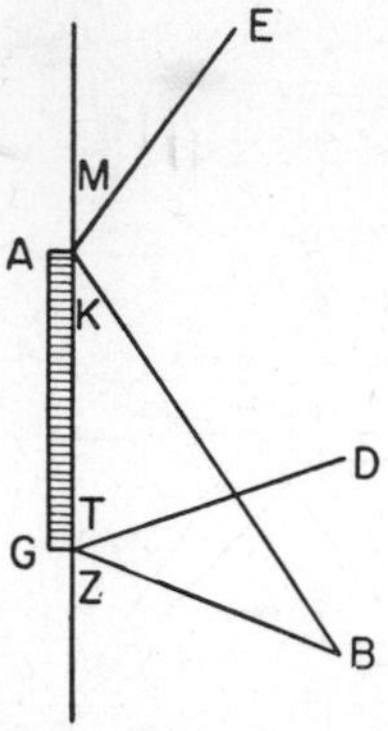

7. In the case of plane mirrors the reflected rays neither will converge nor are parallel.

For let AG be a plane mirror, B the eye, [D and E the objects, GB and BA the incident rays,][1] and GD and AE the reflected rays.

$\therefore \angle Z = \angle T.$

But $\angle Z > \angle K$,

that is, $\angle Z > \angle M$.

$\therefore \angle T > \angle M.$

$\therefore$ GD and AE are neither parallel nor will they meet in the direction of D and E.[2]

10. In the case of concave mirrors, when the eye is situated at the circumference the reflected rays converge.

Let BGA be a concave mirror, and let the eye be placed at B. Let BG and BA be incident rays and GX and AN reflected rays. I say that GX and AN will meet on the side of X and N.

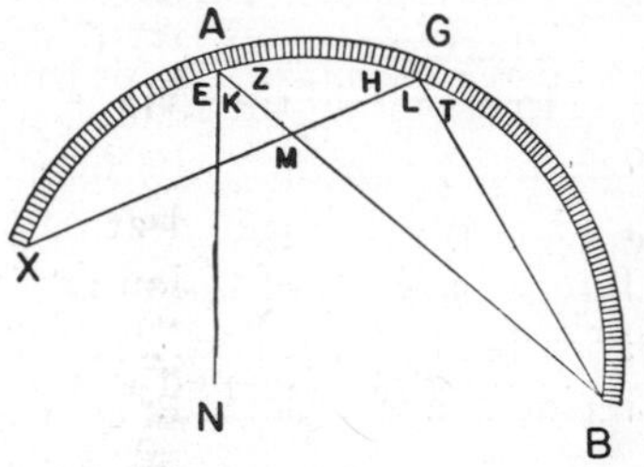

For since arc $AB >$ arc GB,

$\angle Z > \angle T.$

$\therefore \angle E > \angle H.$

$\therefore \angle L > \angle K$ (as remainders).

But $\angle M > \angle L$.

$\therefore \angle M > \angle K$, and, consequently, GX and AN will meet on the side of N and X.

15. It is desired to secure the same effect[3] by another construction.

Let ABG be a right triangle. Bisect BG at T. Let ZH and DE be plane mirrors on lines AG and AB, respectively. Consider TK as an observer with the eye at point T capable of looking into either mirror as desired. And so the problem will be solved.

[1] The material within [ ] translates Schmidt's restoration of a lacuna assumed here.

[2] A similar proof is given for the case of convex mirrors (Prop. VIII).

[3] I.e., the effect of surprising the observer. The reference may, however, be to a combination of mirrors having the same purpose as that described in the previous chapter (14), the *speculum theatrale*.

If one mirror (ZH into which the observer looks) is kept unmoved, while the other (DE, behind the observer) is moved up and down, the ray will reach a point where the image of the heel[1] of the observer will appear in the mirror and he will think that he is flying.

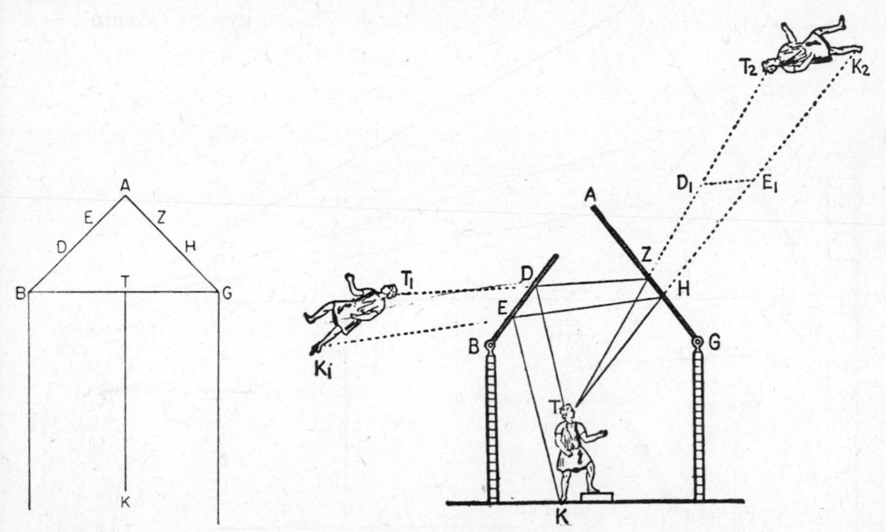

18. To place a mirror so that one approaching it sees neither his own image nor that of another but only the image which we select.

Let AB [p. 268] be the wall where the mirror is to be put and let the mirror be inclined to it at a given angle. If this angle is one-third of a right angle the measurements will be suitable. Let BG be the surface of the mirror, and let BD be perpendicular to AB. D, the point on BD at which the eye is, is so situated that a perpendicular drawn from it to BG falls outside BG. Let this perpendicular be ED. Draw DG to the end G of the mirror and let $\angle EGD = \angle BGH$. If then, a visual ray from the eye D falls on G, the end of the mirror, it will be reflected to H. Now let HN be drawn from H at right angles to DB. Now let DT[2] be another incident ray and draw HT.

$\therefore \angle BTH > \angle ETD$,[3]

and $\angle BTK = \angle GTD$.

$\therefore TK$ intersects HN, as do all rays incident upon the mirror when reflected.

---

[1] K_2, in the figure at the right.

[2] T being any point on BG.

[3] For $\angle BTH > \angle BGH$,
and $\angle ETD < \angle EGD$.
But $\angle BGH = \angle EGD$.
$\therefore \angle BTH > \angle ETD$.

Now let a plane [mirror] *LM* be drawn parallel to mirror *GB* and intersected by a ray reflected from that mirror. Clearly, then, the eye will see only that which lies within *HN*, since all the reflected rays fall within *HN*.

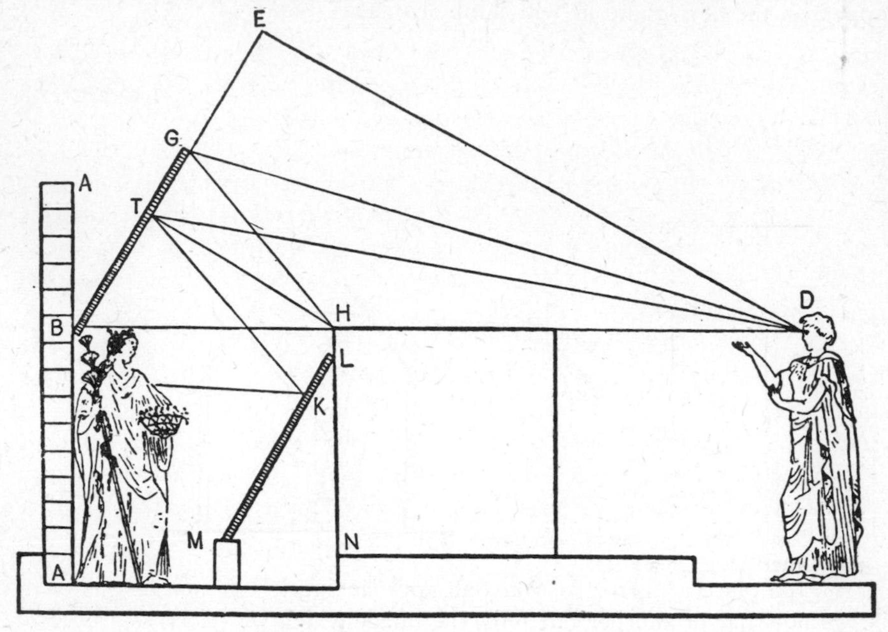

Therefore, if we place whatever object we wish near plane *LM*, those approaching will see not their own image but merely that of the aforesaid object. It will consequently be necessary, as we have said, to place *LM* within *HN* so that the object in question may be between the parallel plane mirrors[1]. . . .

EXPERIMENTAL CONFIRMATION OF LAWS OF REFLECTION

Ptolemy (?), *Optics* III, pp. 60.23–64.28 (Govi) [2]

Now in seeking knowledge in any field we must start with certain general principles, and must make assumptions which are definite and self-evident either from the point of view of their practical effect or of their internal consistency. Only from such assumptions may the subsequent demonstrations be derived.

Now the basic principles required for the study of mirrors are three in number, and they are matters of primary knowledge, knowable in and of

[1] Reading, as Schmidt suggests, *inter plana equidistantia specula*. The rest of the paragraph, which is not given here, deals with the placing of the mirrors in a temple so that the apparition may be seen by one approaching. The figure shows the apparatus. Temple magic is also an important motive in Hero's *Pneumatics* (see pp. 327–329).

[2] On this work see p. 271.

themselves. They are as follows: (1) objects seen in mirrors are seen in the direction of the visual ray which is reflected from the mirror to the object, depending on the position of the eye; (2) images in mirrors appear to be on the perpendicular drawn from the object to the surface of the mirror, and produced; (3) the position of the reflected ray, from the eye to the mirror and from the mirror to the object, is such that each of the two parts contains the point of reflection and makes equal angles with the perpendicular to the mirror at that point.[1]

Now in the case of spherical mirrors what is meant by the perpendicular to the surface at a given point is the line perpendicular at the given point to the plane containing all the lines tangent to the sphere at that point. Hence all perpendiculars to the surface of a sphere must, when produced, pass through the center of the sphere.

The truth of the elementary principles which we have set forth is corroborated by the actual phenomena, as we shall now explain. For in the case of all mirrors we find that if we mark the points on the surface through which images are seen, and cover these points, the images will no longer be visible.[2] When, however, we uncover the points successively and direct our vision toward these uncovered points, both the points and the images in question will be seen in the direction of the visual ray.

Again, if we place long, straight objects at right angles to the surfaces of mirrors and take a position some distance off, both the images of the objects and the objects themselves as actually seen outside the mirror will appear to form a single straight line.[3]

From both these circumstances it follows that the image of an object must appear in the mirror at the intersection of the visual ray and the perpendicular from the object to the mirror.[4] Now these lines lie in the same plane, since they intersect. Again, this plane is perpendicular to the surface of the mirror, since one of the aforesaid lines is perpendicular to that surface. Finally, the visual ray, since it is reflected to the visible object, is in the aforesaid plane, and the perpendicular to the surface of the mirror at the point of reflection is the common boundary of all planes

---

[1] With reference to the figure where MR is the mirror, A the eye, B the object, B' the image, O the point at which the visual ray strikes the mirror, and TO perpendicular to the mirror, the assumptions are: (1) B' lies on AO, (2) B' lies on BP, perpendicular to MR, (3) $\angle TOA = \angle TOB$.

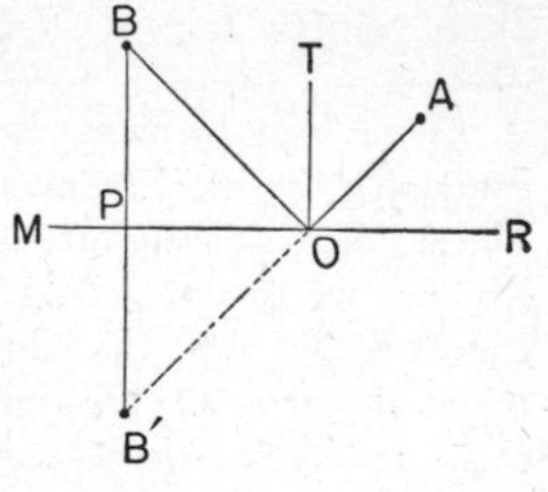

[2] This is a confirmation of the first assumption.

[3] This is a confirmation of the second assumption.

[4] Both, of course, produced. The visual ray referred to is that which, in the first instance, lies along the straight line from eye to image.

containing incident visual rays to the point in question and the corresponding reflected rays. . . .[1]

But this will become clearer and still more obvious and its truth will be amply demonstrated by the following experiment.

Take a round copper disk of moderate size, such as the one illustrated, with center at *A* and both surfaces as even as possible. Have the edges of the circumference well rounded and smoothed. Draw a small circle *BGDE* about center *A* on one surface of the disk, and draw two diameters, *BD* and *GE*, intersecting at right angles. Divide each quadrant of the circle into ninety equal parts. With *B* and *D*, respectively, as centers, and *BA* and *DA* as radii, draw *ZAH* and *TAK*, arcs of two circles. Now take three small, thin bars of iron, squared off and straight. Let one bar remain straight, and smooth one of its sides, making of it a polished mirror. Let the other two bars be curved so that a surface of one is convex and of the other concave over an arc of a circle equal to circle *BGDE*. Polish these two surfaces of these bars so that they may act as two mirrors.

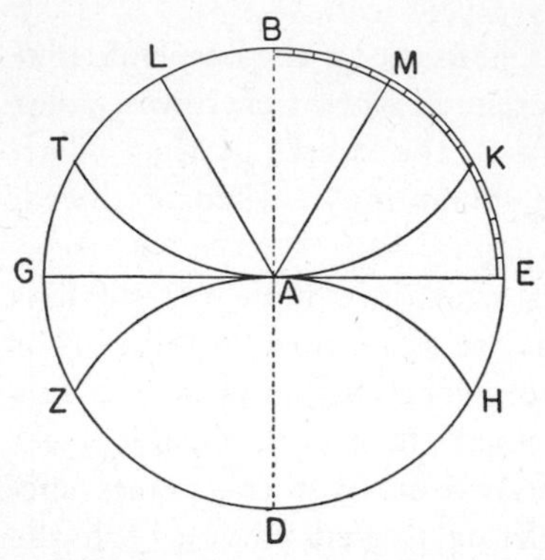

Now take arcs *ZAH* and *TAK* on each of the two curved bars, respectively, draw *BA* in white and *AL* in some other color, and set up a small dioptra[2] upon *AL*, placing the disk in such a way that the line of sight of the dioptra coincides with *AL* passing through *L*. . . . Place the plane mirror on *GAE*, the convex mirror on *ZAH*, and the concave mirror on *TAK*. Place at the common midpoint of the upper edge of the three mirrors a knob protruding from the disk so as to mark its position over point *A*.

Now if we place one eye at the dioptra at *L*, on line *AL*, and look toward the point at which the axis meets the mirrors, and if we then pass some small colored object over the surface of the disk, moving it until it appears to us on the other side of point *A*, which is on the line of vision, then points *L* and *A* and the image of the object in the three mirrors[3]

[1] What may be meant is that the said perpendicular is the boundary between all planes containing the incident rays and all planes containing the reflected rays.

[2] A simple sighting tube is meant.

[3] Are we to understand that the plane mirror stood higher than the concave and the convex higher than the plane, to prevent interference, and that the object was sufficiently tall to appear in all three mirrors so arranged? Or was it intended that the mirrors be used consecutively and the place of the image noted in each case? The experiment is an empirical test of the equality of the angles of incidence and reflection.

will appear to us on a single straight line. If, then, we mark the position of the object on the surface of the disk, that is, the point from which the image of the object is produced in the mirrors, say point *M*, and draw *AM*, we shall find that arc *BM* is always equal to arc *BL*. Angle *LAB* will, consequently, be equal to angle *MAB*, and *BD* will be perpendicular to all the mirrors.

Furthermore, *AL* is the path of the ray from the eye incident upon the surface of the mirror, and *AM* the path of the ray reflected from the mirror's surface to the object. Again, if we place an object of moderate length at *B* and place the eye on line *AB*, produced, the whole will appear upon a single straight line, *AD*.

The truth of the principles which we have assumed is, then, evident from our illustrations, and it may readily be seen that in these cases our reasoning accords with the evidence of our senses. Now it is the nature of a visual ray to proceed in a straight line from its source to all objects which are seen directly. A reflected ray, however, which proceeds from a mirror is not, in general, collinear with the visual ray. Our senses, therefore, must have recourse to an action which is natural and customary, and so we join the reflected ray to the first part of the visual ray, the part before reflection. Thus we have the impression that both parts constitute one straight ray, as if that were actually the case and nothing had happened to the ray. Hence the image of the object will be seen as if it were an object in the direct line of sight.

Refraction

AN INVESTIGATION OF REFRACTION

Ptolemy(?), *Optics* V, pp. 142.1–150.5 (Govi)[1]

Visual rays may be altered in two ways: (1) by reflection, i.e., the rebound from objects, called mirrors, which do not permit of penetration [by the visual ray], and (2) by bending [i.e., refraction] in the case of media

[1] Claudius Ptolemy, to whose work reference has been made (p. 162), was the author of a treatise on optics which is no longer extant in its original form. There is, however, some evidence that an extant Latin translation of a lost Arabic version of a Greek work on optics is to be identified with the otherwise lost *Optics* of Ptolemy, though the ascription cannot be made with certainty.

The original work seems to have been in five books. The first two dealt with the general theory of vision, the third and fourth with the theory of mirrors, and the fifth with the subject of refraction. Great interest attaches to this last book, for here is set forth the experimental procedure for the measurement of angles of refraction corresponding to given angles of incidence for a visual ray passing (*a*) from air to water, (*b*) from air to glass, and (*c*) from water to glass. (The incident ray, it is to be recalled, was treated in Greek geometrical optics as that which passes from the eye toward the reflecting or the refracting surface.) The figures obtained experimentally have evidently been corrected by the author to correspond to a set

which permit of penetration and have a common designation ["transparent"] for the reason that the visual ray penetrates them.

Now in the preceding books we dealt with mirrors, and explained so far as was possible according to the aforementioned principles of the science of optics, the relation of images and objects, and how the object in each case comes to produce the particular image. It remains for us to consider here the illusions involved when the objects are seen in media through which sight penetrates.

Now it has been shown in what precedes: (1) that this type of bending of a visual ray does not take place in all liquids and rare media, but that a definite amount of bending takes place only in the case of those media that have some likeness to the medium from which the visual ray originates, so that penetration may take place,[1] (2) that a visual ray proceeds along a straight line and may be naturally bent only at a surface which forms a boundary between two media of different densities, (3) that the bending takes place not only in the passage from rarer and finer media to denser (as is the case in reflection) but also in the passage from a denser medium to a rarer, and (4) that this type of bending does not take place at equal angles but that the angles, as measured from the perpendicular, have a definite quantitative relationship.[2]

We must now consider the effect of particular increments in the angles under discussion. Let us, however, first point out what refraction has in common with reflection, namely, that in either case the image appears to

---

formula (note the constancy of the "second differences" in the tables below).

The extant Latin version, made in the twelfth century by Eugenio of Sicily, lacks the first book and part of the fifth, as did the copy of the Arabic version on which it was based. Moreover the Latin is in many places obscure and unintelligible. This is due not only to the barbaric style of the translator but to the number of steps by which the translation is removed from the original. Our English version is consequently to be considered in many places as tentative and merely suggestive.

[1] This very obscure passage seems to mean that the process of refraction (as opposed to that of reflection) requires *two* different media, but not so different that one of them does and the other does not allow a visual ray to penetrate.

[2] Precisely what is meant is difficult to say because of the obscurity of the text. It is tempting, however, to interpret the passage as referring to a quantitative relation between angle of incidence and angle of refraction (the incident ray passing from the eye, and the angles being measured from the perpendicular upon the surface dividing the media). From the way in which the results of the observations seem to have been corrected, as the tables given on p. 278 indicate, the relation was of the form $r = ai - bi^2$ (where i is the angle of incidence, r the angle of refraction, and a and b constants depending on the specific media). This does not accord with accurate modern observation as well as does the relation $\sin i/\sin r =$ constant, but it is more accurate than the relation $i/r =$ constant, which seems to have been widely adopted (see the tables, p. 278). Throughout the discussion it is to be remembered that the ray from the eye is considered the incident ray, in contrast with modern usage, in which the ray from the object is generally so named.

be on the prolongation of the straight line forming the initial portion of that ray which passes from the eye and is reflected or refracted to the object.[1] That is, the image appears at the intersection of this ray (produced), which passes from the eye to the reflecting or refracting surface, and the perpendicular from the object to the same surface. It follows, therefore, in the case of refraction, just as in the case of reflection, that the plane passing through this altered ray is perpendicular to the surface at which the alteration takes place.[2]

From this basis follow conclusions relevant to the nature of perception and involving a quantitative relation[3] as we have shown in the passage dealing with the principles of mirrors.

This is quite clear and obvious[4] and we may understand it immediately with the help of a coin placed in a vessel called a *baptistir*. For suppose that the position of the eye is such that the visual ray emanating from it and just passing over the edge of the vessel reaches a point higher than the coin. Then, allowing the coin to remain in its position, pour water gently into the vessel until the ray which just passes over the edge is bent downward and falls upon the coin. The result is that objects not previously seen are then seen along the straight line passing from the eye to a point above the true position of the object. Now the observer will not suppose that the visual ray has been bent toward the objects but that the objects are themselves afloat and are raised toward the ray. The objects, therefore, will appear on the perpendicular drawn from them to the surface of the water, in accordance with the principles set forth above.

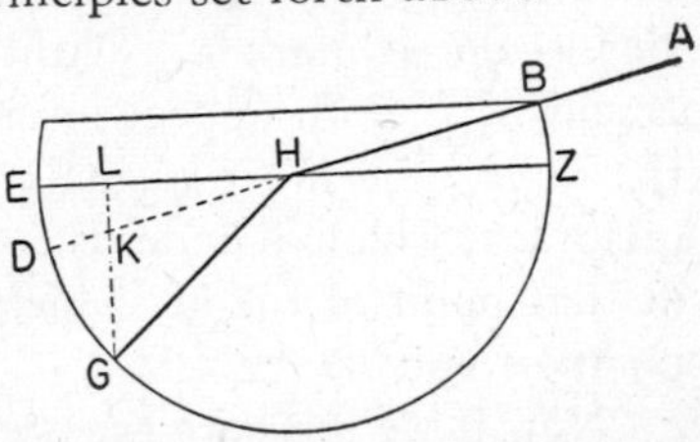

Thus [see fig.] if the eye be placed at point *A* . . . and the coin at point *G* in the lower part of the vessel, the coin will not be seen so long as the vessel is empty. The reason is that the portion of the vessel just below point *B* blocks the visual ray which could proceed directly to the coin. But when enough water is poured into the vessel so that it stands at the

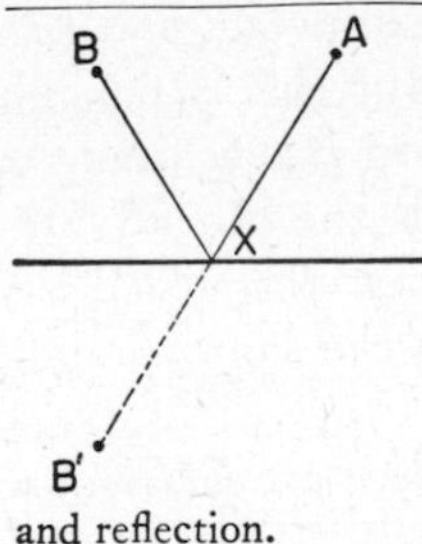

[1] That is, if *A* be the eye, *B* the object, *B′* the image, and *X* the point where the initial portion of the ray which is reflected to the object impinges on the reflecting or refracting surface, the image appears on a prolongation of *AX*.

[2] What seems to be meant is that the plane containing the visual and refracted rays is perpendicular to the refracting surface, just as, in the case of reflection, the plane containing the visual and reflected rays is perpendicular to the reflecting surface.

[3] In the case of reflection, the equality of the angles of incidence and reflection.

[4] The reference seems to be to the rule for the location of the image in the case, now, of refraction.

level *ZHE*, the ray *ABH* is bent in the direction of *GH*, below the prolongation of *AH*. In that case, the position of the coin will appear to be on the perpendicular from *G* to *EH*, that is on *LKG* which meets *AHD* at *K*. The apparent position of the coin, then, will be on the straight line proceeding from the eye and produced so that it passes through *K*, which is above the line that the visual ray actually takes and nearer to the surface of the water. The coin will, therefore, be seen at point *K*.

The amount of refraction which takes place in water and which may be observed is determined by an experiment like that which we performed, with the aid of a copper disk, in examining the laws of mirrors.[1]

On this disk draw a circle [see fig.] *ABGD* with center at *E* and two diameters *AEG* and *BED* intersecting at right angles. Divide each quadrant into ninety equal parts and place over the center a very small colored marker. Then set the disk upright in a small basin and pour into the basin clear water in moderate amount so that the view is not obstructed. Let the surface of the disk, standing perpendicular to the surface of the water, be bisected by the latter, half the circle, and only half, that is, *BGD*, being entirely below the water. Let diameter *AEG* be perpendicular to the surface of the water.

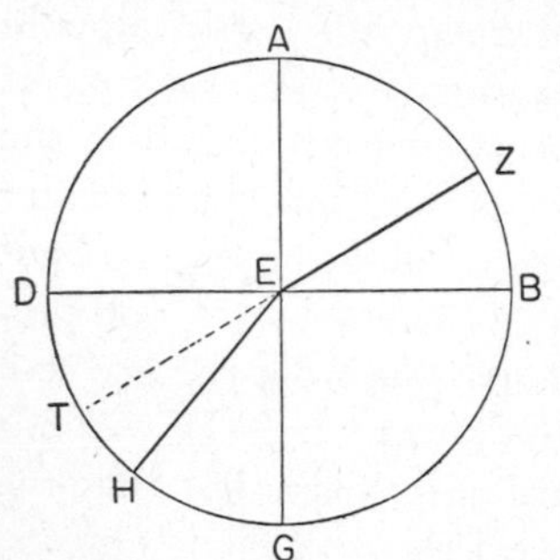

Now take a measured arc, say *AZ*, from point *A*, in one of the two quadrants of the disk which are above the water level. Place over *Z* a small colored marker. With one eye take sightings until the markers at *Z* and at *E* both appear on a straight line proceeding from the eye. At the same time move a small, thin rod along the arc, *GD*, of the opposite quadrant, which is under the water, until the extremity of the rod appears at that point of the arc which is on a prolongation of the line joining the points *Z* and *E*.

Now if we measure the arc between point *G* and the point *H*, at which the rod appears on the aforesaid line, we shall find that this arc, *GH*, will always be smaller than arc *AZ*. Furthermore, when we draw *ZE* and *EH*, angle *AEZ* will always be greater than angle *GEH*. But this is possible only if there is a bending, that is, if ray *ZE* is bent toward *H*, according to the amount by which one of the opposite angles exceeds the other.[2]

If, now, we place the eye along the perpendicular *AE* the visual ray will not be bent but will fall upon *G*, opposite *A* and in the same straight line as *AE*.

In all other positions, however, as arc *AZ* is increased, arc *GH* is also

---

[1] See p. 270, above.

[2] The amount of bending depends on the law connecting $\angle AEZ$ and $\angle GEH$, angles of incidence and refraction, respectively.

increased, but the amount of the bending of the ray will also be progressively greater.[1]

When *AZ* is 10°, *GH* will be about 8°
" " " 20°, " " " 15½°
" " " 30°, " " " 22½°
" " " 40°, " " " 29°
" " " 50°, " " " 35°
" " " 60°, " " " 40½°
" " " 70°, " " " 45½°
" " " 80°, " " " 50° . . .[2]

This is the method by which we have discovered the amount of refraction in the case of water. We have not found any perceptible difference in this respect between waters of different densities.

Now if we make our observation from the relatively dense natural water to the rarer medium, there will be considerable difference in the amount of refraction, corresponding to various increments in the angle of incidence, in the passage of the ray from the denser medium, water, to the rarer. But since it is impossible for us to determine, by an experiment such as that just now described, the amount of refraction which takes place when a visual ray proceeds from a denser to a rarer medium, we have applied the following method of measuring the angles.

Construct a semicylinder of pure glass similar to half the circular disk. Let the base of this semicylinder take the position *TKL* [see fig.] and let its diameter be smaller than that of the aforesaid metal disk. Fit the base of the semicylinder to the disk so that the whole base is fastened to the disk, the common center is at *E*, the diameter *TL* lies along diameter *BD*, and *AE* is perpendicular to the plane side of the glass surface. Hence all lines drawn from *E* to arcs *BGD* and *TKL* will be perpendicular thereto.

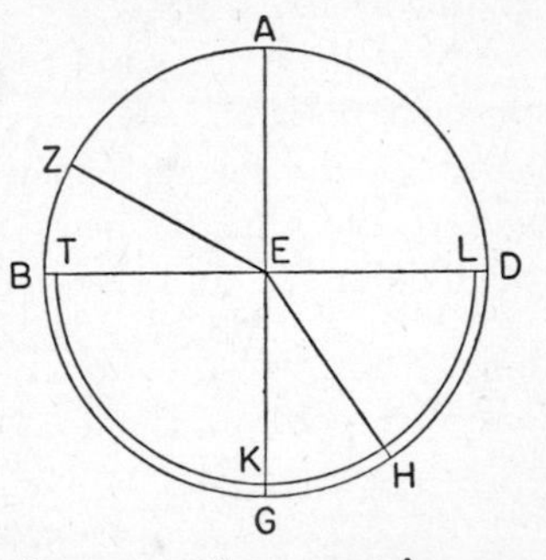

Now we arrange this experiment as we did the preceding experiment, placing a small marker on the glass just over *E*, the midpoint of the straight edge of the base of the semicylinder. We then look, with one eye, along line *AE* toward the edge of the glass, and keep moving an object along that part of the circumference opposite that from which we are observing, until it appears before our eye. Now this object will be found to be over point *G*, for *AEG* is perpendicular both to *TEL* and *TKL*.

---

[1] That is, *GH* increases but at a progressively slower rate than *AZ*.

[2] See the comparisons in Table A, *From Air to Water*, below. Note that r, the angle of refraction, is given by the equation $r = ai - bi^2$, where $a = 0.825$, $b = 0.0025$, and i is the angle of incidence (the incident ray being that which passes from the eye to the refracting surface: modern usage generally calls this the angle of refraction).

Again, if we shift our viewpoint to a point opposite *A* and look in the direction of *GE*, an object moved over the opposite circumference of the disk will come into view on a prolongation of *GE*, that is, above *EA*. For the same reason as before, there will be no bending of the visual ray in its passage from glass to air.[1]

But now take a certain arc measured from *A*, say arc *AZ*, draw *ZE*, coloring it black, and direct the vision along this line until an object moved back and forth behind the glass is seen in the direction of that line. If we place a marker at the point, *H*, reached when *EH* appears collinear with the black line *ZE*, we shall again in this case find that angle *AEZ* is larger than angle *GEH*. Moreover we shall find the excess of the one angle over the other greater than in the case of water, for given angles of incidence.[2]

And again, when the eye is at point *H* on the other side of *E* and looks from *H* in the direction of *HE*, both points will appear to be on precisely the same line as in the preceding case. But since there was a bending of the ray at point *E*, it follows that whether the ray proceeded from the air to the glass, as did *ZE*, and was bent along *EH*, or proceeded from the glass to the air, as did *HE*, and was bent along *EZ*, in either case there was a bending in the direction of *T*. And since the perpendiculars which are drawn from E to arc *TKL* are all similar, they are not bent, whether the rays which they represent are considered as beginning or as ending at *E*.[3]

Now if in this case, too, we wish to find the amount of the refraction in each position, we place the eye successively in each of the positions taken in the former experiment [where the visual ray passed from air to water] and thus we vary the angle made on the disk[4] between the perpendicular *AE* and the visual ray *EZ*. The results are as follows:

When $\angle AEZ$ is 10°, $\angle GEH$ is approximately 7°
" " " 20°, " " 13½°
" " " 30°, " " 19½°
" " " 40°, " " 25°
" " " 50°, " " 30°
" " " 60°, " " 34½°
" " " 70°, " " 38½°
" " " 80°, " " 42° . . .[5]

---

[1] I.e., there is no refraction when the visual ray is perpendicular to the boundary between the media. Thus there is no refraction in the light from a star at the zenith (see p. 282).

[2] I.e., the excess of the angle of incidence over the angle of refraction is greater in the passage of the visual ray from air to glass than for the passage from air to water.

[3] The point seems to be that so long as the ray is within the same medium, glass, there is no bending.

[4] Possibly, "in the air."

[5] See the comparisons in Table B, *From Air to Glass*. Note that the angle of refraction is given by the equation $r = ai - bi^2$, where $a = 0.725$ and $b = 0.0025$.

But the amount of refraction will be less when the glass is placed next to water, since the difference between angles of incidence and refraction in the passage of a visual ray from one of these bodies to the other is not large. For the difference in density between water and glass is less than that between air and water or between air and glass. But we are able again in this case to determine the amount of refraction, as we shall now explain.

Attach a semicylinder of glass [see fig.] to the bronze disk and adjust it so that the center of the straight edge is the same as that of the disk. Again color point *E*, and set up the bronze disk in a basin so that the disk is at right angles to the surface of the water and half under the water. Place the curved side of the glass, *TKL*, above, and pour into the basin an amount of water so that edge *TEL* of the cylinder will be just above the surface of the water.

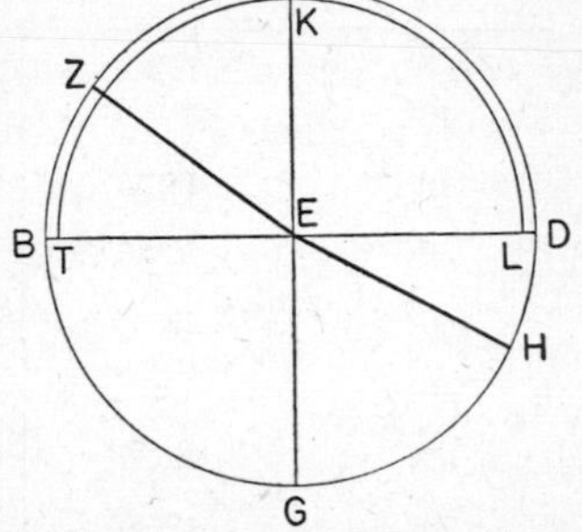

Now take arc *GH* in the less dense medium, that is, in the water, containing, say, 10°. Mark *H* with a small colored marker, and sight it with one eye until an object *Z*, which is being moved over arc *AB*, is seen along the line joining *H* with the marked point *E*. When this has been done, draw the two lines *EH* and *EZ*.

If, then, we wish to measure on arc *AB* the angle subtended in the denser medium, that is, in the glass, as the angle in the water measured from the perpendicular, that is, angle *GEH*, varies, we shall find the following results:

When $\angle GEH$	is	10°,	$\angle AEZ$	is approximately	9½°
"	"	" 20°,	"	"	18½°
"	"	" 30°,	"	"	27°
"	"	" 40°,	"	"	35°
"	"	" 50°,	"	"	42½°
"	"	" 60°,	"	"	49½°
"	"	" 70°,	"	"	56°
"	"	" 80°,	"	"	62° . . . [1]

[1] That is, $r = ai - bi^2$, where $a = 0.975$ and $b = 0.0025$ (Table C, below).

The following tables will serve to compare the theory of refraction as given in our text with that in which the ratio of the angles of incidence and refraction is a constant, and with the modern theory, in which the ratio of the sines of the angles is a constant. See P. Brunet and A. Mieli, *Histoire des sciences: antiquité*, pp. 826–827.

It may be noted that the correction of experimental data to correspond to a set formula is discussed by Kepler in connection with Witelo's (thirteenth century) tables of refraction, which are much the same as those in the present text. Kepler contented himself with the formula $r = ki$ for relatively small angles of incidence. The relation between refraction and

SUMMARY OF THE LAWS OF REFRACTION

Ptolemy(?), *Optics* V, pp. 154.1–156.6 (Govi)

Now it is possible, on the basis of our investigations, to draw general conclusions about this type of refraction. Thus, near the point at which

the velocity of light in different media (which leads to the sine law) was not explicitly stated until the seventeenth century.

Angles of incidence (i)	Angles of refraction (r), according to text	First differences	Second differences	i:r	sin i:sin r
				(on the basis of the angles in the first two columns)	
		A. From Air to Water			
0°	0°				
		8°			
10°	8°		30′	1.25	1.248
		7°30′			
20°	15°30′		30′	1.29	1.270
		7°			
30°	22°30′		30′	1.33	1.308
		6°30′			
40°	29°		30′	1.38	1.369
		6°			
50°	35°		30′	1.43	1.336
		5°30′			
60°	40°30′		30′	1.48	1.333
		5°			
70°	45°30′		30′	1.55	1.329
		4°30′			
80°	50°			1.60	1.286
		B. From Air to Glass			
0°	0°				
		7°			
10°	7°		30′	1.43	1.425
		6°30′			
20°	13°30′		30′	1.48	1.465
		6°			
30°	19°30′		30′	1.54	1.498
		5°30′			
40°	25°		30′	1.60	1.521
		5°			
50°	30°		30′	1.67	1.531
		4°30′			
60°	34°30′		30′	1.74	1.529
		4°			
70°	38°30′		30′	1.82	1.509
		3°30′			
80°	42°			1.91	1.472
		C. From Water to Glass			
0°	0°				
		9°30′			
10°	9°30′		30′	1.07	1.052
		9°			
20°	18°30′		30′	1.09	1.078
		8°30′			
30°	27°		30′	1.11	1.101
		8°			
40°	35°		30′	1.14	1.121
		7°30′			
50°	42°30′		30′	1.18	1.134
		7°			
60°	49°30′		30′	1.22	1.139
		6°30′			
70°	56°		30′	1.25	1.133
		6°			
80°	62°			1.28	1.115

the visual ray bends and at which a perpendicular drawn from any external point reaches the surface that bounds the two aforesaid dissimilar bodies,[1] objects in the denser medium appear larger than they do in the rarer medium (the same position in each medium being preserved). The visual ray passes in this case from the rarer to the denser medium. The opposite will be the case when the passage of the visual ray is from the denser to the rarer medium.[2]

Our proposition is that the *amount* of the refraction is the same in each of the two types of passage but that the two refractions differ in type. For in its passage from a rarer to a denser medium the ray inclines *toward* the perpendicular, whereas in its passage from a denser to a rarer medium it inclines *away* from the perpendicular.[3]

For consider a plate such as we have previously described, with diameter *BD* [see fig.] lying on the surface which divides two dissimilar media. Draw the perpendicular *AEG*. Let a ray inclined to the perpendicular, for example *EH*, make ∠*GEH* with the perpendicular. Now the position of the refracted ray remains exactly the same when the visual ray passes through point *E*[4] and the position of the eye is at point *Z*. For the line beyond the point of refraction, that is *EK*, inclines, in its course, toward the perpendicular,[5] whereas the visible object appears to be along the straight line (*EK*). Again, if the eye is at point *H*, and *EZ* is in the rarer medium (*BAD*), line *ET* (beyond the point of refraction) will incline away from perpendicular *AE*,[6] an action exactly the opposite of that in the former case. That is, the ray is further from the perpendicular than if it were to proceed in a straight line.

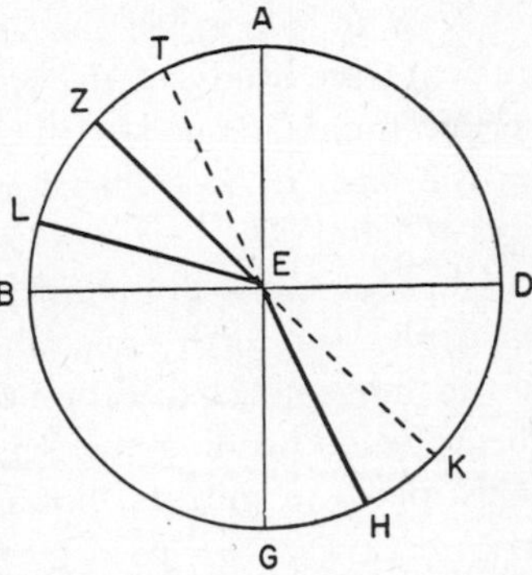

Again when the media differ considerably (in density), so do the angles (of incidence and refraction), and the difference in the angles becomes greater as the density of (the denser) one of the media is increased. For if we assume that semicircle *BAD* is in a rarer and *BGD* in a denser medium, and take angle *AEZ* as it is [see fig.], then, if the medium of section

[1] The boundary between the air and the aether had been referred to, but the sequel considers any two media of different densities.

[2] The point seems to be that if the eye and the object exchange positions the object (now in the rarer medium) appears smaller than it did in the denser medium.

[3] Here we have the general statement of the reversibility of refraction. The experimental evidence is set forth in the following paragraphs.

[4] I.e., originates further back than *E*.

[5] Obviously what is meant is that the visual ray, which if unbent would lie along *EK*, is bent toward the perpendicular.

[6] I.e., the ray, which if unbent would lie along *ET*, is bent away from the perpendicular.

BGD is made denser, the excess of $\angle AEZ$ over $\angle GEH$ will vary with the excess of the density of the new medium over the old.[1] For example, when $\angle AEZ$ in air is a third of a right angle, $\angle GEH$ will, in water, be about a fourth part of a right angle, and, in glass a fifth part of a right angle plus a sixtieth thereof, approximately. In this latter case the amount of the refraction and the excess of the angle of incidence over the angle of refraction will, as the former approaches 90°, be greater, since glass is a denser substance than water.

In the same way, if we take the path of one of the refracted visual rays, say *LEK*, other than that of perpendicular *AE*,

$$AL:AZ > GK:HG$$

and, *alternando*, $AL:GK > AZ:GH$.

Again, *separando*, $LZ:AZ > KH:HG$

and $LZ:KH > AZ:GH$.[2]

Now it is possible for us to understand these various points from a quantitative study of the refractions we have investigated, if we assume certain numbers,[3] and, with their help take up the several changes indicated on the basis of such initial assumptions, as we did in the case of the two arcs *AZ* and *AL*.

But some one in opposition to this may ask why it is that in the first principles set forth—i.e., about the perpendiculars, and the appearance of the image in the direction of the visual ray—there is a similarity between the type of bending just discussed [i.e., refraction] and reflection, as it takes place in mirrors, but there is no such similarity in the measure of

---

[1] Neither the method of measuring relative densities nor the precise way in which the angular differences vary with the differences in densities can be gathered from the example that follows, viz., when $\angle AEZ = 30°$, $\angle GEH = 22\frac{1}{2}°$ in water and $19\frac{1}{2}°$ in glass (see p. 278, above).

[2] If i_1 and i_2 are angles of incidence (the incident visual ray being that which passes from the eye to the refracting surface) with $i_2 > i_1$, and r_1 and r_2 the corresponding angles of refraction, the preceding paragraph gave the result $(i_2 - r_2) > (i_1 - r_1)$. The present paragraph gives the results:

$$i_2 : i_1 > r_2 : r_1,$$
$$i_2 : r_2 > i_1 : r_1,$$
$$(i_2 - i_1) : i_1 > (r_2 - r_1) : r_1,$$

and

$$(i_2 - i_1) : (r_2 - r_1) > i_1 : r_1.$$

But $i_2 : i_1 > r_2 : r_1$ (from which the other results follow immediately) is valid only in the case where the visual ray from the eye passes from the less dense to the denser medium (the angle of incidence being the angle made by the line of vision with the perpendicular), as in all the cases given by the author. Since we, however, consider the light as passing from the object to the eye, and take the angle of incidence as that angle which the ray from the object makes with the perpendicular, it follows that, for us, $i_2 : i_1 > r_2 : r_1$ only in the passage of light from the less refrangent to the more refrangent medium: otherwise, $i_2 : i_1 < r_2 : r_1$.

[3] The point seems to be that the theoretical results set forth above may be better grasped if we assign specific values to i_2 and i_1, and note the corresponding values of r_2 and r_1 (from tables such as those on p. 278).

the angles.[1] The answer, as well as the necessity that things be so, will be found in what we are to set forth. And from this will be seen something even more remarkable, namely, the operation of nature in conserving the activity of force.[2]

ATMOSPHERIC REFRACTION

Ptolemy(?), *Optics* V, pp. 151.1–153.20 (Govi)

Again, it is possible for us to see from the phenomena which I am about to discuss that at the boundary between air and aether there is a bending of the visual ray because of a difference between these two bodies. We find that stars which rise and set seem to incline more toward the north when they are near the horizon and are measured by the instrument used for such measurement.[3] For the circles, parallel to the equator, described upon these stars when they are rising or setting are nearer the north than the circles described upon them when they are in the middle of the heaven.[4] As they draw nearer the horizon they have a greater inclination toward the north. In the case of stars which do not rise and set, their distance from the north pole will be smaller when their position on the meridian is nearer the horizon. For when their position on the meridian is nearer the zenith, the circle parallel to the equator will, at that point, be larger, whereas it is smaller in the other position.[5] This is due to the bending of the visual ray at the surface which divides the air from the aether, a spherical surface, of necessity, whose center is the common center of all the elements and of the earth.

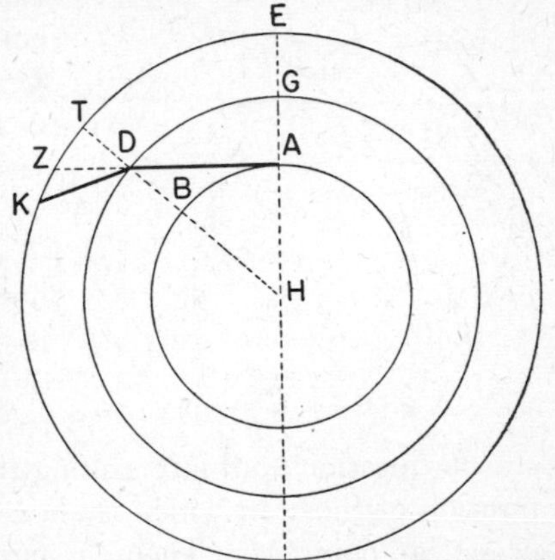

Consider, then, in the first place point *E* [see fig.] as the zenith and the great circles of the various spheres which we have mentioned, circle *AB* on the surface of the earth, circle *GD* on the surface dividing air and aether, and *EZ* which passes through certain stars. Let the center of all the spheres

[1] The angles of incidence and reflection are equal, but the angles of incidence and refraction are unequal.

[2] This statement of a principle of least action or conservation of energy is to be noted, though its application to the problem of refraction is not clearly made by the author in the portion of Book V that is extant.

[3] The dioptra or the astrolabe (see p. 139).

[4] I.e., the small circle (parallel to the celestial equator) that represents the apparent path of a star in its diurnal rotation appears to have a greater declination when taken at the rising or setting of the star in question than, say, midway between the rising and setting.

[5] I.e., the apparent distance from a circumpolar star to the pole is greater at its upper than at its lower transit across the meridian.

be *H*. Draw *EAH*, take point *A* as the position of the eye, and line *ADZ* to meet the boundary common to the horizon and to circle *GD*. Again, let *DT* be perpendicular to the circle, and consider *ADK* as a visual ray bent from point *D* along *KD*. Suppose there is a star at point *K*. Since the visual ray is bent at the surface,[1] toward a position more remote from point *E*, the angle *KDT* (which is in the rarer medium) formed with the line [*HDT*] perpendicular to the bounding surface from which a reflection would be at equal angles, is greater [than ∠*ZDT*]. The stars will therefore be seen from point *A*[2] along line *ADZ*, and the distance of the star from the zenith will appear less than the true distance. For its distance will appear to be arc *EZ* instead of arc *KE*.

The higher, then, the position of the star in the heavens, the smaller will be the difference between its apparent and its real position. If the star is at *E* there is no bending, since the visual ray from point *A* to point *E* is not subject to such bending (refraction), for it is perpendicular to the bounding surface at which any refraction would take place.

These, then, are the preliminary propositions. Let us now consider *ABG* [see fig.] as the circle of the horizon, and *AEZG* as that half of the circle of the meridian which is above the earth, *E* being the zenith and *Z* the visible pole of the celestial sphere. Let *BHD* be the portion above earth of the circle parallel to the celestial equator and passing through certain stars. Let there be a star at point *T* on this circle near the horizon, and let *KETL* be the visible half of the circle which passes through the zenith and through the star at *T*. Since, then, when a star is near the horizon it appears nearer the zenith than it really is,[3] and its apparent position differs from its true position on the great circle [passing through the zenith and the star in question and] intersecting the horizon, it follows that the apparent position of the star, which is above point *T*, will be between *E* and *T*, let us say at point *M*. Then the circle parallel to the celestial equator and passing through *M* will lie further to the north than the circle (parallel to the celestial equator) passing through point *T*, which, in our part of the inhabited world, lies toward the north. When the star has risen to point *H* it is in a position where the bending of the visual ray is insufficient to cause a perceptible difference between the apparent and the true position.[4]

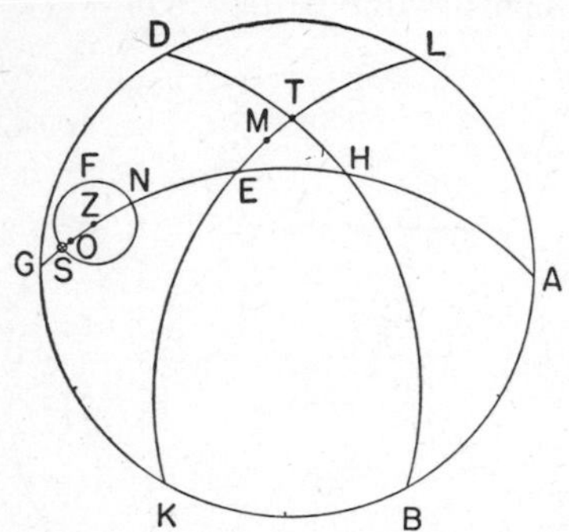

[1] I.e., the surface marking the boundary between aether and air.

[2] The text reads *E*, which can hardly be right.

[3] The text reads "nearer the zenith than to its true position," which is obviously wrong.

[4] At *H* the star is on the meridian, and its angular distance from the zenith is least.

Similarly, take Z as the north celestial pole, and let NSF be the circle parallel to the celestial equator, described by one of the stars which are always above the horizon. When the star is at point S of this circle it will appear nearer point E, the zenith, and will seem to be at about point O. But when the star is at point N there is no difference, or only an imperceptible difference, between the apparent and the real position. And therefore when such a star, in its revolution, is nearer the horizon, its distance from the north pole of the celestial sphere will seem less than the distance from the aforesaid pole when the star in its revolution is nearer the zenith. For arc ZN will be greater than arc OZ. We have therefore seen why stellar phenomena must, because of the bending of visual rays, appear as they do.[1] . . .

Cf. Ptolemy, *Almagest*, p. 13.3–9 (Heiberg)

Now the fact that heavenly bodies appear larger when they are near the horizon is due not to their smaller distance from the horizon but to the vaporous moisture surrounding the earth between our eye and these heavenly bodies.[2] It is the same as when objects immersed in water appear larger and in fact the more deeply immersed the larger.

Cf. Cleomedes, *On the Circular Motion of the Heavenly Bodies* II. 1, pp. 122.15–124.8 (Ziegler)

The sun appears larger to us as it rises and sets, but smaller when it is in mid-heaven, for the reason that in viewing it at the horizon we see it through a thicker layer of air and also one that is more humid, for such is the air next to the earth. But when we view it in mid-heaven, we see it through clearer air. Thus in the latter case the ray issuing from the eye toward the sun is not refracted, but in the former case, where the ray is directed toward the horizon when the sun rises or sets, the ray is necessarily refracted when it encounters the thicker and moister air. Thus the sun appears larger to us, just as the appearance of objects in water is altered because they are not seen in a straight line.[3] Such phenomena are in every case to be considered as due to disturbances of our vision and not connected with the visible objects themselves. It is also said that where it is possible to view the sun from deep wells, its appearance is much larger since it is seen through the humid air of the well. It certainly cannot be said that the sun grows larger when viewed from the bottom of a well and smaller when viewed from the top; but evidently the dimness and dampness of the air in the well cause the sun to appear larger to the observer.

---

[1] The author goes on to indicate the impossibility of measuring astronomical refraction because of the absence of data as to the relative extent of the atmosphere and the aether.

[2] The apparent increase of size of sun and moon at the horizon is largely due to an optical illusion, not to refraction (cf. also p. 123).

[3] But by a refracted ray.

A "PARADOXICAL" ECLIPSE EXPLAINED BY REFRACTION

Cleomedes, *On the Circular Motion of the Heavenly Bodies* II. 6.
Translation of T. L. Heath, *Greek Astronomy*[1]

These facts having been proved with regard to the moon, the argument establishing that the moon suffers eclipse through falling into the earth's shadow would seem to be contradicted by the stories told about a class of eclipses seemingly paradoxical. For some say that an eclipse [sometimes] occurs, even when both the luminaries are seen above the horizon.[2] This should make it clear that [in that case] the moon does not suffer eclipse through falling into the earth's shadow, but in some other way, since, if an eclipse occurs when both sun and moon appear above the horizon, the moon cannot suffer eclipse through falling into the earth's shadow. For the place where the moon is, when both bodies appear above the horizon, is still being lit up by the sun, and the shadow cannot yet be at the place where the moon gives the impression of being eclipsed. Accordingly, if this be the case, we shall be obliged to declare that the cause of the eclipse of the moon is a different one.[3] . . .

Nevertheless, having regard to the many and infinitely various conditions which naturally arise in the air, it would not be impossible that, when the sun has just set, and is under the horizon, we should receive the impression of its not yet having set, if there were cloud of considerable density at the place of setting and the cloud were illuminated by the sun's rays and transmitted to us an image of the sun, or if there were "anthelium." Such images are indeed often seen in the air, especially in the neighbourhood of Pontus. The ray, therefore, proceeding from the eye and meeting the air in a moist and damp condition might be bent, and so might catch the sun although just hidden by the horizon. Even in ordinary life we have observed something similar. For, if a gold ring be thrown into a drinking

---

[1] It is not known when Cleomedes lived. Some place him in the first century B.C., others in the first or second A.D. He was a compiler rather than an original scientist, and his work is important for the wealth of historical information that it contains. His treatise *On the Circular Motion of the Heavenly Bodies*, extant in two books, deals with various astronomical questions and is often controversial, reflecting the Stoic viewpoint against the Epicurean. A passage containing the account of the methods used by Eratosthenes and Posidonius in estimating the size of the earth has been quoted above (p. 149). The present passage discusses possible explanations of a "paradoxical" eclipse of the moon when both sun and moon appear above the horizon. [Edd.]

[2] Pliny refers to an instance of such an eclipse (*Natural History* II. 57) as follows: "And [it was discovered by Hipparchus] why it is that, though after sunrise the eclipsing shadow must be below the earth, it has once happened that the moon was eclipsed in the west while both sun and moon were visible above the horizon."

Such an eclipse was visible in the vicinity of New York City on Nov. 7, 1938. [Edd.]

[3] Cleomedes is skeptical about the reported observation, but considers various explanations before turning to the possibility of refraction as an explanation. [Edd.]

cup or other vessel, then, when the vessel is empty, the object is not visible at a certain suitable distance, since the visual current goes right on in a straight line as it touches the brim of the vessel. But, when the vessel has been filled with water up to the level of the brim, the ring placed in the vessel is now, at the same distance, visible, since the visual current no longer passes straight on past the brim as before, but, as it touches, at the brim, the water which fills the vessel up to the brim, it is thereby bent, and so, passing to the bottom of the vessel, finds the ring there. Something similar, then, might possibly happen in a moist and thoroughly wet condition of the air, namely, that the visual ray should, by being bent, take a direction below the horizon, and there catch the sun just after its setting, and so receive the impression of the sun's being above the horizon. Perhaps, also, some other cause akin to this might sometimes give us the impression of the two bodies being above the horizon, though the sun had already set. But the observed phenomena make it as clear as day that the moon is not eclipsed otherwise than by falling within the earth's shadow.

The Nature and Velocity of Light

Aristotle, *On the Soul* II. 7.[1] Translation of R. D. Hicks (Cambridge, 1907)

And so we shall have first to explain what light is.

There is, then, we assume, something transparent; and by this I mean that which, though visible, is not, properly speaking, visible in itself, but by reason of extrinsic colour. Air, water, and many solid bodies[2] answer to this description. For they are not transparent *quâ* air or *quâ* water, but because there is a certain natural attribute present in both of them which is present also in the eternal body on high. Light is the actuality of this transparent *quâ* transparent.[3] But where the transparent is only potentially present, there darkness is actually. Light is a sort of colour in the transparent when made transparent in actuality by the agency of fire or something resembling the celestial body: for this body also has an attribute which is one and the same with that of fire. What the transparent is, and what light is, has now been stated; namely, that it is neither fire nor body generally nor an effluence from any body[4] (for even then it would still be a sort of body), but the presence of fire or something fiery in the transparent. For it is impossible for two bodies to occupy the same space at the same time.

---

[1] In connection with Aristotelian optical theory note also the discussion of the rainbow (*Meteorologica* III. 4–5) and the comments of T. E. Lones, *Aristotle's Researches in Natural Science* (London, 1912), pp. 36–42, and A. M. Sayili in *Isis* 30 (1939) 65–83. [Edd.]

[2] E.g., glass. [Edd.]

[3] I.e., not as air or as water, but as transparent. [Edd.]

[4] Aristotle is here refuting the view of Empedocles and Plato that light is fire or like fire, and the view of Democritus that light is a corporeal emission from the surface of a body.

Light is held to be contrary to darkness. But darkness is absence from the transparent of the quality above described: so that plainly light is the presence of it. Thus Empedocles and others who propounded the same view are wrong when they represent light as moving in space and arriving at a given point of time between the earth and that which surrounds it without our perceiving its motion. For this contradicts not only the clear evidence of reason, but also the facts of observation: since, though a movement of light might elude observation within a short distance, that it should do so all the way from east to west is too much to assume.[1]

Aristotle, *De Sensu* 6, 446*a*26–*b*3. Translation of J. I. Beare (Oxford, 1908)

Empedocles, for example, says that Light from the Sun arrives first in the intervening space before it comes to the eye, or reaches the Earth. This might plausibly seem to be the case. For whatever is moved [in space], is moved from one place to another; hence there must be a corresponding interval of time also in which it is moved from the one place to the other. But any given time is divisible into parts; so that we should assume a time when the sun's ray was not as yet seen, but was still travelling in the middle space.

Cf. Lucretius VI. 195–204

But we hear the thunder with our ears after our eyes perceive the flash of lightning, because things always reach our ears more slowly than they affect our vision. This may also be seen from the fact that if you perceive someone in the distance cutting down a tall tree with a double edged axe, you see him strike the blow before the sound of it comes to the ears. So also we see the lightning before we hear the thunder clap which arises at the same time and from the same cause as the lightning, being born of the same collision.

Cf. also Pliny, *Natural History* II. 142

It is certain that a lightning flash is seen before the thunder is heard, though they both take place at the same time. And this is not strange for light is swifter than sound.

ACOUSTICS AND MUSICAL THEORY

The Nature of Sound

Archytas, Frag. 1 (Diels)[2]

"Now they [the mathematicians] observed in the first place that there can not be sound without the striking of bodies against one another. . . .

---

[1] Cf. p. 214, above. Having denied both the corporeality and the (spatial) motion of light, Aristotle rejects the notion of a finite velocity. Empedocles and the atomists had held, as we do, that light from the sun reached the intervening spaces between sun and earth before it reached the earth and our eyes. [Edd.]

[2] Porphyrius (third century A.D.) in his commentary on Ptolemy's *Harmonics* quotes the

Now as we are constituted many sounds cannot be heard by us, some because of the feebleness of the blow, others because of the distance from us, still others because of the intensity of the sound. For just as when one seeks to pour a great deal of water into a narrow-necked vessel and none of the water enters, so the very intense sounds fail to penetrate to our hearing. Of the sounds that we do perceive, those that reach us swiftly and violently from the blow seem high pitched, those that reach us slowly and weakly seem to be low pitched.[1] For if one takes a stick and moves it slowly and feebly, the blows produce a low sound, whereas if one strikes swiftly and intensely, the blows produce a high sound. This we may observe not only in the circumstances just described but also when we wish to produce a loud and sharp sound in speaking or singing. In that case we make the sound by increasing the intensity of our breath. Now it is the same as with projectiles. Those that are hurled with greater force are carried far, those that are hurled weakly move only a short distance. For the air gives way more readily before those projectiles that are hurled with great force, but less readily before those hurled weakly. So it is with sounds. Those that are projected by an intense breath are loud and sharp, while those projected with weak breath are soft and low.[2] And we can see this clearly with the help of a most compelling example. For when the same person makes a loud sound we can hear him from a considerable distance, but when he makes a soft sound we cannot hear him even at close range. And in the case of pipes, air blown from the mouth into holes near the mouth produces a sharper note because of the strong pressure. But if the air passes through the holes further from the mouth, the note is lower. Clearly, then, swift motion produces a high-pitched sound and slow motion a low-pitched sound.[3] Now the same thing happens in the case of the bull-roarers that are used in the celebration of the mysteries. Those that are moved slowly give out a low sound, those moved intensely give out a sharp sound. So it is also with the reed. If one blows into it, closing off the bottom, it gives us a low-pitched sound, but if one blows into it using

---

following passage from a no longer extant work of Archytas (see p. 35). The passage is interesting for its connection of sound with motion and its connection of pitch with speed of motion. That there is no clear idea, however, of wave propagation of sound and frequency of vibrations is seen from the statement that high-pitched sounds travel more swiftly than those of low pitch. On the relative velocity of light and sound see the preceding passages.

[1] Cf. Aristotle, *De Sensu* 448*a*19: "Some of those who treat of concords say that the sounds do not reach us at the same time, but only appear to do so."

Theophrastus in Porphyrius's *Commentary on Ptolemy's Harmonics*, p. 64.20 (Düring): "The higher note does not differ [from the lower] in speed, for if it did it would reach the hearing sooner, and there would be no concord. If there is a concord, both notes must have the same speed."

[2] There seems to be some confusion, in Archytas's account, between intensity and pitch.

[3] The inverse relation between the length of the vibrating column of air and the rate of vibration is not very clearly put. Cf. the case of the reed pipe mentioned below.

only half or any other fraction of the reed, a high-pitched sound is produced. For the same breath moves over a long path weakly, but over a shorter path with more intensity."

After further discussion on the proportionality involved in the motion of sound, he [Archytas] concludes his argument by saying "that it is clear from many considerations that high-pitched sounds move more swiftly and low-pitched more slowly."

Aristotle, *On the Soul* II. 8. Translation of R. D. Hicks

There are two sorts of sound, one a sound which is operant, the other potential sound. For some things we say have no sound, as sponge, wool; others, for example, bronze and all things solid and smooth, we say have sound, because they can emit sound, that is, they can produce actual sound between the sonorous body and the organ of hearing. When actual sound occurs it is always of something on something and in something, for it is a blow which produces it. Hence it is impossible that a sound should be produced by a single thing, for, as that which strikes is distinct from that which is struck, that which sounds sounds upon something. And a blow implies spatial motion. As we stated above, it is not concussion of any two things taken at random which constitutes sound. Wool, when struck, emits no sound at all, but bronze does, and so do all smooth and hollow things; bronze emits sound because it is smooth, while hollow things by reverberation produce a series of concussions after the first, that which is set in motion being unable to escape. Further, sound is heard in air, and though more faintly, in water.[1] It is not the air or the water, however, which chiefly determine the production of sound: on the contrary, there must be solid bodies colliding with one another and with the air: and this happens when the air after being struck resists the impact and is not dispersed. Hence the air must be struck quickly and forcibly if it is to give forth sound; for the movement of the striker must be too rapid to allow the air time to disperse: just as would be necessary if one aimed a blow at a heap of sand or a sandwhirl, while it was in rapid motion onwards.

Echo is produced when the air is made to rebound backwards like a ball from some other air which has become a single mass owing to its being within a cavity which confines it and prevents its dispersion. It seems likely that echo is always produced, but is not always distinctly audible: since surely the same thing happens with sound as with light. For light is always being reflected; else light would not be everywhere, but outside the spot where the sun's rays fall there would be darkness. But it is not always reflected in the same way as it is from water or bronze or any other smooth surface; I mean, it does not always produce the shadow, by which we define light.

[1] As a matter of fact, water is better than air as a medium for sound. [Edd.]

Void[1] is rightly stated to be the indispensable condition of hearing. For the air is commonly believed to be a void, and it is the air which causes hearing, when being one and continuous it is set in motion. But, owing to its tendency to disperse, it gives out no sound unless that which is struck is smooth. In that case the air when struck is simultaneously reunited because of the unity of the surface; for a smooth body presents a single surface.

That, then, is resonant which is capable of exciting motion in a mass of air continuously one as far as the ear. There is air naturally attached to the ear. And because the ear is in air, when the external air is set in motion, the air within the ear moves.

[Aristotle], *De Audibilibus* 800*a*1–*b*3, 803*b*18–804*a*8.[2] Translation of T. Loveday and E. S. Forster (Oxford, 1913)

All sounds, whether articulate or inarticulate, are produced by the meeting of bodies with other bodies or of the air with bodies, not because the air assumes certain shapes, as some people think, but because it is set in motion in a way in which, in other cases, bodies are moved, whether by contraction or expansion or compression, or again when it clashes together by an impact from the breath or from the strings of musical instruments. For, when the nearest portion of it is struck by the breath which comes into contact with it, the air is at once driven forcibly on, thrusting forward in like manner the adjoining air, so that the sound travels unaltered in quality as far as the disturbance of the air manages to reach. For, though the disturbance originates at a particular point, yet its force is dispersed over an extending area, like breezes which blow from rivers or from the land. Sounds which happen for any reason to have been stifled where they arise, are dim and misty; but, if they are clear, they travel far and fill all the space around them.

We all breathe in the same air, but the breath and the sounds which we emit differ owing to structural variations of the organs at our disposal, through which the breath must travel in its passage from within—namely the windpipe, the lungs, and the mouth. Now the impact of the breath upon the air and the shapes assumed by the mouth make most difference

---

[1] Used here not in the sense of a vacuum but of a medium like air. [Edd.]

[2] The connection of sound with motion and impulse was made very early in Greek acoustical theory. The importance of the air, or other medium, in transmitting the impulse was also appreciated (see, e.g., the selection from Aristotle, *De Anima*, quoted above.) In the *De Audibilibus*, which seems to be a fragment of a longer work, the explanation is somewhat different from Aristotle's. There is a definite approach to a wave theory, the portions of air communicating the impulse to proximate portions. Further development of a wave theory of sound is found in Boethius (p. 291, below), who in all probability draws upon a Greek source.

The authorship of *De Audibilibus* is doubtful. Some have ascribed the work to Strato of Lampsacus, other to Heraclides of Pontus. [Edd.]

to the voice. This is clearly the case; for indeed all the differences in the kinds of sounds which are produced proceed from this cause, and we find the same people imitating the neighing of horses, the croaking of frogs, the song of the nightingale, the cries of cranes, and practically every other living creature, by means of the same breath and windpipe, merely by expelling the air from the mouth in different ways. Many birds also imitate by these means the cries of other birds which they hear.

As to the lungs, when they are small and inexpansive and hard, they cannot admit the air nor expel it again in large quantities, nor is the impact of the breath strong and vigorous. For, because they are hard and inexpansive and constricted, they do not admit of dilatation to any great extent, nor again can they force out the breath by contracting after wide distension; just as we ourselves cannot produce any effect with bellows, when they have become hard and cannot easily be dilated and closed.

Voices are thin, when the breath that is emitted is small in quantity. Children's voices, therefore, are thin, and those of women and eunuchs, and in like manner those of persons who are enfeebled by disease or over-exertion or want of nourishment; for owing to their weakness they cannot expel the breath in large quantities. The same thing may be seen in the case of stringed instruments; the sounds produced from thin strings are thin and narrow and "fine as hairs," because the impacts upon the air have only a narrow surface of origin. For the sounds that are produced and strike on the ear are of the same quality as the source of movement which gives rise to the impacts; for example, they are spongy or solid, soft or hard, thin or full. For one portion of the air striking upon another portion of the air preserves the quality of the sound, as is the case also in respect of shrillness and depth; for the quick impulsions of the air caused by the impact, quickly succeeding one another, preserve the quality of the voice, as it was in its first origin. Now the impacts upon the air from strings are many and are distinct from one another, but because, owing to the shortness of the intermittence, the ear cannot appreciate the intervals, the sound appears to us to be united and continuous. The same thing is the case with colours; for separate coloured objects appear to join, when they are moved rapidly before our eyes. The same thing happens, too, when two notes form a concord; for owing to the fact that the two notes overlap and include one another and cease at the same moment, the intermediate constituent sounds escape our notice. For in all concords more frequent impacts upon the air are caused by the shriller note, owing to the quickness of its movement; the result is that the last note strikes upon our hearing simultaneously with an earlier sound produced by the slower impact. Thus, because, as has been said, the ear cannot perceive all the constituent sounds, we seem to hear both notes together and continuously.

Cf. Euclid, *Sectio Canonis*, Introduction[1]

If there were complete rest and immobility there would be complete silence. And if there were silence and nothing moved, nothing would be heard. Therefore for anything to be heard there must first be a blow and motion. Now all sounds result from some blow, and a blow cannot take place without a previous motion. Again, some motions are more frequent, others are rarer, and the more frequent produce the higher pitched sounds while the rarer produce the lower pitched. From this it follows necessarily that some sounds are higher pitched, being composed of more frequent and more numerous motions, while others are lower pitched, being composed of rarer and less numerous motions. Hence sounds higher pitched than what is required reach the required pitch by a process of slackening, that is, by a lowering of the amount of motion; while sounds lower pitched than what is required reach the required pitch by a process of tightening, that is, by an increase in the amount of motion. Therefore sounds must be said to consist of parts, since they reach their proper pitch by addition or subtraction. Now all things that consist of parts may be spoken of as in a numerical ratio to one another. And this must be the case also for sounds, which are thus said to stand in a numerical ratio to one another.

Now some numbers are said to be in multiplicate,[2] others in superparticular,[3] and still others in superpartial[4] ratio to one another. Thus sounds must also be said to stand in such numerical ratios to one another. The relation of numbers that are in multiplicate or superparticular ratio may be expressed by a single word.[5]

Now some sounds we know are consonant and others dissonant, the consonant uniting to produce a single blend, the dissonant failing to do so. Thus it is reasonable to say that since consonant sounds join to produce a single blend of tone, they belong to numbers whose relation may be expressed by a single word, the numbers being in multiplicate or superparticular ratio.

Boethius, *De Institutione Musica* I. 3, 8, 14 (Friedlein)[6]

3. Consonance, which is at the basis of all musical compositions, cannot occur without sound, sound cannot be produced without an impulse

[1] This treatise may well be the work of Euclid, though its authenticity has been questioned (see p. 37).

[2] I.e., of the form $n:1$.

[3] I.e., of the form $(n + 1):n$.

[4] I.e., of the form $(n + m):n$.

[5] The meaning of this "single word" or, as it may be rendered, "single category" is not clarified by Euclid.

[6] In addition to his famous *Consolation of Philosophy* Boethius (*ca.* 480–524) wrote works on arithmetic, music, geometry, astronomy, and philosophical commentaries. His scientific

and blow, and an impulse and blow cannot take place unless motion precedes. For if everything remains immobile there can be no striking of object against object so that one is set in motion by the other. That is, if everything remains stationary and without motion, it follows that there is no sound. Therefore sound is defined as a blow upon air, the effect persisting undissolved until the hearing is reached.[1] Now some motions are swifter and others slower, and again some are rarer and others denser.[2] For in the case of a continuous motion, one is aware either of its speed or slowness. But if one moves the hand,[3] it is either with a dense or a rare motion. Now if a motion is slow and of lower frequency (*rarior*), low pitched sounds are of necessity produced by reason of the slowness and low frequency (*raritate*) of the driving force. But if the motions are swift and dense (i.e., of high frequency—*spissi*), high pitched sounds are necessarily produced.

For this reason the same string if tightened gives a high pitched sound, if loosened a low pitched sound. For when it is tauter it delivers a swifter blow and returns more swiftly and strikes the air more frequently and at shorter intervals (*frequentius ac spissius*). But the looser string delivers weak and slow blows of low frequency (*raros*) by reason of its very weakness in striking, and does not vibrate for a longer time.[4] For it is not to be thought that every time a string is struck only one sound is produced, or that there is only one striking [of the air] in these cases.[5] The fact is that the air is struck every time the string as it vibrates delivers a blow. But since the sounds follow swiftly on one another,[6] no interruption is sensed by the ear, and there is the sensation of a single sound, whether low pitched or high. And this is the case even though each sound is really composed of many,[7] the low pitched consisting of slower and less frequent sounds, the high pitched of swift and dense. It is like the case of the cone commonly called a spinning top. If one carefully smoothes it and colors one strip[8] red, or some other color, and then spins it as swiftly as possible,

---

work is a compilation from sources not always clearly understood by him, that on music being based on Nicomachus, Euclid, and Ptolemy. The present passages are of interest because they point to a wave theory of sound with a definite notion of vibrational frequency. Boethius in all probability took the theory from an earlier Greek source. Cf. the preceding passages.

[1] *Percussio aeris indissoluta usque ad auditum*, a literal translation from Nicomachus, *Manual of Harmony* 4.

[2] *rariores, spissiores*. The sequel shows the meanings to be "of lower frequency" and "of higher frequency" in what is essentially a theory of vibrational frequency.

[3] The idea seems to be that of moving the hand back and forth several times.

[4] I.e., to compensate for the fewer impulses.

[5] Perhaps *aeris* is to be read for *in his*.

[6] Literally "since the speeds of the sounds are joined." The meaning is clarified by what follows, especially the analogy in the case of vision.

[7] "Impulses" would have been clearer, but the Latin can hardly sustain that interpretation.

[8] The strip would have to pass through vertex and base.

the whole cone appears covered with red. Not that it is so covered, but the speed prevents the parts that are not red from being seen. But of this more later.

And so, since high pitched sounds are produced by denser and swifter motions and low pitched sounds by rarer and slower motions, clearly it is by an increase, as it were, in the number of motions[1] that increased tension produces a higher pitched sound from a lower, while it is by a diminution in the number of motions that lessened tension produces a lower pitched sound from a higher. For high pitch consists of more motions than low.

But in matters where plurality makes the difference a certain quantitative relationship must be basic. The relation of the few and the many involves the comparison of one number with another. Now of those things that are compared with respect to number, some are equal and others are unequal. Thus sounds too are in some cases equal and in other cases are separated by an interval because of inequality. But in the case of those sounds which are not separated because of inequality, there is no consonance.[2] For a consonance is the concord of sounds differing from one another but united.[3]

8. . . . Now an interval is the distance between a higher pitched and a lower pitched sound. A consonance is the union of a high and a low sound striking the ears pleasantly and uniformly. A dissonance, on the other hand, is the harsh and unpleasant effect on the ear of two sounds mingling with each other. For when they resist union and each strives, as it were, to arrive untouched, and bars the way of the other, then each of them produces an unpleasant sensation on the ear.

14. Let us now speak of the method of hearing.[4] In the case of sounds something of the same sort takes place as when a stone is thrown out and falls into a pool or other calm water. The stone first produces a wave with a very small circumference. Then it causes the waves to spread out in ever wider circles until the motion, growing weaker as the waves spread

---

[1] The context seems to indicate that the reference is to vibrations.

[2] I.e., there is unison.

[3] The Greeks developed various theories of consonances, the mathematical school basing their theories on a "principle of low numbers," i.e., finding the most perfect consonance in the octave (tones corresponding to string lengths in ratio of 2:1), and less perfect in the fifth (3:2), and the fourth (4:3) (see p. 300). There seems to have been considerable difficulty in seeing that increased frequency of vibration does not involve increased speed of propagation of the sound.

[4] This passage is an instance of the much used analogy of concentric waves in a pool of water and the propagation of sound waves. Cf. Chrysippus in Diogenes Laertius VII. 158: "We hear when the air between the object which sounds and the hearer is so struck that it falls upon the ear in spherical waves. We may compare the circular waves that are formed when a stone is cast into a tank of water." See also p. 307, below.

out, finally ceases. The later and larger the wave, the weaker the impulse with which it breaks. Now if there is an object that can block the waves as they grow larger, the motion is at once reversed and forced back,[1] in the same series of waves, to the center from which it originated.

In the same way, then, when air is struck and produces a sound, it impels other air next to it and in a certain way sets a rounded wave of air in motion, and is thus dispersed and strikes simultaneously the hearing of all who are standing around. And the sound is less clear to one who stands further away since the wave of impelled air which comes to him is weaker.

Experiments in Acoustics

Tradition ascribes the discovery of the numerical relation between pitch and string length to Pythagoras. Whether or not this ascription is sound, a certain amount of experiment along these lines seems to have been carried out by the early followers of Pythagoras, and the discussion of these experiments was transmitted by writers on acoustics. That the tradition was overlaid with a legendary element is seen most clearly from the accounts given by Nicomachus and Boethius. But what is most important is that we have in the determination of the numerical relation between pitch and string length perhaps the first application of mathematics to the phenomena of nature. The success of this application undoubtedly confirmed the Pythagoreans in their search for numerical relations in all realms of being.

A great advance was made by the discovery, perhaps by Eudoxus and Archytas about 400 B.C., of the proportionality of pitch to vibrational frequency. But from the point of view of the history of physical theory it is a mistake to minimize, as some have done, the importance of the earlier formulation in terms of string length.

Theon of Smyrna II. 12–13 (pp. 56–61, Hiller)[2]

12. Pythagoras is reputed to have discovered the numerical ratio of sounds that are consonant with one another, namely, the fourth, in the ratio of 4:3, the fifth, in the ratio of 3:2, the octave, in the ratio of 2:1, the octave plus fourth, in the ratio of 8:3 (which is multisuperparticular, for it is $2 + 2/3$), the octave plus fifth, in the ratio of 3:1, the double octave, having the ratio of 4:1; and of the other intervals, those that encompass a tone, in the ratio of 9:8, and those that encompass what is now called a semitone, formerly a diesis, in the ratio of 256 to 243.[3]

He investigated these ratios on the basis of the length and thickness of strings, and also on the basis of the tension obtained by turning the pegs or by the more familiar method of suspending weights from the strings.

---

[1] Reading *retunditur* instead of *rotundatur* of Friedlein's edition.

[2] Theon of Smyrna, who lived probably in the first part of the second century A.D., wrote a treatise on mathematical matters useful to an understanding of Plato. The extant portions deal with arithmetic, astronomy, and music; the portions on plane geometry and stereometry are lost. The work is important chiefly for the light it throws on historical matters.

[3] Since the ratio of a tone is 9:8, the ratio of the semitone should be the geometric mean between 1 and 9/8, *viz.*, $\sqrt{9/8}$. Perhaps to avoid the irrational, the Pythagoreans took the leimma or diesis for certain purposes as a semitone. It is defined as the quotient of the ratio of a fourth by that of a ditone, i.e., $4/3 \div 9^2/8^2 = 256/243$.

And in the case of wind instruments the basis was the diameter of the bore, or the greater or lesser intensity of the breath. Also the bulk and weights of disks[1] and vessels were examined. Now whichever of these criteria is chosen in connection with any one of the aforesaid ratios, other conditions being equal, the consonance which corresponds to the ratio selected will be produced.[2]

For the present let it suffice for us to illustrate by means of the lengths of strings, using the so-called monochord.[3] For if the single string in the monochord is divided into four equal parts, the sound produced by the whole length of the string forms with the sound produced by three quarters of the string (the ratio being 4:3) the consonance of a fourth. Again, the sound produced by the whole string forms with that produced by half the string (the ratio being 2:1) the consonance of an octave. And the sound produced by the whole string forms with that produced by one quarter of the string (the ratio being 4:1) the consonance of a double octave.

Again, the sound produced by three quarters of the string forms with that produced by half the string (the ratio being 3:2) the consonance of a fifth. The sound produced by three quarters of the string forms with that produced by one quarter (the ratio being 3:1) the consonance of octave plus fifth. If the string is divided into 9 equal parts, the sound produced by the whole string and that produced by 8 parts (the ratio being 9:8) encompass the interval of one tone.

All the consonances are contained in the tetractys consisting of 1, 2, 3, and 4. For in these numbers are the consonances of the fourth, the fifth, the octave <the octave plus fifth, and the double octave>, that is, the ratios of 4:3, 3:2, 2:1, 3:1, and 4:1.

---

[1] Cf. Scholium on Plato's *Phaedo* 108D:

"A certain Hippasus constructed 4 bronze disks in such a way that although their diameters were equal, the thickness of the first disk was four-thirds that the second, three-halves that of the third, and double that of the fourth. And when these disks were struck they produced a concord. Now the story goes that when Glaucus heard the sound from these disks, he was the first to try to play on them. And from his playing on them, the expression 'art of Glaucus' is used even now."

[2] These experiments cannot all have been successfully performed. The inverse relation of pitch to string length (and also to string diameter) and to the length of the column of air in wind instruments seems to have been correctly tested. But there is no indication in the ancient sources that pitch varies not with the weight stretching the string, but with the square root thereof. Again, the results of the alleged experiment with "musical glasses" are quite erroneous. That the ratios of pitch really represent ratios of frequency of vibrations seems to be suspected in some of the accounts. Galileo's demonstration of this marks the beginning of modern acoustical science. It was not until the seventeenth century that the absolute frequency of vibrations corresponding to the various pitches was experimentally investigated.

[3] The apparatus consists of a single string, whose effective vibrating length may be varied, and a graduated ruler alongside it.

Some sought to obtain these consonances from weights, others from magnitudes, others from motions [and numbers corresponding thereto],[1] still others from vessels [and magnitudes corresponding thereto].[1] They say that Lasus of Hermione and the Pythagorean Hippasus of Metapontum investigated the motions with regard to speed and slowness, through which the consonances. . . .[2]

Believing <that the consonances depended> on numbers, he obtained these ratios in experiments on vessels. For taking vessels that were in all respects identical he left one empty and half filled another with liquid. On striking each he found that the consonance of an octave was produced.[3] Again, leaving one of the vessels empty, and filling the other to the one-quarter mark, he struck the vessels and found that the consonance of a fourth was produced. The consonance of a fifth was produced when he filled one vessel to the one-third mark.[4] The ratios of the empty spaces in the respective vessels was 2:1 in the case of the octave, 3:2 in the case of the fifth, and 4:3 in the case of the fourth.

A similar investigation is made by the division of the strings, as we have said, not, however, with one string as in the case of the monochord, but with two. Thus, tuning two strings to the same pitch, he stopped one of them by pressing it down at its midpoint and the half string produced with the other [full] string the consonance of an octave. When he stopped off a third part of the string, the remaining two-thirds gave with the other [full] string the consonance of a fifth. Similarly in the case of a fourth, when a fourth part was stopped off on one of the strings, the other three parts produced with the other [full] string that consonance.

He performed the same experiment on the pan-pipes with the same result.

---

[1] The material here bracketed is probably not authentic.

[2] Lasus and Hippasus were members of the early Pythagorean school. Hippasus has sometimes been credited with the discovery of the irrational (see p. 14). There are stories of his expulsion and shipwreck for divulging secrets of the school. It has been suggested that the theory Lasus and Hippasus had referred to was that higher pitched sounds are more quickly propagated than lower pitched (see p. 287).

There seems to be a lacuna in the text of Theon at this point, but the experimenter referred to in what follows is probably Hippasus. For a different view see C. Jan, *Musici Scriptores Graeci*, p. 131.

[3] The results described in this series of supposed experiments are entirely erroneous. Actually the sound obtained by striking the half-filled vessel would be a little lower in pitch. The accord of the octave could be produced by causing the respective columns of air (not the vessels themselves) to vibrate, e.g., by blowing across them, but also in the case of a narrow-necked vessel, by quickly removing a stopper or the finger inserted as a stopper. This seems to be the experimental fact at the basis of the tradition that Theon misinterprets, perhaps because he never repeated the experiment.

[4] The other vessel remaining empty.

Now some produced the consonances by the method of weights, suspending from two strings weights in the aforesaid ratios.[1] Others used the method of lengths, <stopping off different parts> of strings and determining the consonances that are about to be found in strings.

13. . . . that a musical sound (φθόγγος) is the incidence of the voice on a single pitch. For they say that the sound must always be homogeneous without the slightest divergence and not composed of different levels of pitch. Now some voices are high and others low in pitch, and the same is therefore true of musical sounds, of which the high pitched is swift and the low pitched slow.

Now if one should blow into two pipes of equal thickness and bore, provided with holes in the manner of a flute, one pipe being double the other in length, the breath in the case of the half sized flute is thrown back with double the speed (of the other). The consonance of an octave results, the lower sound being that of the longer pipe, the higher sound that of the shorter.

The speed or slowness of the motion is the cause of this. And the same consonances are obtained on a single pipe in accordance with the distances between the holes thereof. For if the pipe be divided into two equal parts, the accord of an octave is produced if one blows first into the whole length and then through the hole dividing the pipe in half. Again, if the pipe is divided into three equal parts and two of the parts are on the side near the mouthpiece and one part away from it,[2] the consonance of a fifth is produced if one blows first into the whole length and then into two-thirds the length. And when there are four divisions, three parts on the side of the mouthpiece and one part away from it, the consonance of a fourth is produced if one blows first into the whole length and then into three-quarters of the length.

Eudoxus and Archytas held that the doctrine of consonances depended on numbers. At the same time they held that the ratios depended on movements and that the swift motion was high pitched[3] since it beat upon and pierced the air continuously and more swiftly, whereas the slower motion was low pitched being less active.

So much for the discovery of the consonances. Let us now return to the discussion of Adrastus. He says that when the instruments have been previously constructed according to these ratios for the purpose of ob-

[1] See p. 295, n. 2.

[2] I.e., separated off by a hole of sufficient size properly placed.

[3] On Eudoxus see p. 36, on Archytas p. 35. It is doubtful whether Archytas himself had a clear notion of the vibration of air in connection with pitch (see p. 287). But it is possible that Eudoxus or his followers developed the idea, and, contrary to the older practice, assigned higher numbers to the higher pitched tones.

taining the consonances, the correctness of the ratios is confirmed by the sense of hearing. Conversely, when the sense of hearing is taken as the starting point, the correctness of the perception is confirmed by the [measurement of the] ratio. . . .[1]

Boethius, *De Institutione Musica* I. 10–11 (Friedlein)

10. This,[2] then, was the chief reason why Pythagoras abandoned the auditory sense as a criterion of judgment and had recourse to the divisions of a measuring rod. He had no faith in a human ear, which suffers change not only naturally but by reason of external accidents and varies with age. He had no confidence in musical instruments because of the great variation and instability to which they were subject. In the case of strings, for example, damper air would weaken the vibrations, drier air strengthen them; a thicker string would produce a lower tone, a thinner string a higher tone, or there might be some other disturbance of the original uniformity. Since the same was true of the other instruments he gave them no consideration, believing as he did that they were worthy of very little confidence. Instead he sought long and ardently for a method by which he might learn the fixed and unalterable measurement of consonances.

Now by a stroke of divine fortune he was passing a metal workers' shop and heard the hammers when struck produce somehow a single concord from their diverse sounds. Surprised to find that which he had long been seeking he went into the shop and after long consideration concluded that it was the variation in the force of those using the hammers that produced the diversity of sounds. To verify this he had the men exchange hammers. But it turned out that the character of the sounds did not depend on the strength of the men but remained the same even after the hammers were exchanged. On noting this he weighed the hammers. Now there happened to be five hammers, and those two which gave the consonance of an octave (diapason) were found to weigh in the ratio of 2 to 1.[3] He took that one which was double the other and found that its weight was four-thirds the weight of a hammer with which it gave the con-

[1] I.e., two sets of experiments were performed. In the one set the senses confirmed the ratios, in the other set the ratios confirmed the senses. This is of interest in connection with the history of experimental method.

[2] The preceding discussion had dealt with the unreliability, from certain viewpoints, of the evidence of the senses, e.g., the auditory sense.

[3] Boethius does not indicate whether it was the lighter or the heavier hammer that sounded the higher note. But in any case the theory is without basis. The proportions are valid only when applied to the length of strings (or pipes), other conditions (thickness, material, tension, etc.) being equal. This case is mentioned below.

The story of the hammers had become part of Pythagorean legend long before Boethius. It is told in other Pythagorean sources, e.g., Nicomachus, *Manual of Harmonics*.

sonance of a fourth (diatessaron). Again he found that this same hammer was three-halves the weight of a hammer with which it gave the consonance of a fifth (diapente). Now the two hammers to which the aforesaid hammers had been shown to bear the ratio of 4 to 3 and 3 to 2, respectively, were found to bear to each other the ratio of 9 to 8. The fifth hammer was rejected, for it made no consonance with the others.

Thus, while it is true that before Pythagoras the musical consonances were called the octave (diapason), fifth (diapente), and fourth (diatessaron), the latter being the smallest consonance, Pythagoras was the first to find, by this method, the proportions involved in these consonances.

To make clearer what has been said, let us suppose, for example, that the weights of the four hammers are represented by the numbers 12, 9, 8, and 6. Then the hammers with weights 12 and 6 gave the consonance of an octave. The hammer of weight 12 gave with that of weight 9 (the ratio being 4 to 3) the consonance of a fourth. The same consonance was given by the hammer of weight 8 with that of weight 6. The hammer of weight 9 gave with that of weight 6 the consonance of a fifth, as did that of weight 12 with that of weight 8. That of weight 9 gave with that of weight 8 (the proportion being 9 to 8) the interval of a tone.

11. On returning home Pythagoras tried to determine by various researches whether the whole theory of consonances could be explained by these proportions. Thus he attached equal weights[1] to strings and judged their consonances by the ear. Again, he varied the procedure by doubling or halving the length of reeds and using the other proportions. In this way he achieved a very considerable degree of certainty.

Often as a means of testing the proportions he would pour cyathi[2] of fixed weight into vessels, and with a bronze or iron rod strike the vessels containing the various weights. He was overjoyed to find no reason to alter his conclusions. He then proceeded to examine the length and thickness of strings. In this way he discovered the [principle of the] monochord of which we shall speak hereafter. The monochord was called *canon*[3] not merely from the wooden ruler by which we measure the length of strings corresponding to a given tone, but because it forms for this type of investigation so definite and precise a standard that no inquirer can be deceived by dubious evidence.

---

[1] Other things being equal, the frequency of vibration varies not with the weight that stretches the string, but with the square root thereof.

[2] A measure of volume (12 cyathi equal about 1 pint). The reference is probably to the pouring of proportionate amounts of liquid into various vessels, but it is impossible that the consonances were obtained by striking the vessels (see p. 296, n. 3).

[3] Latin *regula* (Greek κανών) denotes not only the monochord with the straight edge for measurement, but also the standard or model determined by such measurement.

Musical Intervals

Philolaus, Frag. 6 (Diels)[1]

The octave includes a fourth and a fifth. The interval of a fifth is a full tone greater than that of a fourth. For from *hypate* to *mese* is a fourth, from *mese* to *nete* a fifth, from *nete* to *trite* a fourth, and from *trite* to *hypate* a fifth.[2] Between *mese* and *trite* there is a full tone. Now the ratio of the fourth is 4:3, that of the fifth 3:2, and that of the octave 2:1. Thus the octave includes five full tones and two semitones, the fifth includes three full tones and one semitone, and the fourth includes two full tones and one semitone.

[Aristotle], *Problemata* XIX. 23, 35*a*, 41, 50.[3] Translation of E. S. Forster

23. Why is *hypate* double *nete*?[4] Is it because in the first place, when half the string is struck and when the whole string is struck an accord in the octave is produced? So too with wind instruments, the sound produced through the middle hole and that produced through the whole flute give an accord in the octave. Again, in the reed-pipe an accord in the octave is obtained by doubling the length, and this is how flute-makers produce it. Similarly they obtain a fifth by means of a length in the ratio of 3 to 2. Again, those who construct Pan-pipes stuff wax into the extreme end of the *hypate*-reed, but fill up the *nete*-reed to the middle. Similarly they obtain a fifth by means of a length in the ratio of 3 to 2, and a fourth by means of a length in the ratio of 4 to 3. Further, *hypate* and *nete* on triangular stringed instruments, when they are equally stretched, give an accord in the octave when one is double the other in length.

---

[1] Philolaus, the Pythagorean, may have flourished toward the close of the fifth century B.C. He is thought to have been the first of the Pythagoreans to record the doctrines of that school for publication. Some, however, hold that the surviving fragments have been erroneously ascribed to him and are actually of later date, about 350 B.C. See also pp. 95–97. The present passage is ascribed to Philolaus by Nicomachus and Stobaeus.

[2] These terms, originally names of strings of the lyre, came to represent notes of the scale. Thus, when the tonic is *a*, the four notes mentioned have the following relations:

hypate	*mese*	*trite*	*nete*
e	*a*	*b*	*e'*

Later *paramese* was substituted for *trite* (see p. 302) and *trite* was used to designate other notes.

[3] The *Problemata* consists of questions and answers in mathematics, physics, music, meteorology, physiology, and various other fields, arranged in 38 divisions. It seems, in its present form, to be a compilation from various sources, though the thought is generally Peripatetic. The no longer extant *Problemata* which Aristotle is known to have written is probably an important source. [Edd.]

[4] I.e., double in length of string. *Hypate* is the note of lowest pitch in the scale, *nete* its octave. [Edd.]

35*a*. Why is the accord in the octave the most beautiful of all? Is it because its ratios are contained within integral terms, while those of the others are not so contained? For since *nete* is double *hypate*,[1] as *nete* is two, so *hypate* is one; and as *hypate* is two, *nete* is four; and so on. But *nete* is to *mese*[2] in the ratio of $3/2$ to 1 (for a fifth is in this ratio), and that which is in the ratio of $3/2$ to 1 is not contained within integral terms; for as the lesser number is one, so the greater number is one with the addition of a half, so that it is no longer a comparison of whole numbers, but fractions are left over. The like happens also with the fourth; for the "epitrite" of a term is as great as that term and one third as great again.[3] Or is it because the accord which is made up of both the other two[4] is the most perfect, and because it is the measure[5] of the melody?

41. Why are a double fifth and a double fourth not concordant, whereas a double octave is? Is it because a fifth is in the ratio of 3 to 2, and a fourth in that of 4 to 3? Now in a series of three numbers in the ratio of 3 to 2 or 4 to 3, the two extreme numbers will have no ratio to one another; for neither will they be in a superparticular ratio nor will one be a multiple of the other. But, since the octave is in a ratio of 2 to 1, if it be doubled the extreme numbers would be in a fourfold ratio. So, since a concord is a compound of sounds which are in a proper ratio to one another, and sounds which are at an interval of two octaves from one another are in a ratio to one another (while double fourths and double fifths are not), the sounds constituting the double octave would give a concord (while the others would not) for the reasons given above.[6]

50. Why is it that the sounds produced from two jars of the same size and quality, one empty and the other half-full, give an accord in the octave?[7] Is it because the sound produced from the half-full jar is double that produced from the empty jar? This surely is just what happens in the pipes. For the quicker the movement, the higher seems the note, and

---

[1] I.e., double in frequency of pitch. [Edd.]

[2] *Mese* is the note a fifth below *nete*. [Edd.]

[3] The fourth is in the ratio of $4/3$ to 1. [Edd.]

[4] I.e., the octave. [Edd.]

[5] I.e., the melody remains within the octave. [Edd.]

[6] I.e., given t terms ($t>2$) of the sequence $n, na, na^2, \ldots na^{t-1}$, the ratio of na^{t-1} to n, where a is $3/2$ or $4/3$, is never an integer or a fraction of the form $(m+1)/m$. In the modern tuning of keyboard instruments a compromise is effected, so that, for example, the twelfth fifth above a given note coincides with the seventh octave above that note, though $2^7 \neq (3/2)^{12}$. [Edd.]

[7] If the traditional account is referred to (see p. 296) and it is meant that the striking of the vessels will produce the accord of the octave, the statement is erroneous. Actually the sound produced from the half-filled vessel would be a little lower. Only if the respective columns of air, not the vessels themselves, are caused to vibrate by blowing across them can the accord of the octave be obtained. [Edd.]

in larger spaces the air collects more slowly, and in double the space in double the time, and proportionately in the other spaces. A wine-skin too which is double the size of another gives an accord in the octave with one which is half its size.

Cf. Plutarch, *On Music* 23

For when the lowest[1] string is double[2] the highest, the consonance of an octave results. If we assign twelve units to *nete*, and six to *hypate*, as we indicated before, *paramese*, having the proportion of 3:2 to *hypate*, has nine units. Again, as we said, *mese* has eight units. Thus the most important intervals in music are the fourth, in the ratio of 4:3, the fifth, in the ratio of 3:2, and the octave, in the ratio of 2:1. There is also the fixed ratio of 9:8 in the interval of a full tone. . . . Now *nete* exceeds *mese* by a third part of itself, and *hypate* is exceeded by *paramese* in the same way,[3] so that the excesses have a relationship. For the extremes exceed the inner terms and are exceeded by them by the same parts of themselves, that is, *nete* exceeds and *hypate* is exceeded by *paramese* and *mese* in the same ratios, 4:3 and 3:2. These intervals determine the harmonic progression.[4]

Musical Science and Its Divisions

Aristoxenus, *Elements of Harmonics* II. 32–38.[5] Translation of Henry S. Macran (Oxford, 1902)

We shall now proceed to the consideration of Harmonic and its parts. It is to be observed that in general the subject of our study is the question,

---

[1] I.e., lowest in position, but highest in pitch.

[2] Not in string length but in frequency of vibration. Though there was no satisfactory theory of sound waves, the numerical relations of pitch (with the octave double the base note) came to be applied on the basis of the inverse ratio of string length. Originally, however, the higher number was given to the lower pitched note (see p. 300, n. 4).

[3] I.e., by a third part of *paramese*.

[4] See p. 6. The relations of the notes referred to here are:

hvpate	*mese*	*paramese*	*nete*
e	*a*	*b*	*e'*
6	8	9	12

[5] Aristoxenus of Tarentum, a pupil of Aristotle, was a prolific writer. Apart from an incomplete work on rhythm, an important work on musical theory, *Harmonic Elements*, is extant.

Plato had insisted on numerical relations rather than auditory sensations as the basis of musical science, and was thus opposed to the early Pythagorean experimental tradition in which both elements are united. But the later Pythagoreans seem to have adopted the Platonic view, while Aristoxenus and Ptolemy again sought to unite both elements. Aristoxenus, however, emphasized the auditory to a considerably greater extent than did Ptolemy and is criticized by Ptolemy for excessive empiricism and the lack of a sound mathematical basis. For details of Ptolemy's criticism of Aristoxenus's approximations see I. Düring, *Ptolemaios und Porphyrios über die Musik* (Göteborg, 1934), and an article by L. Boutroux in *Revue générale des sciences* 30(1919)265-274. [Edd.]

In melody of every kind what are the natural laws according to which the voice in ascending or descending places the intervals? For we hold that the voice follows a natural law in its motion, and does not place the intervals at random. And of our answers we endeavour to supply proofs that will be in agreement with the phenomena—in this unlike our predecessors. For some of these introduced extraneous reasoning, and rejecting the senses as inaccurate fabricated rational principles, asserting that height and depth of pitch consist in certain numerical ratios and relative rates of vibration—a theory utterly extraneous to the subject and quite at variance with the phenomena; while others, dispensing with reason and demonstration, confined themselves to isolated dogmatic statements, not being successful either in their enumeration of the mere phenomena. It is our endeavour that the principles which we assume shall without exception be evident to those who understand music, and that we shall advance to our conclusions by strict demonstration.

Our subject-matter then being all melody, whether vocal or instrumental, our method rests in the last resort on an appeal to the two faculties of hearing and intellect. By the former we judge the magnitudes of the intervals, by the latter we contemplate the functions of the notes.[1] We must therefore accustom ourselves to an accurate discrimination of particulars. It is usual in geometrical constructions to use such a phrase as "Let this be a straight line;" but one must not be content with such language of assumption in the case of intervals. The geometrician makes no use of his faculty of sense-perception. He does not in any degree train his sight to discriminate the straight line, the circle, or any other figure, such training belonging rather to the practice of the carpenter, the turner, or some other such handicraftsman. But for the student of musical science accuracy of sense-perception is a fundamental requirement. For if his sense-perception is deficient, it is impossible for him to deal successfully with those questions that lie outside the sphere of sense-perception altogether. This will become clear in the course of our investigation. And we must bear in mind that musical cognition implies the simultaneous cognition of a permanent and of a changeable element, and that this applies without limitation or qualification to every branch of music.[2] To begin with, our perception of the differences of the genera is dependent on the permanence of the containing, and the variation of the intermediate, notes.[3]

[1] I.e., each note has a function in relation to the other notes of a scale. Macran points out, as an analogy, that in our music *b* has the function of a leading note in the key of *c*, that of a dominant in the key of *e*, and that of a tonic in the key of *b*. [Edd.]

[2] There follow ten illustrations of the fact that musical cognition involves the perception of variable elements amid permanent elements. [Edd.]

[3] In the various genera the first and fourth notes of the tetrachord are the same, the second and third varying, so that the intervals are ¼ tone, ¼ tone, 2 tones in the enharmonic, ½

Again, while the magnitude remains constant, we distinguish the interval between *hypate* and *mese* from that between *paramese* and *nete*; here, then, the magnitude is permanent, while the functions of the notes change;[1] similarly, when there are several figures of the same magnitude, as of the fourth, or fifth, or any other; similarly, when the same interval leads or does not lead to modulation, according to its position. Again, in matters of rhythm we find many similar examples. Without any change in the characteristic proportion constituting any one genus of rhythm, the lengths of the feet vary in obedience to the general rate of movement;[2] and while the magnitudes are constant, the quality of the feet undergoes a change;[3] and the same magnitude serves as a foot, and as a combination of feet.[4] Plainly, too, unless there was a permanent quantum to deal with there could be no distinctions as to the methods of dividing it and arranging its parts.[5] And in general, while rhythmical composition employs a rich variety of movements, the movements of the feet by which we note the rhythms are always simple and the same.[6] Such, then, being the nature of music, we must in matters of harmony also accustom both ear and intellect to a correct judgment of the permanent and changeable element alike.

These remarks have exhibited the general character of the science called Harmonic; and of this science there are, as a fact, seven parts. Of these one and the first is to define the *genera*, and to show what are the permanent and what are the changeable elements presupposed by this distinction. None of our predecessors have drawn this distinction at all; nor is this to be wondered at. For they confined their attention to the Enharmonic genus, to the neglect of the other two. Students of instruments, it is true, would not fail to distinguish each genus by ear, but none of them reflected even on the question. At what point does the Enharmonic

---

tone, ½ tone, 1½ tones in the chromatic, and ½ tone, 1 tone, 1 tone in the diatonic, the sum in each case being 2½ tones. [Edd.]

[1] In the eight-note scale, *hypate*, *mese*, *paramese*, and *nete* are the first, fourth, fifth, and eighth notes, respectively. The intervals mentioned in the text are both fourths but their function in the scale is different. [Edd.]

[2] Compare a bar consisting of a half note and two quarter notes with a bar consisting of a quarter note and two eighth notes. The rhythm is the same, but the time differs. [Edd.]

[3] Compare a bar consisting of a dotted half note and three quarter notes with a bar consisting of a whole note and two quarter notes. The time is the same (6/4), the rhythm different. [Edd.]

[4] Compare a bar consisting of a half note and two quarter notes with two bars each consisting of a quarter note and two eighth notes. The magnitude is the same, the number of bars different. [Edd.]

[5] The divisibility of a whole note into two half notes, four quarter notes, etc., implies a permanent unit. [Edd.]

[6] By a rearrangement of the notes within a bar the character of the bar is changed but not its length. [Edd.]

begin to pass into the Chromatic? For their ability to discriminate each genus extended not to all the *shades*, inasmuch as they were not acquainted with all styles of musical composition or trained to exercise a nice discrimination in such distinctions; nor did they even observe that there were certain loci of the notes that alter their position with the change of genus. These reasons sufficiently explain why the genera have not as yet been definitely distinguished; but it is evident that we must supply this deficiency if we are to follow the differences that present themselves in works of musical composition.

Such is the first branch of Harmonic. In the second we shall deal with *intervals*,[1] omitting, to the best of our ability, none of the distinctions to be found in them. The majority of these, one might say, have as yet escaped observation. But we must bear in mind that wherever we come upon a distinction which has been overlooked, and not scientifically considered, we shall there fail to recognize the distinctions in works of melodic composition.

Again, since intervals are not in themselves sufficient to distinguish notes—for every magnitude, without qualification, that an interval can possess is common to several musical functions[2]—the third part of our science will deal with *notes*, their number, and the means of recognizing them; and will consider the question whether they are certain points of pitch, as is vulgarly supposed, or whether they are musical functions, and also what is the meaning of a musical "function." Not one of these questions is clearly conceived by students of the subject.

The fourth part will consider *scales*, firstly as to their number and nature, secondly as to the manner of their construction from intervals and notes. Our predecessors have not regarded this part of the subject in either of these respects. On the one hand, no attention has been devoted to the questions whether intervals are collocated in any order to produce scales, or whether some collocations may not transgress a natural law. On the other hand, the distinctions in scales have not been completely enumerated by any of them. As to the first point, our forerunners simply ignored the distinction between "melodious" and "unmelodious;" as to the second, they either made no attempt at all at enumeration of scale-distinctions, confining their attention to the seven octave scales which they called Harmonies; or if they made the attempt, they fell very short of completeness, like the school of Pythagoras of Zacynthus,[3] and Agenor of

---

[1] E.g., the octave, fourth, fifth, etc. [Edd.]

[2] E.g., in our music the intervals *c* to *f* and *g* to *c* are both fourths, but the functions of these intervals and of the component notes differ in a given key. [Edd.]

[3] Athenaeus (p. 637) gives a description by Artemon of a little-known stringed instrument called the τρίπους, invented by Pythagoras of Zacynthus. [Edd.]

Mitylene.[1] The order that distinguishes the melodious from the unmelodious resembles that which we find in the collocation of letters in language. For it is not every collocation but only certain collocations of any given letters that will produce a syllable.

The fifth part of our science deals with the *keys* in which the scales are placed for the purposes of melody.[2] No explanation has yet been offered of the manner in which those keys are to be found, or of the principle by which one must be guided in enunciating their number. The account of the keys given by the Harmonists closely resembles the observance of the days according to which, for example, the tenth day of the month at Corinth is the fifth at Athens, and the eighth somewhere else. Just in the same way, some of the Harmonists hold that the Hypodorian is the lowest of the keys; that half a tone above lies the Mixolydian; half a tone higher again the Dorian; a tone above the Dorian the Phrygian; likewise a tone above the Phrygian the Lydian. The number is sometimes increased by the addition of the Hypophrygian clarinet at the bottom of the list. Others, again, having regard to the boring of finger-holes on the flutes, assume intervals of three quarter-tones between the three lowest keys, the Hypophrygian, the Hypodorian, and the Dorian; a tone between the Dorian and Phrygian; three quarter-tones again between the Phrygian and Lydian, and the same distance between the Lydian and Mixolydian. But they have not informed us on what principle they have persuaded themselves to this location of the keys. And that the close packing of small intervals is unmelodious and of no practical value whatsoever will be clear in the course of our discussion.

Again, since some melodies are simple, and others contain a modulation, we must treat of modulation,[3] considering first the nature of modulation in the abstract, and how it arises, or in other words, to what modification in the melodic order it owes its existence; secondly, how many modulations there are in all, and at what intervals they occur. On these questions we find no statements by our predecessors with or without proof.

The last section of our science is concerned with the actual *construction*

---

[1] Agenor of Mitylene was the teacher of Isocrates' grandsons. He achieved considerable repute in music. [Edd.]

[2] On the distinctions of the various scales or modes in the Aristoxenian system and the key, or pitch, proper to each of them, see Macran's Introduction. The question of keys and modes is among the thorniest in Greek music. A satisfactory solution of the many problems involved has not yet been found. See R. P. Winnington-Ingram, *Mode in Ancient Greek Music* (Cambridge, 1936) and K. Schlesinger, *The Greek Aulos* (London, 1939). [Edd.]

[3] The modulation with which Aristoxenus is here primarily concerned is thus defined by Bacchius . . . "the transition which a melody makes from one scale into another, by providing for itself a different *mese*." But a different *mese* can mean nothing else than a tonic of different pitch, so this transition means simply modulation into a different key. . . .

of melody.[1] For since in the same notes, indifferent in themselves, we have the choice of numerous melodic forms of every character, it is evident that here we have the practical question of the employment of the notes; and this is what we mean by the construction of melody. The science of harmony having traversed the said sections will find its consummation here.

Applications of Acoustics

Vitruvius, *On Architecture* V. 3, 5, 8. Translation of M. H. Morgan

3. Acoustics of the Theatre

The curved cross-aisles should be constructed in proportionate relation, it is thought, to the height of the theatre, but not higher than the footway of the passage is broad. If they are loftier, they will throw back the voice and drive it away from the upper portion, thus preventing the case-endings of words from reaching with distinct meaning the ears of those who are in the uppermost seats above the cross-aisles. In short, it should be so contrived that a line drawn from the lowest to the highest seat will touch the top edges and angles of all the seats. Thus the voice will meet with no obstruction. . . .

Particular pains must also be taken that the site be not a "deaf" one, but one through which the voice can range with the greatest clearness. This can be brought about if a site is selected where there is no obstruction due to echo.

Voice is a flowing breath of air, perceptible to the hearing by contact. It moves in an endless number of circular rounds, like the innumerably increasing circular waves which appear when a stone is thrown into smooth water,[2] and which keep on spreading indefinitely from the centre unless interrupted by narrow limits, or by some obstruction which prevents such waves from reaching their end in due formation. When they are interrupted by obstructions, the first waves, flowing back, break up the formation of those which follow.

In the same manner the voice executes its movements in concentric circles; but while in the case of water the circles move horizontally on a plane surface, the voice not only proceeds horizontally, but also ascends vertically by regular stages. Therefore, as in the case of the waves formed in the water, so it is in the case of the voice: the first wave, when there is no obstruction to interrupt it, does not break up the second or the following

[1] The other parts of Harmonic science have supplied the material of melody, notes, intervals, and scales; it remains for the composer to make a judicious use of it. The science of the use of musical material is μελοποιία. . . . [Macran then quotes the *Isagoge* of Cleonides (?) to show the functions of diatonic progressions, skips, repetitions, and prolongations in the construction of melodies. Edd.]

[2] See p. 293. [Edd.]

waves, but they all reach the ears of the lowest and highest spectators without an echo.

Hence the ancient architects, following in the footsteps of nature, perfected the ascending rows of seats in theatres from their investigations of the ascending voice, and, by means of the canonical theory of the mathematicians and of the musicians, endeavoured to make every voice uttered on the stage come with greater clearness and sweetness to the ears of the audience. For just as musical instruments are brought to perfection of clearness in the sound of their strings by means of bronze plates or horn ἠχεῖα so the ancients devised methods of increasing the power of the voice in theatres through the application of harmonics.[1]

5. Sounding Vessels in the Theatre

In accordance with the foregoing investigations on mathematical principles, let bronze vessels be made,[2] proportionate to the size of the theatre, and let them be so fashioned that, when touched, they may produce with one another the notes of the fourth, the fifth, and so on up to the double octave. Then, having constructed niches in between the seats of the theatre, let the vessels be arranged in them, in accordance with musical laws, in such a way that they nowhere touch the wall, but have a clear space all around them and room over their tops. They should be set upside down, and be supported on the side facing the stage by wedges not less than half a foot high. Opposite each niche, apertures should be left in the surface of the seat next below, two feet long and half a foot deep.

The arrangement of these vessels, with reference to the situations in which they should be placed, may be described as follows. If the theatre be of no great size, mark out a horizontal range halfway up, and in it construct thirteen arched niches with twelve equal spaces between them, so that of the above mentioned "echea" those which give the note nete hyperbolaeon may be placed first on each side, in the niches which are at the extreme ends; next to the ends and a fourth below in pitch, the note nete diezeugmenon; third, paramese, a fourth below; fourth, nete synhemmenon; fifth, mese, a fourth below; sixth, hypate meson, a fourth below; and in the middle and another fourth below, one vessel giving the note hypate hypaton.[3]

On this principle of arrangement, the voice, uttered from the stage as

---

[1] There follows a summary of the Aristoxenian system of harmonics. [Edd.]

[2] The meaning may be "let bronze vessels be made on mathematical principles," i. e. in mathematical ratios. [Edd.]

[3] The illustration (p. 309) gives the notes to which the different shells resound. For a complete description of the Aristoxenian system, which Vitruvius adopts, see H. S. Macran, *The Harmonics of Aristoxenus*, Introduction. [Edd.]

from a centre, and spreading and striking against the cavities of the different vessels, as it comes in contact with them, will be increased in clearness of sound, and will wake an harmonious note in unison with itself. . . .[1]

Whoever wishes to carry out these principles with ease, has only to consult the scheme at the end of this book, drawn up in accordance with the laws of music. It was left by Aristoxenus, who with great ability and labour classified and arranged in it the different modes. In accordance with it, and by giving heed to these theories, one can easily bring a theatre to perfection, from the point of view of the nature of the voice, so as to give pleasure to the audience.

Somebody will perhaps say that many theatres are built every year in Rome, and that in them no attention at all is paid to these principles; but he will be in error, from the fact that all our public theatres made of wood contain a great deal of boarding, which must be resonant. This may be observed from the behaviour of those who sing to the lyre, who, when they wish to sing a higher key, turn towards the folding doors on the stage, and thus by their aid are reinforced with a sound in harmony with the voice. But when theatres are built of solid materials like masonry, stone, or marble, which cannot be resonant, then the principles of the "echea" must be applied.

If, however, it is asked in what theatre these vessels have been employed, we cannot point to any in Rome itself, but only to those in the districts of Italy and in a good many Greek states. We have also the evidence of Lucius Mummius, who, after destroying the theatre in Corinth, brought its bronze vessels to Rome, and made a dedicatory offering at the temple of Luna with the money obtained from the sale of them. Besides, many skillful architects, in constructing theatres in small towns, have, for lack of means, taken large jars made of clay, but similarly resonant, and have produced very advantageous results by arranging them on the principles described.[2]

---

[1] Vitruvius goes on to expose a more elaborate arrangement of shells for a larger theater. New shells resounding to notes of the various genera (enharmonic, chromatic, and diatonic) in all the tetrachords (from hyperbolaeon to hypaton) are disposed around the theater at various levels. [Edd.]

[2] Until quite recently no substantial advance in practical acoustics had been made since ancient times. Examples of earthenware vessels used presumably for acoustical purposes have been found. [Edd.]

8. Acoustics of the Site of a Theatre

All this having been settled with the greatest pains and skill, we must see to it, with still greater care, that a site has been selected where the voice has a gentle fall, and is not driven back with a recoil so as to convey an indistinct meaning to the ear. There are some places which from their very nature interfere with the course of the voice, as for instance the dissonant, which are termed in Greek κατηχοῦντες; the circumsonant, which with them are named περιηχοῦντες; again the resonant, which are termed ἀντηχοῦντες; and the consonant, which they call συνηχοῦντες. The dissonant are those places in which the first sound uttered that is carried up high, strikes against solid bodies above, and being driven back, checks as it sinks to the bottom the rise of the succeeding sound.

The circumsonant are those in which the voice spreads all round, and then is forced into the middle, where it dissolves, the case-endings are not heard, and it dies away there in sounds of indistinct meaning. The resonant are those in which it comes into contact with some solid substance and recoils, thus producing an echo, and making the terminations of cases sound double. The consonant are those in which it is supported from below, increases as it goes up, and reaches the ears in words which are distinct and clear in tone. Hence, if there has been careful attention in the selection of the site, the effect of the voice, will, through this precaution, be perfectly suited to the purposes of a theatre.

MAGNETISM

Loadstone, which was found in many places in the Greek world, was the subject of widespread interest among the Greeks. It attained considerable importance in folk medicine and magic, and naturalists and philosophers sought to explain its peculiar properties. The artificial magnet was unknown to the Greeks and Romans, and they knew nothing of the principle underlying the compass. In fact they were often unable to distinguish loadstone from various non-magnetic ores. They did, however, observe the chain effects of the magnet (magnetic induction), as well as the action of the magnet through a medium like water, brass, or silver. Without discovering the principle of polarity they noted the repelling as well as the attracting action of the magnet. The use of the magnetic stone in industry (apart from its use for playthings and magical effects) seems to have been merely as a source for iron and for certain pharmaceutical preparations. Thales, Empedocles, Democritus, Diogenes of Apollonia, Plato, Epicurus, and Galen are among those whose explanations of the action of the magnet are referred to in extant sources.

The property of amber (Greek ἤλεκτρον) in attracting and repelling certain substances likewise interested the Greeks, who gave similar explanations of its action.

The numbing action of electric fish, too, was often described on the basis of an analogy with the magnet. See pp. 253, 430.

The Magnet

Pliny, *Natural History* XXXVI. 126–130. Translation of K. C. Bailey, *The Elder Pliny's Chapters on Chemical Subjects* (London, 1929–1932)

Passing from the subject of marbles to other stones with striking properties, who would hesitate to deal first with the magnet?[1] For what phenomenon is more astonishing? Where has nature shown greater audacity? We have already related how she gave rocks a tongue with which to answer man, more, to break in upon his speech. What is harder to move than the frozen stillness of stone? Lo, she has given it hands and feelings. What is more refractory than hard iron? Lo, she has given it feet and character. For iron, the tamer of all substances, is drawn by the magnet, follows some intangible attraction and, as it comes nearer, leaps to meet the magnet, is held, and clings fast in its embrace. So the magnet is given another name, *sideritis*, while some call it *Heraclion*.[2] Nicander is our authority for believing that it was called *magnes* from the man[3] who first found it on Mount Ida (for they may be found everywhere, including Spain), and he is said to have discovered it when the nails in his shoes and the ferrule of his staff adhered to it, as he was pasturing his herds.

Sotacus[4] classifies magnets into five varieties: the first found in Aethiopia, the second from the Magnesia which has a common boundary with Macedonia and lies on one's right hand as one makes for Iolcus from Euboea, the third found at Hyettus in Bœotia, the fourth from the neighbourhood of Alexandria in the Troad, and the fifth from the Magnesia which is in Asia.

The first point of distinction is whether the stone is male or female,[5] the second depends on its colour. Magnets found in the Magnesia near Macedonia are a reddish black colour, while Bœotian magnets are more red than black. The magnets found in the Troad are black[6] and of the female sex, and are therefore without magnetic power, but the worst of all come from Magnesia in Asia. These are white, with a resemblance to pumice, and have no attraction whatever for iron. It is an ascertained

---

[1] There is no doubt that the ancient magnet was the naturally magnetic oxide of iron, ferroso-ferric oxide, or magnetite (Fe_3O_4). This mineral is widely distributed and is usually iron black in colour with a metallic lustre, although, as with most minerals, the colour varies somewhat. Only a comparatively small number of specimens possess magnetic polarity. The most powerful natural magnets of modern times have been found in Siberia and in the Harz. Although the exact composition of the natural magnet was, of course, unknown to the ancients, they realized that it was an ore of iron, for Pliny (XXXIV.147) classes it among the ores of iron, and . . . he speaks of haematite (well known to be an iron ore) as a variety of magnet. . . .

[2] Hesychius says that it [the magnet] was called *Heraclion* from Heraclea, a city in Lydia, and that Sophocles called it a Lydian stone, and others *sideritis*. . . .

[3] The derivation from one or other Magnesia is much more probable, and is supported by Lucretius (VI. 908).

[4] The author of a book on precious stones. Pliny gives him as one of his authorities for this book.

[5] The reference in the next section suggests that a male magnet was one which had powerful magnetic powers, a female magnet one which had little or none. . . .

[6] It has been suggested (e.g., by Stillman, *History of Early Chemistry*) that this black, non-magnetic, substance was pyrolusite, or manganese dioxide. . . .

fact that, the more blue there is in the colour of a magnet, the better it is likely to be. Those from Aethiopia bear off the palm, and sell for their weight in silver. They are found in Zmiris, for so they call the sand-covered region of Aethiopia. In the same district is found the hæmatites magnet which is the colour of blood and which gives, when powdered, a material of a saffron-red hue. Hæmatites has not the same power of attracting iron as the magnet. It is a proof of the Aethiopian origin of a magnet if it attracts other magnets.[1] All these varieties are used, each in its due proportion, in salves for the eyes, and are particularly effective in arresting ophthalmic discharges. A magnet which has been ignited and powdered[2] is used for healing burns.

In Aethiopia also, and not far from the previous locality, is another mountain which bears the stone *theamedes*, which drives away and repels every kind of iron. We have had occasion to mention these characteristics of attraction and repulsion on several occasions.

Cf. Plato, *Ion* 533 D–E.[3] Translation of W. R. M. Lamb (London, 1925)

For, as I was saying just now, this is not an art in you, whereby you speak well on Homer, but a divine power, which moves you like that in the stone which Euripides named a magnet, but most people call "Heraclea stone." For this stone not only attracts iron rings, but also imparts to them a power whereby they in turn are able to do the very same thing as the stone, and attract other rings; so that sometimes there is formed quite a long chain of bits of iron and rings, suspended one from the other; and they all depend for this power on that one stone.[4]

THE EPICUREAN EXPLANATION OF THE MAGNET

Lucretius, *On the Nature of Things* VI. 906–916, 998–1064. Translation of Cyril Bailey

For what follows, I will essay to tell by what law of nature it comes to pass that iron can be attracted by the stone which the Greeks call the mag-

[1] All magnets will attract other magnets if unlike poles are presented to one another.

[2] The product thus obtained would be red ferric oxide (Fe_2O_3).

[3] Socrates is comparing the chain of influence from Muse to poet to rhapsode to audience with the action of a magnet on a chain of objects. [Edd.]

[4] After referring to these phenomena, St. Augustine goes on to say (*On the City of God* XXI. 4):

"But what I heard about this stone from my brother and fellow bishop Severus Milevitanus is even more remarkable. He told how he had seen Bathanarius (formerly Count of Africa, with whom the bishop was then living) take this loadstone and hold it under some silver; he then placed a piece of iron above the silver. Then, as he moved the hand in which he held the stone under the silver, the iron above it also moved. Thus despite the presence of the silver beween the stone and the iron, the brisk movement to and fro of the stone below the silver caused a similar movement of the iron above it, and this without any effect on the silver itself." [Edd.]

net, from the name of its native place, because it has its origin within the boundaries of its native country, the land of the Magnetes. At this stone men marvel; indeed, it often makes a chain of rings all hanging to itself. For sometimes you may see five or more in a hanging chain, and swaying in the light breezes, when one hangs on to the other, clinging to it beneath, and each from the next comes to feel the binding force of the stone; in such penetrating fashion does its force prevail. . . .

Wherefore, when all these things have been surely established and settled for us, laid down in advance and ready for use, for what remains, from them we shall easily give account, and the whole cause will be laid bare, which attracts the force of iron. First of all it must needs be[1] that there stream off this stone very many seeds or an effluence, which, with its blows, parts asunder all the air which has its place between the stone and the iron. When this space is emptied and much room in the middle becomes void, straightway first-beginnings of the iron start forward and fall into the void, all joined together; it comes to pass that the ring itself follows and advances in this way, with its whole body. Nor is anything so closely interlaced in its first particles, all clinging linked together, as the nature of strong iron and its cold roughness. . . .

It comes to pass, too, that the nature of iron retreats from this stone at times, and is wont to flee and follow turn by turn. Further, I have seen Samothracian iron rings even leap up, and at the same time iron filings move in a frenzy inside brass bowls, when this Magnesian stone was placed beneath: so eagerly is the iron seen to desire to flee from the stone. When the brass is placed between, so great a disturbance is brought about because, we may be sure, when the effluence of the brass has seized beforehand and occupied the open passages in the iron, afterwards comes the effluence of the stone, and finds all full in the iron, nor has it a path by which it may stream through as before. And so it is constrained to dash against it and beat with its wave upon the iron texture; and in this way it repels it from itself, and through the brass drives away that which without it it often sucks in.

Herein refrain from wondering that the effluence from this stone has not the power to drive other things in the same way. For in part they stand still by the force of their own weight, as for instance, gold; and partly, because they are of such rare body, that the effluence flies through untouched, they cannot be driven anywhere; among this kind is seen to be the substance of wood. The nature of iron then has its place between the two, and when it has taken in certain tiny bodies of brass, then it comes to pass that the Magnesian stones drive it on with their stream.

[1] In the Epicurean account the creation of a void between the magnet and the iron impels the atoms of iron on the side nearer the magnet to move toward the magnet. This they do, drawing the rest of the iron along. [Edd.]

Cf. Galen, *On the Natural Faculties* I. 14. Translation of A. J. Brock (London, 1916)

Now Epicurus, despite the fact that he employs in his *Physics* elements similar to those of Asclepiades, yet allows that iron is attracted by the lodestone, and chaff by amber. He even tries to give the cause of the phenomenon. His view is that the atoms which flow from the stone are related in shape to those flowing from the iron, and so they become easily interlocked with one another; thus it is that, after colliding with each of the two compact masses (the stone and the iron) they then rebound into the middle and so become entangled with each other, and draw the iron after them. . . . I fail to understand how anybody could believe this. Even if we admit this, the same principle will not explain the fact that, when the iron has another piece brought in contact with it, this becomes attached to it. . . . As a matter of fact, I have seen five writing-stylets of iron attached to one another in a line, only the first one being in contact with the lodestone, and the power being transmitted through it to the others.[1]

APPLIED MECHANICS

The high development of various technological branches in antiquity indicates at least an empirical knowledge of substances and their properties as well as a familiarity with certain mechanical, chemical, metallurgical, and engineering procedures. The shipbuilder is governed by laws of hydrostatics, the dyer by chemical laws, and the bridge builder by mechanical laws. But they need not be aware of a strictly scientific formulation of these laws.

Apart from incidental references scattered throughout all ancient literature, the technological material that is extant in formal or systematic treatment is quite restricted. There are the works of the Greek mechanicians and writers on military affairs, Vitruvius, Frontinus, portions of Pliny's *Natural History*, a few handbooks on farming and surveying, and some brief practical handbooks of chemical procedures. But our chief sources of information for most purposes are the actual archaeological remains.

We have not sought to give here a complete conspectus of Greek and Roman technology but have contented ourselves with a few examples that have a close connection with the theoretic branches already considered. Increasing attention has been paid in recent years to the organization of society in antiquity, and in particular the institution of slavery, as affecting the technological development and possibly also the directions in which science in general developed. Though the discussion of this question lies outside the scope of our Source Book, the material here collected will serve to illustrate some aspects of the problem.

Archimedes and Technology

Plutarch, *Life of Marcellus* 14–17. Translation of Bernadotte Perrin (London, 1917)

14. . . . He [Marcellus] proceeded to attack the city [Syracuse] by land and sea, Appius leading up the land forces, and he himself having a fleet of sixty quinqueremes filled with all sorts of arms and missiles. Moreover, he had erected an engine of artillery on a huge platform supported by

[1] There follows a refutation of Epicurus's explanation of magnetic action. The occasion for the whole discussion is the question of the existence of natural attractions and repulsions between substances. Galen opposes Epicurus's view that it is the mechanical action of certain types of atoms that gives the appearance of an attraction. [Edd.]

eight galleys fastened together, and with this he sailed up to the city wall, confidently relying on the extent and splendour of his equipment and his own great fame. But all this proved to be of no account in the eyes of Archimedes and in comparison with the engines of Archimedes. To these he had by no means devoted himself as work worthy of his serious effort, but most of them were mere accessories of a geometry practised for amusement, since in bygone days Hiero the king had eagerly desired and at last persuaded him to turn his art somewhat from abstract notions to material things, and by applying his philosophy somehow to the needs which make themselves felt, to render it more evident to the common mind.

For the art of mechanics, now so celebrated and admired, was first originated by Eudoxus and Archytas, who embellished geometry with its subtleties, and gave no problems incapable of proof by word and diagram, a support derived from mechanical illustrations that were patent to the senses. For instance, in solving the problem of finding two mean proportional lines, a necessary requisite for many geometrical figures, both mathematicians had recourse to mechanical arrangements, adapting to their purposes certain intermediate portions of curved lines and sections.[1] But Plato was incensed at this, and inveighed against them as corrupters and destroyers of the pure excellence of geometry, which thus turned her back upon the incorporeal things of abstract thought and descended to the things of sense, making use, moreover, of objects which required such mean and manual labour.[2] For this reason mechanics was made entirely distinct from geometry, and being for a long time ignored by philosophers, came to be regarded as one of the military arts.

And yet even Archimedes, who was a kinsman and a friend of King Hiero, wrote to him that with any given force it was possible to move any given weight; and emboldened, as we are told, by the strength of his demonstration, he declared that, if there were another world, and he could go to it, he could move this.[3] Hiero was astonished, and begged him to put his proposition into execution, and show him some great weight moved by a slight force. Archimedes therefore fixed upon a three-masted merchantman of the royal fleet, which had been dragged ashore by the great labours of many men, and after putting on board many passengers and the customary freight, he seated himself at a distance from her, and without any great effort, but quietly setting in motion with his hand a system of compound pulleys, drew her towards him smoothly and evenly, as though she were gliding through the water. Amazed at this, then, and comprehending the power of his art, the king persuaded Archimedes to prepare for

---

[1] See p. 63. [Edd.]

[2] Eutocius gives a solution of the Delian problem ascribed to Plato. Since the solution is as "mechanical" as that of Archytas, some have, on the basis of this passage, considered the ascription erroneous. [Edd.]

[3] An application of the theory of the lever. [Edd.]

him offensive and defensive engines to be used in every kind of siege warfare. These he had never used himself, because he spent the greater part of his life in freedom from war and amid the festal rites of peace; but at the present time his apparatus stood the Syracusans in good stead, and, with the apparatus, its fabricator.

15. When, therefore, the Romans assaulted them by sea and land, the Syracusans were stricken dumb with terror; they thought that nothing could withstand so furious an onset by such forces. But Archimedes began to ply his engines, and shot against the land forces of the assailants all sorts of missiles and immense masses of stones, which came down with incredible din and speed; nothing whatever could ward off their weight, but they knocked down in heaps those who stood in their way, and threw their ranks into confusion. At the same time huge beams were suddenly projected over the ships from the walls, which sank some of them with great weights plunging down from on high; others were seized at the prow by iron claws, or beaks like the beaks of cranes, drawn straight up into the air, and then plunged stern foremost into the depths, or were turned round and round by means of enginery within the city, and dashed upon the steep cliffs that jutted out beneath the wall of the city, with great destruction of the fighting men on board, who perished in the wrecks. Frequently, too, a ship would be lifted out of the water into mid-air, whirled hither and thither as it hung there, a dreadful spectacle, until its crew had been thrown out and hurled in all directions, when it would fall empty upon the walls, or slip away from the clutch that had held it. As for the engine which Marcellus was bringing up on the bridge of ships, and which was called "sambuca" from some resemblance it had to the musical instrument of that name, while it was still some distance off in its approach to the wall, a stone of ten talents' weight was discharged at it, then a second and a third; some of these, falling upon it with great din and surge of wave, crushed the foundation of the engine, shattered its frame-work, and dislodged it from the platform, so that Marcellus, in perplexity, ordered his ships to sail back as fast as they could, and his land forces to retire.

Then, in a council of war, it was decided to come up under the walls while it was still night, if they could; for the ropes which Archimedes used in his engines, since they imported great impetus to the missiles cast, would, they thought, send them flying over their heads, but would be ineffective at close quarters, where there was no space for the cast. Archimedes, however, as it seemed, had long before prepared for such an emergency engines with a range adapted to any interval and missiles of short flight, and, through many small and contiguous openings in the wall, short-range engines called "scorpions" could be brought to bear on objects close at hand without being seen by the enemy.

16. When, therefore, the Romans came up under the walls, thinking themselves unnoticed, once more they encountered a great storm of missiles; huge stones came tumbling down upon them almost perpendicularly, and the wall shot out arrows at them from every point; they therefore retired. And here again, when they were some distance off, missiles darted forth and fell upon them as they were going away, and there was a great slaughter among them; many of their ships, too, were dashed together, and they could not retaliate in any way upon their foes. For Archimedes had built most of his engines close behind the wall, and the Romans seemed to be fighting against the gods, now that countless mischiefs were poured out upon them from an invisible source.

17. However, Marcellus made his escape, and jesting with his own artificers and engineers, "Let us stop," said he, "fighting against this geometrical Briareus, who uses our ships like cups to ladle water from the sea, and has whipped and driven off in disgrace our sambuca, and with the many missiles which he shoots against us all at once, outdoes the hundred-handed monsters of mythology." For in reality all the rest of the Syracusans were but a body for the designs of Archimedes, and his the one soul moving and managing everything; for all other weapons lay idle, and his alone were then employed by the city both in offence and defence. At last the Romans became so fearful that, whenever they saw a bit of rope or a stick of timber projecting a little over the wall, "There it is," they cried, "Archimedes is training some engine upon us," and turned their backs and fled. Seeing this, Marcellus desisted from all fighting and assault, and thenceforth depended on a long siege.

And yet Archimedes possessed such a lofty spirit, so profound a soul, and such a wealth of scientific theory, that although his inventions had won for him a name and fame for superhuman sagacity, he would not consent to leave behind him any treatise on this subject, but regarding the work of an engineer and every art that ministers to the needs of life as ignoble and vulgar, he devoted his earnest efforts only to those studies the subtlety and charm of which are not effected by the claims of necessity. These studies, he thought, are not to be compared with any others; in them the subject matter vies with the demonstration, the former supplying grandeur and beauty, the latter precision and surpassing power. For it is not possible to find in geometry more profound and difficult questions treated in simpler and purer terms. Some attribute this success to his natural endowments; others think it due to excessive labour that everything he did seemed to have been performed without labour and with ease. For no one could by his own efforts discover the proof, and yet as soon as he learns it from him, he thinks he might have discovered it himself; so smooth and rapid is the path by which he leads one to the desired conclusion. And therefore we may not disbelieve the stories told about him, how, under the lasting charm of some familiar and domestic Siren, he forgot even his food and neglected the care of his person; and how, when he was dragged by main force, as he often was, to the place for bathing and anointing his body, he would trace geometrical figures in the ashes, and draw lines with his finger in the oil with which his body was anointed, being possessed by a great delight, and in very truth a captive of the Muses. And although he made many excellent discoveries, he is said to have asked his kinsmen and friends to place over the grave where he should be buried a cylinder enclosing a sphere, with an inscription giving the proportion by which the containing solid exceeds the contained.[1]

On the Construction of Artillery: Application of Empirical Formulae

Philo of Byzantium, *Belopoeica* (*Mechanics* IV), 3. 5–7 (Diels-Schramm)[2]

Now some of the ancients discovered that the diameter of the bore was the basic element, principle, and measurement in the construction of artillery. But it was necessary to determine this diameter not accidentally or haphazardly but by some definite method by which one could also determine the proportionate measurement for all magnitudes on the instrument. But this could not be done except by increasing or decreasing the diameter of the bore and testing the results. And the ancients did not succeed in determining this magnitude by test, because their trials were not conducted on the basis of many different types of performance, but merely in connection with the required performance. But the engineers who came later, noting the errors of their predecessors and the results of subsequent experiments, reduced the principle of construction to a single basic element, viz., the diameter of the circle that receives the twisted skeins. Success in this work was recently achieved by the Alexandrian engineers, who were heavily subsidized by kings eager for fame and interested in the arts.

[1] The surface and volume of the cylinder are to the surface and volume, respectively, of the inscribed sphere as 3:2. This is the result of a series of demonstrations, in which the method of exhaustion plays a decisive part, in the treatise *On the Sphere and Cylinder* (see pp. 45, 60, above). [Edd.]

[2] The treatises of Philo of Byzantium and Hero of Alexandria (see pp. 197, 255) on artillery. are interesting not alone for the evidence they give of technological development in metal and woodworking and in the construction of instruments of high precision, but also in the application of empirical mathematical formulae to the problems of ballistics. The empirical formula of the engineer has often served as a steppingstone to a more satisfactory theoretical treatment. The reduction of all measurements of all parts of the machine to fractions or multiples of the caliber (i.e., the diameter of the opening that receives the skein of gut whose tension furnishes the power) and, in general, the application of mathematics to the problems of engineering are points of greatest interest. In addition to the usual artillery instruments Philo describes one operated by air pressure and one in which the missiles are automatically loaded, a sort of machine gun. The latter, Philo says, was constructed in Rhodes by Dionysius of Alexandria.

For it is not possible to arrive at a complete solution of the problems involved merely by reason and by the methods of mechanics; many discoveries can, in fact, be made only as a result of trial. . . .

Now the goal of artillery construction is the long range shooting of missiles which will land with vigorous impact. It is chiefly on this problem that experiment and investigation have been conducted. I shall, therefore, report to you what I myself learned in Alexandria during a stay which was, for the most part, spent in company with technicians in this field. In Rhodes, too, I became acquainted with not a few master builders and observed that here too the most successful of these engines were constructed pretty closely in accordance with the method about to be described.

The weight of the stone to be hurled is the basis upon which the engine must be constructed. This weight is first reduced to units,[1] the cube root of this quantity extracted, a tenth of this root added to the root, and the result is the number of digits in the diameter of the opening which receives the skein of gut.

Now if the weight (in drachmas) has no rational cube root, it is necessary to make the closest possible approximation. If the approximation is in excess of the true value, one must seek to lessen the tenth part proportionately, and if the approximation is less than the true value, one must, after adding the tenth part, still further increase the result.[2]

Thus the diameters of openings obtained by this method are as follows:

Weight in minas	Diameter of opening in digits*	[Actual calculation of $d = {}^{11}\!/_{10} \sqrt[3]{100m}$
10	11	11
15	12¾	12.592
20	14	13.859
30	15¾	15.864
50	18¾	18.809
60 (= 1 talent)	21	19.988
150 (= 2½ talents)	25	27.128
180 (= 3 talents)	27	28.828]

* The editors, Diels and Schramm, suggest that, in the sixth case, the reading 21 of the manuscripts is a mistake for 20, and that mistakes either of calculation or of copying account for the variation in the last two cases.

[1] The units are drachmas; 100 drachmas = 1 mina. The formula that follows is: $d = {}^{11}\!/_{10} \sqrt[3]{100m}$, where d is the number of digits and m the number of minas. The mina varied at different times and places, being somewhat more than a pound in the Aeginetan and somewhat less than a pound in the Attic system. The digit (Greek *dactylos*) was about three-fourths of an inch.

[2] The meaning, apparently, is that the multiplier must be somewhat less or somewhat more than 11/10, according to whether the approximate value of the cube root exceeds or is less than the true value. On the approximation to the cube root see p. 87, above.

It is also possible, however, after the diameter of the opening has been determined for the smallest of the engines represented in the above figures, that is, the engine which hurls the 10-mina missile, to determine with instruments the corresponding diameter of the others. This is done by the method of duplication of the cube, as I have explained in the first book. . . .[1]

Hero of Alexandria, *Belopoeica*, 30–34 (Diels-Schramm)

Having, therefore, given adequate consideration to the main features of the construction and use of both the *euthytonon* and the *palintonon*,[2] we shall next take up the question of measurements.

The determination of these measurements, be it understood, is based on actual experience. Now the ancients, directing their attention solely to the shape and arrangement, had scarcely any success in securing an effective discharge of the missile, because of their failure to employ the proper ratios. But their successors, by making some parts smaller and other parts larger, produced engines that were correctly proportioned and effective. The size of every individual part of these engines is determined by the diameter of the opening which receives the skein of gut. For this is the basic unit of measure.

Now in the case of the stone-hurler (λιθοβόλος) the size of the opening in question is determined as follows: take the weight in minas of the stone to be hurled, multiply it by 100, take the cube root of this product, multiply this root by 11/10. The result will be the diameter of the opening in digits.

For example, if the stone weighs 80 minas, 100 times 80 is 8,000, of which the cube root is 20. Add a tenth part, 2, and the result is 22, the diameter of the opening in digits. If the cube root cannot be exactly obtained we must approximate it as closely as possible before adding the tenth part.

On the other hand, in the case of the *euthytonon* the diameter of the opening is obtained by taking 1/9 of the length of the arrow to be shot. For example, if the latter be 3 cubits (or 72 digits), the diameter of the opening will be 1/9 as large, or 8 digits.

Now it is possible, given a single diameter, to determine, by the process of doubling the cube,[3] the other dimensions of stone hurling engines. It

---

[1] Philo proceeds to give the method of duplication reported by Eutocius in his *Commentary on Archimedes' Sphere and Cylinder* III², p. 61 (Heiberg). Subsequent chapters of Philo's work involve the reduction of all measurements to calibers and fractions thereof.

[2] The torsion engines called *euthytona* and *palintona* seem to have been distinguished by a single-arched and a double-arched bow, respectively. The *euthytonon* (also called *scorpios*) was the lighter instrument, which discharged arrows only; while the *palintonon*, the heavier instrument, could also discharge heavy stones and was therefore sometimes called *lithobolos*.

[3] More generally, by the process of constructing a cube bearing any requisite ratio to a given cube. The procedure is basically the same as that employed in doubling the cube. On this problem see p. 62.

is also possible, if one engine has proved successful, to determine from that one the measurements of the others, as follows.

Let the basic diameter[1] of the given engine be AB, and let it be required to construct, on the same model, an engine that will hurl a stone three times as heavy as that which the given engine hurls. Now since the projection of the stone depends on the skein of twisted gut, the engine that is to be constructed will require three times as much gut as does the given engine which has AB as the diameter of the opening receiving the gut.[2] Now the diameter of this opening in the new engine must not be chosen haphazardly but must bear the proper ratio to the length of the gut, so that the cylinders formed by the gut are geometrically similar in the case of both engines.

Let CD be the diameter of the opening in question in the case of the engine to be constructed. Now, since the volumes of similar cylinders are to each other as the cubes of the diameters of their bases,

[3]cylinder on AB : cylinder on $CD = AB^3 : CD^3$.

Now let $AB : CD = CD : EZ = EZ : HT$.

Then[4] $AB : HT = AB^3 : CD^3$.

Therefore cylinder on AB : cylinder on $CD = AB : HT$.

But cylinder on $AB = \frac{1}{3}$ cylinder on CD.

Therefore $AB = \frac{1}{3}\, HT$.

Now AB is given, and therefore HT is also given. Again, CD and EZ are two mean proportionals between AB and HT. CD is, therefore, determined.

In the construction of the new engine HT is to be taken three times as large as AB, since the missile in the case of the new engine is three times as heavy as that of the model. CD and EZ will be the two mean proportionals between AB and HT, and CD will be the required diameter of the opening in the new engine.

We shall now explain how, given two straight lines, we may find two mean proportionals between them.[5]

---

[1] I.e., the diameter of the opening that receives the skein of gut. This measurement may be called the "caliber."

[2] That this is the meaning is clear from what follows, where it is assumed that the volume of the cylinder containing the gut (and therefore the amount of gut) must be proportional to the weight of the missile.

[3] By "cylinder on AB" is meant the cylinder having a circular base of diameter AB.

[4] For
$$\frac{AB^3}{CD^3} = \frac{EZ^3}{HT^3} = \frac{CD \cdot EZ \cdot HT}{HT^3},$$
and
$$\frac{CD \cdot EZ}{HT} = AB,$$
wherefore
$$\frac{AB^3}{CD^3} = \frac{AB}{HT}.$$

[5] The demonstration that follows is substantially the same as that given in Hero, *Mechanics* I. 11, in the two Greek versions (Pappus and Eutocius) and in the Arabic version.

Let AB and BC be the two given lines, at right angles to each other, between which the two mean proportionals are to be found. Complete the rectangle $ABCD$, draw AC and BD, and produce DC and DA. Place a ruler at B so that it intersects these produced lines, and let the ruler rotate about point B until the lines drawn from E to the points of intersection (of the ruler and the produced lines) are equal to each other. The ruler will take the position ZBH and the two lines equal to each other will be EZ and EH.

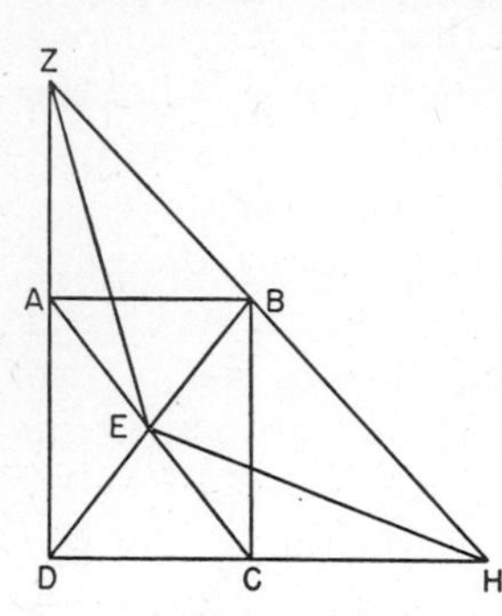

I say that AZ and CH are mean proportionals between AB and BC. And if line AB is the first term, AZ will be the second, CH the third, and BC the fourth.

For since parallelogram $ABCD$ is a rectangle,[1] DE, EA, EB, and EC are all equal.

Since $AE = ED$, and EZ has been drawn,[2] $(DZ \cdot ZA) + AE^2 = EZ^2$.
Also $(DH \cdot HC) + CE^2 = EH^2$.
But $AE = EC$ and $EZ = EH$.
$\therefore DZ \cdot ZA = DH \cdot HC$,
and $DH : DZ = ZA : HC$.
But $DH : DZ = AB : AZ = AZ : CH = CH : CB. \ldots$
That is, AZ and CH are the two mean proportionals between AB and BC.

Vitruvius, *On Architecture* X. 10–12. Translation of M. H. Morgan

10. Catapults or Scorpiones

I shall next explain the symmetrical principles on which scorpiones and ballistae may be constructed, inventions devised for defence against danger, and in the interest of self-preservation.

The proportions of these engines are all computed from the given length of the arrow which the engine is intended to throw, and the size of the holes in the capitals, through which the twisted sinews that hold the arms are stretched, is one ninth of that length.

The height and breadth of the capital itself must then conform to the size of the holes. The boards at the top and bottom of the capital, which are called "peritreti," should be in thickness equal to one hole, and in breadth to one and three quarters, except at their extremities, where they

[1] Reading ὀρθογώνιον (Pappus) rather than διαγώνιον.

[2] Draw also the perpendicular from E to AD.

Then $ZE^2 = [ZA + (AD/2)]^2 + AE^2 - (AD/2)^2$
$= AE^2 + ZA\,(ZA + AD)$
$= AE^2 + ZA \cdot DZ.$

equal one hole and a half. The sideposts on the right and left should be four holes high, excluding the tenons, and five twelfths of a hole thick; the tenons, half a hole. The distance from a sidepost to the hole is one quarter of a hole, and it is also one quarter of a hole from the hole to the post in the middle. The breadth of the post in the middle is equal to one hole and one eighth, the thickness, to one hole.

The opening in the middle post, where the arrow is laid, is equal to one fourth of the hole. The four surrounding corners should have iron plates nailed to their sides and faces or should be studded with bronze pins and nails. The pipe, called *σῦριγξ* in Greek, has a length of nineteen holes. The strips, which some term "cheeks," nailed at the right and left of the pipe, have a length of nineteen holes and a height and thickness of one hole. Two other strips, enclosing the windlass, are nailed on to these, three holes long and half a hole in breadth. The cheek nailed on to them, named the "bench," or by some the "box," and made fast by means of dove-tailed tenons, is one hole thick and seven twelfths of a hole in height. The length of the windlass is equal to. . .[1] holes, the thickness of the windlass to three quarters of a hole.

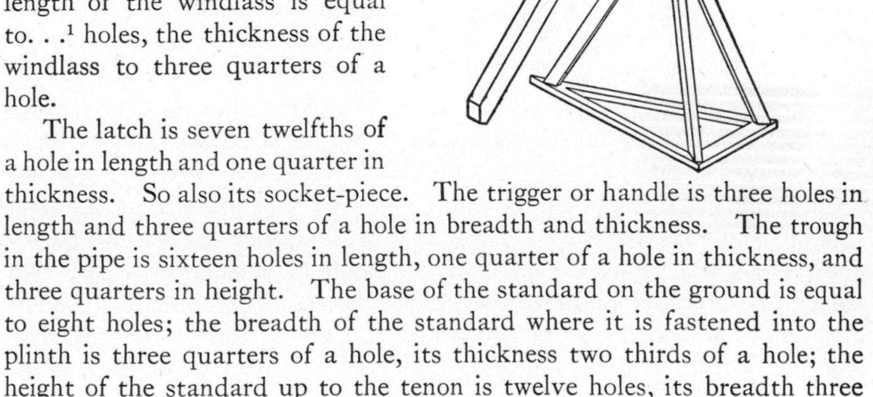

The latch is seven twelfths of a hole in length and one quarter in thickness. So also its socket-piece. The trigger or handle is three holes in length and three quarters of a hole in breadth and thickness. The trough in the pipe is sixteen holes in length, one quarter of a hole in thickness, and three quarters in height. The base of the standard on the ground is equal to eight holes; the breadth of the standard where it is fastened into the plinth is three quarters of a hole, its thickness two thirds of a hole; the height of the standard up to the tenon is twelve holes, its breadth three

[1] The reading here and in numerous other places is very doubtful, chiefly because of the failure of copyists to understand the system of numerical and fractional symbols employed. Some read "four" here. [Edd.]

quarters of a hole, and its thickness two thirds. It has three struts, each nine holes in length, half a hole in breadth, and five twelfths in thickness. The tenon is one hole in length, and the head of the standard one hole and a half in length.

The antefix has the breadth of a hole and one eighth, and the thickness of one hole. The smaller support, which is behind, termed in Greek *ἀντίβασις*, is eight holes long, three quarters of a hole broad, and two thirds thick. Its prop is twelve holes long and has the same breadth and thickness as the smaller support just mentioned. Above the smaller support is its socket-piece, or what is called the "cushion," two and a half holes long, one and a half high, and three quarters of a hole broad. The windlass cup is two and seven twelfths holes long, two thirds of a hole thick, and three quarters broad. The crosspieces with their tenons have the length of... holes, the breadth of three quarters, and the thickness of two thirds of a hole. The length of an arm is seven holes, its thickness at its base two thirds of a hole, and at its end one half a hole; its curvature is equal to two thirds of a hole.

These engines are constructed according to these proportions or with additions or diminutions. For, if the height of the capitals is greater than their width—when they are called "high-tensioned,"—something should be taken from the arms, so that the more the tension is weakened by height of the capitals, the more the strength of the blow is increased by shortness of the arms. But if the capital is less high—when the term "low-tensioned" is used—the arms, on account of their strength, should be made a little longer, so that they may be drawn easily. Just as it takes four men to raise a load with a lever five feet long, and only two men to lift the same load with a ten-foot lever, so the longer the arms, the easier they are to draw, and the shorter, the harder.

I have now spoken of the principles applicable to the parts and proportions of catapults.

11. Ballistae

Ballistae are constructed on varying principles to produce an identical result. Some are worked by handspikes and windlasses, some by blocks and pulleys, others by capstans, others again by means of drums. No ballista, however, is made without regard to the given amount of weight of the stone which the engine is intended to throw. Hence their principle is not easy for everybody, but only for those who have knowledge of the geometrical principles employed in calculation and in multiplication.

For the holes made in the capitals through the openings of which are stretched the strings made of twisted hair, generally women's, or of sinew, are proportionate to the amount of weight in the stone which the ballista is intended to throw, and to the principle of mass, as in catapults the principle is that of the length of the arrow. Therefore, in order that those who do not understand geometry may be prepared beforehand, so as not to be delayed by having to think the matter out at a moment of peril in war, I will set forth what I myself know by experience can be depended upon, and what I have in part gathered from the rules of my teachers, and wherever Greek weights bear a relation to the measures, I shall reduce and explain them so that they will express the same corresponding relation in our weights.

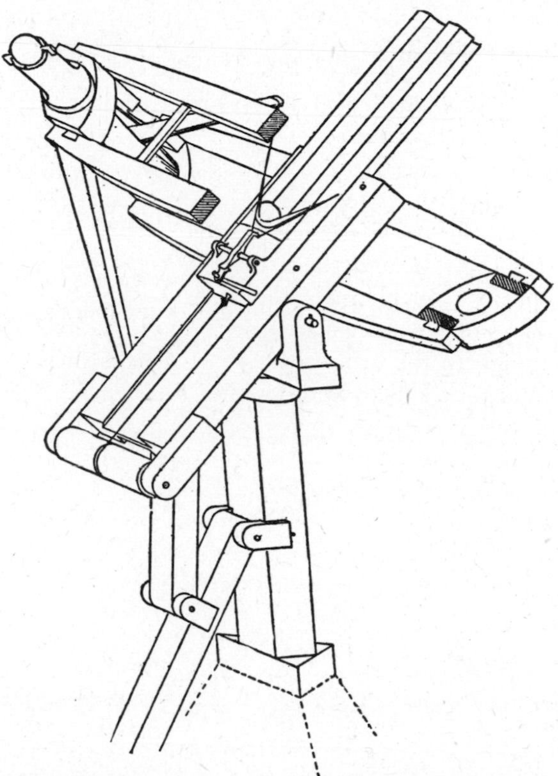

A ballista intended to throw a two-pound stone will have a hole of five digits in its capital;[1] four pounds, six digits, and six pounds, seven digits; ten pounds, eight digits; twenty pounds, ten digits; forty pounds, twelve

[1] Auguste Choisy (*Vitruve*, vol. I, pp. 308 ff.; see also vol. IV, Plate 90) has sought to show that the formula of Vitruvius is of the same form as that of Hero (see p. 319, above). Vitruvius's formula, according to Choisy, is $F = 3.3526 \sqrt[3]{p} + (1/60)\,(60 - p)$, where F is the diameter of the hole in Roman digits and p the weight of the missile in Roman pounds. The second term of the formula is a correction necessary because the empirical factor of proportionality (3.3526) is too small for values of p under 60 and too large for values of p over 60. [Edd.]

and a half digits; sixty pounds, thirteen and a half digits; eighty pounds, fifteen and three quarters digits; one hundred pounds, one foot and one and a half digits; one hundred and twenty pounds, one foot and two digits; one hundred and forty pounds, one foot and three digits; one hundred and sixty pounds, one foot and a quarter; one hundred and eighty pounds, one foot and five digits; two hundred pounds, one foot and six digits; two hundred and forty pounds, one foot and seven digits; two hundred and eighty pounds, one foot and a half; three hundred and twenty pounds, one foot and nine digits; three hundred and sixty pounds, one foot and ten digits. . . .[1]

12. The Stringing and Tuning of Catapults

Beams of very generous length are selected, and upon them are nailed socket-pieces in which windlasses are inserted. Midway along their length the beams are incised and cut away to form framings, and in these cuttings the capitals of the catapults are inserted, and prevented by wedges from moving when the stretching is going on. Then the bronze boxes are inserted into the capitals, and the little iron bolts, which the Greeks call ἐπιζυγίδες, are put in their places in the boxes.

Next, the loops of the strings are put through the holes in the capitals, and passed through to the other side; next, they are put upon the windlasses, and wound round them in order that the strings, stretched out taut on them by means of the handspikes, on being struck by the hand, may respond with the same sound on both sides. Then they are wedged tightly into the holes so that they cannot slacken. So, in the same manner, they are passed through to the other side, and stretched taut on the windlasses by means of the handspikes until they give the same sound. Thus with tight wedging, catapults are tuned to the proper pitch by musical sense of hearing.[2]

Devices Based on Pneumatics

Hero of Alexandria, *Pneumatics* I. 7, 12, 38, 28, 42. Translation of J. G. Greenwood

The Clepsydra

7. Let us now proceed to construct the necessary instruments, beginning with the less important,[3] as from the elements. The following is a contrivance of use in pouring out wine.[4] A hollow globe of bronze is provided, such as *AB* [see fig.], pierced in the lower part with numerous

[1] Vitruvius then goes on to describe the dimensions of the ballista in terms of the "hole," as in the case of the catapult above. [Edd.]

[2] An application of the principle that strings of the same material, breadth, length, and tension have the same pitch. [Edd.]

[3] In the context μικροτέρων would seem to mean "simpler." [Edd.]

[4] This is the clepsydra in its simplest form (see p. 245). Examples of this type of device, dating as far back as the sixth century B.C., have been found. [Edd.]

small holes like a sieve. At the top let there be a tube, *CD*, the upper extremity of which is open, communicating with and soldered into the globe. When it is desired to pour out wine, with one hand grasp the tube *CD* near the mouth *C*, and plunge the globe into the wine until it is wholly immersed. The wine enters through the holes, and the air within, being driven out, passes through the tube *CD*: and if, pressing the thumb on the aperture *C*, you lift the globe out of the wine, the wine contained in the globe will not flow out, as no air can enter to supply the vacuum, for the only entrance is through the mouth *C*, which is closed by the thumb. When, then, we desire to let the wine flow, we remove the finger, and the air, rushing in, fills the vacuum produced. If we again press the finger on the air-hole *C*, there will be no discharge until we once more remove the finger from the vent. We may, in like manner, dip the globe into hot or cold water, and then retain or let out the contents at pleasure, until all the water within is exhausted. If the extremity *C* of the tube *CD* is bent the action will be the same, and it is then easier to stop the orifice with the finger.

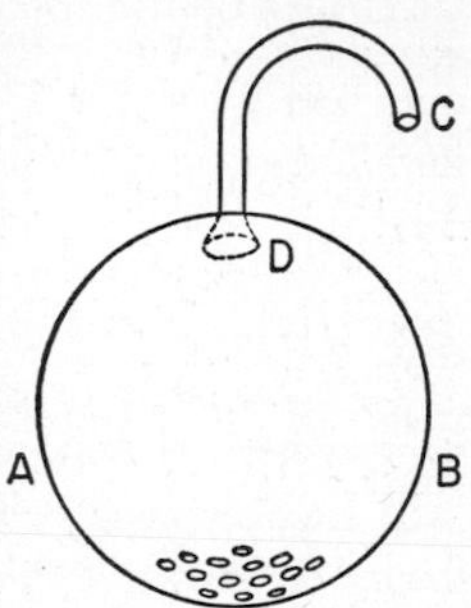

Libations at an Altar Produced by Fire

12. To construct an altar such that, when a fire is raised on it, figures at the side shall offer libations. Let there be a pedestal, *ABCD* [see fig.], on which the figures stand, and also an altar, *EFG*, perfectly air-tight.

The pedestal must also be air-tight, and communicate with the altar at *G*. Through the pedestal insert the tube *HKL*, reaching nearly to the bottom at *L*, and communicating at *H* with a bowl held by one of the figures. Pour liquid into the pedestal through a hole, *M*, which must afterwards be closed. Now if a fire be lighted on the altar *EFG*, the air within it, being rarefied, will descend into the pedestal, and exert pressure on the liquid it contains, which, having no other way of retreat, will pass through the tube *HKL* into the bowl. Thus the figures will pour a libation, and will not cease as long as the fire remains on the altar. When the fire is extinguished, the libation ceases; and as often as the fire is kindled the same will be repeated. The pipe through which the heat is to pass should be broader towards the middle, for it is requisite that the heat, or rather the vapour from it, passing into a broader space, should expand and act with greater force.[1]

Automatic Opening of Temple Doors by Fire on an Altar

38. The construction of a small temple such that, on lighting a fire, the doors shall open spontaneously, and shut again when the fire is extinguished. Let the proposed temple stand on a pedestal, *ABCD* [see fig.], on which lies a small altar, *EA*.

Through the altar insert a tube, *FG*, of which the mouth *F* is within the altar, and the mouth *G* is contained in a globe, *H*, reaching nearly to its centre: the tube must be soldered into the globe, in which a bent siphon, *KLM*, is placed. Let the hinges of the doors be extended downwards and turn freely on pivots in the base *ABCD*; and from the hinges let two chains, running into one, be attached, by means of a pulley, to a hollow vessel, *NX*, which is suspended; while other chains, wound upon the hinges in an opposite direction to the former, and running into one, are attached, by means of a pulley, to a leaden weight, on the descent of which the doors will be shut. Let the outer leg of the siphon *KLM* lead into the suspended vessel; and through a hole, *P*, which must be carefully closed afterwards, pour water into the globe enough to fill one half of it. It will be found that, when the fire has grown hot, the air in the altar becoming heated expands into a larger space; and, passing through the tube *FG* into the globe, it will drive out the liquid contained there through the siphon *KLM* into the suspended vessel, which, descending with its weight, will tighten the chains and open the doors. Again, when the fire is extinguished, the rarefied air will escape through the pores in the side of the globe, and the bent siphon (the extremity of which will be immersed in the water in the suspended vessel) will draw up the liquid in the vessel in order to fill up the void left

[1] Changes of pressure with changes of temperature are made use of in many of the devices described by Hero. [Edd.]

by the particles removed. When the vessel is lightened the weight suspended will preponderate and shut the doors. Some in place of water use quicksilver, as it is heavier than water and is easily disunited by fire.

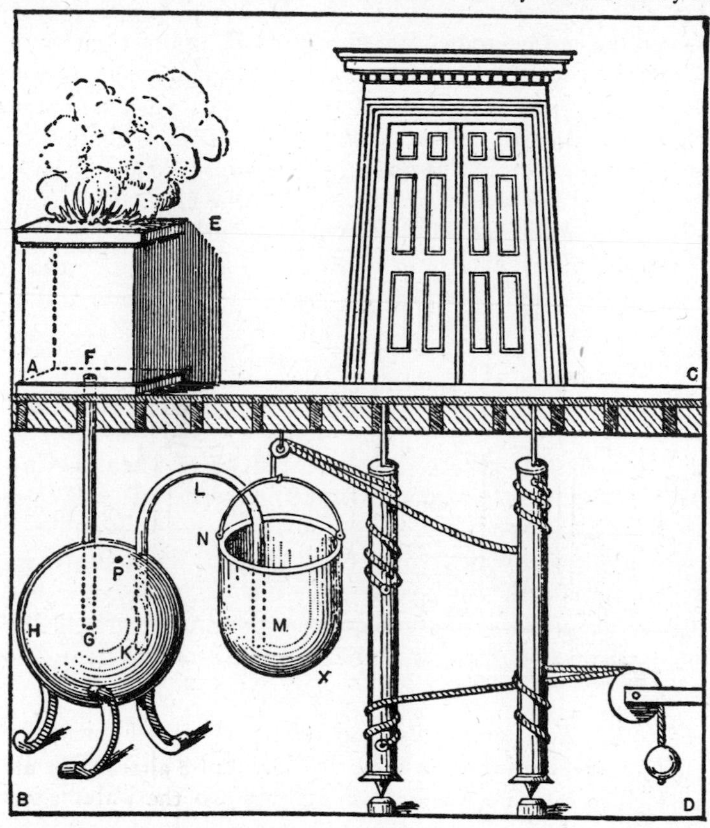

The Fire-Engine[1]

28. The siphons used in conflagrations are made as follows. Take two vessels of bronze, *ABCD*, *EFGH* [see figure, p. 330], having the inner surface bored in a lathe to fit a piston (like the barrels of water-organs), *KL*, *MN*, being the pistons fitted to the boxes. Let the cylinders communicate with each other by means of the tube *XODF*, and be provided with valves, *P*, *R*, such as have been explained above, within the tube *XODF* and opening outwards from the cylinders. In the bases of the cylinders pierce circular

[1] This pump, which was probably invented by Ctesibius, is the simplest type of water pump with valves. A similar device could have been used to pump air and, with a slight rearrangement, to exhaust air. Whether this was done in antiquity is doubtful.

The pump is also described by Philo of Byzantium, Appendix I, 2 (Carra de Vaux) and by Vitruvius X. 7. [Edd.]

apertures, *S*, *T*, covered with polished hemispherical cups, *VQ*, *WY*, through which insert spindles soldered to, or in some way connected with, the bases of the cylinders, and provided with shoulders at the extremities that the cups may not be forced off the spindles. To the centre of the pistons fasten the vertical rods *SE*, *SE*, and attach to these the beam *A′ A′*, working, at its centre, about the stationary pin *D*, and about the pins *B*, *C*,

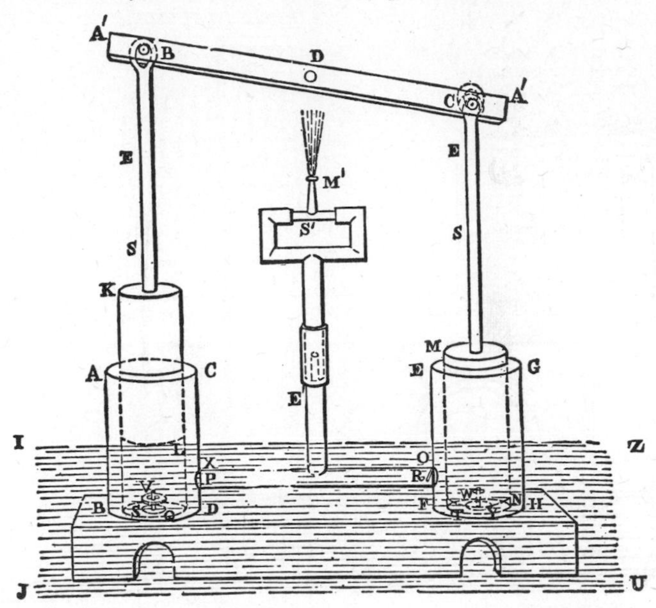

at the rods *SE*, *SE*. Let the vertical tube *S′ E′* communicate with the tube *XODF*, branching into two arms at *S′*, and provided with small pipes through which to force up water, such as were explained above in the description of the machine for producing a water-jet by means of the compressed air. Now, if the cylinders, provided with these additions, be plunged into a vessel containing water, *IJUZ*, and the beam *A′ A′* be made to work at its extremities *A′*, *A′*, which move alternately about the pin *D*, the pistons, as they descend, will drive out the water through the tube *E′ S′* and the revolving mouth *M′*. For when the piston *MN* ascends it opens the aperture *T*, as the cup *WY* rises, and shuts the valve *R*; but when it descends it shuts *T* and opens *R*, through which the water is driven and forced upwards. The action of the other piston, *KL*, is the same. Now the small pipe *M′*, which waves backward and forward, ejects the water to the required height but not in the required direction, unless the whole machine be turned round; which on urgent occasions is a tedious and difficult process. In order, therefore, that the water may be ejected to the spot required, let the tube *E′ S′* consist of two tubes, fitting closely together lengthwise, of which one must be attached to the tube *XODF*, and the other to the part from which the arms branch off at *S′*; and thus, if the upper tube be turned round, by the inclination of the mouthpiece *M′* the stream of water can be forced to any spot we please. The upper joint of the double tube must be secured to the lower, to prevent its being dislodged from the machine by the violence of the water. This may be effected by holdfasts in the shape of the letter *L*, soldered to the upper tube, and sliding on a ring which encircles the lower.

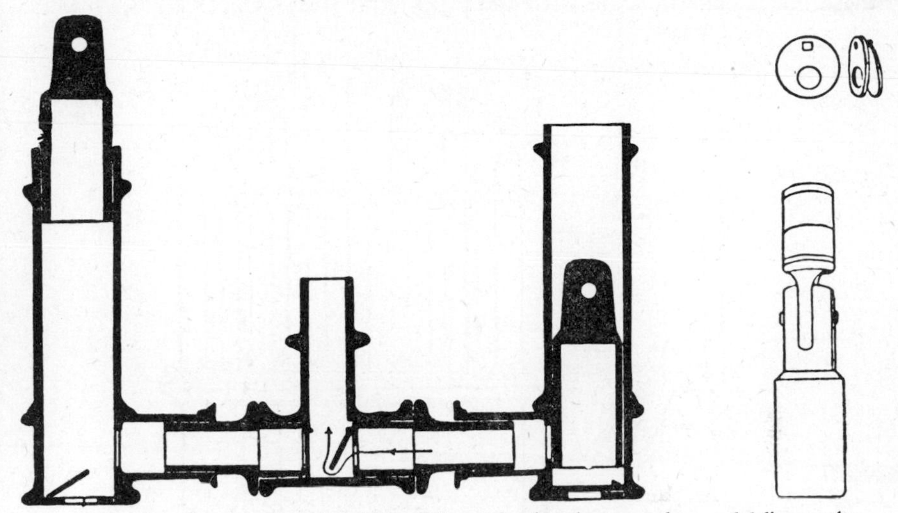

Reconstruction of ancient pump found near Bolsena, showing plungers, valves, and delivery tube.

The Hydraulic Organ[1]

42. The construction of a hydraulic organ. Let *ABCD* [see figure, p. 332] be a smaller altar of bronze containing water. In the water invert a hollow hemisphere, called a *pnigeus*, *EFGH*, which will allow of the passage of the water at the bottom. From the top of this let two tubes ascend above the altar; one of them, *GKLM*, bent without the altar and communicating with a box, *NXOP*, inverted, and having its inner surface made perfectly level to fit a piston. Into this box let the piston *RS* be accurately fitted, that no air may enter by its side; and to the piston attach a rod, *TU*, of great strength. Again, attach to the piston rod another rod, *UQ*, moving about a pin at *U*, and also working like the beam of a lever on the upright rod *WY*, which must be well secured. On the inverted bottom of the

[1] In antiquity the term "hydraulic" (derived from the Greek words for "water" and "flute") always referred specifically to the musical instrument. The invention of this instrument, called *hydraulis*, is ascribed to Ctesibius. An instrument generally similar to that described by Hero is described by Vitruvius X. 8. [Edd.]

box *NXOP* let another smaller box, *Z*, rest, communicating with *NXOP* and closed by a lid above: in the lid is a hole through which the air will enter the box. Place a thin plate under the hole in the lid to close it, upheld by means of four pins passing through holes in the plate, and furnished with heads so that the plate cannot fall off: such a plate is called a valve. Again, let another tube, *FI*, ascend from *FG*, communicating with a transverse tube, *A′ B′*, on which rest the pipes *A*, *A*, *A*, communicating with the tube, and having at the lower extremities small boxes, like those used for money; these boxes communicate with the pipes, and their orifices *B*, *B*, *B*, must

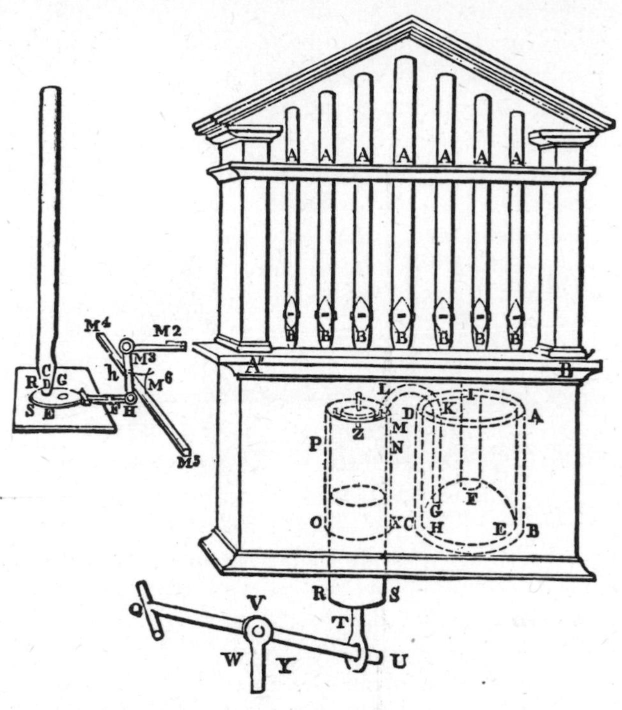

be open. Across these orifices let perforated lids slide, so that, when the lids are pushed home, the holes in them coincide with the holes in the pipes, but, when the lids are drawn outwards the connexion is broken and the pipes are closed. Now, if the transverse beam *UQ* be depressed at *Q*, the piston *RS* will rise and force out the air in the box *NXOP*; the air will close the aperture in the small box *Z* by means of the valve described above, and pass along the tube *MLKG* into the hemisphere: again it will pass out of the hemisphere along the tube *FI* into the transverse tube *A′ B′*, and out of the transverse tube into the pipes, if the apertures in the pipes and in the lids coincide, that is, if the lids, either all, or some of them, have been pushed home.

In order that, when we wish any of the pipes to sound, the corresponding holes may be opened and closed again when we wish the sound to cease, we may employ the following contrivance. Imagine one of the boxes at the extremities of the pipes, *CD*, to be isolated, *D* being its orifice, *E* the communicating pipe, *RS* the lid fitted to it, and *G* the hole in the lid not coinciding with the pipe *E*. Take three jointed bars, *FH*, *HM*, MM^2, of which the bar *FH* is attached to the lid *SF*, while the whole moves about a pin at M^3. Now, if we depress, with the hand, the extremity M^2 towards *D*, the orifice of the box, we shall push the lid inwards, and, when it is in, the aperture in it will coincide with that in the tube. That, when we withdraw the hand, the lid may be spontaneously drawn out and close communication, the following means may be employed. Underneath the boxes let a rod M^4 M^5 run, equal and parallel to the tube A' B', and fix to this slips of horn, elastic and curved, of which M^6, lying opposite to *CD*, is one. A string, fastened to the extremity of the slip of horn, is carried round the extremity *H*, so that, when the lid is pushed out, the string is tightened; if, therefore, we depress the extremity M^2 and drive the lid inwards, the string will forcibly pull the piece of horn and straighten it, but, when the hand is withdrawn, the horn will return again to its original position and draw away the lid from the orifice, so as to destroy the correspondence between the holes. This contrivance having been applied to the box of each pipe, when we require any of the pipes to sound we must depress the corresponding key with the fingers; and when we require any of the sounds to cease, remove the fingers, whereupon the lids will be drawn out and the pipes will cease to sound.

The water is poured into the altar that the superabundant air (I mean, of course, that which is thrust out of the box and forces the water upwards) may be confined in the hemisphere, so that the pipes which are free to sound may always have a supply. The piston, *RS*, when raised, drives the air out of the box into the hemisphere, as has been explained; and when depressed, opens the valve in the small box *Z*. By this means the box is filled with air from without, which the piston, when forced up again, will again drive into the hemisphere. It would be better that the rod *TU* should move about a pivot at *T* also, by means of a single [loop], *R*, which may be fitted into the bottom of the piston, and through which the pivot must pass, that the piston may not be drawn aside, but rise and fall vertically.

Cf. Aristocles in Athenaeus, *Deipnosophistae* 174. Translation of Charles B. Gulick (London, 1928)

The question is debated whether the water-organ belongs to the wind or the stringed instruments.[1] Now Aristoxenus, to be sure, does not know

[1] "Or the percussion instruments" is a plausible addition suggested by H. Diels in view of the sequel. [Edd.]

of it; but it is said that Plato imparted a slight hint of its construction in having made a time-piece for use at night which resembled a water-organ, being a very large water-clock.[1] And in fact the water-organ does look like a water-clock. Therefore it cannot be regarded as a stringed instrument or a percussion instrument, but perhaps may be described as a wind instrument, since wind is forced into it by the water. For the pipes are set low in water, and as the water is briskly agitated by a boy, air is released in the pipes through certain valves which fit into the pipes from one side of the organ to the other, and a pleasant sound is produced.

Philo of Byzantium, *Pneumatics*, Appendix I. 1 (Carra de Vaux)[2]

A Bellows for Lifting Water

A machine for raising water from the bottom of a well by an ingenious device.

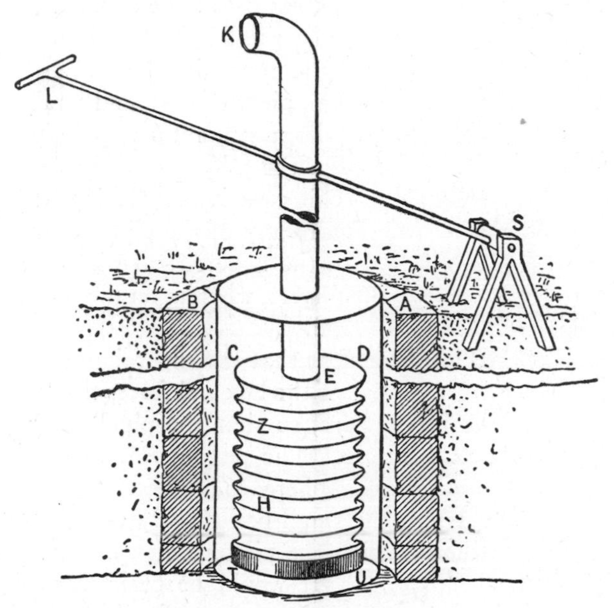

The well must be of the same breadth throughout and uniformly round. The bottom of the well should, if possible, be cemented or covered with boards or other material. Let this well be *AB*. Now construct a cylindrical cover of hard wood exactly fitting the well as a sort of faucet. As it is inserted, it just fills the well without touching the sides and covers the

[1] Diels attempts a reconstruction of this clock (*Sitzungsberichte der Preussischen Akademie der Wissenschaften*, 1915, pp. 824-830). [Edd.]

[2] On Philo of Byzantium see p. 255. The attribution to Philo of this selection is not certain, though evidently favored by Carra de Vaux.

surface of the water completely. Let this cover be *CD*. Now place over this cover a series of hides sewed together and fitted hermetically without a break, similar to a conduit tube. The length of this series of hides should be somewhat greater than the depth of the water. This leather device goes into the water along the cylindrical wall and down to the bottom of the well. There are rings within the device which are so placed around the interior that it closes and opens as it is moved, just as a goldsmith's bellows called *zauqi*. This leather is marked *ZH*. Pierce an opening in the middle of the cover, at *E*, and set up a pipe fitting this opening. The pipe is of copper or some other substance and long enough to emerge from the well. Then, at that end of the leather device which touches the bottom of the well, place a very heavy leaden ring so that when this ring reaches the base of the well it clings firmly to it. This ring is marked *TU*. Now at a point on this pipe, dependent on the depth of the water, attach a wooden shaft which we may call the lever, *L*. Behind the well this shaft is fitted, at its extremity, with two hinges attached to a transverse bar connected with two columns of very solid construction. These hinges turn easily. They are marked *S*.

The upward and downward pressure is exerted at *L*. When the lever is lifted the pipe is necessarily raised together with that which is attached to it, and the leather device is drawn up. When, on the other hand, the lever is depressed, the pipe descends forcefully along with that which is attached to it. The cover then exerts pressure on the water which issues from *K*, the upper opening of the pipe. That is what we wished to explain.

When *L* is depressed toward the brim of the well at side *B*, the cylindrical cover drops and the leather rings *ZH* are drawn together until *E* is near *T*. The water then of necessity must rise through *E* and issue through *K*.[1] That is what we had in mind.

A Mechanical Dove

Aulus Gellius, *Attic Nights* X. 12.8[2]

Many well-known Greeks and the philosopher Favorinus, a very assiduous antiquarian, have definitely asserted that Archytas constructed a wooden model of a dove according to certain mechanical principles, and that the dove actually flew, so delicately balanced was it with weights and propelled by a current of air enclosed and concealed within it. Indeed,

---

[1] There is no explanation of the pneumatic principle involved, the creation of a partial vacuum by the forcing out of the air. This pump without valves is quite primitive in comparison with the pump of Ctesibius described by Hero (see p. 329).

[2] Gellius (*ca.* 130–*ca.* 180) records in the twenty books of his *Attic Nights* miscellaneous information on a wide range of subjects. Favorinus was a contemporary. On Archytas see page 35. The dove is presumably a toy, the wings being moved by a mechanism utilizing air pressure.

in a matter so incredible it is preferable to quote Favorinus' actual language: "Archytas of Tarentum, who among other things was a mechanician, constructed a flying dove of wood. . . ."

A Surveying and Levelling Instrument

The dioptra in its simplest form is a sighting tube. In the form described by Hero it is an instrument or pair of instruments for surveying and levelling, for the measurement of heights and distances of inaccessible places, and for the determination of angles in both terrestrial and astronomical problems. The sighting piece can be set in any position and at any angle. With it is combined a hydraulic level for the determination of the horizontal plane. The tilting of the line of sight to any plane between the horizontal and the vertical and to any position in the plane is made possible by sets of screws and geared wheels. The precise angles involved could, presumably, be read off when the position of the sighting piece was determined. In finding the relative height of two widely separated points a series of intermediate sightings was made with the help of levelling staffs. The latter are also described by Hero. On the use of the dioptra as an astronomical instrument see p. 139, above.

Hero of Alexandria, *On the Dioptra* 3–5, 8, 14, 15 (Schöne)

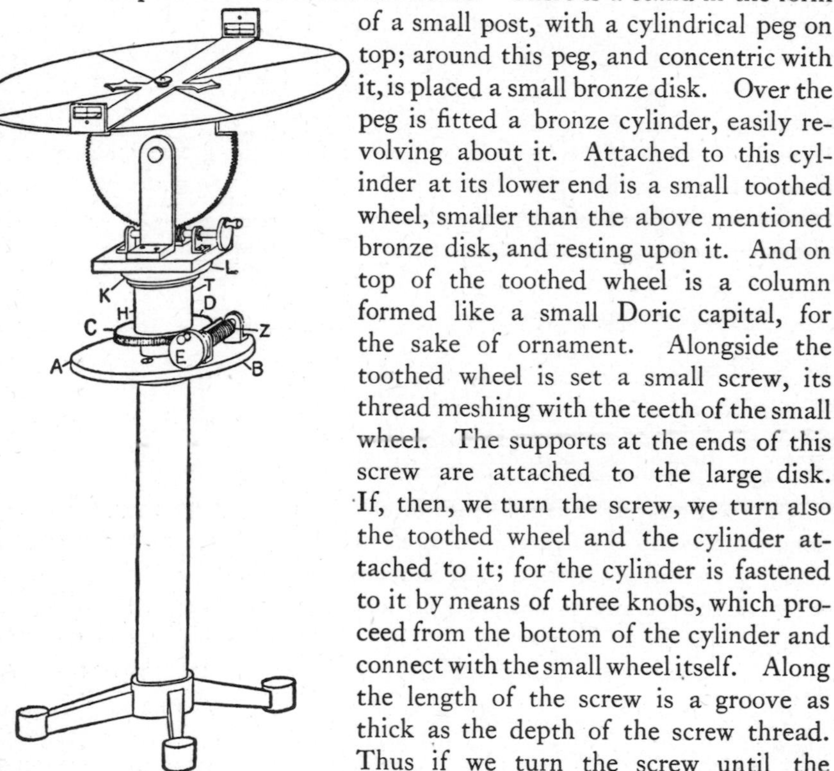

3. The dioptra is constructed as follows. There is a stand in the form of a small post, with a cylindrical peg on top; around this peg, and concentric with it, is placed a small bronze disk. Over the peg is fitted a bronze cylinder, easily revolving about it. Attached to this cylinder at its lower end is a small toothed wheel, smaller than the above mentioned bronze disk, and resting upon it. And on top of the toothed wheel is a column formed like a small Doric capital, for the sake of ornament. Alongside the toothed wheel is set a small screw, its thread meshing with the teeth of the small wheel. The supports at the ends of this screw are attached to the large disk. If, then, we turn the screw, we turn also the toothed wheel and the cylinder attached to it; for the cylinder is fastened to it by means of three knobs, which proceed from the bottom of the cylinder and connect with the small wheel itself. Along the length of the screw is a groove as thick as the depth of the screw thread. Thus if we turn the screw until the groove in it fits into the teeth of the toothed wheel, the toothed wheel will turn automatically. And if we set the toothed wheel as occasion re-

quires and turn the screw enough for the spiral to mesh with the teeth, we render the toothed wheel immovable.

Thus, let *AB* be the disk that is around the cylindrical socket and attached to the stand, *CD* the toothed wheel attached to the cylinder, *EZ* the screw alongside it, and *HT* the cylinder attached to toothed wheel *CD*, on which, as has been mentioned, is the small Doric capital *KL*. Let two bronze supports like rulers be placed upright on the upper surface of the column, just far enough apart to permit of the insertion of a small disk; and on the column between these uprights let there be a revolving screw, whose supports are attached to the above-mentioned column.[1] The large

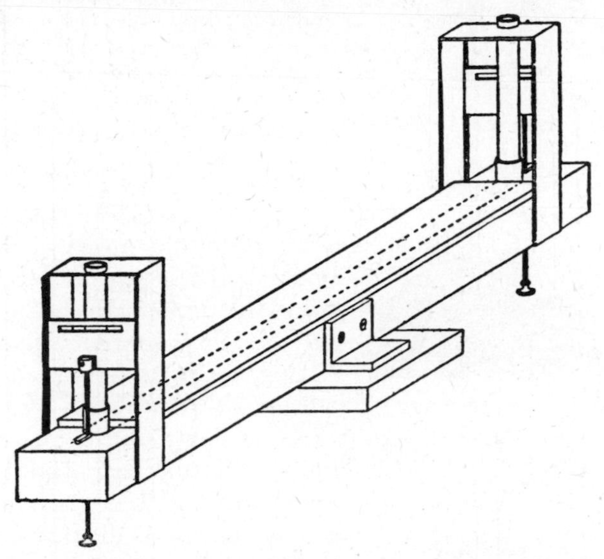

uprights attached to the column go upwards to a height of 4 digits. Across the space between the projections is fitted transversely a beam, about 4 cubits long and sufficiently wide and thick to fit the space. Let the midpoint of this rule be the center of this space.

4. In the upper surface of the beam a round or square groove is cut out, long enough to hold a bronze tube that is about 12 digits shorter than the beam. Two other tubes are set perpendicular to the bronze tubes at its ends, in such a way as to appear to bend up from the bronze tube; these are not more than 2 digits high. Then the bronze tube is covered with a long rule which is fitted into the groove. This rule holds the bronze tube

[1] There is probably a break in the text here. The description of the angle-measurer or theodolite (p. 336) is left incomplete, and what follows refers to the levelling instrument shown on this page. It seems that these two distinct devices could separately fit the same base. In the levelling device each sight is adjusted to the level of water in the cylinder next to it, so that the line of vision is horizontal.

secure and gives the instrument a neater appearance. In each of the two upright tubes is fitted a small glass cylinder wide enough to fit the bronze tube and about 12 digits high. The glass cylinders are sealed all around with wax or some other cementing material to the upright extensions of the tube. In this way, when water is poured into one of the cylinders, none of it can escape. In addition, two small frames are set over the transverse beam at the points where the glass cylinders are set up. Thus

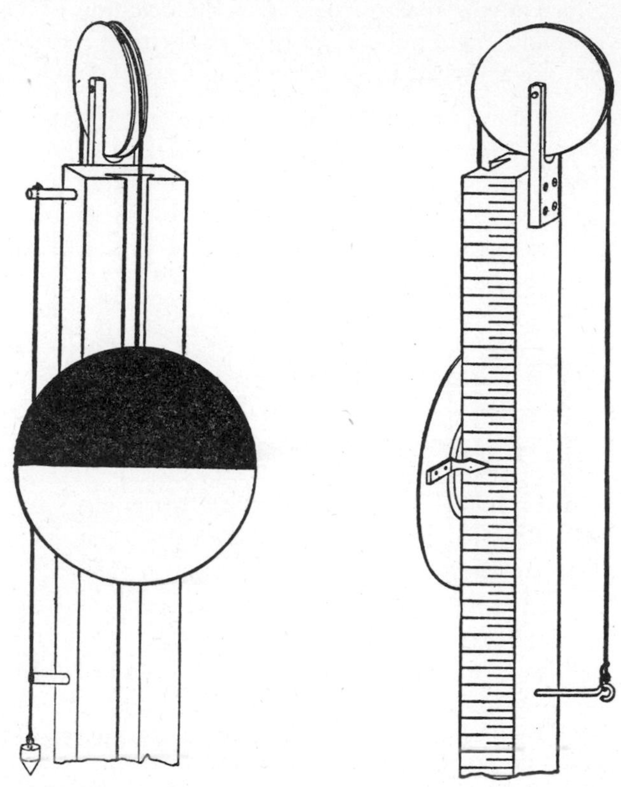

the cylinders pass through these frames and are held secure by them. In these frames are fitted small bronze plates which can move on grooves along the sides of the framework, very close to the glass cylinders. The plates have openings in the center, through which it is possible to take sightings. Small hollow cylinders, half a digit high, are attached to the lower parts of the bronze plates; and fitting into them are bronze rods of the same length as the height of the glass cylinders. These rods pass through an opening in the grooved beam. Screw threads are cut into these rods; and knobs attached to the beam are fitted to these screw threads. Now by turning the lower projections of these rods, we can move the plates with the openings in them upwards and downwards. For the upper end of the rod near the plate will contain a knob which fits into the opening in the hollow cylinder.

5. Now that the construction of the dioptra has been presented, we shall describe the levelling posts and disks that are used with it. There are two posts about 10 cubits long, about 5 digits wide, and about 3 digits thick. In the center of the wider side of each of the two posts is a groove, running along the whole length of the post, its narrow part outside. A tenon, which can easily ride along it without falling out, is fitted into this groove. To this rider is nailed a circular disk 10 or 12 digits in diameter. A straight line is drawn across the face of the disk at right angles to the length of the post. One of the semicircles thus formed is painted white, the other black. A rope, attached to the rider, is brought from the rider over a pulley fixed to the upper part of the post and then down the other side of the post, i.e., the side without the circular disk. Thus by setting the post upright on the ground and pulling the rope from the back, it is possible to raise the disk. If the rope is let go, the disk will drop to the bottom by its own weight. For the disk will have a lead plate nailed to its back, so that it will automatically drop. Now, then, if we slacken the rope, the disk descends and may be stopped at any desired point on the measuring post. Now measure off on the entire length of the post, beginning at the bottom, the divisions into cubits, palms, and digits. At the points of division cut lines into the post on the side to the right of the disk. The disk itself will have on its back an indicator projecting from a point on its diameter and falling along the aforesaid measuring divisions in the side of the post.

The levelling posts will be accurately set up on the ground at right angles [to the horizon] if we proceed as follows. On their sides, where no measuring divisions have been drawn, is inserted a peg about 3 digits long. At the end of this peg, and extending vertically through it, is a hole through which passes a cord that has a weight attached to its end. In the same manner, at the bottom of the post a peg is inserted extending as far out as the hole in the upper peg extends from the post. A vertical line is drawn down the middle of the outer face of the lower peg. When the plumb line coincides with this straight line, the post will stand upright [i.e., at right angles to the horizon].

8. *Given two points, one near us and the other at a distance, to find the distance between them along a horizontal line without approaching the distant point.*

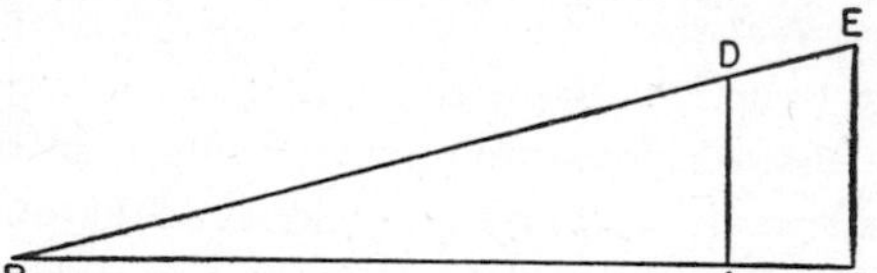

E
D
B
A
C

Let the given points be A and B, A situated near us and B at a distance. Place the dioptra with the semi-circular disk at A. Let the sighting apparatus connected with the circular drum be turned until B is visible.

Then changing my position to the other side of the sighting apparatus I turn the semicircular disk, the other parts remaining stationary,[1] and sight a point, *C*, on the side accessible to us and lying on line *AB*. Then with the help of the dioptra I draw *AD* from *A* perpendicular to *BC* and also a second perpendicular, *CE*, from *C*, taking any point, e.g., *E*, on *CE*. Transferring the dioptra to *E*, I so place the line of sight that *B* is visible through it, and thus I determine on *AD* another point, *D*, collinear with *BE*. Thus we have a triangle *BCE* with *AD* parallel to *CE*. Therefore $CB:BA = CE:AD$. But I can determine the ratio $CE:AD$ by measuring each along a horizontal line, as has been previously indicated. Suppose that we find $CE = 5AD$. Then $BC = 5BA$ and $CA = 4AB$. But I am able to measure the distance *AC* along a horizontal line. Therefore I am able also to measure the distance *AB* along a horizontal line.

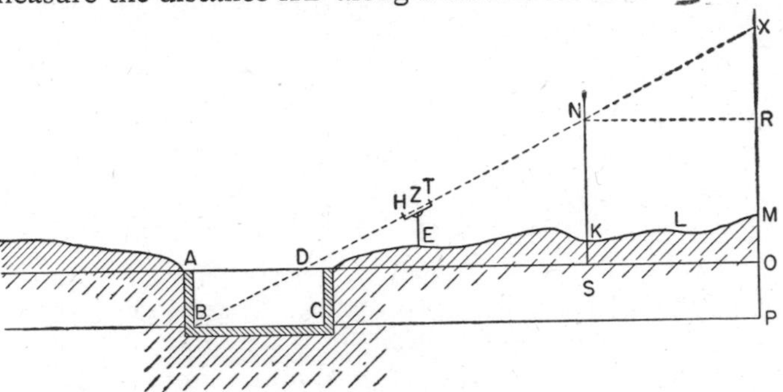

14. *Given a ditch, to find its depth, i.e., the length of the perpendicular from the point at the bottom to the plane passing through us and parallel to the horizon, or, again, to the plane drawn through any other point and parallel to the horizon.*

Let *ABCD* be the given ditch and *B* the point at the bottom. Place the dioptra at *D* or at some other point, e.g., at *E*, the dioptra being *EZ*. Let *HT* be the line along which we take the sighting. Let this be inclined until the point *B* is visible through it. Consider the surface of the ground as lying along line *DEKLM* and the horizontal plane through us as passing through line *ADSO*. Let two measuring posts, *KN* and *MX*, be set up perpendicular to this, and passing through an extension of *HT*, the sighting line. Now let *N* be sighted on *KN* and *X* on *XM*.

Suppose it is required to measure the perpendicular from *B* to the plane drawn through *D* parallel to the horizon, i.e., the perpendicular drawn to line *ADO*. Now this perpendicular which it is required to measure is *BA*.

Consider the plane through *B* parallel to the horizon and passing through *BP* and suppose *XM* extended to *P* and *NK* to *S*. Suppose, further, that *NR* is drawn through *N* parallel to *DO*. Now *NR* is the

---

[1] I.e., keeping the position of line *AB*, which has been determined by sighting.

distance, along a horizontal plane, between points *K* and *M*. It is therefore possible to determine *NR* because *KS* and *MO* can be determined. But *XR* is the excess of *XRO* over *NS*, and can be determined, since it is possible to measure *KS* and *MO* as we did when, by the use of two measuring posts, we determined the perpendicular distance of any given point.[1]

Let us, for example, say that we find *NR* = 4*RX*. Then *BP* = 4*PX*. Now it is possible to determine *BP*, i.e., *AO*. For *AO* or *BP* is the distance in the horizontal plane between *O* and *A*. It is possible, therefore, to determine *XP*, for *XP* = ¼*BP*. But we know the length *XO*, and can consequently obtain *OP*, which is the same as the perpendicular *AB*.

15. *To make a rectilinear tunnel through a mountain, given the location of the openings at each end.*

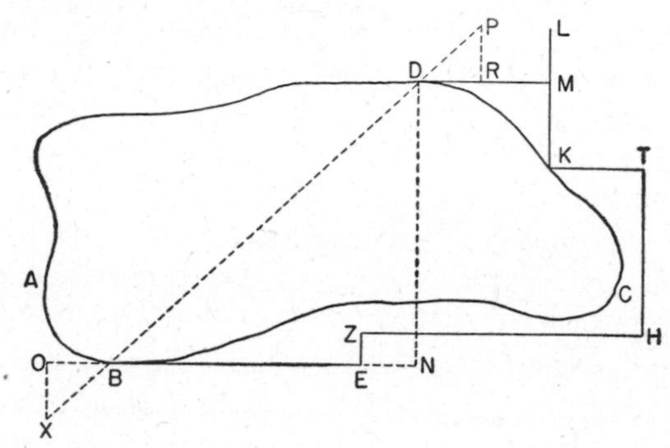

Consider line *ABCD* as representing the base of the mountain and *B* and *D* the openings through which it is required that we build the tunnel. Draw a straight line on the ground from *B*, e.g., *BE*, and from any point thereon, e.g., *E*, draw *EZ* perpendicular to *BE*; again, from any point on *EZ*, e.g., *Z*, draw *ZH* perpendicular to *EZ*; again, from any point on *ZH*, e.g., *H*, draw *HT* perpendicular to *ZH*; and from any point on *HT*, e.g., *T*, draw *TK* perpendicular to *TH*, and, similarly *KL* perpendicular to *TK*. We now move the dioptra along line *KL*,[2] until point *D* comes into sight in a direction perpendicular to *KL* when the dioptra is at *M*. Thus *MD* will be perpendicular to *KL*. Now suppose *EB* extended to *N* and *DN* perpendicular thereto. Now it is possible for us to compute *DN* from *EZ*, *HT*, and *KL*,[3] just as we computed the distance of a given point to another that could not be seen. So, too, we may compute *BN* from *BE*, *ZH*, *TK*, and *MD*.

Suppose, for example, that *BN* = 5*DN*. Now imagine that *BD* is drawn and extended to a point, *X*, and *XO* drawn perpendicular to *BE*.

[1] I.e., its elevation above any given accessible point. The accessible point here is *D*.

[2] I.e., keeping the line of sight always perpendicular to *KL*.

[3] I.e., *KM*, the effective part of *KL*.

Similarly, suppose that BD is extended to a point, P, and PR is a perpendicular to DM. Then $BO = 5OX$ and $DR = 5RP$.

Now if we take any point, e.g., O, on BE, and draw OX at right angles to BO, and make $OX = \frac{1}{5}BO$, then BX will, if produced, pass through D. Again if we make $PR = \frac{1}{5}DR$, DP will, if produced, pass through B.[1]

Thus we shall commence our tunnelling operations along line BX from B and along line DP from D. And, for the rest of the tunnel, we shall proceed by setting our direction line along the determined lines XB or PD or both. If the tunnel is dug in this way the workers [from the opposite ends] will meet.[2]

The Hodometer

Hero of Alexandria, *Dioptra* 34 (Schöne)[3]

The measurement of distances on land by the so-called hodometer is, in my opinion, a subject that appropriately follows the discussion of the dioptra. With the hodometer we can measure such distances without using the laborious and slow method of chain or cord, but merely by the number of revolutions of the wheels of our carriage as we ride. Our predecessors have indicated various methods of performing such measurements, but it will now be possible to pass judgment on these earlier methods as compared with our own.

Build a casing, in the form of a box, to hold the entire mechanism which we are now to describe. At the base of this framework place a bronze [disk] $ABCD$ to which are attached eight[4] little pegs of the kind described. Cut an opening in the base of the framework, and through this opening let a pin [E] attached to the hub of one of the carriage wheels protrude at each revolution of the wheel. Let this pin at each revolution push one of the pegs of the disk in such a way that the next following peg takes the position of the one before, and so on indefinitely. Consequently, when the carriage wheel has made eight revolutions, the disk with the pegs will have made a single revolution.

---

[1] I.e., the hypotenuse of the right triangle determined by OX and BO ($= 5OX$) will be collinear with the (imagined) line BD, as will also the hypotenuse of the right triangle determined by RP and DR ($= 5RP$). The text given by Schöne is slightly different and does not yield a satisfactory sense.

[2] Some time before 500 B.C., the engineer Eupalinus of Megara tunnelled a hill in Samos for the purpose of carrying water through pipes to the city. The tunnel, about 1,000 yards long, is described by Herodotus; it was rediscovered in 1882. The tunnelling operation was conducted from both ends with remarkable accuracy.

[3] This chapter on the hodometer has no connection with the essential subject matter of the *Dioptra* and may have been interpolated, by a later editor, from another work of Hero. Another description of a system of cogged wheels, used in a different connection, has also found its way into this treatise (p. 228, above).

[4] "Eight" is omitted from the Greek text, though in view of what follows it belongs there.

Now fit a screw [*F*] to the center of the aforesaid pegged disk [*ABCD*] and at right angles to it. Let the other end of this screw fit into a crossbeam fastened to the sides of the framework. Next to this screw place a cogged wheel [*G*] with cogs fitting the spiral thread of the screw [*F*] and lying, naturally, at right angles to the base of the framework. Now provide this wheel, too, with an axle [*H*] having its ends revolve in sockets in the sides of the framework.

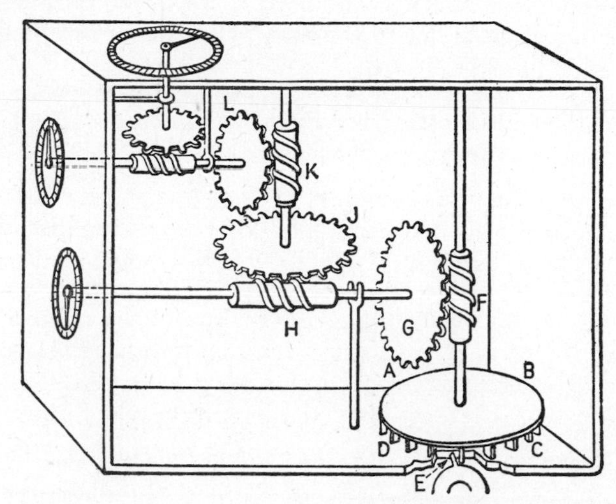

This axle should, in turn, have a spiral thread on one side and thus operate as a screw. Next to this screw place another cogged wheel [*J*], parallel, in this case, to the base and combined with an axle [*K*]. One end of this axle should revolve in a socket at the base of the framework, the other in a crossbeam fastened to the sides of the framework. Provide one part of this axle [*K*] with a screw thread to fit, in turn, the cogs of another wheel [*L*]. The latter wheel is, of course, at right angles to the base of the frame.

Repeat this same procedure as often as desired, or to the extent that the space available within the framework permits. For the larger the number of wheels and screws the larger will be the distances that may be measured. A complete revolution of a given screw will mean the motion of only a single cog's distance in the wheel next to it. Therefore, a single revolution of the screw [*F*] fitting the cogged disk [*ABCD*] involves eight revolutions of the carriage wheel itself, and the motion of only a single cog's distance in the wheel [*G*] next to this screw.

Now suppose that this wheel [*G*] has, say, 30 teeth. In that case a single revolution of the wheel [*G*] under the action of the screw [*F*] will signify 240 revolutions of the wheel of the carriage. Again, if the wheel [*G*] makes one revolution, the screw [*H*] attached to it will also revolve once, whereas the wheel [*J*] next to this screw will move only a single cog's distance. If, again, the wheel [*J*] also has 30 teeth—and it may well have that number and even more—a single revolution of this wheel [*J*] will indicate 7,200 revolutions of the carriage wheel. Now if the carriage wheel has a circumference of 10 cubits, the journey thus measured would be 72,000 cubits or 180 stades. And this result is indicated by the second cogged wheel.[1] But with a greater series of such wheels and an increased number of teeth in each, the length of the journey measured may be increased many fold.

Now the mechanism used need not be capable of measuring a distance much greater than can be traversed by a carriage in a single day. For it is possible each day, after reckoning the distance traversed, to make a fresh start for the recording of the next day's journey.

But since the revolution of each screw does not necessarily produce the precise motion of the adjacent cogs that we may have estimated,[2] we make a test by turning the first screw [*F*] until the cogged wheel [*G*] adjacent to it makes one complete revolution, and we measure the number of revolutions made by the screw [*F*]. Suppose, for example, that the latter revolves 20 times while the adjacent wheel [*G*] is making one revolution. Now this wheel had 30 cogs. Therefore 20 revolutions of the disk [*ABCD*] have produced a motion to the extent of 30 cogs in the wheel [*G*] next to the screw [*F*]. These 20 revolutions involve the turning of 160 cogs [in *ABCD*], which is the number of revolutions of the carriage wheel. The journey, therefore, is 1,600 cubits.

Now if motion to the extent of 30 cogs [in wheel *G*] indicates a journey of 1,600 cubits, the motion of one cog's distance in this wheel [*G*] will measure 53⅓ cubits of the journey. Similarly, when this cogged wheel has begun to revolve and is found to have turned to the extent of 15 cogs, a journey of 800 cubits, or two stades, is indicated. We inscribe, then, upon the center of this cogged wheel [*G*] "53⅓ cubits." We then make the same calculation for the other cogged wheels and inscribe the corresponding numbers. And so, when, in each case, a wheel has turned to the extent of a given number of cogs, we are able to calculate the distance traversed.

To obviate the necessity of opening the casing and examining the position of the cogs of each wheel whenever we wish to ascertain the distance traversed, we shall now show how it is possible to ascertain this distance by the use of dials and indicators on the outer surface of the casing. For while the aforesaid cogged wheels are so placed that they do not themselves touch the sides of the framework, their respective axles should protrude to

[1] I.e., by cogged wheel *J*. This does not include the disk *ABCD*.

[2] We may have estimated that every complete revolution of a screw will produce a motion of a single cog's distance in the cogged wheel next to it. But measurement is necessary to correct whatever error is involved in this estimate.

the outer part of the walls. The protruding parts of these axles should be of square cross section so that they may be fitted with indicators having openings of this form too; thus as the cogged wheel revolves with its axle the indicator will also revolve.

The point of the indicator will, as it revolves, describe a circle on the outside of the corresponding wall. This circle we may divide into the same number of parts as the number of cogs on the wheel within the casing. Now the indicator should be large enough to describe a circle of considerable size. In this way the distances between the points indicating the divisions may be correspondingly large. Upon the circle thus described by the indicator write the same number as was written upon the cogged wheel within. In this way we shall find recorded on the outer surface of the casing the distance traversed.

Cf. Vitruvius, *On Architecture* X. 9. 5–7.[1] Translation of M. H. Morgan

On board ship, also, the same principles may be employed with a few changes. An axle is passed through the sides of the ship, with its ends projecting, and wheels are mounted on them, four feet in diameter, with projecting floatboards fastened round their faces and striking the water. The middle of the axle in the middle of the ship carries a drum with one tooth projecting beyond its circumference. Here a case is placed containing a drum with four hundred teeth at regular intervals, engaging the tooth of the drum that is mounted on the axle, and having also one other tooth fixed to its side and projecting beyond its circumference.

Above, in another case fastened to the former, is a horizontal drum toothed in the same way, and with its teeth engaging the tooth fixed to the side of the drum that is set on edge, so that one of the teeth of the horizontal drum is struck at each revolution of that tooth, and the horizontal drum is thus made to revolve in a circle. Let holes be made in the horizontal drum, in which holes small round stones are to be placed. In the receptacle or case containing that drum, let one hole be opened with a small pipe attached, through which a stone, as soon as the obstruction is removed, falls with a ringing sound into a bronze vessel.

So, when a ship is making headway, whether under oars or under a gale of wind, the floatboards on the wheels will strike against the water and be driven violently back, thus turning the wheels; and they, revolving, will move the axle, and the axle the drum, the tooth of which, as it goes round, strikes one of the teeth of the second drum at each revolution, and makes it turn a little. So, when the floatboards have caused the wheels to revolve four hundred times, this drum, having turned round once, will strike a tooth of the horizontal drum with the tooth that is fixed to its side. Hence,

[1] After describing the hodometer in substantially the same terms as Hero, Vitruvius goes on to describe the method of measuring a journey on water. [Edd.]

every time the turning of the horizontal drum brings a stone to a hole, it will let the stone out through the pipe. Thus by the sound and the number, the length of the voyage will be shown in miles.

A Self-trimming Lamp

Hero of Alexandria, *Pneumatics* I. 34. Translation of J. G. Greenwood

To contrive a self-trimming lamp. Let *AC* be a lamp through the mouth of which is inserted an iron bar, *DE*, capable of sliding freely about the point *E*, and let the wick be wound loosely about the bar. Place near a toothed wheel *F*, moving freely about an axis, its teeth in contact

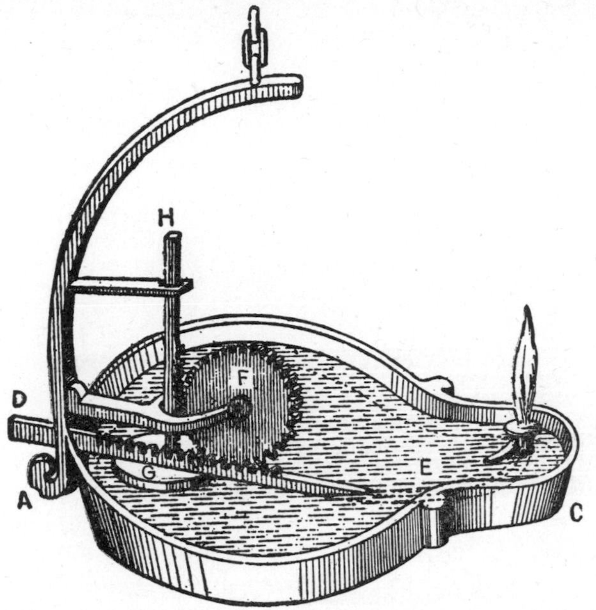

with the iron bar, that, as the wheel revolves, the wick may be pushed on by means of the teeth. Let the opening for the oil be of considerable width, and when the oil is poured in let a small basin float upon it, *G*, to which is attached a perpendicular toothed bar, *H*, the teeth of which fit into the teeth of the wheel. It will be found that, as the oil is consumed, the basin sinks and causes the wheel *F* to revolve by means of the teeth of the bar, and thus the wick is pushed on.

A Case of Cardan's Suspension

Philo of Byzantium, *Pneumatics* 56 (Carra de Vaux)[1]

The construction of an octagonal inkstand, a very ingenious device.—We may make an inkstand which is octagonal, hexagonal, square, pentagonal,

[1] This is possibly an interpolation of late date. It is essentially the device that enables an object, e.g., a ship's compass, to remain level even when its ultimate supporting frame is

or of any other prismatic form in which glass is shaped. This inkstand may be used for writing, no matter which face is uppermost. Regardless of how you place the inkstand it always presents on its upper face an opening to receive the pen, without any ink upsetting. You may insert the pen, obtain ink, and write. Suppose the inkstand is hexagonal, as you see it. In the interior there is a ring on a pivot *AB*; within this ring is another ring on another pivot *CD*. Within this second ring is a cup on a pivot *EZ*. It is this cup which is the inkwell.

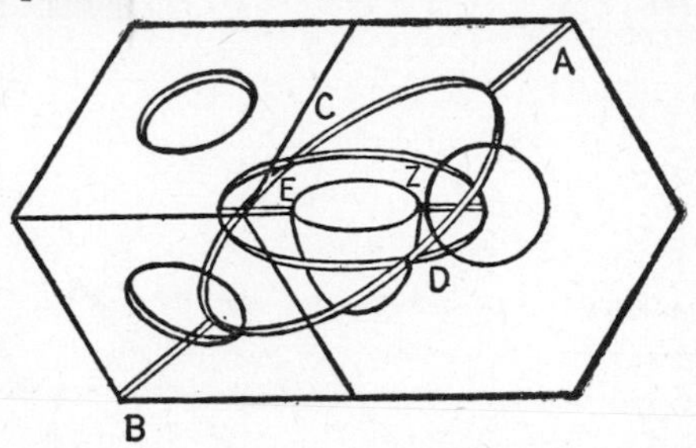

This device, it may be noted, is used among the Jews. Its construction resembles that of a censer which turns without losing its equilibrium. In constructing this inkstand measure accurately and make adjustments carefully so that, when you place it on any side, your pen will enter the top of the inkwell. The figure is appended.

Machines for Lifting

Vitruvius, *On Architecture* X. 2, 4–6.[1] Translation of M. H. Morgan

2. Hoisting Machines

First we shall treat of those machines which are of necessity made ready when temples and public buildings are to be constructed. Two timbers are provided, strong enough for the weight of the load. They are fastened together at the upper end by a bolt, then spread apart at the bottom, and so set up, being kept upright by ropes attached at the upper ends and fixed

---

disturbed. The successive pairs of pivots should be at right angles to each other, though this is not specified in the text.

[1] Marcus Vitruvius Pollio, Roman architect and engineer, flourished, according to the general view of scholars, in the Augustan Age. His treatise *On Architecture*, in ten books, deals with building materials, the rules of construction and decoration of public and private buildings, the water supply of cities, instruments of timekeeping, and machinery. While the chief emphasis is on architecture, there is a discussion of various engineering and technological matters. The work contains important material for a history of technology, as well as some material dealing with the history of pure science.

The treatise of Vitruvius had wide influence during the Renaissance, particularly in connection with the development of neoclassical architecture.

With his chief interest in the applications of science rather than in its theoretical aspects, Vitruvius is typical of the Roman viewpoint as distinguished from that which is customarily associated with Greek science. It should be noted, however, that the material on machines is substantially the same as that found in the treatises of Hero of Alexandria. Both authors seem to have made independent use of earlier Greek sources.

We have also given passages of Vitruvius in the sections on astronomical instruments, applications of acoustics, and the construction of artillery. Vitruvius also describes the water pump and the hydraulic organ of Ctesibius, of which we have given the Heronian descriptions (see pp. 329, 331). [Edd.]

at intervals all round. At the top is fastened a block, which some call a "rechamus." In the block two sheaves are enclosed, turning on axles. The traction rope is carried over the sheave at the top, then let fall and

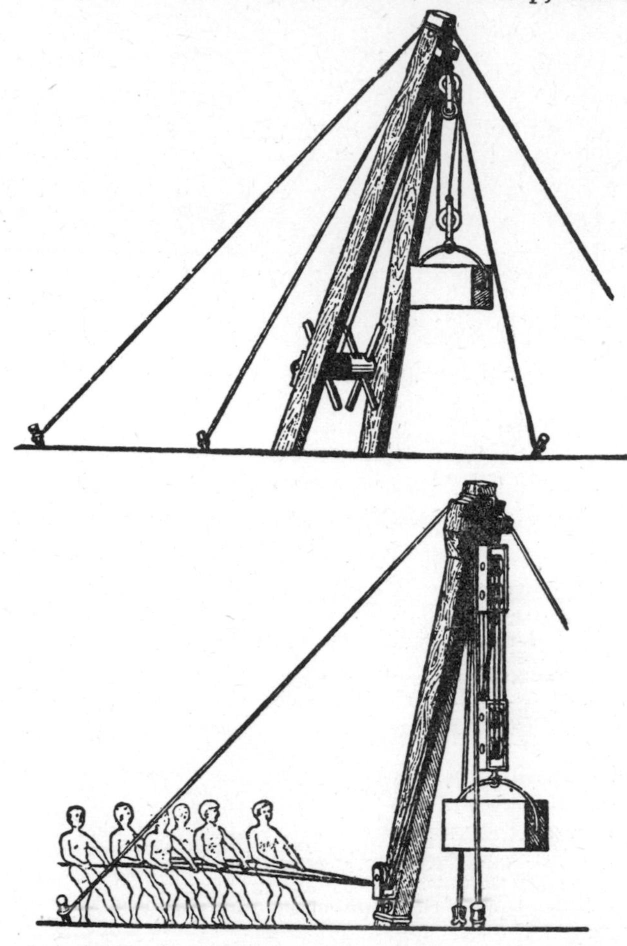

passed round a sheave in a block below. Then it is brought back to a sheave at the bottom of the upper block, and so it goes down to the lower block, where it is fastened through a hole in that block. The other end of the rope is brought back and down between the legs of the machine.

Socket-pieces are nailed to the hinder faces of the squared timbers at the point where they are spread apart, and the ends of the windlass are inserted into them so that the axles may turn freely. Close to each end of the windlass are two holes, so adjusted that handspikes can be fitted into them. To the bottom of the lower block are fastened shears made of iron, whose prongs are brought to bear upon the stones, which have holes bored in them. When one end of the rope is fastened to the windlass, and the latter is turned round by working the handspikes, the rope winds round the windlass, gets taut, and thus it raises the load to the proper height and to its place in the work.

This kind of machinery, revolving with three sheaves, is called a trispast. When there are two sheaves turning in the block beneath and three in the upper, the machine is termed a pentaspast. But if we have to furnish machines for heavier loads, we must use timbers of greater length and thickness, providing them with correspondingly large bolts at the top, and windlasses turning at the bottom. When these are ready, let forestays be attached and left lying slack in front; let the backstays be carried over the shoulders of the machine to some distance, and, if there is nothing to which they can be fastened, sloping piles should be driven, the ground rammed down all round to fix them firmly, and the ropes made fast to them. . . .[1]

4. Engines for Raising Water

I shall now explain the making of the different kinds of engines which have been invented for raising water and will first speak of the tympanum. Although it does not lift the water high, it raises a great quantity very quickly. An axle is fashioned on a lathe or with the compasses, its ends are shod with iron hoops, and it carries round its middle a tympanum made of boards joined together. It rests on posts which have pieces of iron on them under the ends of the axle. In the interior of this tympanum there are eight crosspieces set at intervals, extending from the axle to the circumference of the tympanum, and dividing the space in the tympanum into equal compartments.

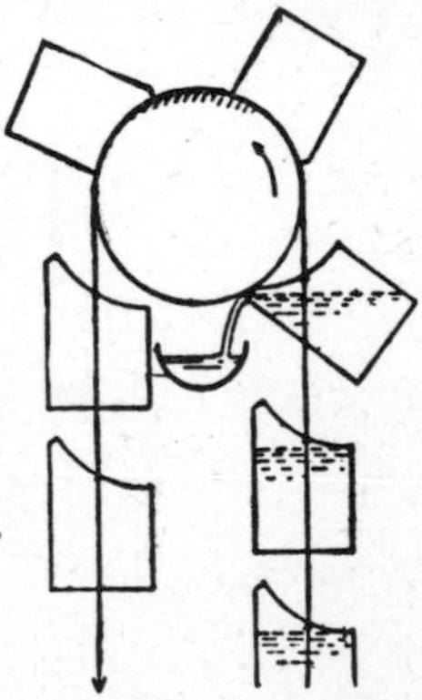

Planks are nailed round the face of it, leaving six-inch apertures to admit the water. At one side of it there are also holes, like those of a dovecot, next to the axle, one for each compartment. After being smeared with pitch like a ship, the thing is turned by the tread of men, and raising the water by means of the apertures in the face of the tympanum, delivers it through the holes next to the axle into a wooden trough set underneath, with a conduit joined to it. Thus, a large quantity of water is furnished for irrigation in gardens, or for supplying the needs of saltworks.

But when it has to be raised higher, the same principle will be modified as follows. A wheel on an axle is to be made, large enough to reach the necessary height. All round the circumference of the wheel there will be cubical boxes, made tight with pitch and wax. So, when the wheel is turned by treading, the boxes, carried up full and again returning to the

[1] Vitruvius proceeds to describe other varieties of hoisting machines, including the single-mast type shown in the second figure on p. 348. [Edd.]

bottom, will of themselves discharge into the reservoir what they have carried up.

But, if it has to be supplied to a place still more high, a double iron chain, which will reach the surface when let down, is passed round the axle of the same wheel, with bronze buckets attached to it, each holding about six pints. The turning of the wheel, winding the chain round the axle, will carry the buckets to the top, and as they pass above the axle they must tip over and deliver into the reservoir what they have carried up.

5. Water Wheels and Water Mills

Wheels on the principles that have been described above are also constructed in rivers. Round their faces floatboards are fixed, which, on being struck by the current of the river, make the wheel turn as they move, and thus, by raising the water in the boxes and bringing it to the top, they accomplish the necessary work through being turned by the mere impulse of the river, without any treading on the part of workmen.

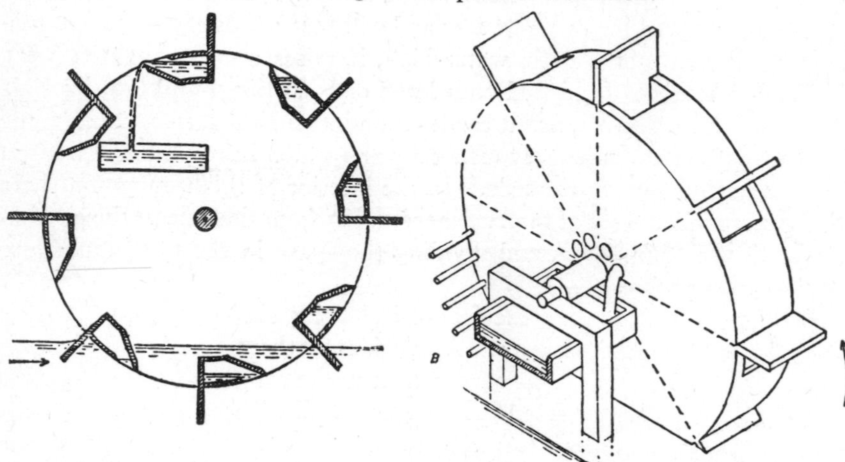

Water mills are turned on the same principle. Everything is the same in them, except that a drum with teeth is fixed into one end of the axle. It is set vertically on its edge, and turns in the same plane with the wheel. Next to this larger drum there is a smaller one, also with teeth, but set horizontally, and this is attached [to the millstone]. Thus the teeth of the drum which is fixed to the axle make the teeth of the horizontal drum move, and cause the mill to turn. A hopper, hanging over this contrivance, supplies the mill with corn, and meal is produced by the same revolution.

6. The Water Screw

There is also the method of the screw, which raises a great quantity of water, but does not carry it as high as does the wheel. The method of constructing it is as follows. A beam is selected, the thickness of which in digits is equivalent to its length in feet. This is made perfectly round. The ends are to be divided off on their circumference with the compass into eight parts, by quadrants and octants, and let the lines be so placed that, if the beam is laid in a horizontal position, the lines on the two ends may perfectly correspond with each other, and intervals of the size of one eighth part of the circumference of the beam may be laid off on the length of it. . . .[1]

[When the screw is completed] the beam is to be set up at an inclination corresponding to that which is produced in drawing the Pythagorean right-angled triangle: that is, let its length be divided into five parts; let three of them denote the height of the head of the screw; thus the distance from the base of the perpendicular to the nozzle of the screw at the bottom will be equal to four of those parts. A figure showing how this ought to be, has been drawn at the end of the book, right in the back.[2]

I have now described as clearly as I could, to make them better known, the principles on which wooden engines for raising water are constructed, and how they get their motion so that they may be of unlimited usefulness through their revolutions.

[1] There follow directions for preparing the screw thread and covering it with planks. Vitruvius mentions only the treadmill as a means of operation, though presumably other means could be used. [Edd.]

[2] This is the so-called "Archimedean screw," perhaps invented by Archimedes when he was in Egypt (Diodorus Siculus 1.34.2, 5.37.3). Series of such screws are supposed to have been used in irrigation and in drying up subterranean streams in mining operations. [Edd.]

CHEMISTRY AND CHEMICAL TECHNOLOGY

Modern chemistry, with its mathematical form and multitude of experiments with instruments of great precision, can hardly be said to have existed before Lavoisier, and in this sense is quite beyond the range of Greek science. We may, however, consider the Greek contribution in this general field from two quite distinct viewpoints. On the one hand we have philosophic reflections on the ultimate constituents of the physical world, on the other hand the technicians' familiarity with various substances and with procedures for the transformation of these substances.

Though the material representing the first viewpoint is generally available in the standard treatises and source books of Greek philosophy, a brief summary of the development may not be out of place here. Practical knowledge of the ways in which different substances could be combined or dissolved into their constituents had existed from time immemorial. The early Greek philosophers sought a theory of matter that would furnish a rational explanation of such combination and dissolution. Thus arose the traditional controversies on monism and pluralism, being and non-being, continuity and discontinuity (e.g., in relation to atomism), and the like. From the debates about the doctrines of Thales, Anaximander, Anaximenes, Heraclitus, the early Pythagoreans and Eleatics, there emerged the system of Empedocles with its four kinds of "roots" or elements, that of Anaxagoras with its infinitude of infinitely divisible "seeds," and then that of Leucippus and Democritus with its indivisible and indestructible atoms.

The influence of Aristotle made the Empedoclean type of theory of the elements the dominant one until the modern rise of atomism. From a consideration of pairs of opposite qualities, hot, cold, dry, and moist, Aristotle derived the pure substances, earth (cold and dry), water (cold and moist), air (warm and moist), and fire (warm and dry), equivalent in name, at least, to the four "roots" of Empedocles. To these Aristotle added a fifth substance, the aether of the heavenly bodies. The Aristotelian theory of matter and form, of the causes, of motion and the prime mover, with which all this doctrine is bound up, need not be entered into here. But it is noteworthy that the underlying unity, at least of non-celestial matter, that appears generally throughout Greek philosophy had an important bearing on the later development of alchemy and chemistry.

It may also be noted that on the question of change and permanence Aristotle's analysis of coming-to-be and passing-away in terms of his concepts of actuality and potentiality served to emphasize the two contrary viewpoints of what we should call "chemical combinations": (1) the theory of the atomists that the atoms in themselves are indestructible and unalterable and that all changes consist merely in the combinations and dissociations of atoms; (2) the theory of Aristotle that in combining to form a compound the constituent substances are neither destroyed nor do they persist, but have a potential type of existence, and that in this sense there may be a real coming-to-be of the compound and passing-away of its constituents, with a converse process at the dissolution of the compound.

Though the Aristotelian doctrines of matter and change triumphed generally in antiquity, the influence of atomism, not only in the form in which Democritus conceived it, but with the modifications on the one hand of Epicurus and his followers and on the other of Strato, should not be overlooked. But it was only in the nineteenth century, when atomism assumed a completely quantitative form, that it served science most successfully.

With reference to the second aspect from which we may consider the ancient Greek contribution to chemistry, the technological aspect, a few remarks may be made. It is hard to point to any particular chemical procedure as specifically a Greek invention. The archaeological evidence and some literary evidence, too, make it quite probable that in this field the Greeks added little to what they borrowed from Egypt and Mesopotamia. In these regions a high state of technological development had been reached in metallurgy, in the production of glass and glazed pottery, and in the art of dyeing fabrics. There are, however, important Greek and Latin writings in this general field from which we shall give selections. While alchemy, insofar as it is non-scientific, is outside the scope of this book, considerable technological material is found in some of the treatises that form the corpus of alchemical literature.

SOME LABORATORY PROCEDURES

Sublimation

Dioscorides, *De Materia Medica* I. 68. 7–8 (Wellmann)

Make the soot of frankincense thus. Bring the grains of frankincense one by one, holding them with a small pair of tongs, to a lamp and place them in a new hollow earthen vessel. Then place over this vessel a bronze cover hollow within and having holes at the middle and carefully wiped clean. Now place on one side or on both[1] small stones four dactyls in height in order to see whether the incense is burning and that there may always be an opening to add other grains. Before the first grain is completely burned out add another, continuing until you think you have a sufficient amount of soot. Always keep wiping the outside of the bronze cover with a sponge dipped in cold water.[2] In this way all the soot clings since the cover is not too warm. Otherwise the soot because of its lightness falls off and is mixed with the ashes of the frankincense. After removing the first soot repeat the process as often as desired, taking away on each occasion the ashes of the burned incense.

Distillation

Dioscorides, *De Materia Medica* I. 72.3; V. 95. 1–2 (Wellmann)[3]

Oil of pitch is obtained by separating off the watery part which rests on top as does whey. This watery part is removed in the cooking of the pitch by the placing of clean wool over it. This wool, becoming moist by the action of the rising steam, is squeezed into a vessel, and this takes place as long as the pitch is cooking. It performs the same functions as liquid pitch.

Quicksilver is prepared from so-called minium, which is also, though

[1] Between cover and vessel, the purpose being to raise the cover.

[2] To aid the condensation.

[3] The two passages here given show two stages in the development of the technique of distillation. For details of the type of apparatus used see F. S. Taylor, "A Survey of Greek Alchemy," *Journal of Hellenic Studies* 50 (1930) 109 ff.

incorrectly, called cinnabar.[1] An iron shell containing cinnabar is placed in an earthen vessel, and over the vessel is placed a flask which is luted all around with clay. A charcoal fire is then kindled under the vessel. Now the soot that clings to the cover after being scraped off and cooled becomes quicksilver. Quicksilver is also found in the form of drops on the ceilings where silver ore is worked. And some say that free quicksilver is found in mines. Quicksilver is kept in vessels of glass, lead, tin, or silver, for it eats into and melts every other material.[2]

The Water Bath

Dioscorides, *De Materia Medica* II. 77.2 (Wellmann)

Fresh softened [marrow] is treated like fat. Water is poured over it, the bones are picked out, and then it is strained through fine linen. It is washed in this manner until the water becomes clear. Next it is melted in a double pot,[3] and the scum which floats on top is skimmed off with a feather. Then it is strained into a mortar. After it becomes solid, the deposit is carefully scraped off, and it [the marrow] is placed in a fresh earthen vessel.

Cf. Theophrastus, *On Odors* 22. Translation of Sir Arthur Hort (London, 1916)

But in all cases the cooking [of spices in making perfumes], whether to produce the astringent quality or to impart the proper odour, is done in vessels standing in water and not in actual contact with the fire; the reason being that the heating must be gentle, and there would be considerable waste if these were in actual contact with the flames; and further the perfume would smell of burning.

METALS AND METALLURGY

Gold and Its Properties

Pliny, *Natural History* XXXIII. 58–65. Translation of K. C. Bailey. *The Elder Pliny's Chapters on Chemical Subjects*

I think that pride of place is given to this substance [gold], not by reason of its colour, for silver is brighter and more like sun-light (that is why it is more commonly used for military standards, since its gleaming is visible at a greater distance), and those who think it is the star-like hue of gold that charms are clearly mistaken, since no special importance is attached to such a colour in gems and other things; nor again is it preferred to the other metals by reason of its heaviness or its malleability, for in both it is

---

[1] See pp. 356ff.

[2] But quicksilver would form amalgams with tin and silver. In fact there is evidence that the reference to lead, tin, and silver is a late interpolation.

[3] What we have here is essentially the double boiler.

surpassed by lead,[1] but because it alone of all substances[2] loses nothing on heating, and survives even conflagrations and the funeral pyre. In fact, the oftener it is heated the better it becomes, and ignition is the test for gold, that on heating till red hot it should still keep its colour. The test is called *obrussa.*

But the best proof of purity is a high melting-point. It is strange too that a substance unsubdued by charcoal, made from the most fiercely-burning wood, is swiftly heated by chaff,[3] and that it is fused with lead to purify it.[4]

Another and more important reason for holding gold in esteem is that it is least worn away by use. On the other hand, lines can be drawn with silver, copper, and lead, and the hands become soiled by the material that wears off. Again, nothing can be beaten into thinner leaves nor divided more finely, for an ounce of it is hammered out into seven hundred and fifty, or even more, gold-leaves measuring four fingers each way.[5] The thickest of gold-leaves are called "leaf of Praeneste," where the image of Fortune, gilded most faithfully, still helps to maintain the name. The next type of leaf is called "quaestorian." In Spain, nuggets of gold are called *striges.*

Above all, it alone is found in nuggets or fine dust. The other metals, after discovery in the mines, have to be perfected by roasting, but gold is gold straight off. Its substance is perfect when found in this way, for the gold in this case is pure; in the other cases which we shall describe it has to be extracted. Finally, neither rust nor verdigris nor anything else can waste its excellence or diminish its weight.

Again, brine and vinegar, the conquerors of matter, make no impression

---

[1] This statement, as it stands, is clearly wrong. Gold is denser than lead and more malleable. The specific gravity of gold is 19.5 and of lead 11.3. . . .

[2] Pliny himself mentions asbestos (XIX. 20). Gold is appreciably volatile at temperatures higher than the melting point (1062° C.).

[3] The use of chaff for the melting of gold is mentioned by Pliny . . . and by Plutarch. . . . It is not as improbable as it seems at first sight that a straw fire should have greater power of fusion than a charcoal fire, for straw contains some of the oxygen necessary for its own combustion, and in a suitable furnace might easily give a flame which would be hot enough for the purpose.

[4] This method is still in use and is termed cupellation. The alloy of gold and lead is heated in air on a bone-ash hearth. The lead is oxidised to litharge, which is removed, partly by blowing from the surface of the molten metal, partly by absorption by the bone-ash. Gold (and silver if present) remains in the cupel.

[5] . . . If we consider 1 *uncia* as 27.28 gms. and the specific gravity of gold as 19.3, and if we take "four fingers" as 7.5 cms., we find that the thickness of one of Pliny's 750 sheets was about 0.000034 cms. If the gold was impure and less dense . . . , the thickness would be a trifle greater. Pliny's gold-leaf was then probably less than four times as thick as the best modern leaf. . . .

on it,[1] and, more than any other metal, it is spun and woven like wool—even without wool.

According to Verrius, Tarquinius Priscus triumphed in a golden tunic, and we ourselves have seen Agrippina, wife of the Emperor Claudius, sitting beside him when he was exhibiting a mimic sea-fight, clad in a cloak woven of gold without other material. Indeed the method of weaving gold into Attalic cloth[2] is old, and was invented by the kings of Asia.

On marble and other objects which cannot be heated white-hot, gold is laid on with white of egg; on wood with a glue called *leucophorum*,[3] made in accordance with a certain formula. Its composition and preparation will be described in its proper place.

It was legal to gild copper by means of quicksilver or at least hydrargyrum,[4] in place of which a fraudulent substitute has been devised, as we shall relate when we are describing their properties.

The copper is first subjected to the violence of fire; then when red-hot, quenched with brine, vinegar, or alum. The brightness of the surface is then used as a test whether the heating has been sufficiently prolonged, and the copper is dried again by heat so that, after polishing with a mixture of pumice and alum, it is fit to receive the gold-leaf laid on with mercury. Alum has purifying properties [in the case of copper] comparable to those already attributed to lead [in the case of gold].

Minium

Pliny, *Natural History* XXXIII. 111–124. Translation of K. C. Bailey

Another thing found in silver mines is minium,[5] still one of the most valued pigments, and formerly held not merely in honour but in veneration by the Romans. . . .

---

[1] Many reagents that will attack gold are now known, the most important being solutions containing a halogen (chlorine, bromine or iodine), and those (such as "aqua regia," a mixture of hydrochloric acid and nitric acid) which are capable of generating a halogen.

[2] It seems clear that Attalic cloth was embroidered with gold, not woven of gold. . . .

[3] Cf. XXXV. 36, where Pliny tells us that this is made by mixing together half a pound of Pontic *sinopis* (red ochre), ten pounds of clear *sil* (yellow ochre), and two pounds of Greek *Melinum* (a white earth, XXXV. 37), and grinding together for twelve days.

[4] See p. 359, below. [Edd.]

[5] It seems to be the general impression that ancient writers in general, and Pliny in particular, were unable to distinguish clearly the various red pigments from one another. A careful examination of par. 111–124 does not confirm this view.

Cinnabaris to Pliny meant "dragon's blood" . . . , and he condemns the popular error by which *minium* was styled *cinnabaris* . . . , tracing the confusion correctly to the Greek word κιννάβαρι, which was used (e.g., by Dioscorides) to indicate mercuric sulphide.

The true *minium* was sulphide of mercury. Its vivid colour (117 and 121), its capacity for recovering its hue after heating (121, though a much clearer account is given by Vitruvius VII. 9.5), the destruction of the red colour by heating with lime (121), its conversion into the

Theophrastus relates that the original discovery of minium was made ninety years before the archonship of Praxibulus (that is, in the three hundred and forty ninth year of our city),[1] by Callias, an Athenian, who hoped at first that gold could be extracted by roasting from the red sand which occurs in silver mines. He adds that, before his time, a hard and sandy variety had been discovered in Spain, and that it has also been found among the Colchians on a certain inaccessible crag, from which they knock it down by throwing javelins at it. This variety, however, was impure, and the best comes from the Cilbian plains beyond Ephesus, where the sand has the colour of the kermes berry.[2] This is ground to a powder, washed, and allowed to subside, the deposit being subjected to further washing.[3] Differences of procedure are found. Some, in preparing the minium, wash only once. Others find that this gives too weak a colour and wash a second time to get a superior product.

I am not surprised at the esteem in which the colour was held. For even in Trojan times red earth[4] was honoured, as we learn from Homer who makes his ships agreeable with it, although he rarely speaks elsewhere of paints and painting. The Greeks call the red earth *miltos* and minium they call *cinnabar*, whence mistakes have arisen on account of the name "Indian cinnabar." For this name is given by them to the gore of the dragon[5] crushed beneath the weight of the dying elephants, a substance in which, as we have already related, the blood of both animals was mingled, and which is the only colour which properly represents blood in painting. This kind of cinnabar is very useful for antidotes and medicines, but, by Hercules, because it also is called cinnabar, doctors employ that minium which we shall shortly show to be a poison. . . .

---

black modification under the action of the light (122), and its use as a source of mercury (123), are all noted.

The *secundarium minium* (119–120) is red lead, probably prepared from cerussite [p. 358, n. 1. Edd.], and Pliny seems to have recognized that the relation between red lead and lead was similar to that between iron rust and iron. . . . No tendency to confusion with kermes (*coccum*) or the red ochres (*rubrica*, *sinopis*, etc.) is evident. The only unquestionable blunder is the reference to roasting *minium* [p. 358, n. 3. Edd.]. . . .

We must conclude that Pliny exhibits remarkably little confusion in dealing with what was for early scientists a very complex subject.

[1] The date would correspond to 405/4 B.C. [Edd.]

[2] The kermes is a red scale-insect common in ever-green oaks. . . . The young insects which are red and worm-like, are collected, dried, and made into a paste with vinegar. Until the sixteenth century, syrup of kermes was used as an astringent medicine.

[3] Such methods of purification of powdered minerals by washing and decantation are in common use still.

[4] Probably our "red ochre," a general term including many earths coloured by ferric oxide.

[5] . . . The "dragon's blood" of the ancients was probably, like the modern article of that name, the resin of certain trees, notably the *Pterocarpus draco*. It has been used in medicine as an astringent.

There is a second kind,[1] which occurs in almost all silver and lead mines and is prepared by burning certain rocks which are mixed with the main vein of ore—not those rocks whose exudation we call quicksilver, for the product obtained by roasting them is also quicksilver—but others found simultaneously. These contain no quicksilver and are recognized only by their leaden colour, for they do not become red until they have been heated in the furnace, after which they are pounded to powder. This is a second-rate kind of minium, known to very few and far inferior to the native "sands" already mentioned. It is used in the company's workshops to adulterate genuine minium. *Syricum*, the manufacture of which will be described in its proper place, may also be employed at will. A calculation, however, made by weighing one against another reveals a minium that has been treated with *Syricum*. In another way, it is convenient for dishonest painters who repeatedly charge their brushes and then dip them in water. The pigment settles and remains at the bottom for the thieves.

Pure minium should have the brilliancy of kermes, but when the second-rate minium is used for mural paintings it experiences rusting, though it is itself the rust of a metal.

In the minium mines of Sisapo the vein of "sand" is pure and contains no silver. It is roasted like gold.[2] It is tested with red-hot gold,[3] for adulterated material darkens, while the pure pigment keeps its colour.[4] I find that lime also is used as an adulterant[5] and that the fraud can be detected in a similar manner, using a plate of hot iron if gold is not available.

Sun and moon injure a minium paint.[6] To prevent this, when the

---

[1] This kind of minium was probably red lead (Pb_3O_4), manufactured from cerussite or native lead carbonate ($PbCO_3$). The latter, as described in this section, has often a colour not unlike that of lead. When roasted, it first loses carbon dioxide and leaves litharge (PbO), which on further roasting (oxidation) gives red lead. The red colour would, therefore, only be obtained after roasting. Cerussite is still found in Spain, and usually occurs associated with galena (argentiferous lead sulphide).

[2] As the *minium* of Sisapo was mercuric sulphide, this statement is mistaken. . . .

[3] The gold is a support during heating, as is the iron mentioned a couple of lines later. We use platinum foil for similar purposes.

[4] Vitruvius (VII. 9.5) gives a more accurate account.

[5] This adulterant is specially mentioned by Vitruvius (VII. 9.5). When heated strongly, a mixture of lime and cinnabar darkens permanently, mercury being set free in accordance with the equation $4CaO + 4\,HgS = 4\,Hg + 3CaS + CaSO_4$. . . . It may be remarked that this test may be misleading in inexperienced hands. On *gentle* heating, the mixture of lime and cinnabar darkens, owing to change of the cinnabar into its black modification, but no chemical change takes place, and the colour is recovered on cooling.

[6] A. H. Church notes that in water-colour painting most vermilions are changed on exposure to solar rays, even in absence of air and moisture. W. D. Bancroft adds that varnish on a picture cuts off ultra-violet light to a great extent and thus protects the picture. . . . The action of the Punic wax and oil was no doubt similar.

wall is dry, hot Punic wax melted with oil should be applied with a brush, and then heated with a brazier of oak-apple charcoal till it sweats. It is rubbed with wax candles and finished with a clean cloth, just as marble statues are polished. Minium-refiners in the factory envelope their faces with loose bladders, which enables them to see without inhaling the fatal dust.[1] Minium finds an application also in inscriptions on rolls, and more vivid letters, even in sepulchral ornamentation, have been produced on gold and marble by its aid.

Of secondary importance is the fact that we have also obtained a substitute for quicksilver—hydrargyrum, consideration of which was postponed somewhat earlier in the book. There are two methods of preparation. The minium may be triturated with vinegar[2] in a copper mortar with a copper pestle, or else it may be put into an iron vessel inside an earthen oven and covered with a lid luted on with clay. It is then heated from beneath, and free use is made of the bellows. Then when the vapour, which condenses on the lid and combines the colour of silver with the fluidity of water, is brushed off, it divides easily into globules and the mobile drops rain down.

Cf. Vitruvius, *On Architecture* VII. 9.5

Minium is adulterated by the admixture of lime. And so if you should want to test its purity, you will have to proceed as follows. Take an iron plate, place the minium on it, and set it over a fire until the plate is glowing hot. Now when the heat has changed the color of the minium to black, take the plate from the fire. If, on cooling, the minium is restored to its original color, this will be proof of its purity. But if it remains black, that is a sign that it has been adulterated.

The Preparation of White Lead

Theophrastus, *On Stones* 56

A piece of lead the size of a brick is placed over vinegar in an earthenware vessel. When the lead has acquired a [rust-like] layer, which usually happens in ten days, they open the vessel and scrape off the decayed part. The process is then repeated again and again until the lead is entirely

---

[1] Continued breathing of either sulphide of mercury or red lead is injurious to health. Dioscorides (V. 109) gives a similar account of the use of bladders, but speaks of the danger of poisoning as occurring in the mines.

[2] The translator has repeated the experiment by grinding cinnabar (HgS) with vinegar and copper turnings.

In the cold the reaction is very slow indeed, but on heating it takes place readily enough, copper-sulphide being formed, while the surface of the copper turnings becomes covered with a copper-mercury amalgam, from which the mercury may be driven off by heating. Pliny follows Theophrastus. . . .

consumed. They take what has been scraped off and keep pulverizing it in a mortar and filtering it. What finally settles to the bottom is white lead.[1]

Cf. Vitruvius, *On Architecture* VIII. 6.10

Water from earthenware pipes is much more wholesome than from lead pipes. For it seems to be spoiled by the lead, since white lead is formed from it. And this is said to be harmful to the human body. Now if what is produced from a substance is harmful, clearly the substance itself is not wholesome.[2]

Imitation of Precious Metals

Leyden Papyrus X, selections.[3] Translation of E. R. Caley, *Journal of Chemical Education* 3 (1926) 1149–1166

1. Purification and hardening of lead.

[1] This is the usual interpretation. But it has been held (see K. C. Bailey on Pliny XXXIV. 175) that *psimynthion* here is an acetate of lead, on the ground that there is no final reaction with carbon dioxide to form white lead, i.e., lead carbonate. See p. 367, below.

[2] This seems to be a reference to lead poisoning. There are also references in the Hippocratic Collection and in Pliny, Nicander, and Celsus to this type of poisoning.

[3] There is a large corpus of alchemical literature in Greek containing the pseudo-Democritean treatises and the treatises and commentaries of Synesius (not the famous bishop), Zosimus, Olympiodorus, and many others. The technical problem of making base metals look like gold had long exercised the Egyptian craftsmen. With the impact of Oriental and Greek philosophy and mysticism upon the native technology, the problem became one of actually transmuting metals into gold. Thus arose alchemy in which Egyptians, Greeks, Syrians, and Arabs participated in turn, and which, in medieval times, extended over the Western as well as the Eastern world.

We have not thought it necessary to include here selections from alchemical works proper, such as form the bulk of Berthelot and Ruelle, *Collection des anciens alchimistes grecs* (Paris, 1888). We have, however, included selections from two documents of this corpus, the Leyden X and Stockholm papyri, which are predominantly concerned with the description of laboratory procedures.

Both papyri were written in the third century A.D. and were brought to light in Egypt in the early part of the nineteenth century. Papyrus Leyden X contains 111 recipes, the great bulk of which have to do with metals and alloys and with methods of imitating precious metals. Others deal with the dyeing of cloth, while still others are identical with sections in the *Materia Medica* of Dioscorides. The 152 recipes of the Stockholm papyrus are devoted mainly to precious stones and methods of imitating them, and to the dyeing of fabrics. The dye required in most cases is a purple to imitate the costly purple of the murex.

The recipes are in the form of brief notes, perhaps originally forming the notebooks of a practising technologist of Roman Egypt, though the procedures recorded in the third-century papyri are doubtless copied, in many cases, from much earlier records. This is further indicated by resemblances to older writings, e.g., to sections of Pliny's *Natural History*. The value of the recipes is very uneven. There is much repetition, and in some cases material seems to have been deliberately inserted to deceive the layman. On the other hand there is revealed a familiarity with a considerable number of alloys and amalgams and their peculiar

Melt it, spread on the surface lamellose alum[1] and copperas reduced to a fine powder and mixed, and it will be hardened.[2]

3. Purification of tin that is put into the alloy of asem.[3]

Take tin purified of any other substance, melt it, let it cool; after having covered it with oil, melt it again; then having crushed together some oil, some bitumen, and some salt, rub it on the metal and melt a third time; after fusion, put aside the tin after having purified it by washing; for it will be like hard silver. Then if you wish to employ it in the manufacture of silver objects, of such a kind that they cannot be found out and which have the hardness of silver, blend 4 parts of silver and 3 parts of tin and the product will become as a silver object.

5. Manufacture of asem.

Tin, 12 drachmas; mercury, 4 drachmas; earth of Chios, 2 drachmas. To the melted tin, add the crushed earth, then the mercury, stir with an iron, and put [the product] in use.[4]

6. Doubling of asem.

One takes: refined copper, 40 drachmas; asem, 8 drachmas; tin in buttons, 40 drachmas; one first melts the copper and after two heatings, the tin; then the asem. When all are softened, remelt several times and cool

---

properties, and, in the art of dyeing, some understanding of mordants and their preparation The methods of dyeing described in the Stockholm papyrus are essentially those used until the advent of our modern coal-tar dyes. Recipes like those of the papyri were transmitted from age to age and many of them are found unchanged in the medieval recipe books.

Difficulties of interpretation abound chiefly because many substances mentioned are unknown to us or known under different names. In fact, the precise notion of a substance that we have today had not yet developed in antiquity. And so we sometimes find the same name applied to more than one substance. In this connection it may be noted that important information of ancient technological procedures has been obtained by actual chemical analysis of metallic and glass objects preserved from antiquity.

A few changes, which seemed advisable, have been made in the translation. [Edd.]

[1] The word alum or rather "alumen" was employed by the ancient writers as a general term to signify a variety of products. Generally, they were impure mixtures of sulfates of iron and aluminum. This must have been widely used for purifying metals. . . .

[2] With a reducing agent present the copperas may be reduced with subsequent hardening. The translator suggests that, here and in many other cases where compounds were employed in the making of alloys, the furnaces were used under reducing conditions, or else it was understood that charcoal or wood was to be placed with the metals being fused. [Edd.]

[3] About a fourth of the recipes of this papyrus deal with the manufacture of asem. The Egyptian word originally denoted an alloy of silver and gold (electrum), but "asemon" came to be used to denote alloys intended to simulate gold or silver, especially the latter. In later alchemical writings the word was used to signify a substance employed in transmuting base metals into gold. [Edd.]

[4] Actually a tin amalgam is formed. The earth of Chios, probably a white clay, might aid in giving the appearance of silver. [Edd.]

by means of the preceding composition. After having augmented the metal by these proceedings, clean it with talc. The tripling is effected by the same procedure, the weights being proportioned in conformity with what has been stated above.[1]

11. Manufacture of asem.

Purify lead carefully with pitch and bitumen, or tin as well; and mix cadmia[2] and litharge[3] in equal parts with the lead, and stir until the alloy is completed and solidifies. It can be used like natural asem.

14. Manufacture of an alloy for a preparation.

Copper, 1 mina, melt and throw on it 2 minas of tin in buttons and work it thus.[4]

15. The coloration of gold.

To color gold to render it fit for use. Misy,[5] salt, and vinegar accruing from the purification of gold; mix it all and throw into the vessel [which contains] the aforementioned gold; let it remain some time, and then having drawn [the gold] from the vessel, heat it upon the coals; then again throw it in the vessel which contains the above-mentioned preparation; do this several times until it becomes fit for use.

16. Augmentation of gold.

To augment gold, take cadmia of Thrace, make the mixture with cadmia in crusts, or that from Galacia.[6]

17. Falsification of gold.

Misy and Sinopian red, equal parts to one part of gold. After the gold has been thrown in the furnace and has become of good color, throw upon it these two ingredients, and removing [the gold] let it cool, and the gold is doubled.[7]

---

[1] What is done is merely to form more asem from the copper and tin. [Edd.]

[2] The word "cadmia" was applied to condensed fumes and smoke gathered from the interior of copper and brass smelters and hence was often a complex mixture of metallic oxides. In a special sense it meant zinc oxide. "Natural asem" was the naturally occurring alloy of gold and silver known as electrum.

[3] The identification of Greek λιθάργυρος with our lead oxide has been questioned. [Edd.]

[4] A typical bronze is formed. [Edd.]

[5] According to Pliny, the "misy" of the ancients was either iron or copper pyrites or oxidation products of these, that is, basic iron or copper sulfates or various mixtures of these salts.

[6] This is apparently the beginning of a longer recipe. No. 17 seems to be the remainder of it. Berthelot has suggested that the title of No. 17 was a comment or gloss erroneously copied into the papyrus by a copyist.

[7] The "augmentation" is merely a case of adulteration with baser metals by the formation of alloys. "Sinopian red" seems to refer to red ochre, mainly ferric oxide. [Edd.]

19. [Manufacture of asem.]

Copper of Cyprus, 4 staters; earth of Samos, 4 staters; lamellose alum, 4 staters; common salt, 2 staters; blackened asem,[1] 2 staters, or if you desire to make it more beautiful, 4 staters. Having melted the copper, spread upon it the earth of Chios and the lamellose alum crushed together, stir until they are mixed; and having melted this asem, pour. Having mixed that which has just been melted with some [wood of] juniper, remove it; before so doing, however, burn it, and extinguish the product in lamellose alum and salt taken in equal parts, with some slimy water slightly thick; and if you wish to finish the work immerse again in the above-mentioned; heat so that [the metal] becomes white. Take care to employ refined copper beforehand, having heated it at the beginning and submitted it to the action of the bellows, until it has rejected its scale and become pure; and then use it as has just been stated.[2]

21. Treatment of hard asem.

A procedure to change black and hard asem into white and soft. Taking some leaves of the castor-oil plant infuse them a day in water; then soak [the metal] in the water before melting and melt twice and sprinkle with aphronitron.[3] And throw alum on the casting; put into use. It possesses quality for it is beautiful.[4]

24. Hardening of tin.

For hardening tin, spread separately [on its surface] lamellose alum and copperas; if, moreover, you have purified the tin as is necessary and have employed the materials previously named, in such a way that they did not escape by flowing away during the heating, you will have Egyptian asem for the manufacture of objects [of jewelry].[5]

25. Gold polish.

For treating gold, in other words for purifying gold and rendering it brilliant: misy, 4 parts; alum, 4 parts; salt, 4 parts. Pulverize with water. And having coated the gold [with it], place it in an earthenware vessel deposited in a furnace and luted with clay, [and heat] until the above-named substances have become molten; then withdraw it and scour carefully.[6]

---

[1] The text is doubtful here. [Edd.]

[2] A copper alloy similar to a bronze is formed. The "poling" of the liquid to remove oxides is noteworthy and is a process still employed. [Edd.]

[3] The word "aphronitron" was applied to a variety of saline efflorescences especially from dry or arid regions. Most probably it was a natural alkali which was essentially sodium carbonate.

[4] This recipe probably does not have the desired effect. [Edd.]

[5] This is a good method for hardening tin by the addition of copper to form a bronze. [Edd.]

[6] The metallic surface is cleaned by the hot bath of molten salts. [Edd.]

26. Purification of silver.

How silver is purified and made brilliant. Take a part of silver and an equal weight of lead; place in a furnace, and keep up the melting until the lead has just been consumed; repeat the operation several times until it becomes brilliant.[1]

27. Coloring in silver.

For silvering objects of copper: tin in sticks, 2 drachmas; mercury, 4 drachmas; earth of Chios, 2 drachmas. Melt the tin, throw on the crushed earth, then the mercury, and stir with an iron and fashion into globules.[2]

28. Manufacture of copper similar to gold.

Crush some cumin; pour on it some water, dilute, and let it stand for three days. On the fourth day shake, and if you wish to use it as a coating mix chrysocolla with it; and the gold will appear.[3]

31. Preparation of chrysocolla.

Solder for gold is prepared thus: copper of Cyprus, 4 parts; asem, 2 parts; gold, 1 part. The copper is first melted, then the asem and finally the gold.

32. To recognize the purity of tin.

After having melted, place some papyrus below it and pour; if the papyrus burns, the tin contains some lead.[4]

34. A procedure for writing in letters of gold.

To write in letters of gold, take some mercury, pour into a clean vessel, and add to it some gold in leaves; when the gold appears dissolved in the mercury, agitate sharply; add a little gum, 1 grain for example, and after letting stand, write in the letters of gold.[5]

36. Manufacture of asem that is black like obsidian.

Asem, 2 parts; lead, 4 parts. Place in an earthen vessel, throw on it a triple weight of unburnt sulfur, and having placed it in the furnace, melt.

---

[1] The silver-oxide coating is converted by this very ingenious method to lead oxide, which is then removed by wiping. [Edd.]

[2] The tin amalgam formed on the surface of the copper gives a silvery appearance. [Edd.]

[3] Or "it will appear gold." The copper is treated with a varnish consisting of gold particles suspended in gum. This is a good method. "Chrysocolla," which generally refers to malachite, seems here to refer to a gold alloy for soldering gold. See No. 31. [Edd.]

[4] An ingenious test. If the metal is pure tin, the melting point is quite low, and the papyrus will not burn. If lead is present, the melting point is raised sufficiently so that the papyrus will burn when the molten metal is poured on it. See also Pliny, *Natural History* XXXIV. 163. [Edd.]

[5] Enough mercury is present so that the gold amalgam remains liquid. The gum acts as a coating, which preserves the writing from wear. [Edd.]

And having withdrawn it from the furnace, beat, and make what you wish. If you wish to make figured objects in beaten or cast metal, then polish and cut. It will not rust.[1]

38. To give objects of copper the appearance of gold.

And neither touch nor rubbing against the touchstone will detect them, but they can serve especially for [the manufacture of] a ring . . .[2] Here is the preparation for this. Gold and lead are ground to a fine powder like flour, 2 parts of lead for 1 of gold; then when mixed, they are incorporated with gum. After that one coats the ring [with this mixture]; then it is heated. One repeats this several times until the object has taken the color. It is difficult to detect [the fraud], because rubbing gives the mark of a gold object, and the heat consumes the lead but not the gold.[3]

43. Testing of gold.

If you wish to test the purity of gold, remelt it and heat it: if it is pure it will keep its color after heating and remain like a piece of money. If it becomes white, it contains silver; if it becomes rougher and harder some copper and tin; if it blackens and softens, lead.[4]

44. Testing of silver.

Heat the silver or melt it, as with gold; and if it remains white and brilliant, it is pure and not false; if it appears black, it contains some lead; if it appears hard and yellow, it contains some copper.

57. [Gilding of silver.]

To gild silver in a durable fashion. Take some mercury and some leaves of gold, and make up into the consistency of wax; taking the vessel of silver, clean it with alum, and taking a little of the waxy material, lay it on with the polisher and let the material fasten itself on. Do this five times. Hold the vessel with a fine linen cloth in order not to soil it. Then taking some embers, prepare some ashes [and with them] smooth the vessel with the polisher, and use it as a genuine tested [gold] vessel.[5]

---

[1] What is actually formed is a mixture of metallic sulfides that is as black as obsidian but does not have its gloss and hardness. [Edd.]

[2] The text is uncertain. [Edd.]

[3] The gold is deposited on the surface and the portion of the lead that is not vaporized sinks down and alloys with the copper. The method is ingenious but probably not permanently effective. [Edd.]

[4] The translator notes that the method described here and in the next recipe, depending, as it does, on the discoloration of the metals by the formation of metallic oxides, would reveal only relatively gross adulterations. [Edd.]

[5] A silver-gold amalgam is formed as a coating over the silver. The text of the last sentence is doubtful; in any case the amalgam will not pass any good test for gold. [Edd.]

PIGMENTS AND DYES

Vitruvius, *On Architecture* VII. 10–12. Translation of M. H. Morgan

Artificial Colours. Black

1. I shall now pass to those substances which by artificial treatment are made to change their composition, and to take on the properties of colours; and first I shall treat of black, the use of which is indispensable in many works, in order that the fixed technical methods for the preparation of that compound may be known.

2. A place is built like a Laconicum, and nicely finished in marble, smoothly polished. In front of it, a small furnace is constructed with vents into the Laconicum, and with a stokehole that can be very carefully closed to prevent the flames from escaping and being wasted. Resin is placed in the furnace. The force of the fire in burning it compels it to give out soot into the Laconicum through the vents, and the soot sticks to the walls and the curved vaulting. It is gathered from them, and some of it is mixed and worked with gum for use as writing ink, while the rest is mixed with size, and used on walls by fresco painters.[1]

3. But if these facilities are not at hand, we must meet the exigency as follows, so that the work may not be hindered by tedious delay. Burn shavings and splinters of pitch pine, and when they turn to charcoal, put them out, and pound them in a mortar with size. This will make a pretty black for fresco painting.

4. Again, if the lees of wine are dried and roasted in an oven, and then ground up with size and applied to a wall, the result will be a colour even more delightful than ordinary black; and the better the wine of which it is made, the better imitation it will give, not only of the colour of ordinary black, but even of that of India ink.

Blue. Burnt Ochre

1. Methods of making blue were first discovered in Alexandria, and afterwards Vestorius set up the making of it at Puzzuoli. The method of obtaining it from the substances of which it has been found to consist is strange enough. Sand and the flowers of natron are brayed together so finely that the product is like meal, and copper is grated by means of coarse files over the mixture, like sawdust, to form a conglomerate. Then it is made into balls by rolling it in the hands and thus bound together for drying. The dry balls are put in an earthen jar, and the jars in an oven.

[1] Any highly carbonaceous material burned in the type of large enclosed furnace described by Vitruvius would produce the required soot. [Edd.]

As soon as the copper and the sand grow hot and unite under the intensity of the fire, they mutually receive each other's sweat, relinquishing their peculiar qualities, and having lost their properties through the intensity of the fire, they are reduced to a blue colour.

2. Burnt ochre, which is very serviceable in stucco work, is made as follows. A clod of good yellow ochre[1] is heated to a glow on a fire. It is then quenched in vinegar, and the result is a purple colour.

White Lead, Verdigris, and Artificial Sandarach

1. It is now in place to describe the preparation of white lead and of verdigris, which with us is called "aeruca." In Rhodes they put shavings in jars, pour vinegar over them, and lay pieces of lead on the shavings; then they cover the jars with lids to prevent evaporation. After a definite time they open them, and find that the pieces of lead have become white lead.[2] In the same way they put in plates of copper and make verdigris,[3] which is called "aeruca."

2. White lead on being heated in an oven changes its colour on the fire, and becomes sandarach.[4] This was discovered as the result of an accidental fire. It is much more serviceable than the natural sandarach dug up in mines.

Dioscorides, *De Materia Medica* V. 98 (Wellmann)

Blue vitriol (chalcanthum) is generically one and the same, for it is a liquid which has been solidified. But it appears in three different states. It appears as a concretion of liquids that filter drop by drop through the roofs of mines. It is for this reason called "stalacton" by those who work in the mines of Cyprus. Another type trickles down abundantly in caves, and is then led off into trenches where it solidifies. This is specifically called "solid" chalcanthum. The third type is called "boiled." This is the most colorless and the weakest kind. It is made in Spain and the method of preparing it is as follows. It is dissolved in water and cooked, then poured off into tanks and left to stand. After a definite number of days it solidifies, dividing into many cube-shaped particles, adhering to one another in clusters.[5]

The type that is heavy, blue, dense, clear, and transparent is considered

---

[1] Probably hydrated ferric oxide. [Edd.]

[2] May the wood shavings be the source of the requisite carbon dioxide, corresponding to the tanbark used in the so-called "Dutch process"? See p. 360, n. 1. [Edd.]

[3] I.e., basic copper acetate. [Edd.]

[4] Sandarach is usually identified with realgar, but here the meaning seems to be red lead, which is of similar appearance. [Edd.]

[5] This observation of crystallization is noteworthy.

the best. Of this type is "stalacton," called by others "lonchoton."[1] Next best is the "solid" type. The "boiled" type seems to be more useful than the others in the preparation of dyes, especially black.[2] Experience shows it to be quite ineffective in medicine.

Stockholm Papyrus, selections.[3] Translation of E. R. Caley, *Journal of Chemical Education* 4 (1927) 979–1002

17. Preparation of emerald.

Take and put so-called topaz[4] stone in liquid alum and leave it there 3 days. Then remove it from this and put it in a small copper vessel in which you have placed pure unadulterated verdigris along with sharp vinegar. Put the cover upon the vessel, close up the cover, and gently keep a fire under the vessel with olive wood for 6 hours, otherwise the longer you maintain the fire, the better and deeper will the stone be—only, as I say, with a gentle fire. Cool and lift the stone out. Its condition will show whether it has become emerald. That is to say, you will observe that a green film has formed upon it. Let it become slowly cooled, however; if not, it soon breaks. Put oil in a small box-tree vessel a sufficient number of days beforehand so that the oil is purified and the product from it can be taken off. Put in the stone and leave it under cover 7 days. On taking out you will have an emerald which resembles the natural ones.[5]

18. Manufacture of a pearl.

Take and grind an easily pulverized stone such as window mica. Take

---

[1] I.e., lance-shaped.

[2] Cf. Pliny, *Natural History* XXXIV. 123–127, on which K. C. Bailey writes:

"There can be no question that the compound whose properties were the origin of the name 'shoemakers' black' was green vitriol or ferrous sulphate (Fe $SO_4.7H_2O$). . . . On the other hand, the connection with copper is, as Pliny notes, indicated by the name *chalcanthos*. . . . Both Pliny and Dioscorides say that the best variety was blue, although there was a kind whose colour was lighter. The blue variety, in its pure form, was blue vitriol or cupric sulphate ($CuSO_4.5H_2O$), and no doubt many intermediate kinds were prepared, for ferrous and cupric sulphates will crystallize together from solution. The confusion has left its trace on chemical nomenclature, for ferrous sulphate is known to this day as 'copperas.' "

[3] See p. 360, n. 3. [Edd.]

[4] The interpretation is doubtful. E. O. von Lippmann has suggested that the reference is to a certain stone found in India (*Chemiker-Zeitung* 27 [1913] 963). [Edd.]

[5] The translator notes that in the imitation of precious stones, as practised in ancient Egypt, the base was so treated as to roughen it and make the surface porous. For this purpose oil, wax, alum, native soda, common salt, vinegar, calcium sulfide, or mixtures of these were used. After the stone was thus corroded a dye was applied. This general method was frequently used in the recipes of the Stockholm papyrus.

Here the stone is etched and green cupric acetate is absorbed into the pores. The green film that forms is quite impermanent, and though the treatment with oil makes the stone appear smoother, the resemblance to the natural gem cannot have been very striking. The method of annealing to prevent fracture of the stone is noteworthy. It is still in use. [Edd.]

gum tragacanth and let it soften for ten days in cow's milk. When it has become soft, dissolve it until it becomes as thick as glue. Melt Tyrian wax; add to this the white of egg and mercury. The mercury should amount to 2 parts and the stone 3 parts, but all remaining substances 1 part each. Mix [the ground mica and the molten wax] and knead the mixture with mercury. Soften the paste in the gum solution and the contents of the hen's egg. Mix all of the liquids in this way with the paste. Then make the pearl that you intend to, according to a pattern. The paste very shortly turns to stone. Make deep round impressions and bore through it while it is moist. Let the pearl thus solidify and polish it highly. If managed properly it will excel the natural.[1]

19. Production of ruby.

The treating of crystal so that it appears like ruby. Take smoky crystal and make the ordinary stone from it. Take and heat it gradually in the dark; and indeed until it appears to you to have the heat within it. Heat it once more in gold-founder's waste. Take and dip the stone in cedar oil mixed with natural sulphur and leave it in the dye, for the purpose of absorption, until morning.[2]

24. Corroding of stones.

A corrosive for any stone. Equal amounts of alum and natron are boiled in an equal amount of water. The small stones are then etched. Previously warm them slightly near the fire and dip them in the corrosive. Do this for a while once to three times while the corrosive boils; dip and leave again three times but no more, so that the small stones do not break.

31. Boiling of stones.

If you wish to make ruby from crystal, which has been worked to any desired end, take and put it in the pan and stir up turpentine balsam and a little pulverized alkanet there until the dye liquid rises; and then take care of the stone.[3]

74. Preparation of verdigris for emerald.

Clean a well-made sheet of Cyprian copper by means of pumice stone and water, dry, and smear it very lightly with a very little oil. Roll it thin and tie a cord around it. Then hang it in a cask with sharp vinegar so that it does not touch the vinegar, and carefully close the cask so that no evaporation takes place. Now if you put it in in the morning, then

---

[1] A white cement will form, but this should be quite distinguishable from a real pearl, besides being impermanent and easily disintegrated. [Edd.]

[2] A colloidal gold suspension reddish in color would probably form on the surface, but, unlike the real ruby, would be impermanent, opaque, and easily abraded. [Edd.]

[3] The dye is dissolved in the balsam, and the balsam coats the crystal. This is essentially lacquering. [Edd.]

scrape off the verdigris carefully in the evening, but if you put it in in the evening, then scrape if off in the morning, and suspend it again until the sheet becomes used up. However, as often as you scrape it off again smear the sheet with oil as explained previously. The vinegar is [thus rendered] unfit for use.[1]

84. A dye liquor for 3 colors.

A dye liquor from which three dye solutions can come. Bruise and mix with water ⅔ of a part of krimnos and 1 part of dyer's alum. Put the wool in and it becomes scarlet red. If it is to be leek-green add ground sulphur with water. If, however, it is to be quince-yellow then add unadulterated natron along with water.

86. For purple.

Boil asphodel and natron, put the wool in it 8 drachmas at a time, and rinse it out. Then take and bruise 1 mina of grape skins, mix these with vinegar and let stand 6 hours. Then boil the mixture and put the wool in.

87. Mordanting.

Boil chalcanthum and scorpiurus and employ for any desired color. These substances, however, also mordant all kinds of stones and skins.[2]

88. Dissolving of alkanet.

Alkanet is dissolved by oil, water, and nuts. The best of all dissolving mediums is, however, camel's urine. For this makes the alkanet dye not only fast, but also durable.

89. Another [recipe].

Bruise alkanet and mix natron with it until it becomes the color of blood. The boiling is done with water. Then dye what you desire. Or else bruise alkanet in the same way with safflower; afterwards put it in and let the blood color be absorbed. And if you bruise alkanet with telis[3] then proceed likewise. Alkanet in company with chalcanthum, however, dyes linen as well as cambric. For with chalcanthum, alkanet red changes into purple.

---

[1] Here we have a clever and effective method for producing basic copper acetate (verdigris). The oil prevents atmospheric oxidation without hindering the reaction of the copper with the acetic acid (see p. 367). [Edd.]

[2] The term "chalcanthum" was used to denote various products of the weathering of iron and copper pyrites and hence was either copper or iron sulfate or mixtures of these salts. The Greek word "scorpiurus" was, according to some, a name given to a sapindaceous plant. [A fairly good mordant is obtained. The iron sulfate will, on boiling, deposit ferric hydroxide in the fibers of the fabrics. This will mordant the dye. Edd.]

[3] Probably our fenugreek. [Edd.]

90. Making purple brilliant.

To make purple brilliant cook alkanet with purging weed and this will dissolve it; or with wild cucumber, purgative cucumber, or hellebore.

93. Mordanting for Sardian purple.

For a mina of wool put in 4 minas of dross of iron and 1 choenix of sour pomegranate; but if not this latter, then use 1 chus of vinegar and 8 chus of water heated over the fire until half of the water has disappeared. Then take the fire away from under it, put the cleaned wool in and leave it there until the water becomes cold. Then take it out, rinse it and it will be mordanted.[1]

94. Mordanting for Sicilian purple.

Put in the kettle 8 chus of water, a half a mina of alum, 1 mina of flowers of copper[2] and 1 mina of gall-nuts. When it boils put in 1 mina of washed wool. When it has boiled two or three times take the wool out. For if you leave it in a longer time, the purple becomes red. Take the wool out, however, rinse it out, and you will have it mordanted.[3]

95. Mordanting and dyeing of genuine purple.

For a stater of wool put in a vessel 5 oboli of alum and 2 kotylae of water. Boil and let it become lukewarm. Leave it until early morning, then take it off and cool it. Then prepare a secondary mordant, putting 8 drachmas of pomegranate blossoms and two kotylae of water in a vessel. Let it boil and put the wool in. However, after you have dipped the wool in several times, lift it out. Add to the pomegranate blossom water about a ball of alumed archil and dye the wool by judging with the eye. If you wish, however, that the purple be dark, add a little chalcanthum and let the wool remain long in it. In another passage it is put in the following way: But if you wish that the purple be dark, then sprinkle natron and a little chalcanthum in the dye bath.

101. Cold dyeing of purple which is done in the true way.

Keep this as a secret because the purple has an extremely beautiful luster. Take scum of woad from the dyer, and a sufficient portion of foreign alkanet of about the same weight as the scum—the scum is very light—and triturate it in the mortar. Thus dissolve the alkanet by grinding in the scum and it will give off its essence. Then take the brilliant color prepared by the dyer—if from kermes it is better, or else from krimnos—heat, and put this liquor into half of the scum in the mortar. Then put

---

[1] An excellent recipe for mordanting. Though wool does not, in general, require a mordant, the method would be effective with other fabrics. [Edd.]

[2] Impure cuprous oxide. [Edd.]

[3] This and the following are good recipes for mordanting. [Edd.]

the wool in and color it unmordanted and you will find it beyond all description.[1]

105. Dyeing in dark blue.

Put about a talent of woad in a tube, which stands in the sun and contains not less than 15 metretes, and pack it in well. Then pour urine in until the liquid rises over the woad and let it be warmed by the sun; but on the following day get the woad ready in such a way that you can tread around in it in the sun until it becomes well moistened. One must do this, however, for 3 days together.[2]

110. Dyeing in bright red purple.

To dye in genuine bright red purple grind archil and take 5 cyathi of the juice for a mina of wool. If you wish a bright tint, mix in ground natron; if you desire a still brighter one, chalcanthum.[3]

111. From the book of Africanus: Preparation of bright red purple.

Take and put the mordanted wool into 1 choenix of krimnos and 4 choenices of archil. Boil these materials, put the wool in, and leave it there until later. Take it out and rinse it with salt water, then with fresh water.[4]

133. Preparing genuine purples.

Iron rust, roasted misy, and pomegranate blossom adapt themselves to mordanting in water and make it possible to give the wool a good deep purple color in 4 hours.[5]

139. Dyeing of colors.

Celandine is a plant root. It dyes a gold color by cold dyeing. Celandine is costly, however. You should accordingly use the root of the pomegranate tree and it will act the same. And if wolf's-milk is boiled and dried it produces yellow. If, however, a little verdigris is mixed with it, it produces green; and safflower blossoms likewise.

153. Making of madder-purple.

After bluing, sprinkle the wool with ashes and trample it down with them in a convenient manner. Then press the liquid out of potter's clay and wash off the blued wool therein. Rinse it in salt water and mordant

---

[1] Dyeing will probably take place in this case. [Edd.]

[2] It appears that the solid woad is simply macerated in the urine mechanically, so that it is finely dispersed and consequently suitable for use in dyeing. [Edd.]

[3] This is simply a case of direct dyeing. The dye does not appear to be fast, according to our standards. [Edd.]

[4] Direct dyeing, as in the preceding recipe, with the use of a mordant. [Edd.]

[5] Iron rust will not dye wool. The misy may accomplish this result, and pomegranate blossoms probably will. [Edd.]

it. You will know if it is sufficiently mordanted when it sinks down in the kettle and the fluid becomes clear. Then heat rain water so that you cannot put your hand in it. Mix roasted, pulverized and sifted madder root, i.e., madder, with white vinegar, a half a mina of madder to a mina of wool, and mix a quarter of a choenix of bean meal with the madder root. Then put these in a kettle and stir up. Then put the wool in; in doing so, stir incessantly and make it uniform. Take it out and rinse it in salt water. If you wish the color to take on a beautiful gloss and not to fade, then brighten it with alum. Rinse the wool out again in salt water; let it dry in the shade and in doing so protect it from smoke.[1]

[1] Another case of direct dyeing. [Edd.]

GEOLOGY AND METEOROLOGY

Descriptions of the earth's surface and its transformations, of the stones and minerals below its surface, of the rivers, oceans, and mountains on its surface, and of the winds, rains, snows, tides, storms, and earthquakes that occur near its surface are found in numerous Greek and Latin authors. Much of this material is included in what the Greeks called "meteorology." The chief extant sources of the various earth sciences are the *Meteorologica* of Aristotle and various other parts of the Aristotelian Corpus, the fragments *On Stones* and *On Wind and Weather Signs* of Theophrastus, Seneca's *Natural Questions*, Pliny's *Natural History*, a large amount of material scattered among the historians, e.g., Herodotus and Diodorus, the geographers, e.g., Ptolemy and Strabo, and the philosophic naturalists. Among the poets, Lucretius is an important source.

PETROGRAPHY

Stones as Distinguished from Metals

Aristotle, *Meteorologica* III. 6, 378*a*18–*b*4.[1] Translation of E. W. Webster

We maintain that there are two exhalations, one vaporous the other smoky, and there correspond two kinds of bodies that originate in the earth, "fossiles" and metals. The heat of the dry exhalation is the cause of all "fossiles." Such are the kinds of stones that cannot be melted, and realgar, and ochre, and ruddle, and sulphur, and the other things of that kind, most "fossiles" being either coloured lye or, like cinnabar, a stone compounded of it. The vaporous exhalation is the cause of all metals, those bodies which are either fusible or malleable such as iron, copper, gold. All these originate from the imprisonment of the vaporous exhalation in the earth, and especially in stones. Their dryness compresses it, and it congeals just as dew or hoar-frost does when it has been separated off, though in the present case the metals are generated before that segregation occurs. Hence, they are water in a sense, and in a sense not. Their matter was that which might have become water, but it can no longer do so: nor are they, like savours, due to a qualitative change in actual water. Copper and gold are not formed like that, but in every case the evaporation congealed before water was formed. Hence, they

[1] In the first three books of the *Meteorologica* Aristotle treats the subjects that were usually included in that branch, (1) phenomena above the earth, e.g., comets, meteors, rain and snow, thunder and lightning, rainbows and halos, and (2) matters of physical geography, e.g., seas, rivers, winds, earthquakes, etc. The fourth book is thought to be by a contemporary and disciple of Aristotle. It is concerned not with meteorology in the usual sense but with certain physical and what we should call chemical processes such as putrefaction, ripening, boiling, broiling, solidification, and liquefaction. [Edd.]

all (except gold) are affected by fire, and they possess an admixture of earth; for they still contain the dry exhalation.[1]

Theophrastus, *On Stones* 1–8[2]

Of the things formed in the ground, some have as their basis water, others earth. Those whose basis is water are the metals, silver, gold, and the like; those whose basis is earth are stone,[3] including the rarer species, and also those types of earth itself which are peculiar by reason of their color, smoothness, density, or some other quality. The discussion on metals has been given in another work. In this treatise we shall now consider stones.

All these [stones] must be considered, to put it simply, to be formed from some pure and homogeneous substance. This substance may come into existence by means of a conflux or a percolation, or, as was observed above, may be separated out in some other manner. The stone may be formed in the various cases by any one of these means. From these [differences stones get] their smoothness, density, brightness, transparency, and other such qualities; and the more uniform and pure each stone is, the more does it have these properties. For, in general, these qualities depend on the perfection of the concretion or coagulation.

The concretion arises, in some substances, from heat, in others from cold. And there is no reason why some stones should not be formed as a result of both these causes (although earth products would all seem to have been formed by fire), since solidification and melting in each case are effected by opposite causes.

Stones show a large number of individual differences. In "earths" most variations are due to color, cohesion, smoothness, density, and the like, while other kinds of variation are rare; but stones show these differ-

---

[1] Here we have Aristotle's attempt to link the origin of stones and metals with his doctrine of the four primary qualities, i.e., hot, cold, moist, and dry. [Edd.]

[2] Though the ancients made widespread use of stones and minerals in various industries, they made little progress in what we call scientific mineralogy and petrology. It was not until the rise of chemistry and crystallography that great advances were made in these branches. The chief writings that have survived from antiquity are a considerable fragment of Theophrastus' treatise *On Stones* and Books 33–37 of Pliny's *Natural History.*

Theophrastus sought to lay a foundation for the classification of stones and minerals. His work treats the material rationally with only an occasional reference to the widespread magical lore on stones such as is found in the later Alexandrian treatises. Though in some cases it is difficult to identify the objects he is describing, because they are not pure substances, the descriptions are usually sufficient to permit identification. The classification according to the reactions to fire is particularly interesting.

The passages given here set forth general considerations about the origin and classification of stones and some discussion of particular varieties.

[3] The reference is to the Aristotelian classification. See the preceding selection.

ences and, in addition, differences due to the faculty of acting [upon other bodies], or of being acted upon or not. For example, they are soluble or not soluble, combustible or not combustible, and so on. In addition, they also show many differences when they are exposed to the action of fire.[1] Further, some are said to have the power of making water appear of their own color, as the emerald; others of turning completely into stone whatever is put into [vessels made of] them;[2] others of attracting; and still others of testing gold and silver, as the so-called Heraclian stone and the Lydian stone. Now the most marvellous and greatest quality of stones (if the account is true) is that of stones which bring forth young.[3] But better known and of wider application is the matter of the workability of stones; for stone may be carved, others turned on a lathe, and others sawn. Yet there are some which no iron instrument will affect at all, and others which are cut with difficulty, or scarcely at all.

Classification of Stones According to the Action of Fire on Them

Theophrastus, *On Stones* 9–19

Some stones are melted by the action of fire and flow in it like metals. For metal-bearing ores melt along with the silver or copper or iron that they contain, either because of the moisture inherent in them or because of the metals themselves. And so pyromachus and millstone[4] melt with the material the kindlers place on them. Some absolutely affirm that all stones are fusible except marble, which burning reduces to lime. But this would seem to be an exaggeration. For many stones are shattered and split while resisting fire, as, for example, potter's clay. And this is understandable, for it is completely dry; for whatever is fusible must be humid and contain considerable moisture. Indeed some stones are said to dry up entirely on exposure to the sun and to become useless unless they are soaked and made wet again, while others, under the same conditions, become softer and more easily broken. It is clear that in both cases moisture is extracted, but the compact stones harden as they dry, while those of loose texture and of a similar [i.e., less compact] structure can be broken down and dissolved.

Some of these breakable stones remain in the fire quite a long time and burn to ashes, like the stones near Binae, which are found in mines and are washed down by the river. For they become kindled when [burning]

[1] See the following selection.

[2] The reference is to "sarcophagus," a type of limestone said to have been found at Assos in Asia Minor and to have the properties here ascribed to it. See K. C. Bailey, *The Elder Pliny's Chapters on Chemical Subjects* II. 251–252.

[3] What stones are referred to is doubtful. The reference may be to some aspect of crystallization. See K. C. Bailey, *op. cit.*, II. 253.

[4] The precise identification of the pyromachus and millstone is uncertain.

coals are placed on them, and they burn so long as any one blows upon them. Then they die down; but they may be made to burn again. For this reason they may be used over a long period of time, but their odor is very pungent and unpleasant.

The stone called "spinus,"[1] which is found in the same mines, burns when broken and piled up and exposed to the sun; and it burns even more readily if it is moistened or sprinkled with water.

The Liparian stone[2] is worn into holes by burning and becomes like pumice-stone, so that it changes both its color and its density. For before it is burnt it is black and smooth and compact. This stone is found in pumice, distributed here and there as in cells, but not continuous. Similarly the pumice in Melos is said to be found in another stone. . . .

There is a stone in Tetras in Sicily, a place opposite Lipara, which is also worn into holes by burning. And there is a large quantity of stone on the promontory called Erineas. Like that found near Binae, this stone gives forth a bituminous odor while burning, and when burnt, it leaves an ash like burnt earth.

Those substances called "anthrax," which are dug up for use are earthy; they kindle and burn like charcoal. They are found in Liguria, where there is also amber, and in Elis, on the way to Olympia through the mountains. They are used by smiths.

Formerly a stone was found in the mines of Scaptesyle which looked like rotten wood; but when oil was poured on it, it burned. And when the oil was burned away, the stone too stopped burning as if it had not been affected.

These then are substantially all the different reactions of stones to fire.

There is another kind of stone of a quite different nature that is entirely incombustible. It is [also] called "anthrax."[3] From it signets are carved; it is red in color, and when held against the sun assumes the glow of a burning coal. It is very valuable; a very small one costs 40 gold staters. It comes from Carthage and Massilia.

The angular stone, including also the hexagonal, which is found near Miletus, is also incombustible. It too, strangely enough, is called anthrax.[4] The same is true also of the diamond. These stones seem to be unlike pumice and ashes in that they never contained any moisture; while pumice and ashes are incombustible and resist the action of fire because the moisture in them has been removed. Indeed, some hold that all pumice results from burning, except for a variety formed by [the concretion of] the foam of the sea.

---

[1] Perhaps a coal of decomposed pyrite or marcasite.

[2] Perhaps obsidian.

[3] The reference is probably to garnet.

[4] Here the reference is probably to a reddish ferruginous quartz.

GEOLOGIC CHANGES: TRANSFORMATIONS OF THE SURFACE OF THE EARTH

Hippolytus, *Refutation of All Heresies* I. 14.5–6 (Diels, *Fragmente der Vorsokratiker* I[5]. 122)

Now Xenophanes[1] holds that the earth is recurrently mingled with the sea and then as time passes is freed again of moisture. He puts forth such proofs as these: that shells are found far inland and on mountains; and that in the quarries of Syracuse imprints of a fish and of seaweed[2] have been found, and in Paros the imprint of a small fry[3] deep in the stone, and in Malta flat slabs [bearing the impressions] of all sorts of fish. He says that the imprint, made long ago when everything was covered with mud, then dried in the mud.

Aristotle, *Meteorologica* I. 14. Translation of E. W. Webster

The lower land [of Egypt] came to be inhabited later than that which lay higher. For the parts that lie nearer to the place where the river is depositing the silt are necessarily marshy for a longer time since the water always lies most in the newly formed land. But in time this land changes its character, and in its turn enjoys a period of prosperity. For these places dry up and come to be in good condition while the places that were formerly well-tempered some day grow excessively dry and deteriorate. This happened to the land of Argos and Mycenae in Greece. In the time of the Trojan wars the Argive land was marshy and could only support a small population, whereas the land of Mycenae was in good condition (and for this reason Mycenae was the superior). But now the opposite is the case, for the reason we have mentioned: the land of Mycenae has become completely dry and barren, while the Argive land that was formerly barren owing to the water has now become fruitful. Now the same process that has taken place in this small district must be supposed to be going on over whole countries and on a large scale. . . .

Since there is necessarily some change in the whole world, but not in the way of coming into existence or perishing (for the universe is permanent), it must be, as we say, that the same places are not for ever moist through the presence of sea and rivers, nor for ever dry. And the facts prove this. The whole land of the Egyptians, whom we take to be the most ancient of men, has evidently gradually come into existence and been produced by the river. This is clear from an observation of the country, and the facts about the Red Sea suffice to prove it too. One of their kings tried to make

---

[1] Xenophanes of Colophon (*ca.* 570–*ca.* 475 B.C.) is primarily known as a philosophical poet and opponent of traditional religion.

[2] Adopting a conjecture of Gomperz. The reading of the manuscripts would give "seals."

[3] Adopting a conjecture of Gronov. The reading of the manuscripts would give "bay-leaf." See also D. W. Thompson, *A Glossary of Greek Fishes*, pp. 95f.

a canal to it (for it would have been of no little advantage to them for the whole region to have become navigable; Sesostris is said to have been the first of the ancient kings to try), but he found that the sea was higher than the land. So he first, and Darius afterwards, stopped making the canal, lest the sea should mix with the river water and spoil it. So it is clear that all this part was once unbroken sea. For the same reason Libya—the country of Ammon—is, strangely enough, lower and hollower than the land to the seaward of it. For it is clear that a barrier of silt was formed and after it lakes and dry land, but in course of time the water that was left behind in the lakes dried up and is now all gone. Again the silting up of the lake Maeotis by the rivers has advanced so much that the limit to the size of the ships which can now sail into it to trade is much lower than it was sixty years ago. Hence it is easy to infer that it, too, like most lakes, was originally produced by the rivers and that it must end by drying up entirely.

Again, this process of silting up causes a continuous current through the Bosporus; and in this case we can directly observe the nature of the process. Whenever the current from the Asiatic shore threw up a sandbank, there first formed a small lake behind it. Later it dried up and a second sandbank formed in front of the first and a second lake. This process went on uniformly and without interruption. Now when this has been repeated often enough, in the course of time the strait must become like a river, and in the end the river itself must dry up.

So it is clear, since there will be no end to time and the world is eternal, that neither the Tanais nor the Nile has always been flowing, but that the region whence they flow was once dry: for their effect may be fulfilled, but time cannot. And this will be equally true of all other rivers. But if rivers come into existence and perish and the same parts of the earth were not always moist, the sea must needs change correspondingly. And if the sea is always advancing in one place and receding in another it is clear that the same parts of the whole earth are not always either sea or land, but that all this changes in the course of time.[1]

The Rise of the Nile

Seneca, *Natural Questions* IV. 2.16–30.[2] Translation of John Clarke (London, 1910)

But I must now go on to inquire into the explanations of the occurrence

[1] For additional material on geologic changes see Herodotus II. 10–12, Strabo I. 3.4, and Ovid, *Metamorphoses* XV. 262 ff. [Edd.]

[2] Lucius Annaeus Seneca (*ca.* 4 B.C.–A.D. 65), son of Lucius(?) Annaeus Seneca, the rhetorician, was born at Cordova in Spain, embarked upon a political career at Rome, and in 48 became tutor of the young Nero. In 65, accused of participation in a conspiracy against the Emperor, he committed suicide.

He was the author of tragedies, moral epistles, and philosophical essays. From the point

of the rise of the Nile in summer;[1] and I will begin with the most ancient of them. Anaxagoras asserts that the snow melting on the peaks of Ethiopia is constantly running down to the Nile. All antiquity shared the same view, which is recorded by Aeschylus, Sophocles, and Euripides. But many proofs make it plain that it is a mistaken one. First of all, the blackened complexion of the people shows that Ethiopia is exceedingly hot. So do the habits of the Troglodytes (cave-dwellers), who for coolness have underground houses. The rocks glow with heat as if a fire had been applied, and that, not only at mid-day, but even toward nightfall. The dusty ground is so hot that no foot of man can endure it. Silver is unsoldered. The joints of statues are melted. No coating of plated metal will stick on. The south wind, too, coming from that tract of country, is the hottest of all winds. None of the animals that go to earth in winter ever hibernates there. Even in midwinter the serpent is seen above ground in the open. At Alexandria, too, which lies far north of this excessive heat, snow does not fall; but the upper regions have not even rain.

How then, I ask, could a district exposed to such broiling heat receive a snowfall sufficient to last through a whole summer? No doubt some of the mountains in Ethiopia, as well as elsewhere, intercept snow; but there can never be a greater fall than in the Alps, or the peaks of Thrace, or the Caucasus. It is in spring, however, or early summer, that the rivers that flow from the European mountains are swollen; subsequently during winter time they decrease. The reason, of course, is that the rains in spring wash off so much of the snow, and the first heat of summer soon scatters the remnants. Neither the Rhine, nor the Rhone, nor the Danube, nor yet the Caystrus is liable to the catastrophe of an overflow in winter; their increase is in summer, though in those northern peaks where they rise the snow lies very deep. The Phasis, too, and the Dnieper would swell during summer if snows had the power of raising the rivers high in spite of the heat of that season. Besides, if this were the cause of the flooding of the Nile, its stream would be fullest in early summer; for that is the period when the snow is deepest and least impaired, and when from its softness

---

of view of the history of science, however, his most significant work is the *Natural Questions*, a treatise extant in seven (though probably composed in eight) books, dealing with a whole range of problems mainly in meteorology and physical geography. In this treatise Seneca sets forth the views of various ancient schools but adheres mainly to those of the Stoic school as formulated by Posidonius. The topics treated include meteoric fires, halos, rainbows, thunder and lightning, terrestrial waters (with a special treatment of the Nile), snow, hail, rain, winds, earthquakes, and comets. [Edd.]

[1] Few scientific questions were as widely discussed in antiquity as the cause of the annual rise of the Nile. The present discussion of Seneca, which ends so abruptly as to suggest that it was never completed or that there is a lacuna in the text, takes up some of the ancient theories. The true cause, in one sense, is the heavy seasonal rainfall in Ethiopia. [Edd.]

the thaw is quickest. The Nile, however, has a regular increase to its stream during four months.[1]

If one may believe Thales, the Etesian winds[2] hinder the descent of the Nile and check its course by driving the sea against its mouths. It is thus beaten back, and returns upon itself. Its rise is not the result of increase: it simply stops through being prevented from discharging, and presently, wherever it can, it bursts out into forbidden ground. Euthymenes of Marseilles[3] bears corroborative testimony: I have, he says, gone a voyage in the Atlantic Sea. It causes an increase in the Nile as long as the Etesian winds observe their season. For at that period the sea is cast up by pressure of the winds. When the winds have fallen, the sea is at rest, and supplies less energy to the Nile in its descent. Further, the taste of that sea is fresh, and its denizens resemble those of the Nile. Now, if the Etesian winds, as alleged, stir up the Nile, why, I should like to know, does its rise begin before them and last after them? Moreover, it does not rise higher in proportion to the violence of their blast. Nor does it swell and fall according as they blow furiously or gently. All which would happen if it derived from them the strength of its increase. Then, again, the Etesian winds beat on the shore of Egypt, and the Nile comes down in their teeth: whereas, if its rise is to be traced to them, the river ought to come from the same quarter as they do.[4] Furthermore, if it flowed out of the sea, its waters would be clear and dark blue, not muddy, as they are. Add to this that Euthymenes' evidence is refuted by a whole crowd of witnesses. At such a time when foreign parts were all unknown, there was opportunity for falsehood: people like Euthymenes had scope for giving currency to travellers' myths. But nowadays the whole coast of the sea beyond Gibraltar is visited by trading vessels: none of the traders tell us that the Nile rises there, or that the sea in the Atlantic tastes differently from what it does elsewhere. The very nature of the sea forbids belief in the story that it is fresh: the freshest water is always lightest, and as such attracted by the sun in evaporation: the residuum, sea, must be salt. Besides, why, on this theory, does the Nile not rise in winter? The sea may be raised at that season by storms too, which are considerably greater than the Etesians; the latter are comparatively moderate in their force. Besides if the source were derived from the Atlantic Ocean, Egypt would be

[1] The occurrence of maximum height at different times for different points along the course of the Nile is not mentioned by Seneca. [Edd.]

[2] The etesian or northerly winds blow with some regularity in the Mediterranean during the summer months. [Edd.]

[3] How this navigator (who may have lived at the close of the sixth century B.C.) came to connect the Atlantic with the Nile is not known, if indeed the error is his. [Edd.]

[4] Perhaps Seneca did not quite understand the argument, which held merely that the winds by blowing contrary to the direction of the river dam it up. [Edd.]

flooded all at once; but, as a matter of fact, the increase is very gradual.

Oenopides of Chios has another explanation: he says that in winter heat is stored up under the ground; that is why caves are then warm, and the water in wells is less cold. The veins of water are dried up by this internal heat, he thinks. In other countries rivers swell through rain: but the Nile, being aided by no rainfall, dwindles during the rainy season of winter, and by and by increases in summer, a season at which the interior of the earth is cold, and the frost returns to the springs. Now, if that were true, rivers in general would increase in summer, and all wells would then have greater abundance of water. Besides, it is not true that there is an increase in the heat underground in winter. Water and caves and wells are warm at that season because they do not admit the frosty air from without. Thus, they do not possess heat, they merely exclude cold. For the same reason they are chilly in summer, because the air heated by the sun is drawn off to a distance, and does not penetrate to them.

The next account is that of Diogenes of Apollonia.[1] It runs thus: The sun attracts moisture; the earth drained of it replenishes its supply in part from the sea, in part from other water. Now, it is impossible that one land should be dry and another overflowing with moisture. The whole earth is full of perforations, and there are paths of intercommunication from part to part. From time to time the dry parts draw upon the moist. Had not the earth some source of supply, it would ere this have been completely drained of its moisture. Well, then, the sun attracts the waves. The localities most affected are the southern. When the earth is parched, it draws to it more moisture. Just as in a lamp the oil flows to the point where it is consumed, so the water inclines toward the place to which the overpowering heat of the burning earth draws it. But where, it may be asked, is it drawn from? Of course, it must be from those northern regions of eternal winter, where there is a superabundance of it. This is why a swift current sets from the Black Sea toward the Lower Sea, without interruption, and not, as in the case of other seas, with alternate flow and ebb of tide; there is always a descending flood in the one direction. Unless this took place, and these routes supplied the means whereby what is lacking may be bestowed on each land, and what is superfluous may be given off, the whole earth would ere now have been either drained or flooded. Now, one would like to ask Diogenes, seeing the deep and all streams are in intercommunication, why the rivers are not everywhere larger in summer. Egypt, he will perhaps tell me, is more baked by the sun, and therefore the Nile rises higher from the extra supply it draws;

[1] Diogenes of Apollonia, a philosopher of the fifth century B.C., is important for his contribution to the biological theory of pneumatism. His description of the vascular system in man is preserved by Aristotle (*History of Animals* III. 2). [Edd.]

but in the other countries, too, the rivers receive some addition. Another question—seeing that every land attracts moisture from other regions, and a greater supply in proportion to its heat, why is any part of the world without moisture? Another—why is the Nile fresh if its water comes from the sea? No river has a fresher and sweeter taste.

Cf. Aëtius IV. 1.4 (Diels, *Fragmente der Vorsokratiker* II^5. 107)

[Democritus holds that] when the snow in the northern parts is melted at the time of the summer solstice and flows away, clouds are formed by the vapors. When the clouds are driven towards the south and towards Egypt by the etesian winds violent storms arise and cause the lakes and the Nile river to be filled.

Cf. Strabo XVII. 1.5

[Posidonius] says that Callisthenes, on the authority of Aristotle, holds that the summer rains are the cause of the overflow [of the Nile].

Cf. Proclus on Plato, *Timaeus*, vol. I, p. 121.8 (Diehl)

Eratosthenes declares it is no longer necessary to inquire as to the cause of the overflow of the Nile, since we know definitely that men have come to the sources of the Nile and have observed the rains there. This confirms Aristotle's conclusion.[1]

Cf. Lucretius VI. 712–737. Translation of Cyril Bailey

The Nile, the river of all Egypt, alone in the world rises, as summer comes, and overflows the plains. It waters Egypt often amid the hot season, either because in summer the north winds, which at that time are said to be the etesian winds, are dead against its mouths; blowing against its stream they check it, and driving the waters upwards fill the channel and make it stop. For without doubt these blasts, which are started from the chill constellations of the pole, are driven full against the stream. The river comes from the south out of the quarter where heat is born, rising among the black races of men of sunburnt colour far inland in the region of mid-day. It may be too that a great heaping up of sand may choke up the mouths as a bar against the opposing waves, when the sea, troubled by the winds, drives the sand within; and in this manner it comes to pass that the river has less free issue, and the waves likewise a less easy downward flow. It may be, too, perhaps that rains occur more at its source at that season, because the etesian blasts of the north winds then drive all the clouds together into those quarters. And, we may suppose, when they have come together driven towards the region of mid-day, there at last the clouds,

[1] See Aristotle, Frag. 246 (Rose).

thrust together upon the high mountains, are massed and violently pressed. Perchance it swells from deep among the high mountains of the Ethiopians, where the sun, traversing all with his melting rays, forces the white snows to run down into the plains.

SEISMOLOGY

Explanation of Earthquakes

Aristotle, *Meteorologica* II. 8. Translation of E. W. Webster

The severest earthquakes take place where the sea is full of currents or the earth spongy and cavernous: so they occur near the Hellespont and in Achaea and Sicily, and those parts of Euboea which correspond to our description—where the sea is supposed to flow in channels below the earth. The hot springs, too, near Aedepsus are due to a cause of this kind. It is the confined character of these places that makes them so liable to earthquakes. A great and therefore violent wind is developed, which would naturally blow away from the earth: but the onrush of the sea in a great mass thrusts it back into the earth. The countries that are spongy below the surface are exposed to earthquakes because they have room for so much wind. . . .

The force wind can have may be gathered not only from what happens in the air (where one might suppose that it owed its power to produce such effects to its volume), but also from what is observed in animal bodies. Tetanus and spasms are motions of wind, and their force is such that the united efforts of many men do not succeed in overcoming the movements of the patients. We must suppose, then (to compare great things with small), that what happens in the earth is just like that.

Our theory has been verified by actual observation in many places. It has been known to happen that an earthquake has continued until the wind that caused it burst through the earth into the air and appeared visibly like a hurricane. This happened lately near Heracleia in Pontus and some time past at the island Hiera, one of the group called the Aeolian islands. Here a portion of the earth swelled up and a lump like a mound rose with a noise: finally it burst, and a great wind came out of it and threw up live cinders and ashes which buried the neighbouring town of Lipara and reached some of the towns in Italy. The spot where this eruption occurred is still to be seen.

Indeed, this must be recognized as the cause of the fire that is generated in the earth: the air is first broken up in small particles and then the wind is beaten about and so catches fire.

A phenomenon in these islands affords further evidence of the fact that winds move below the surface of the earth. When a south wind is going to blow there is a premonitory indication: a sound is heard in the places from which the eruptions issue. This is because the sea is being pushed on from a distance and its advance thrusts back into the earth the wind that was issuing from it. The reason why there is a noise and no earthquake is that the underground spaces are so extensive in proportion to the quantity of the air that is being driven on that the wind slips away into the void beyond. . . .

Subterranean noises, too, are due to the wind; sometimes they portend earthquakes but sometimes they have been heard without any earthquake following. Just as the air gives off various sounds when it is struck, so it does when it strikes other things; for striking involves being struck and so the two cases are the same. The sound precedes the shock because sound is thinner and passes through things more readily than wind. But when the wind is too weak by reason of thinness to cause an earthquake the absence of a shock is due to its filtering through readily, though by striking hard and hollow masses of different shapes it makes various noises, so that the earth sometimes seems to "bellow" as the portent-mongers say.

Water has been known to burst out during an earthquake. But that does not make water the cause of the earthquake. The wind is the efficient cause whether it drives the water along the surface or up from below: just as winds are the causes of waves and not waves of winds. Else we might as well say that earth was the cause; for it is upset in an earthquake, just like water (for effusion is a form of upsetting). No, earth and water are material causes (being patients, not agents); the true cause is the wind.[1]

Seneca, *Natural Questions* VI. 21–23.[2] Translation of John Clarke

21. We Stoics also are convinced that it is only air that can attempt such a feat as the production of an earthquake, for than it nothing in the whole realm of nature is more powerful, more energetic; in absence of it even the elements that are most violent lose their force. It is by air that fire is kindled; if you withdraw wind, water is sluggish. Water becomes impetuous only when the blast tosses it with violence. This force it is that has power to scatter vast spaces of earth, to raise from the depths new mountains, and to set in mid-ocean islands hitherto unseen. Can

[1] Only with the development of geological science has seismology been able, in comparatively recent years, to record much progress. But the ancient theories of the slipping and displacement of earth masses and the part played by subterranean gases are not without interest. [Edd.]

[2] After describing the earthquake in Campania February 5, A.D. 63, Seneca examines various ancient theories of the cause of earthquakes, the part played by subterranean waters, fire, air, etc. He concludes by asserting the view most generally held in antiquity, viz., that the action of air plays the chief role, and he seeks to show precisely how the air acts in various types of earth movements. The modern study of geology and seismology has discovered no foundation for the ancient doctrine. [Edd.]

any one doubt that There and Therasia and this island which in our days under our very eyes rose out of the Aegean Sea, were carried up to the light by the force of air?

Posidonius will have it that there are two different varieties in the movements of the earth, each with its distinctive name. The one is a quaking when the earth is shaken and moves up and down; the other is a tilting when, like a ship, it leans over to one or other side. I am of opinion that there is still a third variety, which we have a special term to denote. Our forefathers had good reason for speaking of a trembling of the earth, for it is unlike either of the other kinds of movement. On such an occasion things are neither all shaken nor all tilted, but they quiver. In a case of this kind no great damage is usually done; while, on the other hand, a tilting is far more destructive than a shock; for unless a contrary movement set in very quickly from the other side to restore the level, downfall follows of necessity.

22. These movements being dissimilar, their causes are likewise different. Let us deal first with the shaking movement. If great loads are being conveyed by a row of many waggons, and the wheels, under the unusual strain, fall into the ruts of the road, one feels the earth shaken. Asclepiodotus[1] has put it on record that on one occasion the fall of a rock that was torn off from the mountain-side caused by the tremor the collapse of some houses in its vicinity. Just the same thing may occur beneath the earth; parts of the over-hanging crags may be loosened and fall with great weight and noise upon the floor of the cavern beneath, and with a violence proportionate to the weight of the mass and the height of the fall. The whole roof of the subterranean valley is disturbed by an occurrence of this kind. It is conceivable, too, that rocks are not always wrenched off by their own weight; when rivers roll over them, the constant moisture weakens the joints of the stone, and day by day bears away part of its fastening, causing abrasion, so to speak, of the skin in which the stone is enclosed. The long waste of ages, through constant daily rubbing, by and by so weakens the fastenings that they cease to be able to sustain their burden. Then blocks of vast size fall down, then the crag hurled headlong will not suffer anything to stand that it strikes in the rebound from its fall, but Comes away with a roar; and all things seem suddenly to rush headlong, as our countryman Virgil says.[2] Such must be the cause of the earthquake that shakes the ground beneath. Now I pass on to the second kind.

23. The earth is naturally full of cavities, containing much empty space. Through these cavities air roams. When an excessive quantity has entered

[1] Asclepiodotus is elsewhere mentioned by Seneca as a pupil of Posidonius and is perhaps to be identified with the author of an extant treatise on military tactics. [Edd.]

[2] *Aeneid* VIII. 525. [Edd.]

and cannot escape it shakes the earth. This explanation is approved by others, too, as mentioned a little above. Perhaps the crowd of witnesses will impress you. The view has the adhesion of Callisthenes,[1] and he is a man not lightly to be set aside. . . . Air, he says, enters the earth by hidden openings under the sea, just as everywhere else. By and by, when the path is blocked by which it had descended, and the resistance of the water in the rear has cut off its retreat, it is borne hither and thither, and encountering itself in its course it undermines the earth. That is the reason why land over against the sea is most frequently harassed by earthquakes; and hence it is that Neptune has been assigned this power of moving the earth. Any one who has learned the elements of Greek knows that he is called among the Greeks Earthshaker (Ἐνοσίχθων).

Cf. Epicurus, *Letter to Pythocles* 105–106.[2] Translation of Cyril Bailey (Oxford, 1926)

Earthquakes may be brought about both because wind is caught up in the earth, so that the earth is dislocated in small masses and is continually shaken, and that causes it to sway. The wind it either takes into itself from outside, or else because masses of ground fall in into cavernous places in the earth and fan into wind the air that is imprisoned in them. And again, earthquakes may be brought about by the actual spreading of the movement which results from the fall of many such masses of ground and the return shock, when the first motion comes into collision with more densely packed bodies of earth. There are also many other ways in which these motions of the earth may be caused.

OCEANOGRAPHY

The Saltness of the Sea

Aristotle, *Meteorologica* II. 3. Translation of E. W. Webster

There is more evidence to prove that saltness is due to the admixture of some substance, besides that which we have adduced. Make a vessel of wax and put it in the sea, fastening its mouth in such a way as to prevent any water getting in. Then the water that percolates through the wax sides of the vessel is sweet,[3] the earthy stuff, the admixture of which makes the water salt, being separated off as it were by a filter. It is this stuff which makes salt water heavy (it weighs more than fresh water) and thick.

[1] Callisthenes of Olynthus, nephew of Aristotle, was a philosopher and historian in Alexander's retinue until he was put to death by the tyrant for complicity in a conspiracy against him. [Edd.]

[2] The Epicurean account seems to draw upon Anaximenes, Anaxagoras, and other early thinkers. Note the typical Epicurean caveat against the single explanation. [Edd.]

[3] This experiment, referred to elsewhere by Aristotle and by other ancient authors, has been tried, without success, in recent times. The earliest reference is in Empedocles (see Frag. A 66, Diels). [Edd.]

The difference in consistency is such that ships with the same cargo very nearly sink in a river when they are quite fit to navigate in the sea.[1] This circumstance has before now caused loss to shippers freighting their ships in a river. That the thicker consistency is due to an admixture of something is proved by the fact that if you make strong brine by the admixture of salt, eggs, even when they are full, float in it. It almost becomes like mud; such a quantity of earthy matter is there in the sea. The same thing is done in salting fish.

Again if, as is fabled, there is a lake in Palestine,[2] such that if you bind a man or beast and throw it in it floats and does not sink, this would bear out what we have said. They say that this lake is so bitter and salt that no fish live in it and that if you soak clothes in it and shake them it cleans them. The following facts all of them support our theory that it is some earthy stuff in the water which makes it salt. In Chaonia there is a spring of brackish water that flows into a neighbouring river which is sweet but contains no fish. The local story is that when Heracles came from Erytheia driving the oxen and gave the inhabitants the choice, they chose salt in preference to fish. They get the salt from the spring. They boil off some of the water and let the rest stand; when it has cooled and the heat and moisture have evaporated together it gives them salt, not in lumps but loose and light like snow. It is weaker than ordinary salt and added freely gives a sweet taste, and it is not as white as salt generally is. Another instance of this is found in Umbria. There is a place there where reeds and rushes grow. They burn some of these, put the ashes into water and boil it off. When a little water is left and has cooled it gives a quantity of salt.

Factors Determining the Extent of the Sea

Lucretius, *On the Nature of Things* VI. 608–622

In the first place they wonder that nature does not make the sea grow larger, in view of the great flow of water into the sea, as all the rivers from all sides pour down. Add to this the passing rains and flitting storms which spatter every sea and soak every land. Add the springs of the sea itself. Yet all these will be to the whole bulk of the sea as the addition of scarcely one drop. Wherefore it is less remarkable that the great sea does not increase. But in addition the sun draws off by its heat a considerable amount. For we see the sun with its burning rays dry out clothes that were dripping wet. And we know that there are many seas extending over mighty surfaces. Therefore, however small an amount of moisture the sun may draw up in any one place, still it will, over so broad a surface, draw heavily from the sea.

---

[1] Cf. p. 239, above. [Edd.]

[2] The Dead Sea. [Edd.]

The Theory of Tides

Pliny, *Natural History* II. 212–217.[1] Translation of John Bostock and H. T. Riley

Much has been said about the nature of waters; but the most wonderful circumstance is the alternate flowing and ebbing of the tides, which exists, indeed, under various forms, but is caused by the sun and the moon.[2] The tide flows twice and ebbs twice between each two risings of the moon, always in the space of twenty-four hours. First, the moon rising with the stars swells out the tide, and after some time, having gained the summit of the heavens, she declines from the meridian and sets, and the tide subsides. Again, after she has set, and moves in the heavens under the earth, as she approaches the meridian, on the opposite side, the tide flows in; after which it recedes until she again rises to us. But the tide of the next day is never at the same time with that of the preceding; as if the planet was in attendance, greedily drinking up the sea, and continually rising in a different place from what she did the day before. The intervals are, however, equal, being always of six hours; not indeed in respect of any particular day or night or place, but equinoctial hours,[3] and therefore they are unequal as estimated by the length of common hours, since a greater number of them fall on some certain days or nights, and they are never equal everywhere except at the equinox. This is a great, most clear, and even divine proof of the dullness of those, who deny that the stars go below the earth and rise up again, and that nature presents the same face in the same states of their rising and setting; for the course of the stars is equally obvious in the one case as in the other, producing the same effect as when it is manifest to the sight.

---

[1] Pliny the Elder (A.D. 23–79) was the author of numerous treatises, of which the *Natural History* in 37 books is alone extant. This encyclopedic work is a summary of a lifetime of reading. Though it represents no original scientific work, the *Natural History* is a mine of historical information, for most of the sources used have since been lost. The treatise was an important textbook during the medieval and renaissance periods, and it would be difficult to exaggerate its influence. Among the subjects treated are astronomy, meteorology, geography, zoology, botany, pharmacology, and mineralogy. Pliny shows little critical ability in dealing with his sources. He transmits truth and error with equal zeal. The story of his death during the eruption of Vesuvius is told in a famous letter of his nephew, the younger Pliny. [Edd.]

[2] The usual explanation of tides in the earlier Greek period was the action of the sun in causing winds or in drawing up the water in some way. In the fourth century B.C. the influence of the moon began to be stressed, and later there were attempts (e.g., by Seleucus and Posidonius) to "explain" the action of the moon and sun. After Pliny and Seneca we have little on the subject until Priscian of Lydia in the sixth and Bede at the end of the seventh century. The passage here cited is interesting as a record of information (and some misinformation) in antiquity. [Edd.]

[3] For civil purposes the hour was generally variable in antiquity, depending on the seasons and the latitude of the place. For astronomical and scientific purposes the invariant equinoctial hour was substituted. [Edd.]

There is a difference in the tides, depending on the moon, of a complicated nature, and, first, as to the period of seven days. For the tides are of moderate height from the new moon to the first quarter; from this time they increase, and are at the highest at the full: they then decrease. On the seventh day they are equal to what they were at the first quarter, and they again increase from the time that she is at first quarter on the other side.[1] At her conjunction with the sun they are equally high, as at the full. When the moon is in the northern hemisphere, and recedes further from the earth, the tides are lower than when, going towards the south, she exercises her influence at a less distance.[2] After an interval of eight years, and the hundredth revolution of the moon, the periods and the heights of the tide return into the same order as at first, this planet always acting upon them; and all these effects are likewise increased by the annual changes of the sun, the tides rising up higher at the equinoxes, and more so at the autumnal than at the vernal; while they are lower about the winter solstice, and still more so at the summer solstice; not indeed precisely at the points of time which I have mentioned, but a few days after; for example, not exactly at the full nor at the new moon, but after them; and not immediately when the moon becomes visible or invisible, or has advanced to the middle of her course, but generally about two hours later than the equinoctial hours; the effect of what is going on in the heavens being felt after a short interval; as we observe with respect to lightning, thunder, and thunderbolts.

But the tides of the ocean cover greater spaces and produce greater inundations than the tides of the other seas; whether it be that the whole of the universe taken together is more full of life than its individual parts, or that the large open space feels more sensibly the power of the planet,[3] as it moves freely about, than when restrained within narrow bounds. On which account neither lakes nor rivers are moved in the same manner. Pytheas of Massilia[4] informs us that in Britain the tide rises 80 cubits. Inland seas are enclosed as in a harbour, but, in some parts of them, there is a more free space which obeys the influence. Among many other examples, the force of the tide will carry us in three days from Italy to Utica,[5] when the sea is tranquil and there is no impulse from the sails. But these motions are more felt about the shores than in the deep parts

---

[1] The account of Posidonius (in Strabo III. 5.8) is similar, but differs in details. [Edd.]

[2] This, as well as the following reference to the eight-year cycle, is incorrect. [Edd.]

[3] I.e., the moon. [Edd.]

[4] Pytheas of Massilia made his famous voyage of exploration to Spain, Gaul, Britain, Germany, and perhaps, too, the Scandinavian countries, toward the end of the fourth century B.C. The reference to a tide of 80 cubits (120 feet) is exaggerated. Perhaps Pytheas was speaking of a storm that raised huge waves. [Edd.]

[5] This would be due to currents rather than tide. [Edd.]

of the seas, as in the body the extremities of the veins feel the pulse, which is the vital spirit, more than the other parts. And in most estuaries, on account of the unequal rising of the stars in each tract, the tides differ from each other, but this respects the period, not the nature of them; as is the case in the Syrtes.

Cf. Aëtius, *Placita* III. 17

How the ebb and flow of tides occur.

... Pytheas of Massilia holds that the flood tides occur as the moon becomes full, and the ebb tides as it wanes. ...

Seleucus, the mathematician, who had written in opposition to Crates, and who was himself a believer in the earth's motion, says that the revolution of the moon resists the rotation[1] of the earth; and since the air between the two bodies is displaced and falls upon the Atlantic Ocean, the sea is accordingly swollen with tide.

Cf. Strabo, *Geography* III. 5.9

Posidonius says that Seleucus, an inhabitant of the region of the Erythraean Sea, holds that the regularity and irregularity of these tides depends on the different positions of the moon on the zodiac; that when the moon is in the equinoctial signs the tides are regular, but when the moon is in the tropical signs there is irregularity both in the height and the speed of the tides; while in the case of any of the other signs, the regularity or irregularity depends on the moon's proximity to the signs mentioned.

METEOROLOGY

Dew and Hoarfrost

Aristotle, *Meteorologica* I. 10. Translation of E. W. Webster

Some of the vapour that is formed by day does not rise high because the ratio of the fire that is raising it to the water that is being raised is small. When this cools and descends at night it is called dew and hoarfrost. When the vapour is frozen before it has condensed to water again it is hoarfrost; and this appears in winter and is commoner in cold places. It is dew when the vapour has condensed into water and the heat is not so great as to dry up the moisture that has been raised, nor the cold sufficient (owing to the warmth of the climate or season) for the vapour itself to freeze. For dew is more commonly found when the season or the place is warm, whereas the opposite, as has been said, is the case with hoarfrost. For obviously vapour is warmer than water, having still the fire that raised it: consequently more cold is needed to freeze it.

[1] In another version: "the rotation and the motion." See p. 109, n. 1.

Both dew and hoarfrost are found when the sky is clear and there is no wind. For the vapour could not be raised unless the sky were clear, and if a wind were blowing it could not condense.

The fact that hoarfrost is not found on mountains contributes to prove that these phenomena occur because the vapor does not rise high. One reason for this is that it rises from hollow and watery places, so that the heat that is raising it, bearing as it were too heavy a burden, cannot lift it to a great height but soon lets it fall again. A second reason is that the motion of the air is more pronounced at a height, and this dissolves a gathering of this kind.

Everywhere, except in Pontus, dew is found with south winds and not with north winds. There the opposite is the case and it is found with north winds and not with south. The reason is the same as that which explains why dew is found in warm weather and not in cold. For the south wind brings warm, and the north, wintry weather. For the north wind is cold and so quenches the heat of the evaporation. But in Pontus the south wind does not bring warmth enough to cause evaporation, whereas the coldness of the north wind concentrates the heat by a sort of recoil, so that there is more evaporation and not less. This is a thing which we can often observe in other places too. Wells, for instance, give off more vapour in a north than in a south wind. Only the north winds quench the heat before any considerable quantity of vapour has gathered, while in a south wind the evaporation is allowed to accumulate.

Water, once formed, does not freeze on the surface of the earth, in the way that it does in the region of the clouds.

Winds as Weather Signs

Theophrastus, *Concerning Weather Signs* 35–37. Translation of Sir Arthur Hort (London, 1916)

The winds which most often come on the top of other winds while these are still blowing are the north wind (*aparktias*), the north-north-east, and the north-west. When however the winds are not dispersed by one another, but die down of their own accord, they change to the next winds on the figure, reckoning from left to right according to the course of the sun. When the south wind begins to blow, it is dry, but it becomes wet before it ceases: so too does the south-east wind. The east wind, coming from the quarter where the sun rises at the equinox, is wet: but it brings the rain in light showers.

The north-east and south-west are the wettest winds; the north, the north-north-east, and the north-east bring hail; snow comes with the north-north-east and north. The south, the west, and the south-east winds bring heat. Some of these have their effect on places which they strike as they come from the sea, others on places which they visit as they come over land. The winds which more than any others make the sky thick with cloud and completely cover it are the north-east and the south-west, especially the former. While the other winds repel the clouds from themselves, the north-east alone attracts them as it blows. Those winds which chiefly bring a clear sky are the north-north-west and the north-west, and next after them the north. Those which most have the character of a hurricane are the north, and north-north-west, and the north-west.

They acquire this character when they fall upon one another as they blow, especially in autumn, but to some extent in spring. Those which are accompanied by lightning are the north-north-west, the north-west, the north and the north-north-east. If at sea a quantity of down is seen blown along, which has come from thistles, it indicates that there will be a great wind. Wind may be expected from any quarter in which a number of shooting stars are seen. If these appear in every quarter alike, it indicates many winds.

BIOLOGY

The rise of Greek scientific thought among the Ionian philosophers and in the schools of Magna Graecia embraced the biological as well as the inorganic sciences. The tradition that many of the early philosophers were also physicians seems to be a reflection of this fact.

The chief extant sources for our knowledge of Greek biology are the works of the Hippocratic Collection, the works of Aristotle, the botanical works of Theophrastus, the pharmacology of Dioscorides, parts of the encyclopedic treatise of Pliny, and the later Greek medical writers, notably Galen.

We find not only the observation, description, and classification of the various forms of plant and animal life, but also attempts to bring the varied phenomena into rational systems. Discussions about mechanism, teleology, evolution, heredity, and the like, are fully developed.

Before considering the special branches of zoology and botany, we have given some passages illustrating the general definition, scope, and method of biology.

In accordance with customary usage we have treated human anatomy and physiology in the section on medicine, rather than in the subdivision "Structure and Function" of zoology.

THE STUDY OF BIOLOGY

Aristotle, *On the Parts of Animals* I. 5.[1] Translation of William Ogle (Oxford, 1911)

Of things constituted by nature some are ungenerated, imperishable, and eternal, while others are subject to generation and decay. The former are excellent beyond compare and divine, but less accessible to knowledge. The evidence that might throw light on them, and on the problems which we long to solve respecting them, is furnished but scantily by sensation; whereas respecting perishable plants and animals we have abundant information, living as we do in their midst, and ample data may be collected concerning all their various kinds, if only we are willing to take sufficient pains. Both departments, however, have their special charm. The scanty conceptions to which we can attain of celestial things give us, from their excellence, more pleasure than all our knowledge of the world in which we live; just as a half glimpse of persons that we love is more delightful than a leisurely view of other things, whatever their number and dimensions. On the other hand, in certitude and in completeness our knowledge of terrestrial things has the advantage. Moreover, their greater nearness and affinity to us balances somewhat the loftier interest of the heavenly things that are the objects of the higher philosophy. Having already treated of the celestial world, as far as our conjectures could reach, we proceed to treat of animals, without omitting, to the best of our ability, any member of the kingdom, however ignoble. . . . Every realm of nature is marvellous: and

[1] On Aristotle's zoological works see p. 400, n. 1. [Edd.]

as Heraclitus, when the strangers who came to visit him found him warming himself at the furnace in the kitchen and hesitated to go in, is reported to have bidden them not to be afraid to enter, as even in that kitchen divinities were present, so we should venture on the study of every kind of animal without distaste; for each and all will reveal to us something natural and something beautiful. Absence of haphazard and conduciveness of everything to an end are to be found in Nature's works in the highest degree, and the resultant end of her generations and combinations is a form of the beautiful.

If any person thinks the examination of the rest of the animal kingdom an unworthy task, he must hold in like disesteem the study of man. For no one can look at the primordia of the human frame—blood, flesh, bones, vessels, and the like—without much repugnance. Moreover, when any one of the parts or structures, be it which it may, is under discussion, it must not be supposed that it is its material composition to which attention is being directed or which is the object of the discussion, but the relation of such part to the total form.

NATURAL SELECTION

The origin of life in its various forms, and of man in particular, occupied the Greek thinkers from earliest times. The idea that man somehow arose from the earth or evolved from other animals is often encountered. Anaximander seems to have held that the first manifestation of animal life was marine life, that changes of structure occurred as the animals moved to dry land, and that man thus evolved from the fish. The long period of nursing required by the human infant before it becomes self-sufficient was cited by Anaximander as proof of the fact that man arose from other species, for he could not have survived if he had always been so helpless in infancy. With Empedocles we have the definite idea of chance combinations of organs arising and dying out because of their lack of adaptation.

We have not given the fragmentary sources for the early philosophic views on this subject because their interpretation is difficult and by no means certain. But in Aristotle and Lucretius we have clear statements on the question of natural selection and its role in the development of living forms. In Aristotle we have an attempt to refute the Empedoclean doctrine of chance or purposeless variations in nature. This is in line with Aristotle's view of the preponderating part played by purpose in the workings of nature. Only rarely, according to him, is this purpose frustrated, as in a chance monstrous birth. On the other hand, the passage from Lucretius gives us a view of evolution based on chance combinations, in line with the doctrine of Empedocles and Epicurus.

Purpose in Nature

Aristotle, *Physics* II. 8. Translation of R. P. Hardie and R. K. Gaye

We must explain then (1) that Nature belongs to the class of causes which act for the sake of something; (2) about the necessary and its place in physical problems, for all writers ascribe things to this cause, arguing that since the hot and the cold, etc., are of such and such a kind, therefore certain things *necessarily* are and come to be—and if they mention any other cause (one his "friendship and strife,"[1] another his "mind"[2]), it is only to touch on it, and then good-bye to it.

A difficulty presents itself: why should not nature work, not for the sake of something nor because it is better so, but just as the sky rains, not in order to make the corn grow, but of necessity? What is drawn up must cool, and what has been cooled must become water and descend, the result of this being that the corn grows. Similarly if a man's crop is spoiled on the threshing floor, the rain did not fall for the sake of this—in order that the crop might be spoiled—but that result just followed. Why then should it not be the same with the parts in nature, e.g., that our teeth should come up *of necessity*—the front teeth sharp, fitted for tearing, the molars broad and useful for grinding down the food—since they did not arise for this end, but it was merely a coincident result; and so with all other parts in which we suppose that there is purpose? Wherever then all the parts came about just what they would have been if they had come to be for an end, such things survived, being organized spontaneously in a fitting way; whereas those which grew otherwise perished and continue to perish,[3] as Empedocles says his "man-faced ox-progeny" did.

Such are the arguments (and others of the kind) which may cause difficulty on this point. Yet it is impossible that this should be the true view. For teeth and all other natural things either invariably or normally come about in a given way; but of not one of the results of chance or spontaneity is this true. We do not ascribe to chance or mere coincidence the frequency of rain in winter, but frequent rain in summer we do; nor heat in the dog-days, but only if we have it in winter. If then, it is agreed that things are either the result of coincidence or for an end, and these cannot be the result of coincidence or spontaneity, it follows that they must be for an end; and that such things are all due to nature even the champions of the theory which is before us would agree. Therefore action for an end is present in things which come to be and are by nature.

Further, where a series has a completion, all the preceding steps are for the sake of that. Now surely as in intelligent action, so in nature; and as in nature, so it is in each action, if nothing interferes. Now intelligent action is for the sake of an end; therefore the nature of things also is so. Thus if a house, e.g., had been a thing made by nature, it would have been made in the same way as it is now by art; and if things made by nature were made also by art, they would come to be in the same way as by nature. Each step then in the series is for the sake of the next; and generally art

---

[1] Empedocles.

[2] Anaxagoras.

[3] This passage is quoted by Darwin on the first page of his *Origin of Species* without any realization, however, that it is the view not of Aristotle but of Empedocles, and that Aristotle proceeds to refute it. [Edd.]

partly completes what nature cannot bring to a finish, and partly imitates her. If, therefore, artificial products are for the sake of an end, so clearly also are natural products. The relation of the later to the earlier terms of the series is the same in both.

This is most obvious in the animals other than man: they make things neither by art nor after inquiry or deliberation. Wherefore people discuss whether it is by intelligence or by some other faculty that these creatures work—spiders, ants, and the like. By gradual advance in this direction we come to see clearly that in plants, too, that is produced which is conducive to the end—leaves, e.g., grow to provide shade for the fruit. If then it is both by nature and for an end that the swallow makes its nest and the spider its web, and plants grow leaves for the sake of the fruit and send their roots down (not up) for the sake of nourishment, it is plain that this kind of cause is operative in things which come to be and are by nature. And since "nature" means two things, the matter and the form, of which the latter is the end, and since all the rest is for the sake of the end, the form must be the cause in the sense of "that for the sake of which."

Now mistakes come to pass even in the operations of art: the grammarian makes a mistake in writing and the doctor pours out the wrong dose. Hence clearly mistakes are possible in the operations of nature also. If then in art there are cases in which what is rightly produced serves a purpose, and if where mistakes occur there was a purpose in what was attempted, only it was not attained, so must it be also in natural products, and monstrosities will be failures in the purposive effort. Thus in the original combinations the "ox-progeny" if they failed to reach a determinate end must have arisen through the corruption of some principle corresponding to what is now the seed.

Further, seed must have come into being first, and not straightway the animals: the words "whole-natured first . . ." must have meant seed.

Again, in plants too we find the relation of means to end, though the degree of organization is less. Were there then in plants also "olive-headed vine-progeny," like the "man-headed ox-progeny" or not? An absurd suggestion; yet there must have been, if there were such things among animals.

Moreover, among the seeds anything must have come to be at random. But the person who asserts this entirely does away with "nature" and what exists "by nature." For those things are natural which, by a continuous movement originated from an internal principle, arrive at some completion: the same completion is not reached from every principle; nor any chance completion, but always the tendency in each is towards the same end, if there is no impediment.

The end and the means towards it may come about by chance. We say, for instance, that a stranger has come by chance, paid the ransom, and gone away, when he does so as if he had come for that purpose, though it was not for that that he came. This is incidental, for chance is an incidental cause, as I remarked before. But when an event takes place always or for the most part, it is not incidental or by chance. In natural products the sequence is invariable, if there is no impediment.

It is absurd to suppose that purpose is not present because we do not observe the agent deliberating. Art does not deliberate. If the ship-building art were in the wood, it would produce the same results *by nature*. If, therefore, purpose is present in art, it is present also in nature. The best illustration is a doctor doctoring himself: nature is like that.

It is plain then that nature is a cause, a cause that operates for a purpose.

Epicurean Evolutionary Theory

Lucretius, *On the Nature of Things* V. 783–877. Translation of Cyril Bailey

First of all the earth gave birth to the tribes of herbage and bright verdure all around the hills and over all the plains, the flowering fields gleamed in their green hue, and thereafter the diverse trees were started with loose rein on their great race of growing through the air. Even as down and hair and bristles are first formed on the limbs of four-footed beasts and the body of fowls strong of wing, so then the newborn earth raised up herbage and shrubs first, and thereafter produced the races of mortal things, many races born in many ways by diverse means. For neither can living animals have fallen from the sky nor the beasts of the earth have issued forth from the salt pools. It remains that rightly has the earth won the name of mother, since out of the earth all things are produced. And even now many animals spring forth from the earth, formed by the rains and the warm heat of the sun; wherefore we may wonder the less, if then more animals and greater were born, reaching their full growth when earth and air were fresh. First of all the tribe of winged fowls and the diverse birds left their eggs, hatched out in the spring season, as now in the summer the grasshoppers of their own will leave their smooth shells, seeking life and livelihood. Then it was that the earth first gave birth to the race of mortal things. For much heat and moisture abounded then in the fields; thereby, wherever a suitable spot or place was afforded, there grew up wombs, clinging to the earth by their roots; and when in the fullness of time the age of the little ones, fleeing moisture and eager for air, had opened them, nature would turn to that place the pores in the earth and constrain them to give forth from their opened veins a sap, most like to milk; even as now every woman, when she has brought forth, is filled with sweet milk, because all the current of her nourishment is turned towards her paps. The earth furnished food for the young, the warmth raiment, the grass a couch rich in much soft down. But the youth of the world called not into being hard frosts nor exceeding heat nor winds of mighty violence: for all things grow and come to their strength in like degrees.

Wherefore, again and again, rightly has the earth won, rightly does she keep the name of mother, since she herself formed the race of men, and almost at a fixed time brought forth every animal which ranges madly everywhere on the mighty mountains, and with them the fowls of the air with their diverse forms. But because she must needs come to some end of child-bearing, she ceased, like a woman worn with the lapse of age. For time changes the nature of the whole world, and one state after another must needs overtake all things, nor does anything abide like itself: all things change their abode, nature alters all things and constrains them to turn. For one thing rots away and grows faint and feeble with age, thereon another grows up and issues from its place of scorn. So then time changes the nature of the whole world, and one state after another overtakes the earth, so that it cannot bear what it did, but can bear what it did not of old.

And many monsters[1] too earth then essayed to create, born with strange faces and strange limbs, the man-woman, between the two, yet not either, sundered from both sexes, some things bereft of feet, or in turn robbed of hands, things too found dumb without mouths, or blind without eyes, or locked through the whole body by the clinging of the limbs, so that they could not do anything or move towards any side or avoid calamity or take what they needed. All other monsters and prodigies of this sort she would create; all in vain, since nature forbade their increase, nor could they reach the coveted bloom of age nor find food nor join in the work of Venus. For we see that many happenings must be united for things, that they may be able to beget and propagate their races; first that they may have food, and then a way whereby birth-giving seeds may pass through their frames, and issue from their slackened limbs; and that woman may be joined with man, they must needs each have means whereby they can interchange mutual joys.

And it must needs be that many races of living things then perished and could not beget and propagate their offspring. For whatever animals you see feeding on the breath of life, either their craft or bravery, aye or their swiftness has protected and preserved their kind from the beginning of their being. And many there are, which by their usefulness are commended

[1] It is to be noted that the type of evolution described by Lucretius in this and the following paragraph does not involve a development from lower to higher forms through the mechanism of heredity. In fact, for Lucretius (and Empedocles, whom he seems to be following) the functioning of heredity and sexual reproduction entered only after earth herself, through chance groupings of atoms, had produced an organism having characteristics that might make for survival. The vast majority of chance groupings being incapable of survival, had, of course, disappeared. [Edd.]

to us, and so abide, trusted to our tutelage. First of all the fierce race of lions, that savage stock, their bravery has protected, foxes their cunning, and deer their fleet foot. But the lightly-sleeping minds of dogs with their loyal heart, and all the race which is born of the seed of beasts of burden, and withal the fleecy flocks and the horned herds, are all trusted to the tutelage of men, Memmius. For eagerly did they flee the wild beasts and ensue peace and bounteous fodder gained without toil of theirs, which we grant them as a reward because of their usefulness. But those to whom nature granted none of these things, neither that they might live on by themselves of their own might, nor do us any useful service, for which we might suffer their kind to feed and be kept safe under our defence, you may know that these fell a prey and spoil to others, all entangled in the fateful trammels of their own being, until nature brought their kind to destruction.

ZOOLOGY

General Methodology

Aristotle, *On the Parts of Animals* I. 5.[1] Translation of William Ogle

The course of exposition must be first to state the attributes common to whole groups of animals, and then to attempt to give their explanation. Many groups, as already noticed, present common attributes, that is to say, in some cases absolutely identical affections, and absolutely identical

[1] Of the zoological treatises of Aristotle the most important are the *History of Animals* (nine books), *On the Generation of Animals* (five books), and *On the Parts of Animals* (four books). Shorter works are *On the Motion of Animals* and *On the Progression of Animals* (one book each). From the point of view of the wealth of observation, exactness of description, careful ordering and classification, and depth of insight into the general questions at the basis of biological science, Aristotle's works in this branch are among the most significant in the entire history of science. Here we have discussions of the anatomy, physiology, generation, embryology, and habits of hundreds of species of animals and classification of these species along lines that have much in common with those we use today. Not only is all this preserved in great detail but basic questions such as the possibility of spontaneous generation, criteria for distinguishing plant from animal life, heredity and the nature of the generative process are discussed with great penetration.

Generally speaking the *History of Animals* outlines the observed facts, whereas the treatise *On the Parts* tries to explain these facts within the framework of a teleological system of nature. Thus the *History* stresses structure, the *Parts* stresses function. Approximately 500 different animals are mentioned.

The statements of Aristotle are generally accurate and sometimes indicate truths that were subsequently forgotten and not rediscovered until the modern period. There are, however, numerous errors. It seems clear that Aristotle drew largely on the work and reports of others and that he was not always fortunate in his choice of authority. In this connection it is possible that some of the reports brought back from the campaigns of Alexander were already utilized by Aristotle in his zoological works, as they certainly were by Theophrastus in his botanical works. An excellent analysis of the factual content of Aristotle's scientific work, particularly on the biological side, is given by T. E. Lones, *Aristotle's Researches in Natural Science* (London, 1912). [Edd.]

organs—feet, feathers, scales, and the like; while in other groups the affections and organs are only so far identical as that they are analogous. For instance, some groups have lungs, others have no lung, but an organ analogous to a lung in its place; some have blood, others have no blood, but a fluid analogous to blood, and with the same office.[1] To treat of the common attributes in connexion with each individual group would involve, as already suggested, useless iteration. For many groups have common attributes. So much for this topic.

As every instrument and every bodily member subserves some partial end, that is to say, some special action, so the whole body must be destined to minister to some plenary sphere of action. Thus the saw is made for sawing, for sawing is a function, and not sawing for the saw. Similarly, the body too must somehow or other be made for the soul, and each part of it for some subordinate function, to which it is adapted.

We have, then, first to describe the common functions, common, that is, to the whole animal kingdom, or to certain large groups, or to the members of a species. In other words, we have to describe the attributes common to all animals, or to assemblages, like the class of birds, of closely allied groups differentiated by gradation, or to groups like Man not differentiated into subordinate groups. In the first case the common attributes may be called analogous, in the second generic, in the third specific.

When a function is ancillary to another, a like relation manifestly obtains between the organs which discharge these functions; and similarly, if one function is prior to and the end of another, their respective organs will stand to each other in the same relation. Thirdly, the existence of these parts involves that of other things as their necessary consequents.

Instances of what I mean by functions and affections are Reproduction, Growth, Copulation, Waking, Sleep, Locomotion, and other similar vital actions. Instances of what I mean by parts are Nose, Eye, Face, and other so-called members or limbs, and also the more elementary parts of which these are made. So much for the method to be pursued. Let us now try to set forth the causes of all vital phenomena, whether universal or particular, and in so doing let us follow that order of exposition which conforms, as we have indicated, to the order of nature.

Animals and Plants

Aristotle, *History of Animals* VIII. 1.[2] Translation of D. W. Thompson (Oxford, 1910)

Nature proceeds little by little from things lifeless to animal life in such a way that it is impossible to determine the exact line of demarcation,

[1] For the Aristotelian division of animals into Sanguineous (i.e., red-blooded) and Bloodless, see the Table, pp. 405–406. [Edd.]

[2] Aristotle was keenly aware of the absence of a sharp dividing line between the plant and the animal life, and indeed, between the animate and the inanimate. This and the following

nor on which side thereof an intermediate form should lie. Thus, next after lifeless things in the upward scale comes the plant, and of plants one will differ from another as to its amount of apparent vitality; and, in a word, the whole genus of plants, whilst it is devoid of life as compared with an animal, is endowed with life as compared with other corporeal entities. Indeed, as we just remarked, there is observed in plants a continuous scale of ascent towards the animal. So, in the sea, there are certain objects concerning which one would be at a loss to determine whether they be animal or vegetable.[1] For instance, certain of these objects are fairly rooted, and in several cases perish if detached;[2] thus the pinna[3] is rooted to a particular spot, and the solen (or razor-shell) cannot survive withdrawal from its burrow.[4] Indeed, broadly speaking, the entire genus of testaceans have a resemblance to vegetables, if they be contrasted with such animals as are capable of progression.

In regard to sensibility, some animals give no indication whatsoever of it, whilst others indicate it but indistinctly. Further, the substance of some of these intermediate creatures is fleshlike, as is the case with the so-called tethya (or ascidians)[5] and the acalephae (or sea-anemones);[6] but the sponge is in every respect like a vegetable. And so throughout the entire animal scale there is a graduated differentiation in amount of vitality and in capacity for motion.

A similar statement holds good with regard to habits of life. Thus of plants that spring from seed the one function seems to be the reproduction of their own particular species, and the sphere of action with certain animals is similarly limited. The faculty of reproduction, then, is common to all alike. If sensibility be superadded, then their lives will differ from one another in respect to sexual intercourse through the varying amount of pleasure derived therefrom, and also in regard to modes of parturition and ways of rearing their young. Some animals, like plants, simply procreate their own species at definite seasons; other animals busy themselves also in procuring food for their young, and after they are

---

passage recall some cases near the border between plant and animal life. Aristotle's criteria, e.g., capability of sensation, have, of course, given way, but it is noteworthy that even now no single criterion has proved entirely satisfactory in classifying certain microscopic forms of life. [Edd.]

[1] Elsewhere Aristotle speaks of the sea nettle as resembling a plant in certain ways. [Edd.]

[2] See p. 403, below. [Edd.]

[3] A type of bivalve mollusk. [Edd.]

[4] Elsewhere Aristotle describes how the solen burrows more deeply in the sand when it hears a noise. Thompson refers to the method of withdrawing the solen from its burrow with an iron pike of the sort still used in the Adriatic for the capture of razor fish. [Edd.]

[5] See p. 403, n. 4. [Edd.]

[6] A subgroup of the family of invertebrates known as coelenterates. The subgroup includes the jellyfishes, hydroids, and allied species. [Edd.]

reared quit them and have no further dealings with them; other animals are more intelligent and endowed with memory, and they live with their offspring for a longer period and on a more social footing.

The life of animals, then, may be divided into two acts—procreation and feeding; for on these two acts all their interests and life concentrate. Their food depends chiefly on the substance of which they are severally constituted; for the source of their growth in all cases will be this substance. And whatsoever is in conformity with nature is pleasant, and all animals pursue pleasure in keeping with their nature.

Aristotle, *On the Parts of Animals* IV. 5. Translation of William Ogle

The Ascidians differ but slightly from plants, and yet have more of an animal nature than the sponges, which are virtually plants and nothing more. For nature passes from lifeless objects to animals in such unbroken sequence, interposing between them beings which live and yet are not animals, that scarcely any difference seems to exist between two neighbouring groups owing to their close proximity.

A sponge,[1] then, as already said, in these respects completely resembles a plant, that throughout its life it is attached to a rock, and that when separated from this it dies. Slightly different from the sponges are the so-called Holothurias[2] and the sea-lungs, as also sundry other sea-animals that resemble them. For these are free and unattached. Yet they have no feeling, and their life is simply that of a plant separated from the ground. For even among land-plants there are some that are independent of the soil, and that spring up and grow, either upon other plants, or even entirely free. Such, for example, is the plant which is found on Parnassus, and which some call the Epipetrum.[3] This you may hang up on a peg and it will yet live for a considerable time. Sometimes it is a matter of doubt whether a given organism should be classed with plants or with animals. The Ascidians, for instance, and the like so far resemble plants as that they never live free and unattached,[4] but, on the other hand, inasmuch as they have a certain flesh-like substance, they must be supposed to possess some degree of sensibility.

Cf. [Aristotle], *On Plants* I. 1. Translation of E. S. Forster (Oxford, 1913)

Plato says that whatsoever takes food desires food, and feels pleasure in satiety and pain when it is hungry, and that these dispositions do not

[1] The question whether sponges were plants or animals was debated until the nineteenth century. [Edd.]

[2] What Aristotle included in this class is doubtful, for the sea-cucumbers are said to give definite indications of feeling. [Edd.]

[3] Probably a Sedum. There is an English species, *S. telephium*, which has gained the popular name "livelong" from its persistent vitality after being pulled up from the ground.

[4] Aristotle's Tethya, or Ascidians, are not Tunicata generally, but only the simple solitary Ascidians, which are always sessile.

occur without the accompaniment of sensation. The view of Plato in thus holding that plants have sensation and desire was marvellous enough; but Anaxagoras[1] and Democritus and Empedocles declared that they possessed intellect and intelligence.

Classification of Animals

Aristotle, *History of Animals* I. 1.[2] Translation of D. W. Thompson

Of animals, some resemble one another in all their parts, while others have parts wherein they differ. Sometimes the parts are identical in form or species, as, for instance, one man's nose or eye resembles another man's nose or eye, flesh flesh, and bone bone; and in like manner with a horse, and with all other animals which we reckon to be of one and the same species: for as the whole is to the whole, so each to each are the parts severally.[3] In other cases the parts are identical, save only for a difference in the way of excess or defect, as is the case in such animals as are of one and the same genus. By "genus" I mean, for instance, Bird or Fish, for each of these is subject to difference in respect of its genus, and there are many species of fishes and of birds.

Within the limits of genera, most of the parts as a rule exhibit differences through contrast of the property or accident, such as colour and shape, to which they are subject: in that some are more and some in a less degree the subject of the same property or accident; and also in the way of multitude or fewness, magnitude or parvitude, in short in the way of excess or defect. Thus in some the texture of the flesh is soft, in others firm; some have a long bill, others a short one; some have abundance of feathers, others have only a small quantity. It happens further that some have parts that others have not: for instance, some have spurs and others

[1] Anaxagoras also held that plants breathe ([Aristotle], *On Plants* 816*b*26). [Edd.]

[2] Here we have the basic characteristic of Aristotle's system of classification. In opposition to the Platonic system of dichotomies, Aristotle preferred the more natural and popular division into groups (e.g., birds, fishes) and subgroups.

In determining whether two animals belong to what we should call different species of the same genus or to different genera, Aristotle inquires whether the differences are differences of "excess" and "defect" or of "analogy." Birds may differ with respect to the number of their feathers or the length of their bills, etc. These are differences of excess or defect, and the birds are held by Aristotle to be different species in the same genus. Where, however, the organs of two animals are similar only by analogy, e.g., scales for feathers, or hoofs for nails, the animals belong to different genera. There is no one passage in which Aristotle gives his complete classification of animals and, in fact, from different viewpoints different classifications are suggested. There are differences not merely of structure, but of physiology, generation, etc. Aristotle is quite aware of the difficulties of classification, the existence of apparently isolated species, and of intermediate species.

In connection with the next passage we have given a general scheme of Aristotle's classification of animals in relation to generation. [Edd.]

[3] In this case the animals are alike in species, but differ only as individuals. [Edd.]

not, some have crests and others not; but as general rule, most parts and those that go to make up the bulk of the body are either identical with one another, or differ from one another in the way of contrast and of excess and defect.[1] For "the more" and "the less" may be represented as "excess" or "defect."

Once again, we may have to do with animals whose parts are neither identical in form nor yet identical save for differences in the way of excess or defect:[2] but they are the same only in the way of analogy, as, for instance, bone is only analogous to fish-bone, nail to hoof, hand to claw, and scale to feather; for what the feather is in a bird, the scale is in a fish.

CLASSIFICATION OF ANIMALS IN RELATION TO GENERATION

Aristotle, *On the Generation of Animals* II. 1. Translation of Arthur Platt (Oxford, 1910)[3]

Some animals bring to perfection and produce into the world a creature

---

[1] In this case the animals are alike in genus, but differ in species. [Edd.]

[2] In this case the animals are of different genera. [Edd.]

[3] The attempt to classify animals on the basis of differences in mode of generation is typical of Aristotle in that it goes beyond that which is merely obvious to the eye and takes a more fundamental criterion. The following Table (reprinted from Charles Singer, *A Short History of Biology*, p. 42 [Oxford, 1931]) gives a general scheme of the Aristotelian classification of animals on the basis of generation. The material is derived from various passages in Aristotle. [Edd.]

Enaima = *Vertebrates*

(*Having red blood and either viviparous or oviparous*)

Viviparous in the internal sense.		1. Man.
		2. Cetaceans.
		3. Viviparous quadrupeds.
		(*a*) Ruminants with cutting teeth in lower jaw only, and with cloven hoofs=sheep, oxen, etc.
		(*b*) Solid-hoofed animals=horses, asses, etc.
		(*c*) Other viviparous quadrupeds.
Oviparous though sometimes externally viviparous.	With perfect eggs.	4. Birds.
		(*a*) Birds of prey with talons.
		(*b*) Swimmers with webbed feet.
		(*c*) Pigeons, doves, etc.
		(*d*) Swifts, martins, etc.
		(*e*) Other birds.
		5. Oviparous quadrupeds=amphibians and most reptiles.
		6. Serpents.
	With imperfect eggs.	7. Fishes.
		(*a*) Selachians=cartilaginous fishes ("Galeos" an exception).
		(*b*) Other fishes.

like themselves, as all those which bring their young into the world alive;[1] others produce something undeveloped which has not yet acquired its own form; in this latter division the sanguinea lay[2] eggs, the bloodless animals either lay an egg or give birth to a scolex. The difference between egg and scolex is this: an egg is that from a part of which the young comes into being, the rest being nutriment for it; but the whole of a scolex is developed into the whole of the young animal.[3] Of the vivipara, which bring into the world an animal like themselves, some are internally viviparous (as men, horses, cattle, and of marine animals dolphins and the other cetacea);[4] others first lay eggs within themselves, and only after this are externally viviparous (as the cartilaginous fishes). Among the ovipara some produce the egg in a perfect condition (as birds and all oviparous quadrupeds and footless animals, e.g., lizards and tortoises and most snakes;[5] for the eggs of all these do not increase when once laid). The eggs of others are imperfect; such as those of fishes, crustaceans, and cephalopods, for their eggs increase after being produced.[6]

All the vivipara are sanguineous, and the sanguinea are either viviparous or oviparous, except those which are altogether infertile.[7] Among bloodless[8] animals the insects produce a scolex, alike those that are generated by copulation and those that copulate themselves though not so generated. For there are some insects of this sort, which though they come

---

Anaima = *Invertebrates*

(*Without red blood and viviparous, vermiparous, budding, or spontaneous*)

With perfect eggs.	8. Cephalopods. 9. Crustaceans.
With eggs of special type.	10. Insects, spiders, scorpions, etc.
With generative slime, buds, or spontaneous generation.	11. Molluscs (except Cephalopods), Echinoderms, etc.
With spontaneous generation.	12. Sponges, Coelenterates, etc.

[1] The vivipara, including the cartilaginous fishes.

[2] Birds and reptiles.

[3] Cf. the modern distinction between meroblastic and holoblastic yolks.

[4] The exclusion of the cetacea, i.e., whales, dolphins, and porpoises, from the class of fishes is noteworthy. Aristotle remarks elsewhere that these animals have lungs and blowholes and suckle their young. Though he indicates that this latter condition is true only of viviparous animals, he does not define a class of mammals. [Edd.]

[5] The exception among the snakes is the viper.

[6] The increase in size is however only due to imbibition of water; there is no further development of the egg. . . .

[7] Mules.

[8] See the table. "Bloodless," in this sense, means lacking in red blood. The class is roughly equivalent to our invertebrates, while the Sanguinea are roughly equivalent to our vertebrates. Thus Aristotle says (*History of Animals* 516*b*22): "all sanguineous animals have a backbone . . . composed either of bone or of spine." [Edd.]

into being by spontaneous generation are yet male and female; from their union something is produced, only it is imperfect; the reason of this has been previously stated.

These classes admit of much cross-division. Not all bipeds are viviparous (for birds are oviparous), nor are they all oviparous (for man is viviparous), nor are all quadrupeds oviparous (for horses, cattle, and countless others are viviparous), nor are they all viviparous (for lizards, crocodiles, and many others lay eggs). Nor does the presence or absence of feet make the difference between them, for not only are some footless animals viviparous, as vipers and the cartilaginous fishes, while others are oviparous, as the other fishes and serpents, but also among those which have feet many are oviparous and many viviparous, as the quadrupeds above mentioned. And some which have feet, as man, and some which have not, as the whale and dolphin, are internally viviparous. By this character then it is not possible to divide them,[1] nor is any of the locomotive organs the cause of this difference, but it is those animals which are more perfect in their nature and participate in a purer element[2] which are viviparous, for nothing is internally viviparous unless it receive and breathe out air. But the more perfect are those which are hotter in their nature and have more moisture and are not earthy in their composition. And the measure of natural heat is the lung when it has blood in it, for generally those animals which have a lung are hotter than those which have not, and in the former class again those whose lung is not spongy nor solid nor containing only a little blood, but soft and full of blood. And as the animal is perfect but the egg and the scolex are imperfect, so the perfect is naturally produced from the more perfect. If animals are hotter as shown by their possessing a lung but drier in their nature, or are colder but have more moisture, then they either lay a perfect egg or are viviparous after laying an egg within themselves. For birds and scaly reptiles because of their heat produce a perfect egg, but because of their dryness it is only an egg; the cartilaginous fishes have less heat than these but more moisture, so that they are intermediate, for they are both oviparous and viviparous within themselves,[3] the former because they are cold, the latter because of their moisture; for moisture is vivifying, whereas dryness is furthest removed from what has

[1] A hit at Plato's "dichotomy" of animals which *was* based on this character. Aristotle's own test, the condition in which the young are produced, is of course vastly superior, foreshadowing the doctrine that embryology affords the most important evidence of affinities. And he refuses to consider even that as the sole test. . . .

[2] . . . The internally viviparous animals, including the cetacea, are air-breathers; the cartilaginous or elasmobranch fishes are only externally viviparous, laying eggs first within themselves.

[3] From comparison of I. 8 it seems that this should mean "oviparous internally and viviparous externally." [Aristotle had not seen the ovum of what we call mammals. Edd.]

life. Since they have neither feathers nor scales such as either reptiles or other fishes have, all which are signs rather of a dry and earthy nature, the egg they produce is soft; for the earthy matter does not come to the surface in their eggs any more than in themselves. This is why they lay eggs in themselves, for if the egg were laid externally it would be destroyed, having no protection.

Animals that are cold and rather dry than moist also lay eggs, but the egg is imperfect; at the same time, because they are of an earthy nature and the egg they produce is imperfect, therefore it has a hard integument that it may be preserved by the protection of the shell-like covering. Hence fishes,[1] because they are scaly, and crustacea, because they are of an earthy nature, lay eggs with a hard integument.

The cephalopods, having themselves bodies of a sticky nature, preserve in the same way the imperfect eggs they lay, for they deposit a quantity of sticky material about the embryo.[2]

All insects produce a scolex. Now all the insects are bloodless, wherefore all creatures that produce a scolex from themselves are so.[3] But we cannot say simply that all bloodless animals produce a scolex, for the classes overlap one another, (1) the insects, (2) the animals that produce a scolex, (3) those that lay their egg imperfect, as the scaly fishes, the crustacea, and the cephalopods. I say that these form a gradation, for the eggs of these latter resemble a scolex, in that they increase after oviposition, and the scolex of insects again as it develops resembles an egg; how so we shall explain later.

We must observe how rightly Nature orders generation in regular gradation. The more perfect and hotter animals produce their young perfect in respect of quality (in respect of quantity this is so with no animal, for the young always increase in size after birth), and these generate living animals within themselves from the first. The second class do not generate perfect animals within themselves from the first (for they are only viviparous after first laying eggs), but still they are externally viviparous. The third class do not produce a perfect animal, but an egg, and this egg is perfect. Those whose nature is still colder than these produce an egg, but an imperfect one, which is perfected outside the body, as the class of scaly fishes, the crustacea, and the cephalopods. The fifth and coldest class does not even lay an egg from itself;[4] but so far as the young ever attain to this condition at all, it

[1] . . . The teleosteans, the vast majority of fish, opposed to the cartilaginous elasmobranchs. The latter have no scales like those of the teleosteans. The crustacea are "earthy," as shown by their exo-skeletons. . . .

[2] I.e., about their eggs.

[3] Since *only* insects produce a scolex. [Edd.]

[4] "From itself," because the scolex *does* later develop into an egg, i.e., the pupa, according to Aristotle, but the parent does not lay this pupa ἐξ αὑτοῦ.

is outside the body of the parent, as has been said already. For insects produce a scolex first; the scolex after developing becomes egg-like (for the so-called chrysalis or pupa is equivalent to an egg); then from this it is that a perfect animal comes into being, reaching the end of its development in the second change.

Some animals then, as said before, do not come into being from semen, but all the sanguinea do so which are generated by copulation, the male emitting semen into the female; when this has entered into her the young are formed and assume their peculiar character, some within the animals themselves when they are viviparous, others in eggs.

DIFFICULTY OF COMPLETE CLASSIFICATION

Aristotle, *History of Animals* IV. 7. Translation of D. W. Thompson

Furthermore, there are some strange creatures to be found in the sea, which from their rarity we are unable to classify. Experienced fishermen affirm, some that they have at times seen in the sea animals like sticks, black, rounded, and of the same thickness throughout; others that they have seen creatures resembling shields, red in colour, and furnished with fins packed close together; and others that they have seen creatures resembling the male organ in shape and size, with a pair of fins in the place of the testicles, and they aver that on one occasion a creature of this description was brought up on the end of a night-line.[1]

STRUCTURE AND FUNCTION

STRUCTURAL DIFFERENCES IN THE VARIOUS GENERA

Aristotle, *History of Animals* II. 1. Translation of D. W. Thompson

With regard to animals in general, some parts or organs are common to all, as has been said, and some are common only to particular genera; the parts, moreover, are identical with or different from one another on the lines already repeatedly laid down. For as a general rule all animals that are generically distinct have the majority of their parts or organs different in form or species; and some of them they have only analogically similar and diverse in kind or genus, while they have others that are alike in kind but specifically diverse; and many parts or organs exist in some animals, but not in others.

For instance, viviparous quadrupeds have all a head and a neck, and all the parts or organs of the head, but they differ each from other in the shapes of the parts. The lion has its neck composed of one single bone instead of vertebrae; but, when dissected, the animal is found in all internal characters to resemble the dog.

---

[1] The identification of these three species is not clear. D. W. Thompson believes the last to be *Gastropteron meckelii*. [Edd.]

The quadrupedal vivipara instead of arms have forelegs. This is true of all quadrupeds, but such of them as have toes have, practically speaking, organs analogous to hands; at all events, they use these fore-limbs for many purposes as hands. . . .

The fore-limbs then serve more or less the purpose of hands in quadrupeds, with the exception of the elephant. This latter animal has its toes somewhat indistinctly defined, and its front legs are much bigger than its hinder ones; it is five-toed, and has short ankles to its hind feet. But it has a nose such in properties and such in size as to allow of its using the same for a hand. For it eats and drinks by lifting up its food with the aid of this organ into its mouth, and with the same organ it lifts up articles to the driver on its back; with this organ it can pluck up trees by the roots, and when walking through water it spouts the water up by means of it; and this organ is capable of being crooked or coiled at the tip, but not of flexing like a joint, for it is composed of gristle.

MEMBRANES

Aristotle, *History of Animals* III. 13. Translation of D. W. Thompson

In all sanguineous animals membranes are found. And membrane resembles a thin close-textured skin, but its qualities are different, as it admits neither of cleavage nor of extension. Membrane envelops each one of the bones and each one of the viscera, both in the larger and the smaller animals; though in the smaller animals the membranes are indiscernible from their extreme tenuity and minuteness. The largest of all the membranes are the two that surround the brain, and of these two the one that lines the bony skull is stronger and thicker than the one that envelops the brain;[1] next in order of magnitude comes the membrane that encloses the heart. If membrane be bared and cut asunder it will not grow together again, and the bone thus stripped of its membrane mortifies.

BONES

Aristotle, *On the Parts of Animals* II. 9. Translation of William Ogle

There is a resemblance between the osseous and the vascular systems; for each has a central part in which it begins, and each forms a continuous whole. For no bone in the body exists as a separate thing in itself, but each is either a portion of what may be considered a continuous whole, or at any rate is linked with the rest by contact and by attachments; so that nature may use adjoining bones either as though they were actually continuous and formed a single bone, or, for purposes of flexure, as though they

[1] The *dura mater* and *pia mater*, respectively.

were two and distinct. And similarly no blood-vessel has in itself a separate individuality; but they all form parts of one whole. For an isolated bone, if such there were, would in the first place be unable to perform the office for the sake of which bones exist; for, were it discontinuous and separated from the rest by a gap, it would be perfectly unable to produce either flexure or extension; nor only so, but it would actually be injurious, acting like a thorn or an arrow lodged in the flesh. Similarly if a vessel were isolated, and not continuous with the vascular centre, it would be unable to retain the blood within it in a proper state. For it is the warmth derived from this centre that hinders the blood from coagulating; indeed, the blood, when withdrawn from its influence, becomes manifestly putrid. Now the centre or origin of the blood-vessels is the heart, and the centre or origin of the bones, in all animals that have bones, is what is called the chine. . . .

Now the bones of viviparous animals, of such, that is, as are not merely externally but also internally viviparous,[1] vary but little from each other in point of strength, which in all of them is considerable. For the Vivipara in their bodily proportions are far above other animals, and many of them occasionally grow to an enormous size, as is the case in Libya and in hot and dry countries generally. But the greater the bulk of an animal, the stronger, the bigger, and the harder are the supports which it requires; and comparing the big animals with each other, this requirement will be most marked in those that live a life of rapine. Thus it is that the bones of males are harder than those of females; and the bones of flesh-eaters, that get their food by fighting, are harder than those of Herbivora. Of this the Lion is an example; for so hard are its bones, that, when struck, they give off sparks, as though they were stones. It may be mentioned also that the Dolphin, inasmuch as it is viviparous, is provided with bones and not with fish-spines.

In those sanguineous animals, on the other hand, that are oviparous, the bones present successive slight variations of character. Thus in Birds there are bones, but these are not so strong as the bones of the Vivipara. Then come the Oviparous fishes, where there is no bone, but merely fish-spine. In the Serpents too the bones have the character of fish-spine, excepting in the very large species, where the solid foundation of the body requires to be stronger, in order that the animal itself may be strong, the same reason prevailing as in the case of the Vivipara. Lastly, in the Selachia, as they are called, the fish-spines are replaced by cartilage. For it is necessary that the movements of these animals shall be of an undulating character; and this again requires the framework that supports the body to be made of a pliable and not of a brittle substance. Moreover, in these

[1] This is to exclude the ovo-viviparous Selachia, whose bones are cartilaginous. [Edd.]

Selachia[1] nature has used all the earthy matter on the skin; and she is unable to allot to many different parts one and the same superfluity of material. Even in viviparous animals many of the bones are cartilaginous. This happens in those parts where it is to the advantage of the surrounding flesh that its solid base shall be soft and mucilaginous. Such, for instance, is the case with the ears and nostrils; for in projecting parts, such as these, brittle substances would soon get broken. Cartilage and bone are indeed fundamentally the same thing, the differences between them being merely matters of degree. Thus neither cartilage nor bone, when once cut off, grows again. Now the cartilages of these land animals are without marrow, that is, without any distinctly separate marrow. For the marrow, which in bones is distinctly separate, is here mixed up with the whole mass, and gives a soft and mucilaginous consistence to the cartilage. But in the Selachia the chine, though it is cartilaginous, yet contains marrow; for here it stands in the stead of a bone.

THE STOMACH IN RUMINANTS

Aristotle, *History of Animals* II. 17.[2] Translation of D. W. Thompson

All the afore-mentioned animals have a stomach, and one similarly situated, that is to say, situated directly under the midriff; and they have a gut connected therewith and closing at the outlet of the residuum and at what is termed the "rectum." However, animals present diversities in the structure of their stomachs. In the first place, of the viviparous quadrupeds, such of the horned animals as are not equally furnished with teeth in both jaws are furnished with four such chambers. These animals, by the way, are those that are said to chew the cud. In these animals the oesophagus extends from the mouth downwards along the lung, from the midriff to the big stomach (or paunch); and this stomach is rough inside and semi-partitioned. And connected with it near to the entry of the oesophagus is what from its appearance is termed the "reticulum" (or honeycomb bag); for outside it is like the stomach, but inside it resembles a netted cap; and the reticulum is a great deal smaller than the stoma Connected with this is the "echinus" (or many-plies), rough inside and laminated, and of about the same size as the reticulum. Next after this comes what is called the "enystrum" (or abomasum), larger and longer than the echinus, furnished inside with numerous folds or ridges, large and smooth. After all this comes the gut.[3]

---

[1] The skin of the fishes called Selachia by Aristotle is studded with numerous tubercles, granules, or spines, of bony matter; a peculiarity designated as "placoid" by modern ichthyologists.

[2] Aristotle's description of the stomach in ruminants is quite accurate and was the result of careful dissection either by him or by his authorities. [Edd.]

[3] Cf. Aristotle, *On the Parts of Animals* III. 14. Translation of William Ogle: "For all

Such is the stomach of those quadrupeds that are horned and have an unsymmetrical dentition;[1] and these animals differ one from another in the shape and size of the parts, and in the fact of the oesophagus reaching the stomach centralwise in some cases and sideways in others. Animals that are furnished equally with teeth in both jaws have one stomach; as man, the pig, the dog, the bear, the lion, the wolf.

THE MECHANISM OF LOCOMOTION IN ANIMALS

Aristotle, *On the Progression of Animals* 9. Translation of A. S. L. Farquharson (Oxford, 1912)

What follows will explain that if there were no point at rest flexion and straightening would be impossible. Flexion is a change from a right line to an arc or an angle, straightening a change from either of these to a right line. Now in all such changes the flexion or the straightening must be relative to one point. Moreover, without flexion there could not be walking or swimming or flying. For since limbed creatures stand and take their weight alternately on one or other of the opposite legs, if one be thrust forward the other must of necessity be bent. For the opposite limbs are naturally of equal length, and the one which is under the weight must be a kind of perpendicular at right angles to the ground.

When the one leg is advanced it becomes the hypotenuse of a right-angled triangle. Its square then is equal to the square on the other side together with the square on the base. As the legs then are equal, the one at rest must bend either at the knee or, if there were any kneeless animal which walked, at some other articulation.[2] The following experiment exhibits the fact. If a man were to walk parallel to a wall in sunshine, the line described [by the shadow of his head] would be not straight but zigzag, becoming lower as he bends, and higher when he stands and lifts himself up.

It is, indeed, possible to move oneself even if the leg be not bent, in the way in which children crawl. This was the old though erroneous account of the movement of elephants. But these kinds of movements involve a

animals that have horns, the sheep for instance, the ox, the goat, the deer, and the like, have several stomachs. For since the mouth, owing to its lack of teeth, only imperfectly performs its office as regards the food, this multiplicity of stomachs is intended to make up for its shortcomings; the several cavities receiving the food one from the other in succession; the first taking the unreduced substances, the second the same when somewhat reduced, the third when reduction is complete, and the fourth when the whole has become a smooth pulp. Such is the reason why there is this multiplicity of parts and cavities in animals with such dentition. The names given to the several cavities are the paunch, the honey-comb bag, the manyplies, and the reed."

[1] Elsewhere Aristotle indicates that those animals ruminate which lack upper front teeth, e.g., kine, sheep, goats. [Edd.]

[2] I.e., the advancing leg cannot be the hypotenuse of a right-angled triangle and still be equal to the other leg. The backward leg must therefore be bent. [Edd.]

flexion in the shoulders or in the hips. Nothing at any rate could walk upright continuously and securely without flexions at the knee, but would have to move like men in the wrestling schools who crawl forward through the sand on their knees. For the upper part of the upright creature is long so that its leg has to be correspondingly long; in consequence there must be flexion. For since a stationary position is perpendicular, if that which moves cannot bend it will either fall forward as the right angle becomes acute or will not be able to progress. For if one leg is at right angles to the ground and the other is advanced, the latter will be at once equal and greater. For it will be equal to the stationary leg and also equivalent to the hypotenuse of a right-angled triangle. That which goes forward therefore must bend, and while bending one, extend the other leg simultaneously, so as to incline forward and make a stride and still remain above the perpendicular; for the legs form an isosceles triangle, and the head sinks lower when it is perpendicularly above the base on which it stands.

Of limbless animals, some progress by undulations (and this happens in two ways, either they undulate on the ground, like snakes, or up and down, like caterpillars), and undulation is a flexion; others by a telescopic action, like what are called earthworms and leeches. These go forward, first one part leading and then drawing the whole of the rest of the body up to this, and so they change from place to place. It is plain too that if the two curves were not greater than the one line which subtends them undulating animals could not move themselves; when the flexure is extended they would not have moved forward at all if the flexure or arc were equal to the chord subtended; as it is, it reaches further when it is straightened out, and then this part stays still and it draws up what is left behind.

In all the changes described that which moves now extends itself in a straight line to progress, and now is hooped; it straightens itself in its leading part, and is hooped in what follows behind. Even jumping animals all make a flexion in the part of the body which is underneath, and after this fashion, make their leaps. So too flying and swimming things progress, the one straightening and bending their wings to fly, the other their fins to swim. Of the latter some have four fins, others which are rather long, for example eels, have only two. These swim by substituting a flexion of the rest of their body for the [missing] pair of fins to complete the movement, as we have said before. Flat fish use two fins, and the flat of their body as a substitute for the absent pair of fins. Quite flat fish, like the Ray, produce their swimming movement with the actual fins and with the two extremes or semicircles of their body, bending and straightening themselves alternately.[1]

---

[1] Aristotle is not entirely correct in ascribing to all flat fish the mode of swimming observed in the ray. [Edd.]

Generation and Embryology

ARISTOTLE'S THEORY OF SEXUAL GENERATION

Aristotle, *On the Generation of Animals* I. 21.[1] Translation of Arthur Platt

At the same time the answer to the next question we have to investigate is clear from these considerations, I mean how it is that the male contributes to generation and how it is that the semen from the male is the cause of the offspring. Does it exist in the body of the embryo as a part of it from the first, mingling with the material which comes from the female? Or does the semen communicate nothing to the material body of the embryo but only to the power and movement in it? For this power is that which acts and makes, while that which is made and receives the form is the residue of the secretion in the female.[2] Now the latter alternative appears to be the right one both *a priori* and in view of the facts. For, if we consider the question on general grounds, we find that, whenever one thing is made from two of which one is active and the other passive, the active agent does not exist in that which is made;[3] and, still more generally, the same applies when one thing moves and another is moved; the moving thing does not exist in that which is moved. But the female, as female, is passive, and the male, as male, is active, and the principle of the movement comes from him. Therefore, if we take the highest genera under which they each fall, the one being active and motive and the other passive and moved, that one thing which is produced comes from them only in the sense in which a bed comes into being from the carpenter and the wood, or in which a ball comes into being from the wax and the form. It is plain then that it is not necessary that anything at all should come away from the male,[4] and if anything does come away it does not follow that this gives rise to the em-

---

[1] On Aristotle as an embryologist see Joseph Needham, *A History of Embryology* (Cambridge University Press, 1934), pp. 19–41. Among Aristotle's contributions Needham lists the introduction of the comparative method into embryology, the distinction between primary and secondary sexual characteristics, the putting of the origin of sex determination to the very beginning of embryonic development, the association of the phenomena of regeneration with the embryonic state, the formulation of the antithesis between preformation and epigenesis, the foreshadowing of the theory of recapitulation, and the assignment of the correct functions to the placenta and the umbilical cord.

It is only since the invention of the microscope that the structure and function of ova and spermatozoa have been susceptible of scientific investigation. Aristotle's philosophical argument, however, has been included here because of its profound influence on biological history. [Edd.]

[2] I.e., as much of the catamenia as is not discharged.

[3] E.g., a man, the active agent, makes a boat out of wood, the passive material; the man does not exist in the boat as a part of it.

[4] In view of this statement it is not surprising to find Aristotle giving credence to the myth that hen partridges may be impregnated by the scent of the cock (*Hist. An.* VI. 2). [Edd.]

bryo, as being in the embryo, but only as that which imparts the motion and as the form; so the medical art cures the patient.

This *a priori* argument is confirmed by the facts. For it is for this reason that some males which unite with the female do not, it appears, insert any part of themselves into the female, but on the contrary the female inserts a part of herself into the male; this occurs in some insects.[1] For the effect produced by the semen in the female (in the case of those animals whose males do insert a part) is produced in the case of these insects by the heat and power in the male animal itself when the female inserts that part of herself which receives the secretion. And therefore such animals remain united a long time, and when they are separated the young are produced quickly. For the union lasts until that which is analogous to the semen has done its work, and when they separate the female produces the embryo quickly; for the young is imperfect inasmuch as all such creatures give birth to scoleces.

What occurs in birds and oviparous fishes is the greatest proof that neither does the semen come from all parts of the male nor does he emit anything of such a nature as to exist within that which is generated, as part of the material embryo,[2] but that he only makes a living creature by the power which resides in the semen (as we said in the case of those insects whose females insert a part of themselves into the male). For if a hen-bird is in process of producing wind-eggs and is then trodden by the cock before the egg has begun to whiten and while it is all still yellow,[3] then they become fertile instead of being wind-eggs. And if while it is still yellow she be trodden by another cock, the whole brood of chicks turn out like the second cock. Hence some of those who are anxious to rear fine birds act thus; they change the cocks for the first and second treading, not as if they thought that the semen is mingled with the egg or exists in it, or that it comes from all parts of the cock; for if it did it would have come from both cocks, so that the chick would have all its parts doubled. But it is by its force that the semen of the male gives a certain quality to the material and the nutriment in the female, for the second semen added to the first can produce this effect by heat and concoction, as the egg acquires nutriment so long as it is growing.[4]

The same conclusion is to be drawn from the generation of oviparous

[1] Aristotle frequently repeats this assertion, but it is quite unfounded. [Edd.]

[2] The matter of the embryo was entirely contributed by the female, according to Aristotle. [Edd.]

[3] I.e., while the egg consists merely of the yolk in the vitelline membrane, when it first begins its passage through the oviduct. The "white of egg" is deposited round it as it gets further on in the oviduct.

[4] Clearly the chickens inherit from the second cock only if the first has failed to fertilize the ova. But Aristotle seems to be reporting the actual notions of the poultrymen. [Edd.]

fishes. When the female has laid her eggs, the male sprinkles the milt over them, and those eggs are fertilized which it reaches, but not the others; this shows that the male does not contribute anything to the quantity but only to the quality of the embryo.

From what has been said it is plain that the semen does not come from the whole of the body of the male in those animals which emit it, and that the contribution of the female to the generative product is not the same as that of the male, but the male contributes the principle of movement and the female the material. This is why the female does not produce offspring by herself, for she needs a principle, i.e., something to begin the movement in the embryo and to define the form it is to assume. Yet in some animals, as birds, the nature of the female unassisted can generate to a certain extent, for they do form something, only it is incomplete; I mean the so-called wind-eggs.

ARGUMENTS AGAINST PANGENESIS

Aristotle, *On the Generation of Animals* I. 17–18. Translation of Arthur Platt

17. . . . Now it is thought that all animals are generated out of semen, and that the semen comes from the parents. Wherefore it is part of the same inquiry to ask whether both male and female produce it or only one of them, and to ask whether it comes from the whole of the body[1] or not from the whole; for if the latter is true it is reasonable to suppose that it does not come from both parents either.[2] Accordingly, since some say that it comes from the whole of the body, we must investigate this question first.

The proofs from which it can be argued that the semen comes from each and every part of the body may be reduced to four. First, the intensity of the pleasure of coition; for the same state of feeling is more pleasant if multiplied, and that which affects all the parts is multiplied as compared to that which affects only one or a few. Secondly, the alleged fact that mutilations are inherited, for they argue that since the parent is deficient in this part the semen does not come from thence, and the result is that the corresponding part is not formed in the offspring. Thirdly, the resemblances to the parents, for the young are born like them part for part as well as in the whole body; if then the coming of the semen from the whole

[1] This was a Hippocratic view, held also by Democritus, and very similar to Darwin's famous "pangenesis."... [See p. 489, below. Edd.]

[2] . . . Aristotle's arguments against pangenesis, if we may adopt the Darwinian term for the similar though not identical theory, are of varying validity, but he is certainly right in the main. . . . On the other hand, when Aristotle denies that the female contributes semen, though he may be right in a certain sense, he is in truth fundamentally wrong, for his position is that the mother does not contribute anything resembling the semen of the male, whereas the ovum is just as important as the spermatozoon and carries inheritance equally with it.

body is cause of the resemblance of the whole, so the parts would be like because it comes from each of the parts. Fourthly, it would seem to be reasonable to say that as there is some first thing from which the whole arises, so it is also with each of the parts, and therefore if semen or seed is cause of the whole so each of the parts would have a seed peculiar to itself. And these opinions are plausibly supported by such evidence as that children are born with a likeness to their parents, not only in congenital but also in acquired characteristics; for before now, when the parents have had scars, the children have been born with a mark in the form of the scar in the same place, and there was a case at Chalcedon where the father had a brand on his arm and the letter was marked on the child, only confused and not clearly articulated.[1] That is pretty much the evidence on which some believe that the semen comes from all the body.

18. On examining the question, however, the opposite appears more likely, for it is not hard to refute the above arguments and the view involves impossibilities. First, then, the resemblance of children to parents is no proof that the semen comes from the whole body, because the resemblance is found also in voice, nails, hair, and way of moving, from which nothing comes.[2] And men generate before they yet have certain characters, such as a beard or grey hair. Further, children are like their more remote ancestors from whom nothing has come, for the resemblances recur at an interval of many generations. . . .

We may also ask whether the semen comes from each of the homogeneous parts only, such as flesh and bone and sinew, or also from the heterogeneous, such as face and hands. For if (1) from the former only, we object that the resemblance exists rather in the heterogeneous parts, such as face and hands and feet; if then it is not because of the semen coming from all parts that children resemble their parents in *these*, what is there to stop the homogeneous parts also from being like for some other reason than this? If (2) the semen comes from the heterogeneous alone, then it does not come from all parts; but it is more fitting that it should come from the homogeneous parts, for they are prior to the heterogeneous which are composed of them; and as children are born like their parents in face and hands, so they are, necessarily, in flesh and nails.[3] If (3) the semen comes from both, what would be the manner of generation? For the heterogeneous parts are composed of the homogeneous, so that to come from the former would be to come from the latter and from their composition. . . .[4]

---

[1] Legends of this sort have always been popular. . . .

[2] Aristotle assumes this for hair and nails because they have no blood-vessels, I suppose.

[3] Since face and hands are composed of flesh and nails and other homogeneous parts.

[4] There follows a long series of additional arguments to meet the third and fourth arguments of those who believe in pangenesis. [Edd.]

Again, the cuttings from a plant bear seed; clearly, therefore, even before they were cut from the parent plant, they bore their fruit from their own mass alone, and the seed did not come from *all* the plant.

But the greatest proof of all is derived from observations we have sufficiently established on insects. For, if not in all, at least in most of these, the female in the act of copulation inserts a part of herself into the male. This, as we said before, is the way they copulate, for the females manifestly insert this from below into the males above, not in all cases, but in most of those observed.[1] Hence it seems clear that, when the males do emit semen, then also the cause of the generation[2] is not its coming from all the body, but something else which must be investigated hereafter. For even if it were true that it comes from all the *body*, as they say, they ought not to claim that it comes from all *parts* of it, but only from the creative part—from the workman, so to say, not the material he works in. Instead of that, they talk as if one were to say that the semen comes from the shoes, for, generally speaking, if a son is like his father, the shoes he wears are like his father's shoes.[3]

As to the vehemence of pleasure in sexual intercourse, it is not because the semen comes from all the body, but because there is a strong friction (wherefore if this intercourse is often repeated the pleasure is diminished in the persons concerned). Moreover, the pleasure is at the end of the act, but it ought, on the theory, to be in each of the parts, and not at the same time, but sooner in some and later in others.[4]

If mutilated young are born of mutilated parents,[5] it is for the same reason as that for which they are like them. And the young of mutilated parents are not always mutilated, just as they are not always like their parents; the cause of this must be inquired into later, for this problem is the same as that.

Again, if the female does not produce semen, it is reasonable to suppose it does not come from all the body of the male either.[6] Conversely, if it does not come from all the male it is not unreasonable to suppose that it does not come from the female, but that the female is cause of the generation in some other way. Into this we must next inquire, since it is plain that the semen is not secreted from all the parts.

---

[1] See p. 416, n. 1, above. [Edd.]

[2] *Sic*; but what have we to do in this argument with the cause of *generation*? Should it not be the cause of the *resemblance*?

[3] In modern language, the germ-cell creates the body which it wears. This body clothes the germ-cells as shoes again clothe the body.

[4] Because the semen would have further to travel from some than from others.

[5] Modern science simply denies the fact *in toto*.

[6] Because the young resemble the mother as well as the father.

THE PLACENTAL SHARK

Aristotle, *History of Animals* VI. 10. Translation of D. W. Thompson

The so-called smooth shark has its eggs in betwixt the wombs like the dog-fish; these eggs shift into each of the two horns of the womb and descend, and the young develop with the navel-string attached to the womb, so that, as the egg-substance gets used up, the embryo is sustained to all appearance just as in the case of quadrupeds. The navel-string is long and adheres to the under part of the womb (each navel-string being attached as it were by a sucker), and also to the centre of the embryo in the place where the liver is situated. If the embryo be cut open, even though it has the egg-substance no longer, the food inside is egg-like in appearance. Each embryo, as in the case of quadrupeds, is provided with a chorion and separate membranes. When young the embryo has its head upwards, but downwards when it gets strong and is completed in form.[1]

SPONTANEOUS GENERATION

Aristotle, *History of Animals* V. 1; VI. 15, 16.[2] Translation of D. W. Thompson

Now there is one property that animals are found to have in common with plants. For some plants are generated from the seed of plants, whilst other plants are self-generated through the formation of some elemental principle similar to a seed; and of these latter plants some derive their nutriment from the ground, whilst others grow inside other plants, as is mentioned, by the way, in my treatise on Botany. So with animals, some spring from parent animals according to their kind, whilst others grow spontaneously and not from kindred stock, and of these instances of spontaneous generation some come from putrefying earth or vegetable matter, as is the case with a number of insects, while others are spontaneously generated in the inside of animals out of the secretions of their several organs.

---

[1] Aristotle frequently describes Selachia and is generally, though not entirely, correct in classifying them as externally viviparous. His reference to a shark (or dogfish) which has a sort of placenta is particularly remarkable. The accuracy of his description was not fully appreciated until the nineteenth century, when Johannes Müller "rediscovered" certain types of Selachia having structures that function as placentas. Thompson, *Glossary of Greek Fishes*, p. 42, notes similar observations in the sixteenth and seventeenth centuries. [Edd.]

[2] Aristotle, in common with other Greek thinkers, held that animate beings may be generated from inanimate matter. Spontaneous generation was ascribed to certain species of flowerless plants, insects, and fishes. The present passages are illustrative. Cf. also p. 459. The demonstration, within the last century, that what had been supposed to be lifeless matter actually harbored abundant microscopic life caused the rejection of the theory of spontaneous generation. But the larger issues implicit in this theory, the interrelation of the animate and the inanimate and the possibility of the creation of life *in vitro*, are by no means closed. [Edd.]

VI. 15. The great majority of fish, then, as has been stated, proceed from eggs. However, there are some fish that proceed from mud and sand, even of those kinds that proceed also from pairing and the egg. . . . Some writers actually aver that mullet all grow spontaneously. In this assertion they are mistaken, for the female of the fish is found provided with spawn, and the male with milt. However, there is a species of mullet that grows spontaneously out of mud and sand.

From the facts above enumerated it is quite proved that certain fishes come spontaneously into existence not being derived from eggs or from copulation. Such fish as are neither oviparous nor viviparous arise all from one of two sources, from mud, or from sand and from decayed matter that rises thence as a scum; for instance, the so-called froth of the small fry comes out of sandy ground. This fry is incapable of growth and of propagating its kind; after living for a while it dies away and another creature takes its place, and so, with short intervals excepted, it may be said to last the whole year through.

16. Eels are not the issue of pairing, neither are they oviparous;[1] nor was an eel ever found supplied with milt or spawn, nor are they when cut open found to have within them passages for spawn or for eggs. In point of fact, this entire species of blooded animals proceeds neither from pairing nor from the egg.

There can be no doubt that the case is so. For in some standing pools, after the water has been drained off and the mud has been dredged away, the eels appear again after a fall of rain. In time of drought they do not appear even in stagnant ponds, for the simple reason that their existence and sustenance is derived from rain-water.

There is no doubt, then, that they proceed neither from pairing nor from an egg. Some writers, however, are of opinion that they generate their kind, because in some eels little worms are found, from which they suppose that eels are derived. But this opinion is not founded on fact. Eels are derived from the so-called "earth's guts" that grow spontaneously in mud and in humid ground; in fact, eels have at times been seen to emerge out of such earthworms, and on other occasions have been rendered visible

[1] It was only in comparatively recent times that the details of the generation of the eel were discovered. The eel is not the *direct* issue of ova or pairing. The *leptocephalus brevirostris*, a small fish, previously thought to be a separate species, is the larval form of the eel. The cycle involves a migration on the part of the eels to far distant waters, the development of organs of reproduction in the course of the migration, the generation of the larval form, and the transformation thereof into eels after remigration. The fact that the leptocephalus in question is today called "casentula" by Sicilian fishermen has evoked D. W. Thompson's suggestion that an ancient form of this word lurks beneath the γῆς ἔντερα (from the supposed Doric γᾶς ἔντερα) of the text—here translated "earth's guts." [Edd.]

when the earthworms were laid open by either scraping or cutting. Such earthworms are found both in the sea and in rivers, especially where there is decayed matter: in the sea in places where seaweed abounds, and in rivers and marshes near to the edge; for it is near to the water's edge that sun-heat has its chief power and produces putrefaction. So much for the generation of the eel.

THE DEVELOPMENT OF THE CHICK

Aristotle, *History of Animals* VI. 3.[1] Translation of D. W. Thompson

Generation from the egg proceeds in an identical manner with all birds, but the full periods from conception to birth differ, as has been said. With the common hen after three days and three nights there is the first indication of the embryo; with larger birds the interval being longer, with smaller birds shorter. Meanwhile the yolk comes into being, rising towards the sharp end, where the primal element of the egg is situated, and where the egg gets hatched;[2] and the heart appears, like a speck of blood, in the white of the egg. This point beats and moves as though endowed with life, and from it two vein-ducts with blood in them trend in a convoluted course [as the egg-substance goes on growing, towards each of the two circumjacent integuments];[3] and a membrane carrying bloody fibres now envelops the yolk, leading off from the vein-ducts. A little afterwards the body is differentiated, at first very small and white. The head is clearly distinguished, and in it the eyes, swollen out to a great extent. This condition of the eyes lasts on for a good while, as it is only by degrees that they diminish in size and collapse. At the outset the under portion of the body appears insignificant in comparison with the upper portion. Of the two ducts that lead from the heart, the one proceeds towards the circumjacent integument, and the other, like a navel-string, towards the yolk. The life-element of the chick is in the white of the egg, and the nutriment comes through the navel-string out of the yolk.[4]

When the egg is now ten days old the chick and all its parts are distinctly visible. The head is still larger than the rest of its body, and the eyes

---

[1] The authenticity of the whole section is suspected by the translator, who has included within [ ] the parts that are especially doubtful. Regardless of the question of authorship the passage is important from a methodological viewpoint. [Edd.]

[2] "It is now known that the development of the embryo commences in the germinal spot or disc, situated on one side of the yolk. In consequence of the yolk opposite the germinal spot being denser than that on the side of the spot, this remains uppermost, however the egg may be rotated. . . ." (T. E. Lones, *Aristotle's Researches in Natural Science*, p. 204). [Edd.]

[3] . . . What are afterwards described as the two integuments (allantois and yolk-sac) are not yet differentiated on the third day.

[4] Just the opposite view is given in the Hippocratic treatise *On the Nature of the Child* 30: "The bird arises from the yolk of the egg, but it has its sustenance and growth from the white." [Edd.]

larger than the head, but still devoid of vision. The eyes, if removed about this time, are found to be larger than beans, and black; if the cuticle be peeled off them there is a white and cold liquid inside, quite glittering in the sunlight, but there is no hard substance whatsoever. Such is the condition of the head and eyes. At this time also the larger internal organs are visible, as also the stomach and the arrangement of the viscera; and the veins that seem to proceed from the heart are now close to the navel. From the navel there stretch a pair of veins; one[1] towards the membrane that envelops the yolk (and, by the way, the yolk is now liquid, or more so than is normal), and the other[2] towards that membrane which envelops collectively the membrane wherein the chick lies, the membrane of the yolk, and the intervening liquid. [For, as the chick grows, little by little one part of the yolk goes upward, and another part downward, and the white liquid is between them; and the white of the egg is underneath the lower part of the yolk, as it was at the outset.] On the tenth day the white is at the extreme outer surface, reduced in amount, glutinous, firm in substance, and sallow in colour.

The disposition of the several constituent parts is as follows. First and outermost comes the membrane of the egg, not that of the shell, but underneath it. Inside this membrane[3] is a white liquid; then comes the chick, and a membrane round about it,[4] separating it off so as to keep the chick free from the liquid; next after the chick comes the yolk, into which one of the two veins was described as leading, the other one leading into the enveloping white substance. [A membrane with a liquid resembling serum envelops the entire structure. Then comes another membrane right round the embryo, as has been described, separating it off against the liquid. Underneath this comes the yolk, enveloped in another membrane (into which yolk proceeds the navel-string that leads from the heart and the big vein), so as to keep the embryo free of both liquids.]

About the twentieth day, if you open the egg and touch the chick, it moves inside and chirps; and it is already coming to be covered with down, when, after the twentieth day is past, the chick begins to break the shell. The head is situated over the right leg close to the flank, and the wing is placed over the head; and about this time is plain to be seen the membrane resembling an after-birth that comes next after the outermost membrane of the shell, into which membrane the one of the navel-strings was described as leading (and, by the way, the chick in its entirety is now within it), and so also is the other membrane resembling an after-birth, namely, that surrounding the yolk, into which the second navel-string was described

[1] The vitelline vein and artery.

[2] The allantoic vein and artery.

[3] The allantois.

[4] The amnion.

as leading; and both of them were described as being connected with the heart and the big vein. At this conjuncture the navel-string that leads to the outer after-birth collapses and becomes detached from the chick, and the membrane that leads into the yolk is fastened on to the thin gut of the creature, and by this time a considerable amount of the yolk is inside the chick and a yellow sediment is in its stomach. About this time it discharges residuum[1] in the direction of the outer after-birth, and has residuum inside its stomach; and the outer residuum is white [and there comes a white substance inside]. By and by the yolk, diminishing gradually in size, at length becomes entirely used up and comprehended within the chick (so that, ten days after hatching, if you cut open the chick, a small remnant of the yolk is still left in connexion with the gut), but it is detached from the navel, and there is nothing in the interval between, but it has been used up entirely. During the period above referred to the chick sleeps, wakes up, makes a move and looks up and chirps; and the heart and the navel together palpitate as though the creature were respiring. So much as to generation from the egg in the case of birds.

Birds lay some eggs that are unfruitful, even eggs that are the result of copulation, and no life comes from such eggs by incubation; and this phenomenon is observed especially with pigeons.

Twin eggs have two yolks. In some twin eggs a thin partition of white intervenes to prevent the yolks mixing with each other, but some twin eggs are unprovided with such partition, and the yolks run into one another. There are some hens that lay nothing but twin eggs, and in their case the phenomenon regarding the yolks has been observed. For instance, a hen has been known to lay eighteen eggs, and to hatch twins out of them all, except those that were wind-eggs; the rest were fertile (though, by the way, one of the twins is always bigger than the other), but the eighteenth was abnormal or monstrous.

Cf. Hippocratic Collection, *On the Nature of the Child* 29

The embryo is in a membrane and at the center of the embryo is the umbilicus. Now the embryo draws breath to itself and then breathes out. There are membranes connected with the umbilicus. And if you use the evidence I am going to set forth, you will find that the nature of the embryo is in all other respects, from beginning to end, just as I have described it in these discourses.

For if you take twenty or more eggs and put them under two or more hens for hatching, and each day from the second to the last, that is, the day of the hatching, take one egg, break it open and examine it, you will find everything as I have described it, in so far as the nature of a bird may

[1] The urates of the allantoic fluid.

be compared with that of man. You will find that the membranes stretch from the umbilicus, and that all the other things I spoke of in connection with the human embryo are similarly in a bird's egg, from beginning to end. And if one has not yet seen it, he will be surprised at finding an umbilicus in a bird's egg.

Habits of Animals[1]

BEES

Aristotle, *History of Animals* IX. 40.[2] Translation of D. W. Thompson

Of insects, there is a genus that has no one name that comprehends all the species, though all the species are akin to one another in form; it consists of all the insects that construct a honeycomb; to wit, the bee, and all the insects that resemble it in form. There are nine varieties, of which six are gregarious—the bee, the king-bee,[3] the drone-bee, the annual wasp, and, furthermore, the anthrene (or hornet), and the tenthredo (or ground-wasp); three are solitary—the smaller siren, of a dun colour, the larger siren, black and speckled, and the third, the largest of all, that is called the humble-bee. Now ants never go a-hunting, but gather up what is ready to hand; the spider makes nothing, and lays up no store, but simply goes a-hunting for its food; while the bee—for we shall by and by treat of nine varieties—does not go a-hunting, but constructs its food out of gathered material and stores it away, for honey is the bee's food. This fact is shown by the bee-keepers' attempt to remove the combs; for the bees, when they are fumigated, and are suffering great distress from the process, then devour the honey most ravenously, whereas at other times they are never observed to be so greedy, but apparently are thrifty and disposed to lay by for their future sustenance. They have also another food which is called bee-bread; this is scarcer than honey and has a sweet fig-like taste; this they carry as they do the wax on their legs.

Very remarkable diversity is observed in their methods of working and their general habits. When the hive has been delivered to them clean and empty, they build their waxen cells, bringing in the juice of all kinds of flowers and the "tears" or exuding sap of trees, such as willows and elms and such others as are particularly given to the exudation of gum. With this material they besmear the ground-work, to provide against attacks of other creatures; the bee-keepers call this stuff "stop-wax." They

[1] See also the section on Animal Psychology, p. 533, below. [Edd.]

[2] In addition to this fine passage on the economy of the hive, Aristotle has a less successful discussion of the generation of bees (*On the Generation of Animals* III. 10). The importance of beekeeping in antiquity is evidenced by the extensive observations and literary records on the subject (see Pliny, *Natural History* XI; Vergil, *Georgics* IV; Varro, *Res Rusticae* III). [Edd.]

[3] We call this the queen bee. [Edd.]

also with the same material narrow by side-building the entrances to the hive if they are too wide. They first build cells for themselves; then for the so-called kings and the drones; for themselves they are always building, for the kings only when the brood of young is numerous, and cells for the drones they build if a superabundance of honey should suggest their doing so. They build the royal cells next to their own, and they are of small[1] bulk; the drones' cells they build near by, and these latter are less in bulk than the bees' cells. They begin building the combs downwards from the top of the hive, and go down and down building many combs connected together until they reach the bottom. The cells, both those for the honey and those also for the grubs, are double-doored; for two cells are ranged about a single base, one pointing one way and one the other, after the manner of a double (or hour-glass-shaped) goblet. The cells that lie at the commencement of the combs and are attached to the hives, to the extent of two or three concentric circular rows, are small and devoid of honey; the cells that are well filled with honey are most thoroughly luted with wax. At the entry to the hive the aperture of the doorway is smeared with mitys; this substance is a deep black, and is a sort of dross or residual by-product of wax; it has a pungent odour, and is a cure for bruises and suppurating sores. The greasy stuff that comes next is pitch-wax; it has a less pungent odour and is less medicinal than the mitys. Some say that the drones construct combs by themselves in the same hive and in the same comb that they share with the bees; but that they make no honey, but subsist, they and their grubs also, on the honey made by the bees. The drones, as a rule, keep inside the hive; when they go out of doors, they soar up in the air in a stream, whirling round and round in a kind of gymnastic exercise; when this is over, they come inside the hive and feed to repletion ravenously. The kings never quit the hive, except in conjunction with the entire swarm, either for food or for any other reason. They say that, if a young swarm go astray, it will turn back upon its route and by the aid of scent seek out its leader. It is said that if he is unable to fly he is carried by the swarm, and that if he dies the swarm perishes; and that, if this swarm outlives the king for a while and constructs combs, no honey is produced and the bees soon die out. Bees scramble up the stalks of flowers and rapidly gather the bees-wax with their front legs; the front legs wipe it off on to the middle legs, and these pass it on to the hollow curves of the hind-legs; when thus laden, they fly away home, and one may see plainly that their load is a heavy one. On each expedition the bee does not fly from a flower of one kind

[1] I.e., small collectively, for the queen cells are, individually, the largest of all. So, too, the cells of the drones are individually larger but collectively smaller than those of the workers. [Edd.]

to a flower of another, but flies from one violet, say, to another violet, and never meddles with another flower until it has got back to the hive; on reaching the hive they throw off their load, and each bee on his return is accompanied by three or four companions. One cannot well tell what is the substance they gather, nor the exact process of their work. Their mode of gathering wax has been observed on olive-trees, as owing to the thickness of the leaves the bees remain stationary for a considerable while. After this work is over, they attend to the grubs. There is nothing to prevent grubs, honey, and drones being all found in one and the same comb. As long as the leader is alive, the drones are said to be produced apart by themselves; if he be no longer living, they are said to be reared by the bees in their own cells, and under these circumstances to become more spirited: for this reason they are called "sting-drones," not that they really have stings, but that they have the wish, without the power, to use such weapons. The cells for the drones are larger than the others; sometimes the bees construct cells for the drones apart, but usually they put them in amongst their own; and when this is the case the bee-keepers cut the drone-cells out of the combs. . . . Each comb is of one kind only: that is, it contains either bees only, or grubs only, or drones only; if it happen, however, that they make in one and the same comb all these kinds of cells, each separate kind will be built in a continuous row right through. The long bees build uneven combs, with the lids of the cells protuberant, like those of the anthrene; grubs and everything else have no fixed places, but are put anywhere; from these bees come inferior kings, a large quantity of drones, and the so-called robber-bee; they produce either no honey at all, or honey in very small quantities. Bees brood over the combs and so mature them; if they fail to do so, the combs are said to go bad and to get covered with a sort of spider's web. If they can keep brooding over the part undamaged, the damaged part simply eats itself away; if they cannot so brood, the entire comb perishes; in the damaged combs small worms are engendered, which take on wings and fly away. When the combs keep settling down, the bees restore the level surface, and put props underneath the combs to give themselves free passage-room; for if such free passage be lacking they cannot brood, and the cobwebs come on. When the robber-bee and the drone appear, not only do they do no work themselves, but they actually damage the work of the other bees; if they are caught in the act, they are killed by the working-bees. These bees also kill without mercy most of their kings, and especially kings of the inferior sort; and this they do for fear a multiplicity of kings should lead to a dismemberment of the hive. They kill them especially when the hive is deficient in grubs, and a swarm is not intended to take place; under these circumstances they destroy the cells of the kings if they have been prepared, on the ground that these kings are always ready to lead out swarms. They destroy also the combs of the drones if a failure in the honey supply be threatening and the hive runs short of provisions; under such circumstances they fight desperately with all who try to take their honey, and eject from the hive all the resident drones; and oftentimes the drones are to be seen sitting apart in the hive. The little bees fight vigorously with the long kind, and try to banish them from the hives; if they succeed, the hive will be unusually productive, but if the bigger bees get left mistresses of the field they pass the time in idleness, and do no good at all but die out before the autumn. Whenever the working-bees kill an enemy they try to do so out of doors; and whenever one of their own body dies, they carry the dead bee out of doors also. The so-called robber-bees spoil their own combs, and, if they can do so unnoticed, enter and spoil the combs of other bees; if they are caught in the act they are put to death. It is no easy task for them to escape detection, for there are sentinels on guard at every entry; and, even if they do escape detection on entering, afterwards from a surfeit of food they cannot fly, but go rolling about in front of the hive, so that their chances of escape are small indeed. The kings are never themselves seen outside the hive except with a swarm in flight: during which time all the other bees cluster around them. When the flight of a swarm is imminent, a monotonous and quite peculiar sound made by all the bees is heard for several days, and for two or three days in advance a few bees are seen flying round the hive; it has never as yet been ascertained, owing to the difficulty of the observation, whether or no the king is among these. When they have swarmed, they fly away and separate off to each of the kings; if a small swarm happens to settle near a large one, it will shift to join this large one, and if the king whom they have abandoned follows them, they put him to death. So much for the quitting of the hive and the swarm-flight. Separate detachments of bees are told off for diverse operations; that is, some carry flower-produce, others carry water, others smooth and arrange the combs. A bee carries water when it is rearing grubs. No bee ever settles on the flesh of any creature, or ever eats animal food. They have no fixed date for commencing work; but when their provender is forthcoming and they are in comfortable trim, and by preference in summer, they set to work, and when the weather is fine they work incessantly. The bee, when quite young and in fact only three days old, after shedding its chrysalis-case, begins to work if it be well fed. When a swarm is settling, some bees detach themselves in seach of food and return back to the swarm. In hives that are in good condition the production of young bees is discontinued only for the forty days that follow the winter solstice. When the grubs are grown, the bees put food beside them and cover them with a coating of wax; and, as soon as the grub is strong enough, he of his own accord breaks the lid and comes out. Creatures that make their appearance in hives and spoil the combs the working-bees clear out, but the other bees from sheer laziness look with indifference on damage done to their produce. When the bee-masters take out the combs, they leave enough food behind for winter use; if it be sufficient in quantity, the occupants of the hive will survive; if it be insufficient, then, if the weather be rough, they die on the spot, but if it be fair, they fly away and desert the hive. They feed on honey summer and winter; but they store up another article of food resembling wax in hardness, which by some is called sandarace, or bee-bread. . . . The elder bees do the indoor work, and are rough and hairy from staying indoors; the young bees do the outer carrying, and are comparatively smooth. They kill the drones also when in their work they are confined for room; the drones, by the way, live in the innermost recess of the hive. On one occasion, when a hive was in poor condition, some of the occupants assailed a foreign hive; proving victorious in a combat they took to carrying off the honey; when the bee-keeper tried to kill them, the other bees came out and tried to beat off the enemy but made no attempt to sting the man. The diseases that chiefly attack prosperous hives are first of all the clerus—this consists in a growth of little worms on the floor, from which, as they develop, a kind of cobweb grows over the entire hive, and the combs decay; another diseased condition is indicated in a lassitude on the part of the bees and in malodorousness of the hive. Bees feed on thyme; and the white thyme is better than the red. In summer the place for the hive should be cool, and in winter warm. They are very apt to fall sick if the plant they are at work on be mildewed. In a high wind they carry a stone by way of ballast to steady them. If a stream be near at hand, they drink from it and from it only, but before they drink they first deposit their load; if there be no water near at hand, they disgorge their honey as they drink elsewhere, and at once make off to work. There are two seasons for making honey, spring and autumn; the spring honey is sweeter, whiter, and in every way better than the autumn honey. Superior honey comes from fresh comb, and from young shoots; the red honey is inferior, and owes its inferiority to the comb in which it is deposited, just as wine is apt to be spoiled by its cask; consequently, one should have it looked to and dried. When the thyme is in flower and the comb is full, the honey does not harden. The honey that is golden in hue is excellent. White honey does not come from thyme pure and simple; it is good as a salve for sore eyes and wounds. Poor honey always floats on the surface and should be skimmed off; the fine clear honey rests below. When the floral world is in full bloom, then they make wax; consequently you must then take the wax out of the hive, for they go to work on new wax at once. The flowers from which they gather honey are as follows: the spindle-tree, the melilot-clover, king's-spear, myrtle, flowering-reed, withy, and broom. When they work at thyme, they mix in water before sealing up the comb. . . . They all either fly to a distance to discharge their excrement or make the discharge into one single comb.

MARINE ANIMALS

Aristotle, *History of Animals* IX. 37. Translation of D. W. Thompson

In marine creatures, also, one may observe many ingenious devices adapted to the circumstances of their lives.[1] For the accounts commonly given of the so-called fishing-frog are quite true; as are also those given of the torpedo. The fishing-frog has a set of filaments that project in front of its eyes; they are long and thin like hairs, and are round at the tips; they lie on either side, and are used as baits. Accordingly, when the animal stirs up a place full of sand and mud and conceals itself therein, it raises the filaments, and, when the little fish strike against them, it draws them in underneath into its mouth. The torpedo[2] narcotizes the creatures that it wants to catch, overpowering them by the power of shock that is resident in its body, and feeds upon them; it also hides in the sand and mud, and catches all the creatures that swim in its way and come under its narcotizing influence. This phenomenon has been actually observed in operation. The sting-ray also conceals itself, but not exactly in the same way. That the creatures get their living by this means is obvious from the fact that, whereas they are peculiarly inactive, they are often caught with mullets in their interior, the swiftest of fishes. Furthermore, the fishing-frog is usually thin when he is caught after losing the tips of his filaments, and the torpedo is known to cause a numbness even in human beings. Again, the hake, the ray, the flat-fish, and the angel-fish burrow in the sand, and after concealing themselves angle with the filaments in their mouths, that fishermen call their fishing rods, and the little creatures on which they feed swim up to the filaments, taking them for bits of seaweed such as they feed upon.

[1] It is difficult to separate the consideration of habits from the structure of animals. Since the present passage is concerned with the use of structural features in the life of the organism it is included here. See also the section on Animal Psychology, below. [Edd.]

[2] The numerous references to electric fish in ancient authors are collected by P. Kellaway in *Bulletin of the History of Medicine* 30 (1946) 117–133. Scribonius Largus and others employ the numbing action of such fish as a therapeutic measure in headache and gout. According to Charles Singer (*Studies in the History and Method of Science* II. 49–50) "the first to suggest the true nature of the shock of electric fish was Musschenbroek in 1762, who compared their shocks to those of a Leyden jar." See also Thompson, *Glossary of Greek Fishes*, pp. 169 ff.] [Edd.]

Wherever an anthias-fish is seen, there will be no dangerous creatures in the vicinity, and sponge-divers will dive in security, and they call these signal-fishes "holy-fish." It is a sort of perpetual coincidence, like the

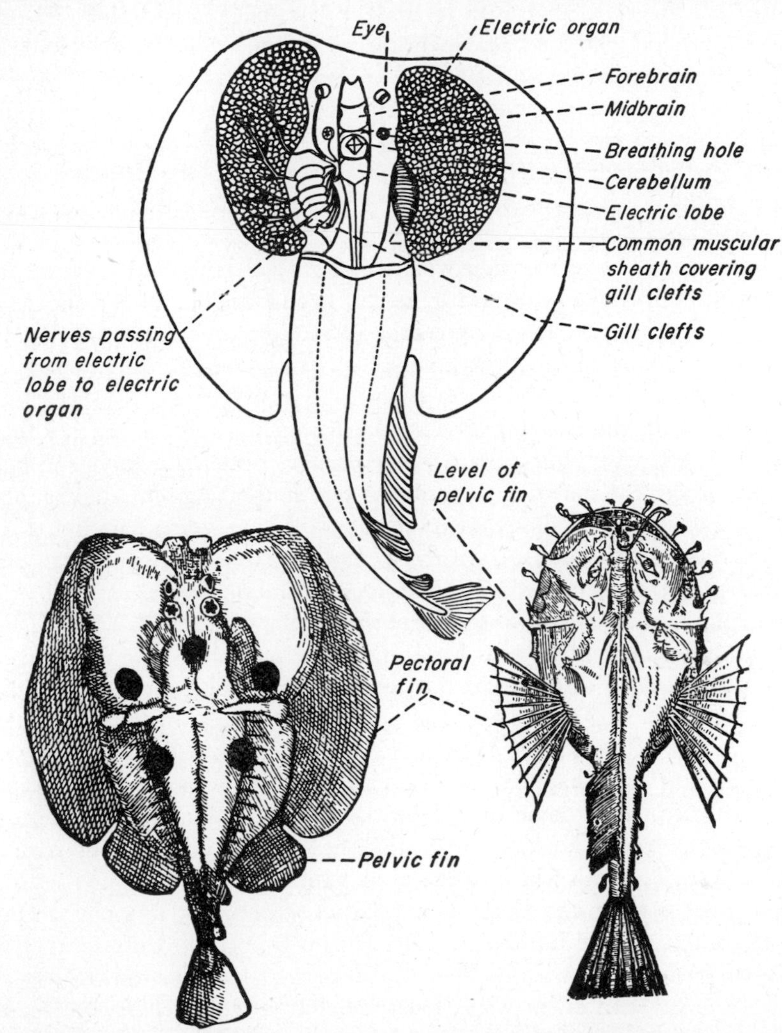

Top and lower left: torpedo-fish. Lower right: angler-fish (fishing-frog), showing filaments. From Charles Singer, *A Short History of Biology*, p. 23 (Oxford, Clarendon Press, 1931).

fact that wherever snails are present you may be sure there is neither pig nor partridge in the neighbourhood; for both pig and partridge eat up the snails.

The sea-serpent resembles the conger in colour and shape, but is of lesser bulk and more rapid in its movements. If it be caught and thrown away, it will bore a hole with its snout and burrow rapidly in the sand; its snout, by the way, is sharper than that of ordinary serpents. The so-called sea-scolopendra,[1] after swallowing the hook, turns itself inside out until it ejects it, and then it again turns itself outside in. The sea-scolopendra, like the land-scolopendra, will come to a savoury bait; the creature does not bite with its teeth, but stings by contact with its entire body, like the so-called sea-nettle. The so-called fox-shark, when it finds it has swallowed the hook, tries to get rid of it as the scolopendra does, but not in the same way; in other words, it runs up the fishing-line, and bites it off short; it is caught in some districts in deep and rapid waters, with night-lines.

The bonitos swarm together when they espy a dangerous creature, and the largest of them swim round it, and if it touches one of the shoal they try to repel it; they have strong teeth. Amongst other large fish, a lamia-shark, after falling in amongst a shoal, has been seen to be covered with wounds.

Of river-fish, the male of the sheat-fish is remarkably attentive to the young. The female after parturition goes away; the male stays and keeps on guard where the spawn is most abundant, contenting himself with keeping off all other little fishes that might steal the spawn or fry, and this he does for forty or fifty days, until the young are sufficiently grown to make away from the other fishes for themselves. The fishermen can tell where he is on guard: for, in warding off the little fishes, he makes a rush in the water and gives utterance to a kind of muttering noise. He is so earnest in the performance of his parental duties that the fishermen at times, if the eggs be attached to the roots of water-plants deep in the water, drag them into as shallow a place as possible; the male fish will still keep by the young, and, if it so happen, will be caught by the hook when snapping at the little fish that come by; if, however, he be sensible by experience of the danger of the hook, he will still keep by his charge, and with his extremely strong teeth will bite the hook in pieces. . . .[2]

The nautilus (or argonaut) is a poulpe or octopus, but one peculiar both in its nature and its habits. It rises up from deep water and swims on the surface; it rises with its shell down-turned in order that it may rise the more easily and swim with it empty, but after reaching the surface it shifts the position of the shell. In between its feelers it has a certain amount of web-growth, resembling the substance between the toes of web-

[1] The identification of this animal is uncertain. [Edd.]

[2] Cf. *History of Animals* VI. 14, 568*b*15–17. This whole account seems to have lacked confirmation until Louis Agassiz, in the middle of the nineteenth century, found in America types of Siluridae of which the males stood guard over the eggs. He then showed that the glanis of the Achelous (presumably the habitat of the species described by Aristotle) differed in this respect from the species found in the other rivers of Europe. [Edd.]

footed birds; only that with these latter the substance is thick, while with the nautilus it is thin and like a spider's web. It uses this structure, when a breeze is blowing, for a sail, and lets down some of its feelers alongside as rudder-oars. If it be frightened, it fills its shell with water and sinks.[1] With regard to the mode of generation and the growth of the shell knowledge from observation is not yet satisfactory; the shell, however, does not appear to be there from the beginning, but to grow in their case as in that of other shell-fish; neither is it ascertained for certain whether the animal can live when stripped of the shell.

Ecology

EFFECT OF LOCATION AND CLIMATE

Aristotle, *History of Animals* VIII. 18, 19, 28. Translation of D. W. Thompson

18. Animals do not all thrive at the same seasons, nor do they thrive alike during all extremes of weather. Further, animals of diverse species are in a diverse way healthy or sickly at certain seasons; and, in point of fact, some animals have ailments that are unknown to others. Birds thrive in times of drought, both in their general health and in regard to parturition, and this is especially the case with the cushat; fishes, however, with a few exceptions, thrive best in rainy weather; on the contrary, rainy seasons are bad for birds—and so by the way is much drinking—and drought is bad for fishes. . . .

19. The majority of fishes, as has been stated, thrive best in rainy seasons. Not only have they food in greater abundance at this time, but in a general way rain is wholesome for them just as it is for vegetation—for, by the way, kitchen vegetables, though artificially watered, derive benefit from rain; and the same remark applies even to reeds that grow in marshes, as they hardly grow at all without a rainfall. That rain is good for fishes may be inferred from the fact that most fishes migrate to the Euxine for the summer; for owing to the number of the rivers that discharge into this sea its water is exceptionally fresh, and the rivers bring down a large supply of food. Besides, a great number of fishes, such as the bonito and the mullet, swim up the rivers and thrive in the rivers and marshes. The sea-gudgeon also fattens in the rivers, and, as a rule, countries abounding in lagoons furnish unusually excellent fish. While most fishes, then, are benefited by rain they are chiefly benefited by summer rain; or we may state the case thus, that rain is good for fishes in spring, summer, and autumn, and fine dry weather in winter. As a general rule what is good for men is good for fishes also. . . .

---

[1] The description of the nautilus's use of its membrane and the other details of its locomotion are incorrect. [Edd.]

Particular places suit particular fishes; some are naturally fishes of the shore, and some of the deep sea, and some are at home in one or the other of these regions, and others are common to the two and are at home in both. Some fishes will thrive in one particular spot, and in that spot only. As a general rule it may be said that places abounding in weeds are wholesome; at all events, fishes caught in such places are exceptionally fat: that is, such fishes as inhabit all sorts of localities as well. The fact is that weed-eating fishes find abundance of their special food in such localities, and carnivorous fish find an unusually large number of smaller fish.

28. Variety in animal life may be produced by variety of locality: thus in one place an animal will not be found at all, in another it will be small, or short-lived, or will not thrive. Sometimes this sort of difference is observed in closely adjacent districts. Thus, in the territory of Miletus, in one district cicadas are found while there are none in the district close adjoining; and in Cephalenia there is a river on one side of which the cicada is found and not on the other. In Pordoselene there is a public road on one side of which the weasel is found but not on the other. In Boeotia the mole is found in great abundance in the neighbourhood of Orchomenus, but there are none in Lebadia though it is in the immediate vicinity, and if a mole be transported from the one district to the other it will refuse to burrow in the soil. The hare cannot live in Ithaca if introduced there; in fact it will be found dead, turned towards the point of the beach where it was landed. The horseman-ant is not found in Sicily; the croaking frog has only recently appeared in the neighbourhood of Cyrene. In the whole of Libya there is neither wild boar, nor stag, nor wild goat; and in India, according to Ctesias—no very good authority, by the way—there are no swine, wild or tame, but animals that are devoid of blood and such as go into hiding or go torpid are all of immense size there. In the Euxine there are no small molluscs nor testaceans, except a few here and there; but in the Red Sea all the testaceans are exceedingly large. . . .

In Egypt animals, as a rule, are larger than their congeners in Greece, as the cow and the sheep; but some are less, as the dog, the wolf, the hare, the fox, the raven, and the hawk; others are of pretty much the same size, as the crow and the goat. The difference, where it exists, is attributed to the food, as being abundant in one case and insufficient in another, for instance, for the wolf and the hawk; for provision is scanty for the carnivorous animals, small birds being scarce; food is scanty also for the hare and for all frugivorous animals, because neither the nuts nor the fruit last long.

In many places the climate will account for peculiarities; thus in Illyria, Thrace, and Epirus the ass is small, and in Gaul and in Scythia the ass is not found at all, owing to the coldness of the climate of these countries.

Animal Pathology

DISEASES OF VARIOUS ANIMALS

The prevailing agricultural economy in the ancient world contributed to the interest in animal diseases and their cure. Not only zoologists, but also writers on agriculture, e.g., Cato, Varro, Columella, and Palladius, deal with the diseases and treatment of farm animals. There are also some special treatises (e.g., Vegetius, in the fourth century) and later collections treating especially of horse medicine, the *Corpus Hippiatricorum Graecorum*, and the *Mulomedicina Chironis*. Noteworthy in all these treatises is the emphasis on the understanding of symptoms and on the prevention of disease (segregation of sick animals), and the absence of any magical element.

Aristotle, *History of Animals* VIII. 19–25. Translation of D. W. Thompson

19. . . . The tunny and the sword-fish are infested with a parasite[1] about the rising of the Dog-star; that is to say, about this time both these fishes have a grub beside their fins that is nicknamed the "gadfly." It resembles the scorpion in shape, and is about the size of the spider. So acute is the pain it inflicts that the sword-fish will often leap as high out of the water as a dolphin; in fact it sometimes leaps over the bulwarks of a vessel and falls back on the deck. . . .

20. . . . To shell-fish in general drought is unwholesome. During dry weather they decrease in size and degenerate in quality; and it is during such weather that the red scallop is found in more than usual abundance. In the Pyrrhaean Strait the clam was exterminated, partly by the dredging-machine used in their capture, and partly by long-continued droughts. Rainy weather is wholesome to the generality of shell-fish, owing to the fact that the sea-water then becomes exceptionally sweet. In the Euxine, owing to the coldness of the climate, shell-fish are not found: nor yet in rivers, excepting a few bivalves here and there. Univalves, by the way, are very apt to freeze to death in extremely cold weather. . . .

21. [2]To turn to quadrupeds, the pig suffers from three diseases, one of which is called branchos,[3] a disease attended with swellings about the wind-pipe and the jaws. It may break out in any part of the body; very often it attacks the foot, and occasionally the ear; the neighbouring parts also soon rot, and the decay goes on until it reaches the lungs, when the animal succumbs. The disease develops with great rapidity, and the moment it sets in the animal gives up eating. The swineherds know but one way to cure it, namely, by complete excision, when they detect the first signs of the

[1] These parasites are parasitic copepods. . . .

[2] The following passages on veterinary medicine have been held to be a post-Aristotelian interpolation. [Edd.]

[3] The translator believes that at least two diseases have been confused here, the fatal disease of anthrax, and the generally curable foot-and-mouth disease. [Edd.]

disease. There are two other diseases, which are both alike termed craurus.[1] The one is attended with pain and heaviness in the head, and this is the commoner of the two, the other with diarrhoea. The latter is incurable, the former is treated by applying wine fomentations to the snout and rinsing the nostrils with wine. Even this disease is very hard to cure; it has been known to kill within three or four days. The animal is chiefly subject to branchos[2] when it gets extremely fat, and when the heat has brought a good supply of figs. The treatment is to feed on mashed mulberries, to give repeated warm baths, and to lance the under part of the tongue.

Pigs with flabby flesh are subject to measles[3] about the legs, neck, and shoulders, for the pimples develop chiefly in these parts. If the pimples are few in number the flesh is comparatively sweet, but if they be numerous it gets watery and flaccid. The symptoms of measles are obvious, for the pimples show chiefly on the under side of the tongue, and if you pluck the bristles off the chine the skin will appear suffused with blood, and further the animal will be unable to keep its hind-feet at rest. Pigs never take this disease while they are mere sucklings. The pimples may be got rid of by feeding on a kind of spelt called tiphe; and this spelt, by the way, is very good for ordinary food. The best food for rearing and fattening pigs is chickpeas and figs, but the one thing essential is to vary the food as much as possible, for this animal, like animals in general, delights in a change of diet; and it is said that one kind of food blows the animal out, that another superinduces flesh, and that another puts on fat, and that acorns, though liked by the animal, render the flesh flaccid. Besides, if a sow eats acorns in great quantities, it will miscarry, as is also the case with the ewe; and, indeed, the miscarriage is more certain in the case of the ewe than in the case of the sow. The pig is the only animal known to be subject to measles.

22. Dogs suffer from three diseases; rabies, quinsy, and sore feet. Rabies drives the animal mad, and any animal whatever, excepting man, will take the disease if bitten by a dog so afflicted; the disease is fatal to the dog itself, and to any animal it may bite, man excepted. Quinsy[4] also is fatal to dogs; and only a few dogs recover from disease of the feet. The camel, like the dog, is subject to rabies. The elephant, which is reputed to enjoy immunity from all other illnesses, is occasionally subject to flatulency.

23. Cattle in herds are liable to two diseases, foot-sickness[5] and craurus.[6]

[1] The two varieties of κραῦρος cannot be identified with any approach to certainty. . . .

[2] Here, according to the translator, the foot-and-mouth disease seems to be meant. [Edd.]

[3] The "measles" here alluded to are the cysticercus-cysts of the tape-worm. . . .

[4] This κυνάγχη may very possibly be "distemper."

[5] Perhaps foot-and-mouth disease. [Edd.]

[6] Perhaps here a type of pneumonia. [Edd.]

In the former their feet suffer from eruptions, but the animal recovers from the disease without even the loss of the hoof. It is found of service to smear the horny parts with warm pitch. In craurus, the breath comes warm at short intervals; in fact, craurus in cattle answers to fever in man. The symptoms of the disease are drooping of the ears and disinclination for food. The animal soon succumbs and when the carcase is opened the lungs are found to be rotten.

24. Horses out at pasture are free from all diseases excepting disease of the feet. From this disease they sometimes lose their hooves: but after losing them they grow them soon again, for as one hoof is decaying it is being replaced by another. Symptoms of the malady are a sinking in and wrinkling of the lip in the middle under the nostrils, and in the case of the male, a twitching of the right testicle.

Stall-reared horses are subject to very numerous forms of disease. They are liable to a disease called "eileus." Under this disease the animal trails its hind-legs under its belly so far forward as almost to fall back on its haunches; if it goes without food for several days and turns rabid, it may be of service to draw blood, or to castrate the male.[1] The animal is subject also to tetanus: the veins get rigid, as also the head and neck, and the animal walks with its legs stretched out straight. The horse suffers also from abscesses. Another painful illness afflicts them called the "barley-surfeit." The symptoms are a softening of the palate and the heat of the breath; the animal may recover through the strength of its constitution, but no formal remedies are of any avail.

25. The ass suffers chiefly from one particular disease which they call "melis."[2] It arises first in the head, and a clammy humour runs down the nostrils, thick and red; if it stays in the head the animal may recover, but if it descends into the lungs the animal will die. Of all animals of its kind it is the least capable of enduring extreme cold, which circumstance will account for the fact that the animal is not found on the shores of the Euxine, nor in Scythia.

BOTANY

Since the major part of human food and drink, medicines and perfumes, and very much of the material of primitive war and industry have been derived directly from various plants, some knowledge of the latter goes back to prehistoric times. Nor have narrowly practical considerations ever completely excluded other types of interest such as the magical, the religious, or the purely noetic satisfaction of curiosity which has led so many to become amateur

[1] εἴλευς probably includes cases of ordinary colic; but the description further suggests the disease known as "staggers"... in which disease the legs are contorted into curious positions.

[2] Perhaps glanders, though the translator notes that this is almost invariably fatal in the ass. [Edd.]

naturalists. The resultant botanical common knowledge finds expression in Greek and Latin as in other literatures, e.g., in Hesiod's *Works and Days* and Vergil's *Georgics*.

In his botanical treatises, Theophrastus freely utilized the work of many predecessors. The latter may be divided into three groups: (1) rhizotomists (root cutters), who gathered plants for various purposes, including the preparation of medicines, perfumes, and poisons; (2) agronomists, who wrote on agriculture in works classically known as "geoponics" or "georgics"; and (3) naturalistic philosophers.

Amidst much folklore and some superstition, the work of the first two of these groups seems to have contained much of the keenly intelligent observations of nature so characteristic of the Greeks. In the writings of Theophrastus we have references to druggists such as Thrasyas of Mantineia and naturalists such as Androtion and Cleidemus, indicating that these men had a good deal of shrewd knowledge of the life of various trees, shrubs, and herbs and knew about the influence of climate, soil, moisture, manure, and cultivation on the growth and diseases of plants.

In dealing with the theory of nature, the Greek philosophers could not well avoid reference to the character and significance of plant life. The early Ionians seem to have emphasized the importance of fluids in the life of plants as well as of animals, and this may have led to the medical doctrine of the four humors. If Menestor of Sybaris represents Pythagorean botany, it emphasized the importance of heat and cold in plant life. Thus he explained the curling or twisting of ivy as due to heat (Theophrastus, *On the Causes of Plants* I. 21.6). The emphasis on sunlight was also expressed by Anaxagoras in the dictum that while the earth was the mother of plants the sun was their father. Besides other speculation on the sex and growth of trees and the symmetry of pores in the leaves, Empedocles suggested certain serviceable analogies between vegetative and animal organic functions, in which he was followed by some of the Hippocratic writers and by Aristotle. While it has been doubted that Aristotle himself composed botanical treatises, his extant works contain numerous botanical passages. These passages together indicate a general philosophy of plant life but no detailed study such as we find in his zoological treatises. In general he is interested in function rather than structure, the parts of plants being treated as organs. Vegetative life or function is confined to nutrition, growth, and reproduction; plants have no organs for locomotion or perception. The root in plants corresponds to the mouth of animals—the food of the former being earth and water. Growth in a plant as in an animal does not go on only in parts but each plant grows or decays as a whole. Plants reproduce themselves and continue their form, not only through seeds but also from parts of the organism such as roots or twigs.

This philosophy or general outlook underlies the two treatises of Theophrastus, *On the History of Plants* (9 books) and *On the Causes of Plants* (6 books). These books, however, are devoted to more detailed studies of a wide range of specific plants and utilize material not known to Aristotle, especially in regard to the botany of Persia and India made available by the Macedonian conquest of those countries. (See Hugo Bretzl, *Botanische Forschungen des Alexanderzuges*.) Theophrastus does not hesitate to correct the views of his predecessor as head of the Peripatetic School, for instance, in denying that roots always go down into the soil. In general, these two botanical treatises are markedly free from the characteristic habit of Aristotle of developing his ideas through criticism of his predecessors. The works of Theophrastus are the high-water mark of scientific botany in antiquity. One has to come down to the sixteenth century to find anything of comparable merit.

Though the extant Pseudo-Aristotelian treatise *On Plants*, which may be the work of Nicolaus of Damascus in the first century B.C., shows the persistence of purely theoretic interest in botany, the ancient Greek and Roman writers after Theophrastus were concerned mainly with its application to agriculture and pharmacy or materia medica. A number of plants are, so far as we know, mentioned for the first time by Nicander of Colophon and the Romans

Varro and Columella. These authors, however, seem to have drawn much of their botanical material directly or indirectly from Theophrastus. This is undoubtedly true of Pliny, who devoted Books XII to XXVII of his *Natural History* to botany. But it has been held that Pliny was the first to write of the distinction between growth buds and fruit buds. (On this see E. L. Greene, *Landmarks of Botanical History*, p. 159.) Of considerable importance in the history of botany is the treatise of Dioscorides *On Materia Medica*, which contains descriptions of some six hundred plants and an attempt at a rational classification of them. According to Greene (*op. cit.*, p. 154), Dioscorides anticipated the moderns in recognizing the natural affiliation of different species, for instance, in putting the herb *Ebulus* and the treelike *Sambucus* into one genus. Galen emphasized the medical importance of botanical knowledge and seems to have had a first-hand acquaintance with a large number of plants, distrusting merely bookish descriptions of them.

ORGANOGRAPHY

THE PARTS OF PLANTS

Theophrastus, *History of Plants* I. 1.1.–2.6.[1] Translation of Arthur Hort (London, 1916)

1. In considering the distinctive characters of plants and their nature generally, one must take into account their parts, their qualities, the ways in which their life originates, and the course which it follows in each case: (conduct and activities we do not find in them, as we do in animals). Now the differences in the way in which their life originates, in their qualities, and in their life history are comparatively easy to observe and are simpler, while those shown in their "parts" present more complexity. Indeed it has not even been satisfactorily determined what ought and what ought not to be called "parts," and some difficulty is involved in making the distinction.

Now it appears that by a "part," seeing that it is something which belongs to the plant's characteristic nature, we mean something which is permanent either absolutely or when once it has appeared (like those parts of animals which remain for a time undeveloped)—permanent, that is, unless it be lost by disease, age, or mutilation. However, some of the parts of plants are such that their existence is limited to a year, for instance,

---

[1] Theophrastus of Eresus (*ca.* 372–*ca.*288 B.C.) was a close associate of Aristotle and his successor as leader of the Peripatetic School. In addition to the botanical works, his famous *Characters* and an introductory essay on metaphysics are extant. We also have considerable fragments of his works *On the Senses*, *On Stones*, *On Fire*, *On Odors*, *On Winds*, and *On Weather Signs*, and many smaller fragments on other scientific and philosophic subjects. A great deal of what later doxographers wrote goes back, directly or indirectly, to Theophrastus' work on the views of the philosophers from Thales to Plato.

In contrast with the dogmatism of Aristotle on metaphysical problems, Theophrastus in his metaphysical essay shows a hesitant and a questioning attitude. A good deal of the material in the botanical works consists of minute and accurate description quite independent of metaphysical considerations. Thus he prefers to begin the *History of Plants* with a decription of the essential organs of the most developed plants (trees) rather than with a formal definition of what constitutes a plant, giving "the various and manifold" character of the latter (I. 1.10) as the source of the difficulty. [Edd.]

flower, "catkin,"[1] leaf, fruit, in fact all those parts which are antecedent to the fruit itself or else appear along with it. Also the new shoot itself must be included with these; for trees always make fresh growth every year alike in the parts above ground and in those which pertain to the roots. . . .

But perhaps we should not expect to find in plants a complete correspondence with animals in regard to those things which concern reproduction any more than in other respects; and so we should reckon as "parts" even those things to which the plant gives birth, for instance their fruits, although we do not so reckon the unborn young of animals. . . .

Again many plants shed their parts every year, even as stags shed their horns, birds which hibernate their feathers, four-footed beasts their hair: so that it is not strange that the parts of plants should not be permanent, especially as what thus occurs in animals and the shedding of leaves in plants are analogous processes.

In like manner the parts concerned with reproduction are not permanent in plants; for even in animals there are things which are separated from the parent when the young is born, and there are other[2] things which are cleansed away, as though neither of these belonged to the animal's essential nature. And so, too, it appears to be with the growth of plants; for, of course, growth leads up to reproduction as the completion of the process.

And in general, as we have said, we must not assume that in all respects there is complete correspondence between plants and animals. . . . It is waste of time to take great pains to make comparisons where that is impossible, and in so doing we may lose sight also of our proper subject of enquiry. The enquiry into plants, to put it generally, may either take account of the external parts and the form of the plant generally, or else of their internal parts; the latter method corresponds to the study of animals by dissection. . . .

But, before we attempt to speak about each, we must make a list of the parts themselves. Now the primary and most important parts, which are also common to most, are these—root, stem, branch, twig; these are the parts into which we might divide the plant, regarding them as members, corresponding to the members of animals: for each of these is distinct in character from the rest, and together they make up the whole.

The root is that by which the plant draws its nourishment, the stem that to which it is conducted. And by the "stem" I mean that part which grows above ground and is single;[3] for that is the part which occurs most generally both in annuals and in long-lived plants; and in the case of trees

[1] I.e., the male inflorescence of some trees; the term [βρύον] is of course wider than "catkin."

[2] I.e., the embryo is not the only thing derived from the parent animal which is not a "part" of it; there is also the food-supply produced with the young, and the after-birth.

[3] I.e., before it begins to divide.

it is called the "trunk." By "branches" I mean the parts which split off from the stem and are called by some "boughs."[1] By "twig" I mean the growth which springs from the branch regarded as a single whole, and especially such an annual growth.

Now these parts belong more particularly to trees. The stem, however, as has been said, is more general, though not all plants possess even this, for instance, some herbaceous plants are stemless; others again have it, not permanently, but as an annual growth, including some whose roots live beyond the year. In fact your plant is a thing various and manifold, and so it is difficult to describe in general terms: in proof whereof we have the fact that we cannot here seize on any universal character which is common to all, as a mouth and a stomach are common to all animals; whereas in plants some characters are the same in all, merely in the sense that all have analogous characters, while others correspond otherwise. For not all plants have root, stem, branch, twig, leaf, flower or fruit, or again bark, core, fibres or veins; for instance, fungi and truffles; and yet these and such like characters belong to a plant's essential nature. However, as has been said, these characters belong especially to trees, and our classification of characters belongs more particularly to these; and it is right to make these the standard in treating of the others.

Trees moreover show forth fairly well the other features also which distinguish plants; for they exhibit differences in the number or fewness of these which they possess, as to the closeness or openness of their growth, as to their being single or divided, and in other like respects. Moreover each of the characters mentioned is not "composed of like parts;"[2] by which I mean that though any given part of the root or trunk is composed of the same elements as the whole, yet the part so taken is not itself called "trunk," but "a portion of a trunk." The case is the same with the members of an animal's body; to wit, any part of the leg or arm is composed of the same elements as the whole, yet it does not bear the same name (as it does in the case of flesh or bone);[3] it has no special name. Nor again have subdivisions of any of those other organic parts[4] which are uniform special names, subdivisions of all such being nameless. But the subdivisions of those parts[5] which are compound have names, as have those of the foot, hand, and head, for instance, toe, finger, nose or eye. Such then are the largest[6] parts of the plant.

---

[1] Or "knots."

[2] There is no exact English equivalent for ὁμοιομερές, which denotes a whole composed of parts, each of which is, as it were, a miniature of the whole. . . .

[3] I.e., any part taken of flesh or bone may be called "flesh" or "bone."

[4] E.g., the bark. . . .

[5] E.g., the fruit. . . .

[6] I.e., the "compound" parts.

2. Again there are the things of which such parts are composed, namely bark, wood, and core (in the case of those plants which have it), and these are all "composed of like parts." Further, there are the things which are even prior to these, from which they are derived—sap, fibre, veins, flesh: for these are elementary substances—unless one should prefer to call them the active principles of the elements; and they are common to all the parts of the plant. Thus the essence and entire material of plants consist in these.

Again there are other as it were annual parts, which help towards the production of the fruit, as leaf, flower, stalk (that is, the part by which the leaf and the fruit are attached to the plant), and again tendril, "catkin" (in those plants that have them). And in all cases there is the seed which belongs to the fruit: by "fruit" is meant the seed or seeds, together with the seed-vessel. Besides these there are in some cases peculiar parts, such as the gall in the oak, or the tendril in the vine.

In the case of trees we may thus distinguish the annual parts, while it is plain that in annual plants *all* the parts are annual; for the end of their being is attained when the fruit is produced. And with those plants which bear fruit annually, those which take two years (such as celery and certain others) and those which have fruit on them for a longer time—with all these the stem will correspond to the plant's length of life: for plants develop a stem at whatever time they are about to bear seed, seeing that the stem exists for the sake of the seed.

Let this suffice for the definition of these parts: and now we must endeavour to say what each of the parts just mentioned is, giving a general and typical description.

The sap is obvious: some call it simply in all cases "juice," as does Menestor[1] among others: others, in the case of some plants give it no special name, while in some they call it "juice," and in others "gum." Fibre and "veins" have no special names in relation to plants, but, because of the resemblance, borrow the names of the corresponding parts of animals. It may be however that, not only these things, but the world of plants generally, exhibits also other differences as compared with animals: for, as we have said, the world of plants is manifold. However, since it is by the help of the better known that we must pursue the unknown, and better known are the things which are larger and plainer to our senses, it is clear that it is right to speak of these things in the way indicated: for then in dealing with the less known things we shall be making these better known things our standard, and shall ask how far and in what manner comparison is possible in each case. And when we have taken the parts,[2] we must

[1] Menestor of Sybaris, a Pythagorean philosopher, is often mentioned by Theophrastus. He made hot and cold the basic principles in plant life. His relation to Empedocles, who emphasized similar principles in connection with animals, is not certain. [Edd.]

[2] E.g., the root, as such.

next take the differences which they exhibit,[1] for thus will their essential nature become plain, and at the same time the general differences between one kind of plant and another. . . .

Every plant, like every animal, has a certain amount of moisture and warmth which essentially belong to it; and if these fall short, age and decay, while, if they fail altogether, death and withering ensue. . . .

There are also other internal characters, which in themselves have no special name, but, because of their resemblance, have names analogous to those of the parts of animals. Thus plants have what corresponds to muscle; and this quasi-muscle is continuous, fissile, long; moreover no other growth starts from it either branching from the side or in continuation of it. Again plants have veins: these in other respects resemble the "muscle,"[2] but they are longer and thicker, and have side-growths and contain moisture.

ROOTS

Theophrastus, *History of Plants* I. 6, 7; III. 18.[3] Translation of Arthur Hort

I. 6. Plants differ in their roots, some having many long roots, as fig, oak, plane; for the roots of these, if they have room, run to any length. Others again have few roots, as pomegranate and apple, others a single root, as silver-fir and fir; these have a single root in the sense that they have one long one which runs deep, and a number of small ones branching from this. Even in some of those which have more than a single root the middle root is the largest and goes deep, for instance, in the almond; in the olive this central root is small, while the others are larger and, as it were, spread out crab-wise. Again the roots of some are mostly stout, of some of various degrees of stoutness, as those of bay and olive; and of some they are all slender, as those of the vine. Roots also differ in degree of smoothness and in density. For the roots of all plants are less dense than the parts above ground, but the density varies in different kinds, as also does the woodiness. Some are fibrous, as those of the silver-fir, some fleshier, as those of the oak, some are as it were branched and tassel-like, as those of the olive; and this is because they have a large number of fine small roots close together; for all in fact produce these from their large roots, but they are not so closely matted nor so numerous in some cases as in others.

Again some plants are deep-rooting, as the oak, and some have surface roots, as olive, pomegranate, apple, cypress. Again some roots are straight

---

[1] E.g., the different forms which roots assume.

[2] Fibre.

[3] While interested in describing the different forms of the root, Theophrastus insists on its functional definition as the organ for absorbing nourishment (I. 6.8). Hence not all parts of a plant underground are roots (I. 6.11), and parts of ivy and mistletoe are real roots when they not only support the plant but also take in food (III. 18.10). [Edd.]

and uniform, others crooked and crossing one another. For this comes to pass not merely on account of the situation because they cannot find a straight course; it may also belong to the natural character of the plant, as in the bay and the olive; while the fig and such like become crooked because they can not find a straight course.

All roots have core, just as the stems and branches do, which is to be expected, as all these parts are made of the same materials. Some roots again have side-growths shooting upwards, as those of the vine and pomegranate, while some have no side-growth, as those of silver-fir, cypress, and fir. The same differences are found in under-shrubs and herbaceous plants and the rest, except that some have no roots at all, as truffle, mushroom, bullfist, "thunder-truffle." . . .

It is not right to call all that which is underground "root," since in that case the stalk of purse-tassels and that of long onion and in general any part which is underground would be a root, and so would the truffle, the plant which some call puff-ball, the *uingon*, and all other underground plants. Whereas none of these is a root; for we must base our definition on natural function and not on position.

However, it may be that this is a true account and yet that such things are roots no less; but in that case we distinguish two different kinds of root . . . one[1] getting its nourishment from the other,[2] though the fleshy roots too themselves seem to draw nourishment.

I. 7. The roots of all plants seem to grow earlier than the parts above ground . . . . But no root goes down further than the sun reaches, since it is the heat which induces growth. Nevertheless the nature of the soil, if it is light, open, and porous, contributes greatly to deep rooting. . . .

Also young plants, provided that they have reached their prime, root deeper and have longer roots than old plants; for the roots decay along with the plant's body. . . .

The character and function of the roots of the "Indian fig" (banyan) are peculiar, for this plant sends out roots from the shoots till it has a hold on the ground and roots again; and so there comes to be a continuous circle of roots around the tree, not connected with the main stem, but at a distance from it.

III. 18. . . .[Ivy] regularly puts forth roots from the shoots between the leaves, by means of which it gets a hold of trees and walls. . . . Wherefore also by withdrawing and drinking up the moisture it starves its host, while, if it is cut off below, it is able to survive and live.

---

[1] I.e., the fleshy root (tuber, etc.)

[2] I.e., the fibrous root (root proper).

FLOWERS

Theophrastus, *History of Plants* I. 13. Translation of Arthur Hort

For the present let so much be clear, that in all the parts of plants there are numerous differences shown in a variety of ways. Thus of flowers some are downy, as that of the vine, mulberry, and ivy, some are "leafy,"[1] as in almond, apple, pear, plum. Again some of the flowers are conspicuous, while that of the olive, though it is "leafy," is inconspicuous. Again it is in annual and herbaceous plants alike that we find some leafy, some downy. All plants again have flowers either of two colours or of one; most of the flowers of trees are of one colour and white, that of the pomegranate being almost the only one which is red, while that of some almonds is reddish. The flower of no other cultivated trees is gay nor of two colours, though it may be so with some uncultivated trees, as with the flower of silver-fir, for its flower is of saffron colour; and so with the flowers of those trees by the ocean which have, they say, the colour of roses.

However, among annuals, most are of this character—their flowers are two-coloured and twofold.[2] I mean by "twofold" that the plant has another flower inside the flower, in the middle, as with rose, lily, violet. Some flowers again consist of a single "leaf,"[3] having merely an indication of more, as that of bindweed. For in the flower of this the separate "leaves" are not distinct; nor is it so in the lower part of the narcissus, but there are angular projections[4] from the edges. And the flower of the olive is nearly of the same character.

But there are also differences in the way of growth and the position of the flower; some plants have it close above the fruit, as vine and olive; in the latter, when the flowers drop off, they are seen to have a hole through them, and this men take for a sign whether the tree has blossomed well; for if the flower is burnt up or sodden, it sheds the fruit along with itself, and so there is no hole through it. The majority of flowers have the fruit-case in the middle of them, or it may be, the flower is on the top of the fruit-case, as in pomegranate, apple, pear, plum, and myrtle, and among the under-shrubs, in the rose and in many of the coronary plants. For these have their seeds below, beneath the flower, and this is most obvious in the rose because of the size of the seed-vessel. In some cases[5] again the flower is on top of the actual seeds, as in pine-thistle, safflower, and all thistle-like plants; for these have a flower attached to each seed. So too with some herbaceous plants, as *anthemon*, and among pot-herbs, with cucumber,

---

[1] I.e., petaloid.

[2] I.e., corolla and stamens, etc.

[3] I.e., are gamopetalous (or gamosepalous).

[4] I.e., something resembling separate "leaves" (petals or sepals).

[5] I.e., composites.

gourd, and bottle-gourd; all these have their flowers attached on top of the fruits, and the flowers persist for a long time while the fruits are developing.

In some other plants the attachment is peculiar, as in ivy and mulberry; in these the flower is closely attached to the whole[1] fruit-case; it is not however set above it, nor in a seed-vessel that envelops each separately, but it occurs in the middle part of the structure—except that in some cases it is not easily recognized because it is downy.

Again some flowers are sterile, as in cucumbers those which grow at the ends of the shoot, and that is why men pluck them off, for they hinder the growth of the cucumber. And they say that in the citron those flowers which have a kind of distaff[2] growing in the middle are fruitful, but those that have it not are sterile. And we must consider whether it occurs also in any other flowering plants that they produce sterile flowers, whether apart from the fertile flowers or not. For some kinds of vine and pomegranate certainly are unable to mature their fruit, and do not produce anything beyond the flower.

Classification

METHODS AND LIMITATIONS IN CLASSIFICATION

Theophrastus, *History of Plants*, I. 3–4. Translation of Arthur Hort

3. Now since our study becomes more illuminating if we distinguish different kinds, it is well to follow this plan where it is possible. The first and most important classes, those which comprise all or nearly all plants, are tree, shrub, under-shrub, herb.

A tree is a thing which springs from the root with a single stem, having knots and several branches, and it cannot easily be uprooted; for instance olive, fig, vine. A shrub is a thing which rises from the root with many branches; for instance, bramble, Christ's thorn. An under-shrub is a thing which rises from the root with many stems as well as many branches; for instance, savory, rue. A herb is a thing which comes up from the root with its leaves and has no main stem, and the seed is borne on the stem; for instance, corn and pot-herbs.

These definitions however must be taken and accepted as applying generally and on the whole. For in the case of some plants it might seem that our definitions overlap; and some under cultivation appear to become different and depart from their essential nature, for instance, mallow when it grows tall and becomes tree-like. For this comes to pass in no long time, not more than six or seven months, so that in length and thickness the plant becomes as great as a spear, and men accordingly use it as a walking-

[1] I.e., compound.

[2] I.e., the pistil.

stick, and after a longer period the result of cultivation is proportionately greater. So too is it with the beets; they also increase in stature under cultivation, and so still more do chaste-tree, Christ's thorn, ivy, so that, as is generally admitted, these become trees, and yet they belong to the class of shrubs. On the other hand the myrtle, unless it is pruned, turns into a shrub, and so does filbert: indeed this last appears to bear better and more abundant fruit, if one leaves a good many of its branches untouched, since it is by nature like a shrub. Again neither the apple nor the pomegranate nor the pear would seem to be a tree of a single stem, nor indeed any of the trees which have side stems from the roots, but they acquire the character of a tree when the other stems are removed. However some trees men even leave with their numerous stems because of their slenderness, for instance, the pomegranate and the apple, and they leave the stems of the olive and the fig cut short.[1]

Indeed it might be suggested that we should classify in some cases simply by size, and in some cases by comparative robustness or length of life. For of under-shrubs and those of the pot-herb class some have only one stem and come as it were to have the character of a tree, such as cabbage and rue: wherefore some call these "tree-herbs;" and in fact all or most of the pot-herb class, when they have been long in the ground, acquire a sort of branches, and the whole plant comes to have a tree-like shape, though it is shorter lived than a tree.

For these reasons then, as we are saying, one must not make a too precise definition; we should make our definitions typical. For we must make our distinctions too on the same principle, as those between wild and cultivated plants, fruit-bearing and fruitless, flowering and flowerless, evergreen and deciduous. Thus the distinction between wild and cultivated seems to be due simply to cultivation, since, as Hippon[2] remarks, any plant may be either wild or cultivated according as it receives or does not receive attention. Again the distinctions between fruitless and fruit-bearing, flowering and flowerless, seem to be due to position and the climate of the district. And so too with the distinction between deciduous and evergreen. Thus they say that in the district of Elephantine neither vines nor figs lose their leaves.

Nevertheless we are bound to use such distinctions.[3] For there is a certain common character alike in trees, shrubs, under-shrubs, and herbs. Wherefore, when one mentions the causes also, one must take account of

---

[1] I.e., so that the tree comes to look like a shrub from the growth of fresh shoots after cutting. . . .

[2] Philosopher of the fifth century B.C. [Edd.]

[3] . . . The sense seems to be: Though these "secondary" distinctions are not entirely satisfactory, yet (if we look to the *causes* of different characters), they are indispensable, since they are due to causes which affect all the four classes of our "primary" distinction.

all alike, not giving separate definitions for each class, it being reasonable to suppose that the causes too are common to all. And in fact there seems to be some natural difference from the first in the case of wild and cultivated, seeing that some plants cannot live under the conditions of those grown in cultivated ground, and do not submit to cultivation at all, but deteriorate under it; for instance, silver-fir, fir, holly, and in general those which affect cold snowy country; and the same is also true of some of the under-shrubs and herbs, such as caper and lupin. Now in using the terms "cultivated" and "wild" [1]we must make these[2] on the one hand our standard, and on the other that which is in the truest sense "cultivated." Now Man, if he is not the only thing to which this name is strictly appropriate, is at least that to which it most applies.[3]

4. Again the differences, both between the plants as wholes and between their parts, may be seen in the appearance itself of the plant. I mean differences such as those in size, hardness, smoothness or their opposites, as seen in bark, leaves, and the other parts; also, in general, differences as to comeliness or its opposite and as to the production of good or of inferior fruit. For the wild kinds appear to bear more fruit, for instance, the wild pear and wild olive, but the cultivated plants better fruit, having even flavours which are sweeter and pleasanter and in general better blended, if one may so say.

These then, as has been said, are differences of natural character, as it were, and still more so are those between fruitless and fruitful, deciduous and evergreen plants, and the like. But with all the differences in all these cases we must take into account the locality, and indeed it is hardly possible to do otherwise. Such differences[4] would seem to give us a kind of division into classes, for instance, between that of aquatic plants and that of plants of the dry land, corresponding to the division which we make in the case of animals. For there are some plants which cannot live except in wet; and again these are distinguished from one another by their fondness for different kinds of wetness; so that some grow in marshes, others in lakes, others in rivers, others even in the sea, smaller ones in our own sea, larger ones in the Red Sea. Some again, one may say, are lovers of very wet places,[5] or plants of the marshes, such as the willow and the plane. Others again cannot live at all in water, but seek out dry places; and of the smaller sorts there are some that prefer the shore.

However, if one should wish to be precise, one would find that even of

---

[1] I.e., we must take the *extreme* cases.

[2] I.e., plants which entirely refuse cultivation.

[3] The translator considers this last sentence an irrelevant gloss. [Edd.]

[4] I.e., as to locality.

[5] I.e., though not actually living *in* water.

these some are impartial and as it were amphibious, such as tamarisk, willow, alder, and that others even of those which are admitted to be plants of the dry land sometimes live in the sea,[1] as palm, squill, asphodel. But to consider all these exceptions and, in general, to consider in such a manner is not the right way to proceed. For in such matters too nature certainly does not thus go by any hard and fast law. Our distinctions therefore and the study of plants in general must be understood accordingly. To return—these plants as well as all others will be found to differ, as has been said, both in the shape of the whole and in the differences between the parts, either as to having or not having certain parts, or as to having a greater or less number of parts, or as to having them differently arranged, or because of other differences such as we have already mentioned. And it is perhaps also proper to take into account the situation in which each plant naturally grows or does not grow. For this is an important distinction, and especially characteristic of plants, because they are united to the ground and not free from it like animals.[2]

SUBCLASSES OF UNDERSHRUBS

Theophrastus, *History of Plants* VI. 1. Translation of Arthur Hort

For the cultivated kinds of this class [of undershrubs] are not numerous, and consist almost entirely of coronary plants, as rose, gilliflower, carnation, sweet marjoram, martagon, lily, to which may be added tufted thyme, bergamot-mint, calamint, southernwood. For all these are woody and have small leaves; wherefore they are classed as undershrubs. This class covers also pot-herbs, such as cabbage, rue, and others like them. Of these it is perhaps more appropriate to speak under their proper designation, that is, when we come to make mention of coronary plants and pot-herbs. Now let us first speak of the wild kinds. Of these are several classes and subdivisions, which we must distinguish by the characteristics of each subdivision as well as by those of each class taken as a whole.

The most important difference distinguishing class from class which one could find is that between the spineless and the spinous kinds. Again under each of these two heads there are many differences distinguishing kinds and forms, of which we must endeavour to speak severally.

Of spinous kinds some just consist of spines, as asparagus and *skorpios*; for these have no leaves except their spines. Then there are the spinous-leaved plants, as thistle, eryngo, safflower; these and the like have their spines on the leaves, whence their name. Others again have leaves as well

[1] Presumably as being sometimes found on the shore below high-water mark.

[2] This circumstance leads Theophrastus to classify plants not only as above (viz., trees, shrubs, undershrubs, and herbs), but also as land and water plants as well as flowering and flowerless. [Edd.]

as their spines, as rest-harrow, caltrop, and *pheos*, which some call *stoibe*. Caltrop is also spinous-fruited, having spines on the fruit-vessel. Wherefore this peculiarity marks it off from almost all other plants; though many trees and shrubs have spines on the shoots, as wild pear, pomegranate, Christ's thorn, bramble, rose, caper. Such are the general distinctions which may be made among spinous plants.

With spineless plants it is not possible to make such "generic" distinctions; for the variation of the leaves in size and shape is endless, and the differences are not clearly marked;[1] but we must try to distinguish on another principle. There are many classes of such plants and they differ widely, as rock-rose, bryony, madder, privet, *kneoron*, marjoram, savory, *sphakos* (sage), *elelisphakos* (salvia), horehound, *konyza*, balm, and others like these; and in addition to these we have the plants with a ferula-like stem or with a stem composed of fibre, as fennel, horse-fennel, *narthekia* (ferula), *narthex* (ferula), and the plant called by some wolf's bane, and others like these. All these, as well as any other ferula-like plants, may be placed in the class of undershrubs.

Propagation: Germination and Growth

Theophrastus, *History of Plants* II. 1.1–2.6. Translation of Arthur Hort

1. The ways in which trees and plants in general originate are these: spontaneous growth,[2] growth from seed, from a root, from a piece torn off, from a branch or twig, from the trunk itself; or again from small pieces into which the wood is cut up (for some trees can be produced even in this manner). Of these methods spontaneous growth comes first, one may say, but growth from seed or root would seem most natural; indeed these methods too may be called spontaneous; wherefore they are found even in wild kinds, while the remaining methods depend on human skill or at least on human choice.

However all plants start in one or other of these ways, and most of them in more than one. Thus the olive is grown in all the ways mentioned, except from a twig; for an olive-twig will not grow if it is set in the ground, as a fig or pomegranate will grow from their young shoots. Not but what some say that cases have been known in which, when a stake of olive-wood was planted to support ivy, it actually lived along with it and became a tree; but such an instance is a rare exception, while the other methods of growth are in most cases the natural ones. The fig grows in all the ways mentioned, except from root-stock and cleft wood; apple and pear grow also from branches, but rarely. However, it appears that most, if not practically all, trees may grow from branches, if these are smooth, young, and

---

[1] I.e., there is a gradation.

[2] See p. 459, below. [Edd.]

vigorous. But the other methods, one may say, are more natural, and we must reckon what may occasionally occur as a mere possibility.

In fact there are quite few plants which grow and are brought into being more easily from the upper parts, as the vine is grown from branches; for this, though it cannot be grown from the "head," yet can be grown from the branch, as can all similar trees and undershrubs, for instance, as it appears, rue, gilliflower, bergamot-mint, tufted thyme, calamint. So the commonest ways of growth with all plants are from a piece torn off or from seed; for all plants that have seeds grow also from seed. And they say that the bay too grows from a piece torn off, if one takes off the young shoots and plants them; but it is necessary that the piece torn off should have part of the root or stock attached to it. However, the pomegranate and "spring apple" will grow even without this, and a slip of almond grows if it is planted. The olive grows, one may say, in more ways than any other plant; it grows from a piece of the trunk or of the stock, from the root, from a twig, and from a stake, as has been said. Of other plants the myrtle also can be propagated in several ways; for this too grows from pieces of wood and also from pieces of the stock. It is necessary, however, with this, as with the olive, to cut up the wood into pieces not less than a span long and not to strip off the bark.

Trees then grow and come into being in the above-mentioned ways; for as to methods of grafting and inoculation, these are, as it were, combinations of different kinds of trees; or at all events these are methods of growth of a quite different class and must be treated of at a later stage.

2. Of under-shrubs and herbaceous plants the greater part grow from seed or a root, and some in both ways; some of them also grow from cuttings, as has been said, while roses and lilies grow from pieces of the stems, as also does dog's-tooth grass. Lilies and roses also grow when the whole stem is set. Most peculiar is the method of growth from an exudation; for it appears that the lily grows in this way too, when the exudation that has been produced has dried up. They say the same of alexanders, for this too produces an exudation. There is a certain reed also which grows if one cuts it in lengths from joint to joint, and sets them sideways, burying it in dung and soil. Again they say that plants which have a bulbous root are peculiar in their way of growing from the root.

The capacity for growth being shown in so many ways, most trees, as was said before, originate in several ways; but some come only from seed, as silver-fir, fir, Aleppo pine, and in general all those that bear cones; also the date-palm, except that in Babylon it may be that, as some say, they take cuttings from it. The cypress in most regions grows from seed but in Crete from the trunk also, for instance in the hill country about Tarra; for there grows the cypress which they clip, and when cut it shoots in every possible way, from the part which has been cut, from the ground, from the middle, and from the upper parts; and occasionally, but rarely, it shoots from the roots also.

About the oak accounts differ; some say it only grows from seed, some from the root also, but not vigorously, others again that it grows from the trunk itself, when this is cut. But no tree grows from a piece torn off or from a root except those which make side-growths.

However in all the trees which have several methods of originating the quickest method and that which promotes the most vigorous growth is from a piece torn off, or still better from a sucker, if this is taken from the root. And, while all the trees which are propagated thus or by some kind of slip seem to be alike in their fruits to the original tree, those raised from the fruit, where this method of growing is also possible, are nearly all inferior, while some quite lose the character of their kind, as vine, apple, fig, pomegranate, pear. As for the fig, no cultivated kind is raised from its seed, but either the ordinary wild fig or some wild kind is the result, and this often differs in colour from the parent; a black fig gives a white, and conversely. Again, the seed of an excellent vine produces a degenerate result, which is often of quite a different kind; and at times this is not a cultivated kind at all, but a wild one of such a character that it does not ripen its fruit; with others again the result is that the seedlings do not even mature fruit, but only get as far as flowering.

Again the stones of the olive give a wild olive, and the seeds of a sweet pomegranate give a degenerate kind, while the stoneless kind gives a hard sort and often an acid fruit. So also is it with seedlings of pears and apples; pears give a poor sort of wild pears, apples produce an inferior kind which is acid instead of sweet; quince produces wild quince. Almond again raised from seed is inferior in taste and in being hard instead of soft; and this is why men bid us graft on to the almond, even when it is fully grown, or, failing that, frequently plant the offsets.

The oak also deteriorates from seed; at least many persons having raised trees from acorns of the oak at Pyrrha could not produce one like the parent tree. On the other hand they say that bay and myrtle sometimes improve by seeding, though usually they degenerate and do not even keep their colour, but red fruit gives black—as happened with the tree in Antandros; and frequently seed of a "female" cypress produces a "male" tree. The date-palm seems to be about the most constant of these trees, when raised from seed, and also the "cone-bearing pine" (stone-pine) and the "lice-bearing pine." So much for degeneration in cultivated trees; among wild kinds it is plain that more in proportion degenerate from seed, since the parent trees are stronger. For the contrary[1] would be very strange, seeing that degenerate forms are found even in cultivated trees,[2]

[1] I.e., that they should improve from seed.

[2] Whereas wild trees are produced *only* from seed.

and among these only in those which are raised from seed. (As a general rule these are degenerate, though men may in some cases effect a change[1] by cultivation.)

Cf. Hippocratic Collection, *On the Nature of the Child* 23–24[2]

23. But as to the development from shoots, trees arise from trees in the following way. The shoot has a wound in its lower part, that nearer the earth, where it was broken off from a tree. It is from there that the roots are put forth. And they are put forth in the following way. When the shoot, planted in the ground, draws moisture from the earth, it swells and contains pneuma, whereas this is not true of the part above the ground. The pneuma and the moisture in the lower part of the shoot cause the quality which is heaviest to condense. There is then a breaking out below and tender roots are produced there. And when the shoot takes hold below, it draws moisture through the roots and distributes it to the part above the ground. Then the upper part too becomes swollen and contains pneuma. And all that quality which is light in the plant, having been condensed and having become leaves, sprouts. There is, then, growth upwards as well as downwards.

Thus, in the matter of their development plants growing from seed differ from those growing from shoots. For in the case of seed, first the leaf arises from the seed and then roots are thrust downward, but the shoot first takes root and then produces leaves. Now the reason is that there is an abundance of moisture in the seed itself, and since the seed is entirely in the ground there is at the beginning sufficient nourishment for the leaf. By this the leaf will be nourished until roots are put forth.

But this is not the case with shoots, for the shoot does not come from another shoot from which the foliage can in the first instance be nourished, but the shoot is itself like a tree, being also for the most part above the ground, so that it could not be filled with moisture above the ground unless a great force coming upwards from below carried moisture to it. And it is necessary for the shoot first to obtain nourishment for itself from the earth by roots, and then, having thus drawn the nourishment from the earth, to transmit it upwards and to put forth leaves for budding and growth.

24. When the plant grows it branches out by reason of a necessity of which I shall speak. When an excess of moisture is drawn up from the

[1] I.e., improve a degenerate seedling.

[2] The Hippocratic treatise *On the Nature of the Child* seems to belong to the same group as *On Generation* and *On Diseases* IV, works which employ experimental methods and frequently draw analogies between plant and animal life to arrive at general theories of generation and heredity.

earth and comes to the plant, the latter breaks out because of this excess at the point where it is greatest, and at that point there is a branching out. There is also growth both in thickness and in height and depth for the reason that the lower part of the earth is warm in winter and cold in summer. . . .

THE GERMINATION OF SEEDS

Theophrastus, *History of Plants* VIII. 2.1–6. Translation of Arthur Hort

In germinating, some of these plants produce their root and their leaves from the same point, some separately, from either end of the seed.[1] Wheat, barley, one-seeded wheat, and in general all the cereals produce them from either end, in a manner corresponding to the position of the seed in the ear, the root growing from the stout lower part, the shoot from the upper part; but the part corresponding to the root and that corresponding to the stem form a single continuous whole. Beans and other leguminous plants do not grow in the same manner, but they produce the root and the stem from the same point, namely, the point at which the seed is attached to the pod, which, it is plain, is a sort of starting point of fresh growth. In some cases there is also a formation resembling the *penis*, as in beans, chick-peas, and especially in lupins; from this the root grows downwards, the leaf and the stem upwards.

There are then these different ways of germinating; but a point in which all these plants agree is that they all send out their roots at the place where the seed is attached to the pod or ear, whereas the contrary is the case with the seeds of certain trees, as almond, hazel, acorn, and the like. And in all these plants the root begins to grow a little before the stem; whereas in certain trees the bud first begins to grow within the seed itself, and, as it increases in size, the seeds split—for all such seeds are in a manner in two halves, and those of leguminous plants again all plainly have two halves and are double—and then the root is immediately thrust out; but in cereals, since the seeds are in one piece, this does not occur, but the root grows a little before the bud.

Barley and wheat come up with a single leaf, but peas, beans, and chick-peas with several. All the leguminous plants have a single woody root, and also slender side-roots springing from this. The chick-pea is about the deepest rooting of these, and sometimes it has side-roots; but wheat, barley and the other cereals have a number of fine roots, wherefore they are matted together. Again all such plants have many branches and many stems. And there is a sort of contrast between these two classes; the leguminous plants, which have a single root, have many side-growths above from the

[1] The passage indicates differences in germination between monocotyledenous and dicotyledenous plants, as we call them, cereals generally belonging to the former and leguminous plants to the latter group. [Edd.]

stem—all except beans; while the cereals, which have many roots, send up many shoots, but these have no side-shoots—except such sorts of wheat as are called *sitanias* and *krithanias* ("barley-wheat").

During the winter cereals remain in the blade, but, as the season begins to smile, they send up a stem from the midst and it becomes jointed. And it comes to pass that the ear also at once appears in the third, or in some cases in the fourth joint, though it is not distinctly seen in the mass of growth (the whole stem contains more joints than three or four), so that it must be formed at the same time that the straw grows or but a little later; though it does not become conspicuous till it has first swollen and formed in the sheath, and by that time its size makes its development visible.

Four or five days after being set free[1] wheat and barley flower and remain in bloom for a like number of days; those who put the period at the longest say that the bloom is shed in seven days. On the other hand the flowering period of leguminous plants lasts a long time; that of vetch and chick-pea is longer than that of most, but that of the bean is far longer than that of any of them; they say that it is in bloom for forty days; some, however, give this period absolutely, others say that at different times different parts are in flower, since the whole plant does not flower at once. For plants with an ear bloom all at once, but plants with pods and all leguminous plants bloom part at a time; the lower parts bloom first, and, when this bloom has fallen, the part next above it, and so on up to the top. Wherefore, at the time when some of the vetches are gathered, the lower seeds have already fallen, while the upper ones are still quite green.

After the flowering is over wheat and barley develop and mature in about forty days; one-seeded wheat and other such plants take about the same time. So too, they say, does the bean, which blooms and matures in a like number of days: but the others take fewer, and fewest of all the chick-pea, since, as some say, it takes only forty days from the time when it is sown to that when it is mature; and in any case it is clear that the plant as a whole develops very rapidly. Millet, sesame, Italian millet, and the summer crops in general, it is fairly well agreed, take the same number of days, that is, forty; though some say that they take less.

GRAFTING AND BUDDING

Theophrastus, *On the Causes of Plants* I. 6.1–2. Translation of R. E. Dengler (Philadelphia, 1927)

It remains to speak of some other of the methods of propagation, such as the matter of grafting and budding. The account is simple and has practically been told before; the ingrafted part uses the other as an ordinary plant uses the ground. "Budding" also is really a kind of planting; it is

[1] I.e., from the sheath. . . .

not simply a mere mechanical juxtaposition, but it is clear that in this process both the germinating and the supporting force is the vital moisture. The "eye" having this moisture is suited to the other part, and the part having the nourishment grows the bud as though it were its own.

All such operations are attended by ready growth because the nourishment is, as it were, predigested, which is particularly true of budding; for in that process the nourishment is purest and similar to the sap right in the stems of fruits. It is always easy to grow like kinds on like; the "eye" is practically something growing on its kind. The reciprocal relation, too, is easily understood—especially in plants whose bark is similar, for there is the least need of adaptation between congeners—and the process amounts practically to a mere transposition.

Cf. Hippocratic Collection, *On the Nature of the Child* 26

Now if buds are taken from trees and inserted into other trees, and, becoming trees themselves, live in these other trees and bear fruit, the fruit is unlike that of the trees upon which they have been grafted. This comes about as follows. First the bud sprouts, for it had nourishment originally from the tree from which it was taken and then in that upon which it was grafted. When it has sprouted in this way it sends out tender roots to the tree and at first derives nourishment from the moisture of this tree upon which it has been grafted. Then after a time it sends roots into the ground through the tree upon which it has been grafted and derives nourishment from the soil, drawing up moisture. It is from this source that it is fed. Hence one should not be surprised to find that grafts bear fruit different from that of the trees upon which they are grafted. For they live from the earth.

SOWING AND GERMINATION OF POTHERBS

Theophrastus, *History of Plants* VII. 1.1–2.4. Translation of Arthur Hort

First we must speak of the class of pot-herbs, beginning with the cultivated kinds, since it happens that these are better known than the wild kinds.

There are three seed-times for all things grown in gardens, at which men sow the various herbs, distinguishing by the season. One is the "winter" seed-time, another the "summer," and the third is that which falls between these, coming after the winter solstice. These terms, however, are given in regard not to the sowing, but to the growth and use of each kind; for the actual sowing takes place, one might almost say, at the opposite seasons. Thus, the "winter" period begins after the summer solstice in the month Metageitnion,[1] in which they sow cabbage, radish,

[1] July.

turnip, and what are called "secondary crops," that is to say, beet, lettuce, rocket, monk's rhubarb, mustard, coriander, dill, cress; and this is also called the "first" period of cultivation. The second period begins after the winter solstice in the month Gamelion,[1] in which they scatter or plant the seed of leeks, celery, long onion, orach. The third period, which is called the "summer" period, begins in the month Munychion[2]: in this are sown cucumber, gourd, blite, basil, purslane, savory. Moreover, they make several sowings of the same herb at each season, as of radish, basil, and the others. And at all the periods are sown the "secondary crops."

Not all herbs germinate within the same time, but some are quicker, others slower, namely, those which germinate with difficulty. The speediest are basil, blite, rocket, and of those sown for winter use, radish; for these germinate in about three days. Lettuce takes four or five, cucumber and gourd about five or six, or as some say, seven; however, cucumber is earlier and quicker than the others. Purslane takes a longer time, dill four days, cress and mustard five. Beet in summer takes six days, in winter ten, orach takes eight, and turnip ten. Leek and long onion do not take the same time, but the former nineteen to twenty days, the latter ten to twelve. Coriander germinates with difficulty; indeed fresh seed will not come up at all unless it is moistened. Savory and marjoram take more than thirty days; but celery germinates with the greatest difficulty of all; for those who make the time comparatively short say forty days, and others fifty, and that too, at whichever period it is sown, for some sow it as a "secondary crop" at all the periods.

Generally speaking, those herbs which are sown at more than one season do not mature faster in the summer. Howbeit it is strange if the season and the state of the atmosphere do not contribute at all to quicker growth, and if, when there is an unfavourable cold season and the atmosphere is cloudy, these conditions do not tend to make growth slower, seeing that, when stormy or fair weather follows the sowing, germination is slower or quicker accordingly. And there is another thing which makes a difference as to the raising of the various herbs; germination begins earlier in sunny places which have an even temperature.

As a matter of fact, to speak roundly, the causes of such differences must be found in several different circumstances, in the seeds themselves, in the ground, in the state of the atmosphere, and in the season at which each is sown, according as it is stormy or fair. However, it is a point for consideration with which herbs the time of sowing makes a difference and with which it makes none; thus it is said that radish germinates on the third day whether it be sown in summer or in winter, while beet, as has been said,

[1] January.
[2] April.

behaves differently according to the season. Anyway such are and are said to be the seasons of germination in each case.

Another thing which makes a difference as to the rapidity with which the seeds germinate is their age; for some herbs come up quicker from fresh seed, as leek, long onion, cucumber, gourd; (some even soak the seed of cucumber first in milk or water, to make it germinate quicker). Some come up quicker from old seed, as celery, beet, cress, savory, coriander, marjoram (unless indeed they are raised from fresh seed in the manner which we have mentioned). There is, they say, a singular feature about beet; the seed does not all germinate at once, but some of it not for some time, some even in the next or in the third year; wherefore it is said that little comes up from much seed.

ARTIFICIAL FERTILIZATION

Theophrastus, *History of Plants* II. 8.4.[1] Translation of Arthur Hort

With dates it is helpful to bring the male to the female; for it is the male which causes the fruit to persist and ripen, and this process some call, by analogy, "the use of the wild fruit." The process is thus performed: when the male palm is in flower, they at once cut off the spathe on which the flower is, just as it is, and shake the bloom with the flower and the dust over the fruit of the female, and, if this is done to it, it retains the fruit and does not shed it. In the case both of the fig and of the date it appears that the "male" renders aid to the "female,"—for the fruit-bearing tree is called "female,"—but while in the latter case there is a union of the two sexes, in the former the result is brought about somewhat differently.[2]

Cf. Theophrastus, *On the Causes of Plants* II. 9.15

The case of the date palm is not the same [as that of the fig] but bears a certain resemblance to it, wherefore they call the process "caprification." For the flower, down, and dust from the male join in producing, through the agency of heat and other power, a dryness and a freedom of respiration. Because of this the fruit persists. Similar, in a sense, is what takes place in the case of fishes, when the male sprinkles semen upon the deposited eggs. Indeed it is possible to draw analogies from quite different types.

---

[1] Generally speaking, the ancients knew nothing of the sexuality of plants, references to "males" and "females" usually being to sterile and fertile varieties. In the case of the date palms, in which the male and female organs are on different trees, the method of artificial fertilization here described was known to the Babylonians and Egyptians long before Theophrastus, though he seems to have correctly grasped the sexual nature of the process, as the analogy in *On the Causes of Plants* II. 9.15 indicates. Actual experiments to determine the sexuality of plants by the segregation of male and female flowers both in dioecious and in monoecious species were not performed until the end of the seventeenth century. [Edd.]

[2] I.e., by hanging fruiting branches of the wild fig near the cultivated fig so that pollen may be brought from the former to the latter by the gall insect. See *On the Causes of Plants* II. 9.6. [Edd.]

SPONTANEOUS GENERATION

Theophrastus, *On the Causes of Plants* I. 5.[1] Translation of R. E. Dengler

Spontaneous generation, to put the matter simply, takes place in smaller plants, especially in those that are annuals and herbaceous. But still it occasionally occurs too in larger plants whenever there is rainy weather or some peculiar condition of air or soil; for thus it is said that the silphium[2] sprang up in Libya when a murky and heavy sort of wet weather condition occurred, and that the timber growth which is now there has come from some similar reason or other; for it was not there in former times.

The rains also produce certain decompositions and changes—when the moisture penetrates deeply—and they can nourish and increase the resulting growths under the warming and drying influence of the sun. Many believe that animals also come into being in the same way.[3]

But if, in truth, the air also supplies seeds, picking them up and carrying them about, as Anaxagoras says, then this fact is much more likely to be the explanation; for it would produce other examples of beginnings and nourishings. Moreover the rivers and the gathering together and breaking forth abroad of waters purvey seed from everywhere, both of trees and shrubby plants; wherefore also changes in the courses of rivers make many places wooded that before were woodless.

Such growths would not appear spontaneous, but rather as though sown or planted. Of the sterile sorts, one might the rather expect growth of their own accord, as they are neither planted nor grown from seed. If, however, they come from neither of these latter processes, then their origin must be spontaneous. But this may possibly not be true, at least for a good many of them; it may rather be that all the stages of development of their seeds escape our observation, just as was said in *Historia* about willow and elm. Moreover the development of seed is hidden also in many of the smaller herbaceous plants, as we said, about thyme and the others, that

---

[1] The possibility of spontaneous, i.e., seedless, generation or, in general, the propagation of plants from lifeless matter is often mentioned by Theophrastus, but he almost always indicates that seeds may be so small as not to be visible and that in any case the problem must be further investigated. The present passage is typical. Theophrastus correctly describes (*On the Causes of Plants* II. 12.9) the propagation of mistletoe in opposition to those who held the development to be without seed. But cf. Theophrastus, *On the Causes of Plants* I. 1.2:

"It is therefore clear that generation by seed is common to all plants. Now if some may be generated in two ways, from seed or spontaneously, that is not any more remarkable than the case of animals which propagate from their own kind and also from the earth." [Edd.]

[2] Silphium played a prominent role in the economic life of the cities of Cyrenaica, where it was cultivated. It was also discussed by Pliny and Dioscorides and is depicted on coins, e.g., those of Cyrene. It had become quite rare by the time of Strabo and thereafter disappeared completely. Modern botanists have been unable to identify the plant with any known species. [Edd.]

[3] See p. 420. [Edd.]

they were not apparent to the sight, but that their power was evident, since the flowers when sown propagate the plants.

It happens, also, that the seeds of some trees are small and hard to see, for instance, those of cypress. For it is not the entire globular fruit of this tree that is the seed, but the light and, as it were, branlike slight object which grows in the midst of the fruit, and which flutters away when the fruit breaks open; and so it requires some experienced person to collect it, one who can observe narrowly the time of maturity of the fruit and who can recognize the seed itself.

This happens to be true of many other trees, especially in all such as grow close together in natural forests and on the mountains. For one might not easily suppose that this densely crowded condition of growth would persist as a result of their growing up spontaneously; but one of the two things must occur, either growth from the root or growth from seeds.

The woodsmen say that some of the congeners of these trees are sterile. It is likely either that the seeds escape their notice, or that, because the nourishment has been spent on other parts, these trees have actually become sterile, just as in the case of luxuriant vines and whatever plants have a like experience.

But when this phenomenon is found in fruit-bearing and flowering plants, what prevents it occurring in others not fruit-bearing? Let this be given merely as our opinion; more accurate investigation must be made of the subject and the matter of spontaneous generation must be thoroughly inquired into. To sum it up simply: this phenomenon is bound to occur when the earth is thoroughly warmed and when the collected mixture is changed by the sun, as we see also in the case of animals.

Ecology

EFFECTS OF SITUATION, CLIMATE, AND CULTIVATION

Theophrastus, *History of Plants* VIII. 7.6–7; II. 2.7–12. Translation of Arthur Hort

VIII. 7.6. For growth and nourishment the climate is the most important factor, and in general the character of the season as a whole; for when rain, fair weather, and storms occur opportunely, all crops bear well and are fruitful, even if they be in soil which is impregnated with salt or poor. Wherefore there is an apt proverbial saying that "it is the year which bears and not the field."

But the soil also makes much difference, according as it is fat or light, well watered or parched, and it also makes quite as much difference what sort of air and of winds prevails in that region; for some soils, though light and poor, produce a good crop because the land has a fair aspect in regard to sea breezes. But, as has been repeatedly said already, the same breeze has not this effect in all places; some places are suited by a west, some by a north, some by a south wind.

Again the working of the soil and above all that which is done before the sowing has an important effect; for when the soil is well worked it bears easily. Also dung is helpful by warming and ripening the soil, for manured land gets the start by as much as twenty days of that which has not been manured. However, manure is not good for all crops; and further it is beneficial not only to corn and the like but to most other things, except fern, which they say it destroys if it is put on. (Fern is also destroyed if sheep lie on it, and, as some say, lucerne is destroyed by their dung and urine.)

II. 2.7. . .In some places, as at Philippi, the soil seems to produce plants which resemble their parent; on the other hand a few kinds in some few places seem to undergo a change, so that wild seed gives a cultivated form, or a poor form one actually better. We have heard that this occurs, but only with the pomegranate, in Egypt and Cilicia; in Egypt a tree of the acid kind both from seeds and from cuttings produces one whose fruit has a sort of sweet taste, while about Soli in Cilicia near the river Pinaros (where the battle with Darius was fought) all those pomegranates raised from seed are without stones.

If anyone were to plant our palm at Babylon, it is reasonable to expect that it would become fruitful and like the palms of that country. And so would it be with any other country which has fruits that are congenial to that particular locality; for the locality is more important than cultivation and tendance. A proof of this is the fact that things transplanted thence become unfruitful, and in some cases refuse to grow altogether.

There are also modifications due to feeding and attention of other kinds, which cause the wild to become cultivated, or again cause some cultivated kinds to go wild, such as pomegranate and almond. Some say that wheat has been known to be produced from barley, and barley from wheat, or again both growing on the same stool; but these accounts should be taken as fabulous. Anyhow those things which do change in this manner do so spontaneously,[1] and the alteration is due to a change of position (as we said happens with pomegranates in Egypt and Cilicia), and not to any particular method of cultivation.

So too is it when fruit-bearing trees become unfruitful, for instance the *persion* when moved from Egypt, the date-palm when planted in Hellas, or the tree which is called "poplar" in Crete, if anyone should transplant it. Some again say that the sorb becomes unfruitful if it comes into a very warm position, since it is by nature cold-loving. It is reasonable to suppose

[1] I.e., cultivation has nothing to do with it.

that both results follow because the natural circumstances are reversed, seeing that some things entirely refuse to grow when their place is changed. Such are the modifications due to position.

As to those due to method of culture, the changes which occur in things grown from seed are as was said; (for with things so grown also the changes are of all kinds). Under cultivation the pomegranate and the almond change character, the pomegranate if it receives pig-manure and a great deal of river water, the almond if one inserts a peg and removes for some time the gum which exudes and gives the other attention required. In like manner plainly some wild things become cultivated and some cultivated things become wild; for the one kind of change is due to cultivation, the other to neglect: however, it might be said that this is not a change but a natural development towards a better or an inferior form; (for that it is not possible to make a wild olive, pear, or fig into a cultivated olive, pear, or fig). As to that indeed which is said to occur in the case of the wild olive, that if the tree is transplanted with its top-growth entirely cut off, it produces "coarse olives," this is no very great change. However, it can make no difference which way[1] one takes this.

PLANTS OF RIVERS, MARSHES, AND LAKES

Theophrastus, *History of Plants* IV. 8.1–4. Translation of Arthur Hort

Next we must speak of plants which live in rivers, marshes, and lakes. Of these there are three classes, trees, plants of "herbaceous" character, and plants growing in clumps. By "herbaceous" I mean here such plants as the marsh celery and the like; by "plants growing in clumps" I mean reeds, galingale, *phleo*, rush, sedge—which are common to almost all rivers and such situations.

And in some such places are found brambles, Christ's thorn, and other trees, such as willow, abele, plane. Some of these are water plants to the extent of being submerged, while some project a little from the water; of some again the roots and a small part of the stem are under water, but the rest of the body is altogether above it. This is the case with willow, alder, plane, lime, and all water-loving trees.

These too are common to almost all rivers, for they grow even in the Nile. However, the plane is not abundant by rivers, while the abele is even more scarce, and the manna-ash and ash are commonest. At any rate, of those that grow in Egypt the list is too long to enumerate separately; however, to speak generally, they are all edible and have sweet flavours. But they differ in sweetness, and we may distinguish also three as the most useful for food, namely, the papyrus, the plant called *sari*, and the plant which they call *mnasion*.

[1] I.e., whether nature or man is said to cause the admitted change.

The papyrus does not grow in deep water, but only in a depth of about two cubits and sometimes shallower. The thickness of the root is that of the wrist of a stalwart man, and the length above four cubits; it grows above the ground itself, throwing down slender matted roots into the mud, and producing above the stalks which give it its name "papyrus"; these are three-cornered and about ten cubits long, having a plume which is useless and weak, and no fruit whatever; and these stalks the plant sends up at many points.

Plant Pathology

DISEASES CAUSED CHIEFLY BY WEATHER

Theophrastus, *History of Plants* IV. 14.2–14. Translation of Arthur Hort

General diseases are those of being worm-eaten, of being sun-scorched, and rot. All trees, it may be said, have worms, but some less, as fig and apple, some more, as pear. Speaking generally, those least liable to be worm-eaten are those which have a bitter acrid juice, and these are also less liable to sun-scorch. Moreover this occurs more commonly in young trees than in those which have come to their strength, and most of all it occurs in the fig and the vine.

The olive, in addition to having worms (which destroy the fig too by breeding in it), produces also a "knot" (whick some call a fungus, others a bark-blister), and it resembles the effect of sun-scorch. Also sometimes young olives are destroyed by excessive fruitfulness. The fig is also liable to scab, and to snails which cling to it. However, this does not happen to figs everywhere, but it appears that, as with animals, diseases are dependent on local conditions; for in some parts, as about Aineia, the figs do not get scab.

The fig is also often a victim to rot and to *krados*. It is called rot when the roots turn black, it is called *krados* when the branches do so; for some call the branches *kradoi* (instead of *kladoi*), whence the name is transferred to the disease. The wild fig does not suffer from *krados*, rot, or scab, nor does it get so worm-eaten in its roots as the cultivated tree; indeed some wild figs do not even shed their early fruit—not even if they are grafted into a cultivated tree.

Scab chiefly occurs when there is not much rain after the rising of the Pleiad; if rain is abundant, the scab is washed off, and at such times it comes to pass that both the spring and the winter figs drop off. Of the worms found in fig-trees some have their origin in the tree, some are produced in it by the creature called the "horned worm"; but they all turn into the "horned worm"; and they make a shrill noise. The fig also becomes diseased if there is heavy rain; for then the parts towards the root and the root itself become, as it were, sodden, and this they call "bark-shedding."[1] The vine suffers from over-luxuriance; this, as well as sun-scorch, specially happens to it either when the young shoots are cut by winds, or when it has suffered from bad cultivation, or, thirdly, when it has been pruned upwards.

The vine becomes a "shedder," a condition which some call "casting of the fruit," if the tree is snowed upon at the time when the blossom falls, or else when it becomes over lusty; what happens is that the unripe grapes drop off, and those that remain on the tree are small. Some trees also contract disease from frost, for instance, the vine; for then the eyes of the vine that was pruned early become abortive; and this also happens from excessive heat, for the vine seeks regularity in these conditions too, as in its nourishment. And in general anything is dangerous which is contrary to the normal course of things.

Moreover the wounds and blows inflicted by men who dig about the vines render them less able to bear the alternations of heat and cold; for then the tree is weak owing to the wounding and to the strain put upon it, and falls an easy prey to excess of heat and cold. Indeed, as some think, most diseases may be said to be due to a blow; for that even the diseases known as "sun-scorch" and "rot" occur because the roots have suffered in this way. In fact they think that there are only these two diseases; but there is not general agreement on this point. . . .

Some mutilations, however, do not cause destruction of the whole tree, but only produce barrenness; for instance, if one takes away the top of the Aleppo pine or the date-palm, the tree in both cases appears to become barren but not to be altogether destroyed.

There are also diseases of the fruits themselves, which occur if the winds and rains do not come in due season. For it comes to pass that sometimes trees, figs, for example, shed their fruit when rain does or does not come, and sometimes the fruit is spoilt by being rotted and so choked off, or again by being unduly dried up. It is worst of all for some trees, as olive and vine, if rain falls on them as they are dropping their blossom; for then the fruit, having no strength, drops also.

In Miletus the vines at the time of flowering are eaten by caterpillars, some of which devour the flowers, others, a different kind, the leaves; and they strip the tree; these appear if there is a south wind and sunny weather; if the heat overtakes them, the trees split.

About Taras the olives always show much fruit, but most of it perishes at the time when the blossom falls. Such are the drawbacks special to particular regions.

There is also another disease incident to the olive, which is called cob-web; for this forms on the tree and destroys the fruit. Certain hot winds

[1] I.e., shedding of the "bark" of the roots. . . .

also scorch both olive, vine-cluster, and other fruits. And the fruits of some get worm-eaten, as olive, pear, apple, medlar, pomegranate. Now the worm which infests the olive, if it appears below the skin, destroys the fruit; but if it devours the stone it is beneficial. And it is prevented from appearing under the skin if there is rain after the rising of Arcturus. Worms also occur in the fruit which ripens on the tree, and these are more harmful as affecting the yield of oil. Indeed these worms seem to be altogether rotten; wherefore they appear when there is a south wind and particularly in damp places. The *knips* also occurs in certain trees, as the oak and fig, and it appears that it forms from the moisture which collects under the bark, which is sweet to the taste. Worms also occur in some pot-herbs, as also do caterpillars, though the origin of these is of course different.

Such are in general the diseases, and the plants in which they occur. Moreover there are certain affections due to season or situation which are likely to destroy the plant, but which one would not call diseases: I mean such affections as freezing and what some call "scorching."

DISEASES OF CEREALS AND PULSES

Theophrastus, *History of Plants* VIII. 10. Translation of Arthur Hort

As to diseases of seeds—some are common to all, as rust, some are peculiar to certain kinds; thus chick-pea is alone subject to rot and to being eaten by caterpillars and by spiders; and some seeds are eaten by other small creatures. Some again are liable to canker and mildew, as cummin. But creatures which do not come from the plant itself but from without do not do so much harm; thus the *kantharis* is a visitor among wheat, the *phalangion* in vetches, and other pests in other crops.

Generally speaking, cereals are more liable to rust than pulses, and among these barley is more liable to it than wheat; while of barleys some kinds are more liable than others, and most of all, it may be said, the kind called "Achillean." Moreover the position and character of the land make no small difference in this respect; for lands which are exposed to the wind and elevated are not liable to rust, or less so, while those that lie low and are not exposed to wind are more so. And rust occurs chiefly at the full moon. Again wheat and barley are destroyed by winds, if they are caught by them either when in flower, or when the flower has just fallen and they are weak; and this applies specially to barley, indeed it occurs when the grain is already ripening, if the winds are violent and last a long time; for they dry up and parch the grain, which some call being "wind-bitten." Also a hot sun after cloudy weather destroys both, and wheat more than barley, so that the ear is not even conspicuous, since it is empty.

Wheat is also destroyed by grubs; sometimes they eat the roots, as soon as they appear, sometimes they do their work when by reason of drought the ear cannot be formed; for at such times the grub is engendered, and eats the haulm as it is becoming unrolled; it eats right up to the ear and then, having consumed it, perishes. And, if it has entirely eaten it, the wheat itself perishes; if, however, it has only eaten one side of the haulm and the plant has succeeded in forming the ear, half the ear withers away, but the other half remains sound. However, it is not everywhere that the wheat is so affected; for instance, this does not occur in Thessaly, but only in certain regions, as in Libya and at Lelanton in Euboea.

Grubs occur also in *okhros*, *lathyros*, and peas, whenever these crops get too much rain and then hot weather supervènes; and caterpillars occur in chick-peas under the same conditions. All these pests perish, when they have exhausted their food, whether the fruit in which they occur be green or dry, just as wood-worms do and the grubs found in beans and other plants, as was said of the pests found in growing trees and in felled timber. But the creature called "horned worm" is an exception. Now in regard to all these pests the position makes a great difference, as might be expected. For the climate, it need hardly be said, makes a difference according as it is hot or cold, moist or dry; and it was the climate which gave rise to these pests; wherefore they are not always found even in places in which they ordinarily occur.[1]

[1] I.e., because the atmospheric conditions are not always favourable to the pest.

MEDICINE

Though reason and observation were not absent from pre-Hellenic medicine, it was among the Greeks that rational systems of medicine developed largely free from magical and religious elements and based only on natural causes. An impressive corpus of Greek and Latin medical writings is fortunately extant recording this development.

Scientific medicine can be said to begin with the incorporation of biological and medical problems within the framework of the early Greek naturalistic systems of philosophy, in the sixth and fifth centuries B.C. The development was rapid. Centers of medical instruction arose at Cos, Cnidus, Sicily, southern Italy, and the mainland of Greece, from which there issued works on all sorts of matters connected with medicine. How these works came to be gathered together and ascribed to a fifth-century physician, Hippocrates of Cos, and which of them, if any, he himself wrote, are questions that cannot be answered with any certainty.

With the rapid development of the special sciences in the early Alexandrian period, we find a keen interest in what we call "scientific medicine." For the first time human cadavers were regularly and systematically dissected. This fact, together with the refinement of mechanical instruments, contributed to the advance of surgery. Experiments in physiology were conducted. At the same time there was an increasing divergence between different methodologic viewpoints in medicine; there were differences of opinion even on the worth of scientific investigation in relation to the art of healing. Medical sects developed stressing one or another factor or principle, Dogmatists and Empiricists, and, in the Roman period, Methodists. Despite theoretical divergences, however, physicians continued as always to show eclectic tendencies in their practical activities. These unifying tendencies later became implicit in the medical theories of such synthesizers as Rufus and Galen in the second century.

With the gradual abandonment of the systematic dissection of human cadavers, medical science tended to become static and in some respects to retrograde. It continued, however, to be studied at Alexandria and in numerous other centers and was subsequently inherited by the Arabs, who in turn helped to pass it on to the West.

Of extant Greek medical writings the earliest are the Hippocratic Collection, containing some seventy works of quite diverse character mainly from the fifth and fourth centuries B.C. The treatises of Aristotle and Theophrastus, though not specifically medical, are of prime importance for the history of anatomy, physiology, and pharmacology. Our next important extant source is Celsus' encyclopedic treatise on medicine, written in Latin in the Augustan Age. The works of Pliny and Dioscorides, especially the latter, help us to reconstruct the history of pharmacology. The great medical works of the earlier Alexandrian period, e.g., those of Herophilus and Erasistratus, are lost, but we are to some extent able to reconstruct the tradition with the help of the extant works of subsequent centuries. From the second century, works of Rufus of Ephesus, Aretaeus of Cappadocia, Soranus of Ephesus, and Galen have survived. The works of Galen, which included much polemic against the medicine of his day, in later centuries came to represent the supreme embodiment of Greek and Roman medical knowledge. Later medical writers generally confined themselves to excerpting, summarizing, translating, or commenting on their predecessors, and this is the case with Oribasius in the fourth century, Caelius Aurelianus (translator of Soranus) in the fifth, Aëtius of Amida and Alexander of Tralles in the sixth, and Paul of Aegina in the seventh. These works, it may be noted, remain entirely in the tradition of rational medicine.

We have collected a few passages to illustrate the history of Greek medicine and to give some indication of the methods used and results obtained in the various branches of medical science and its applications. Since our concern is with the scientific aspects of medicine we do not deal with the many interesting phases of the medical profession or the development of temple medicine and the cults of healing gods.

THE DEVELOPMENT OF MEDICINE

Celsus, *On Medicine*, Introduction 8–35.[1] Translation of W. G. Spencer (London, 1935)

It was . . . Hippocrates of Cos,[2] a man first and foremost worthy to be remembered, notable both for professional skill and for eloquence, who separated this branch of learning from the study of philosophy.[3] After him Diocles of Carystus,[4] next Praxagoras[5] and Chrysippus,[6] then Herophilus[7] and Erasistratus,[8] so practised this art that they made advances even towards various methods of treatment.

During the same times the Art of Medicine was divided into three parts:

---

[1] The treatise *On Medicine* written at the beginning of the first century seems to have formed one part of an encyclopedic work dealing with agriculture, military arts, rhetoric, philosophy, law, and medicine. The part on medicine, which alone is extant, gives a systematic and generally competent exposition of current knowledge about dietetics and regimen, pathology, therapeutics, pharmacology, and surgery, all in the compass of a few hundred pages. There are also important historical discussions such as that quoted here. In his eclecticism, his avoidance of sectarian disputation, and his desire to set forth a system of medicine in encyclopedic form, Celsus shows tendencies that are more typical of the Roman than of the Greek viewpoint. Whether Celsus was an original author or a translator, and whether he was a physician or a layman, are questions into which we need not enter. [Edd.]

[2] Little is known of the life of Hippocrates. He flourished at the end of the fifth century B.C. and achieved a great reputation as a physician, which grew after his death. The treatises that in some manner came to be associated with his name, though their authorship is certainly diverse, deal with almost every branch of medical art and science, and include both technical and non-technical discussions. The books of the Hippocratic Collection are, with very few exceptions, free from magical or mystical tendencies. [Edd.]

[3] Celsus' historical account cannot be taken literally. The medical art had developed quite apart from philosophy, and the stage represented in the Homeric poems is, in certain branches, not considerably different from that found in Hippocratic times. Celsus may be referring to the conflict between the relative claims of theory and practice, a perpetual conflict in medical history, evidenced, e.g., in the work *On Ancient Medicine* of the Hippocratic Collection. [Edd.]

[4] Diocles of Carystus, probably a follower of Aristotle, wrote on various branches of biology and medicine and attained great fame in antiquity. Later writers give some fragmentary clues to Diocles' treatment of pharmacology, anatomy, pathology, diet and regimen, general therapy, and embryology. [Edd.]

[5] Praxagoras of Cos flourished in the latter part of the fourth century B.C. The extant fragments of his work indicate that he recognized the connection of brain and spinal cord and established the importance of the pulse in medicine. He espoused a humoral pathology and viewed the arteries as channels for the passage of air. Tradition makes him the teacher of Herophilus. [Edd.]

[6] The reference here is probably to the Cnidian physician who was a teacher of Erasistratus.

one being that which cures through diet, another through medicaments, and the third by the hand. The Greeks termed the first *Διαιτητική*, the second *Φαρμακευτική*, the third *Χειρουργία*. But of that part which cures diseases by diet those who were by far the most famous authorities, endeavouring to go more deeply into things, claimed for themselves also a knowledge of nature, without which it seemed that the Art of Medicine would be stunted and weak. After them first of all Serapion,[9] declaring that this kind of reasoning method was in no way pertinent to Medicine, based it only upon practice and upon experience. He was followed by Apollonius and Glaucias, and somewhat later Heraclides of Tarentum,[10] and other men of no small note, who in accordance with what they professed called themselves Empirici [or Experimentalists]. Thus this Art of Medicine which treats by diet was also divided into two parts, some claiming an Art based upon speculation, others on practice alone. But after those mentioned above no one troubled about anything except what tra-

---

He seems to have made radical departures from the customary practice of his day, e.g., in his avoidance of phlebotomy (in which he was followed by Erasistratus). It is generally held that there was another Cnidian physician named Chrysippus, grandfather of the above. [Edd.]

[7] Herophilus, one of the most famous physicians of antiquity, was born at Chalcedon in the latter part of the fourth century B.C., and taught and practiced medicine at Alexandria under Ptolemy I. He made important anatomical investigations of the eye, the brain and nervous system, the vascular system, and the genital organs. He also wrote on obstetrics and gynecology. He had an elaborate quantitative theory of the pulse. Unlike his contemporary Erasistratus he retained the humoral theory of his predecessors and emphasized the importance of drugs, not, however, to the neglect of diet and regimen. [Edd.]

[8] Erasistratus of Ceos, a younger contemporary of Herophilus at Alexandria, is best known for his anatomical and physiological researches. In the fragments of his work, which alone are preserved, one reads of his distinction between motor and sensory nerves, of his use of the quantitative experiment (p. 480), and of post-mortem investigation of the effect of disease on the organs. He developed theories of digestion, the flow of blood and pneuma, and the causes of disease. He opposed the humoral theory and sought mechanical explanations ultimately based on a type of atomism and a principle of suction. In therapy he relied on dietetics rather than drugs and opposed the emphasis on phlebotomy. He invented an S-shaped catheter. [Edd.]

[9] Serapion of Alexandria, an important figure in the Empiric Sect, lived in the first half of the second century B.C. Among his writings were works on therapeutics and on the medical sects. [Edd.]

[10] Heraclides of Tarentum, who lived in the first half of the first century B.C., was one of the most important physicians of the Empiric Sect and highly esteemed by Galen. He wrote on pharmacology (see p. 510), therapy, and dietetics, as well as Hippocratic commentaries and a treatise on the Empiric Sect.

Glaucias, an Empiric physician contemporary with Heraclides, wrote Hippocratic commentaries and glosses and also a work on pharmacology.

The Apollonius here referred to is probably the Empiric physician of Antioch in the second century B.C. He and his son, also an Empiric, are mentioned by Galen. He is to be distinguished from Apollonius of Citium (p. 521) and the Herophilean Apollonius Mys of Alexandria (first century B.C.). [Edd.]

dition had handed down to him until Asclepiades[1] changed in large measure the way of curing. Of his successors, Themison,[2] late in life, diverged from Asclepiades in some respects. And it is through these men in particular that this health-giving profession of ours has grown up. . . .

They, then, who profess a reasoned theory of medicine[3] propound as requisites, first, a knowledge of hidden causes involving diseases, next, of evident causes, after these of natural actions also, and lastly of the internal parts.

They term hidden the causes concerning which inquiry is made into the principles composing our bodies, what makes for and what against health. For they believe it impossible for one who is ignorant of the origin of diseases to learn how to treat them suitably. They say that it does not admit of doubt that there is need for differences in treatment, if, as certain of the professors of philosophy have stated, some excess, or some deficiency, among the four elements,[4] creates adverse health; or, if all the fault is in the humours, as was the view of Herophilus; or in the breath, according to Hippocrates; or if blood is transfused into those blood-vessels which are fitted for pneuma, and excites inflammation which the Greeks term φλεγμονή, and that inflammation effects such a disturbance as there is in fever, which was taught by Erasistratus; or if little bodies by being brought to a standstill in passing through invisible pores block the passage, as Asclepiades contended—his will be the right way of treatment, who has not failed to see the primary origin of the cause. They do not deny that experience is also necessary; but they say it is impossible to arrive at what should be done unless through some course of reasoning. For the older men, they say, did not cram the sick anyhow, but reasoned out what might be especially suitable, and then put to the test of experience what conjecture of a sort had previously led up to. Again they say that it makes no matter whether by now most remedies have been well explored already . . . if, nevertheless, they started from a reasoned theory; and that in fact this has also been done in many instances. Frequently, too, novel classes of disease occur about which hitherto practice has disclosed nothing, and so it is necessary to consider how such have commenced, without which no one

[1] Asclepiades of Bithynia (*ca.* 124–*ca.* 50 B.C.) achieved great fame as a physician in Rome. He opposed the humoral theory and professed mechanistic and atomistic theories, holding that health depended on the proper movement of the bodily corpuscles. The basis of his mild therapy was regimen and diet rather than drugs. [Edd.]

[2] Themison of Laodicea was a most important link between Asclepiades and the Methodist School of medicine. Themison developed a system of general types of disease, which was to become basic in Methodism. [Edd.]

[3] I.e., the Dogmatists. [Edd.]

[4] I.e., the contraries, hot and cold, dry and moist, which were connected both with the four Empedoclean elements and with the system of humors. See pp. 352, 488. [Edd.]

among mortals can possibly find out whether this rather than that remedy should be used; this is the reason why they investigate the occult causes.

But they call evident those causes concerning which they inquire, as to whether heat or cold, hunger or surfeit, or such like, has brought about the commencement of the disease; for they say that he will be the one to counter the malady who is not ignorant of its origin.

Further, they term natural actions of the body, those by which we draw in and emit breath, take in and digest food and drink, as also those actions through which food and drink are distributed into every part of the members. Moreover, they also inquire why our blood-vessels now subside, now swell up; what is the explanation of sleep and wakefulness: for without knowledge of these they hold that no one can encounter or remedy the diseases which spring up in connexion with them. Among these natural actions digestion seems of most importance, so they give it their chief attention. Some following Erasistratus hold that in the belly the food is ground up; others, following Plistonicus, a pupil of Praxagoras, that it putrefies; others believe with Hippocrates, that the food is cooked up by heat. In addition there are the followers of Asclepiades, who propound that all such notions are vain and superfluous, that there is no concoction at all, but that the material is transmitted throughout the body, crude as swallowed.[1] And on these points there is little agreement indeed among them; but what does follow is that a different food is to be given to patients according as this or that view is true. For if it is ground up inside, that food should be selected which can be ground up the most readily; if it putrefies, that which does so most expeditiously; if heat concocts it, that which most excites heat. But none of these points need be inquired into if there be no concoction but such things be taken which persist most in the state in which they were when swallowed. In the same way, when breathing is laboured, when sleep or wakefulness disturbs, they deem him able to remedy it who has understood beforehand how these same natural actions happen.

Moreover, as pains, and also various kinds of diseases, arise in the more internal parts, they hold that no one can apply remedies for these who is ignorant about the parts themselves; hence it becomes necessary to lay open the bodies of the dead and to scrutinize their viscera and intestines. They hold that Herophilus and Erasistratus did this in the best way by far, when they laid open men whilst alive[2]—criminals received out of prison from the kings—and whilst these were still breathing, observed parts which beforehand nature had concealed, their position, colour, shape, size, arrangement, hardness, softness, smoothness, relation, processes and depressions of each,

[1] I.e., in a fine solution but not altered by any digestive process. [Edd.]

[2] The Aristotelian and Hippocratic writings give examples of vivisection of animals. Here for the first time we have a reference to human vivisection. [Edd.]

and whether any part is inserted into or is received into another. For when pain occurs internally, neither is it possible for one to learn what hurts the patient, unless he has acquainted himself with the position of each organ or intestine; nor can a diseased portion of the body be treated by one who does not know what that portion is. When a man's viscera are exposed in a wound, he who is ignorant of the colour of a part in health may be unable to recognize which part is intact, and which part damaged; thus he cannot even relieve the damaged part. External remedies too can be applied more aptly by one acquainted with the position, shape and size of the internal organs, and like reasonings hold good in all the instances mentioned above. Nor is it, as most people say, cruel that in the execution of criminals, and but a few of them, we should seek remedies for innocent people of all future ages.[1]

On the other hand, those who take the name of Empirici from their experience do indeed accept evident causes as necessary; but they contend that inquiry about obscure causes and natural actions is superfluous, because nature is not to be comprehended. That nature cannot be comprehended is in fact patent, they say, from the disagreement among those who discuss such matters; for on this question there is no agreement, either among professors of philosophy or among actual medical practitioners. Why, then, should anyone believe rather in Hippocrates than in Herophilus, why in him rather than in Asclepiades? If one wants to be guided by reasoning, they go on, the reasoning of all of them can appear not improbable; if by method of treatment, all of them have restored sick folk to health: therefore one ought not to derogate from anyone's credit, either in argument or in authority. Even philosophers would have become the greatest of medical practitioners, if reasoning from theory could have made them so; as it is, they have words in plenty, and no knowledge of healing at all. They also say that the methods of practice differ according to the nature of localities, and that one method is required in Rome, another in Egypt, another in Gaul; but that if the causes which produce diseases were everywhere the same, the same remedies should be used everywhere; that often, too, the causes are apparent, as, for example, of ophthalmia, or of wounds, yet such causes do not disclose the treatment: that if the evident cause does not supply the knowledge, much less can a cause which is in doubt yield it. Since, therefore, the cause is as uncertain as it is incomprehensible, protection is to be sought rather from the ascertained and ex-

[1] Celsus gives his own view (Introduction 74–75; translation of W. G. Spencer): "But to lay open the bodies of men whilst still alive is as cruel as it is needless; that of the dead is a necessity for learners, who should know positions and relations, which the dead body exhibits better than does a living and wounded man. As for the remainder, which can only be learnt from the living, actual practice will demonstrate it in the course of treating the wounded in a somewhat slower yet much milder way."

plored, as in all the rest of the Arts, that is, from what experience has taught in the actual course of treatment: for even a farmer, or a pilot, is made not by disputation but by practice. That such speculations are not pertinent to the Art of Medicine may be learned from the fact that men may hold different opinions on these matters, yet conduct their patients to recovery all the same. This has happened, not because they deduced lines of healing from obscure causes, nor from the natural actions, concerning which different opinions were held, but from experiences of what had previously succeeded. Even in its beginnings, they add, the Art of Medicine was not deduced from such questionings, but from experience; for of the sick who were without doctors, some in the first days of illness, longing for food, took it forthwith; others, owing to distaste, abstained; and the illness was more alleviated in those who abstained. Again, some partook of food whilst actually under the fever, some a little before, others after its remission, and it went best with those who did so after the fever had ended; and similarly some at the beginning adopted at once a rather full diet, others a scanty one, and those were made worse who had eaten plentifully. When this and the like happened day after day, careful men noted what generally answered the better, and then began to prescribe the same for their patients. Thus sprang up the Art of Medicine, which, from the frequent recovery of some and the death of others, distinguished between the pernicious and the salutary.[1]

THE RATIONAL SPIRIT OF GREEK MEDICINE

Hippocratic Collection, *On the Sacred Disease* 5, 21.[2] Translation of W. H. S. Jones

5. But this disease [epilepsy] is in my opinion no more divine than any other; it has the same nature as other diseases, and the cause that gives rise to individual diseases. It is also curable, no less than other illnesses, unless by long lapse of time it be so ingrained as to be more powerful than the remedies that are applied. Its origin, like that of other diseases, lies in heredity. For if a phlegmatic parent has a phlegmatic child, a bilious parent a bilious child, a consumptive parent a consumptive child, and a

[1] Celsus goes on to indicate his own intermediate position, rejecting the extreme claims of Dogmatists and Empiricists while seeking to combine the elements of truth in both their systems. [Edd.]

[2] The Hippocratic work *On the Sacred Disease* holds that the cause and treatment of seizures must be rationally considered. This opposition to popular demonology and magic and the assertion that no part of nature is more divine than any other are typical of the best in Greek medicine. The author attributes seizures to the flow of phlegm from the head and the consequent stoppage of the flow of air through the vessels. This explanation and the suggested treatment by the proper combination of hot, cold, moist, and dry are a reflection of the humoral view of man's constitution. Various reasons were given in antiquity for the popular term "sacred disease," e.g., that a disease so strange and powerful must have been sent by the gods, or that it affected the most divine part (the brain) of man. [Edd.]

splenetic parent a splenetic child, there is nothing to prevent some of the children suffering from this disease when one or the other parent suffered from it; for the seed comes from every part of the body,[1] healthy seed from the healthy parts, diseased seed from the diseased parts. Another strong proof that this disease is no more divine than any other is that it affects the naturally phlegmatic, but does not attack the bilious. Yet, if it were more divine than others, this disease ought to have attacked all equally, without making any difference between bilious and phlegmatic.

21. This disease styled sacred comes from the same causes as others, from the things that come to and go from the body, from cold, sun, and from the changing restlessness of winds. These things are divine. So that there is no need to put the disease in a special class and to consider it more divine than the others; they are all divine and all human. Each has a nature and power of its own; none is hopeless or incapable of treatment. Most are cured by the same things as caused them. One thing is food for one thing, and another for another, though occasionally it does it harm. So the physician must know how, by distinguishing the seasons for individual things, he may assign to one thing nutriment and growth, and to another diminution and harm. For in this disease as in all others it is necessary, not to increase the illness, but to wear it down by applying to each what is most hostile to it, not that to which it is accustomed. For what is customary gives vigour and increase; what is hostile causes weakness and decay. Whoever knows how to cause in men by regimen moist or dry, hot or cold, he can cure this disease also, if he distinguish the seasons for useful treatment, without having recourse to purifications and magic.

ANATOMY AND PHYSIOLOGY[2]

Bones of the Arm and Hand

Celsus, *On Medicine* VIII. 1.18–22. Translation of W. G. Spencer

From this point [i.e., the juncture of the collar-bone with the flat bone of the shoulder-blades] begins the humerus, which at both ends is swollen out, and is there soft, without marrow and cartilaginous; in the middle cylindrical, hard, containing marrow; and slightly curved both forwards and outwards. Now its front part is that on the side of the chest, its back, that on the side of the shoulder-blades; its inner part that which faces the side, its outer away from the side. It will be clear in later chapters that

[1] See p. 417. [Edd.]

[2] We have placed in the section on Zoology: Structure and Function (p. 409) passages dealing with anatomy and physiology where the emphasis is not specifically on man. We have included here, however, passages dealing with human anatomy and those in which the operation upon an animal is primarily for the purpose of drawing conclusions about human structure and physiology. The difference in treatment between the encyclopedist Celsus and the independent researcher Galen is noteworthy. [Edd.]

this applies to all joints. Now the upper head of the humerus is more rounded than any other bone hitherto described and is inserted by a small excrescence into the top of the wide bone of the shoulder-blades, and the greater part of it is held fast by sinews outside its socket.

The humerus at its lower end has two processes, between which the bone is hollowed out even more than at its extremities. This furnishes a seat for the forearm, which consists of two bones. The radius, which the Greeks call *cercis*, is the uppermost and shorter; at its beginning it is thinner, with a round and slightly hollowed head which receives a small protuberance of the humerus; and it is kept in place there by sinews and cartilage. The ulna is further back and longer and at first larger, and at its upper extremity is inserted by two outstanding prominences into the hollow of the humerus which, as I said above, is between the two processes. At their upper ends the two bones of the forearm are bound together, then they gradually separate, to come together again at the wrist, but with an alteration in size; since there the radius is the larger whilst the ulna is quite small. Further the radius as it enlarges into its cartilaginous extremity is hollowed out at its tip. The ulna is rounded at the extremity, and projects a little at one part. And, to avoid repetition, it should not be overlooked that most bones turn into cartilage at their ends, and that all joints are bounded by it, for movement would be impossible unless apposition were smooth, nor could they be united with flesh and sinews unless some such intermediary material formed the connection.

Turning to the hand, the first part of the palm consists of many minute bones, of which the number is uncertain, but all are oblong and triangular and are connected together on some plan since the upper angle of one alternates with the base of another; therefore they appear like one bone which is slightly concave. Now two small bones project from the hand and are fitted into the hollow of the radius; and at the other end five straight bones directed towards the fingers complete the palm; from these spring the fingers themselves, each composed of three bones;[1] and all are similarly formed. A lower bone is hollowed out at its top to admit a small protuberance from an upper bone, and sinews keep them in place; from them grow nails which become hard, and thus these adhere by their roots to flesh rather than to bone.

Passages of the Eye, Nose, and Ear

Celsus, *On Medicine* VIII. 1.5–6. Translation of W. G. Spencer

Now the largest passages leading into the head are those of the eyes, next the nostrils, then those of the ears. Those of the eyes lead direct and without branching to the brain. The two nasal passages are separated by

[1] If there are five metacarpal bones Celsus cannot properly assert that the thumb has three bones. [Edd.]

an intermediate bone. These begin at the eyebrows and eye-corners, and their structure is for almost a third part bony, then changes into cartilage, and the nearer they get to the mouth the more soft and fleshy their structure becomes. Now these passages are single between the highest and lowest part of the nostrils, but there they each break up into two branches, one set from the nostrils to the throat for expiration and inspiration, the other leading to the brain and split up in its last part into numerous small channels through which we get our sense of smell. In the ear the passage is also at first straight and single, but as it goes further becomes tortuous. And close to the brain this too is divided into numerous fine passages, which give the faculty of hearing.[1]

The Coats of the Eye

Rufus of Ephesus, *On the Names of the Parts of the Body* 153[2]

Of the coats of the eye the first, which is visible, is called "horn-like" [cornea]; the second "grape-like" and "chorioid." It is called "grape-like" because the part lying under the cornea is like a grape in its external smoothness and internal roughness; it is called "chorioid" because the part under the white, being full of veins, resembles the chorion that encloses the foetus. The third coat encloses a vitreous humor; because of its fineness its ancient name was "arachnoid" [i.e., like a spider's web]. But since Herophilus[3] likens it to a drawn net, some call it net-like [retiform; cf. retina], others call it glass-like [vitreous] from its humor. The fourth coat encloses the crystalline humor. It was at first without name, but later was called lentiform because of its shape, and crystalloid because of the humor it contains.

The Muscles Moving the Forearm

Galen, *Anatomical Procedures* I. 11 (II. 272 Kühn).[4] Translation of O. Temkin and W. L. Straus, Jr., *Bulletin of the History of Medicine* 19 (1946) 171–176

You will not be able to observe accurately the movements of the forearm towards the upper arm, in flexion or in extension, before you lay the whole

[1] External auditory meatus, middle ear, and internal ear (labyrinth) are here distinguished. [Edd.]

[2] Rufus of Ephesus practiced medicine at Rome in the reign of Trajan. Among his works that are preserved are those *On the Names of the Parts of the Body* and *On the Interrogation of the Patient* (see p. 499n.). There are also extant, mainly in Latin versions, other works on individual diseases and special branches of anatomy and therapy. Rufus emphasized the importance of anatomy and embraced the eclectic tendencies of his time.

[3] Herophilus's success in coining or giving currency to anatomical names is reflected in such modern terms as "torcular," "calamus scriptorius," and "duodenum," which are the Latin equivalents of the terms he used.

[4] Galen's excellent account of the muscular system is largely based on his dissection of the Barbary ape. [Edd.]

upper arm bare of all the surrounding muscles. So let us do it, keeping in mind that we said that the muscle lying upon the radius[1] reaches up to the humerus and that the muscle underneath this one, viz. the muscle that inserts into the metacarpus before the index and middlefinger,[2] also reaches up to the humerus a little distance away. As I said, it is better to preserve the heads of these muscles, or at least that of the one lying upon the radius. For it will only become clearly visible to you when you have laid bare the foremost muscle of the upper arm.[3] You will expose this muscle by paying attention to the following two marks: the vein that runs along the whole upper arm[4] and is called the "shoulder" vein, and the muscle that occupies the upper part of the shoulder[5]—perhaps it would be better to say that "forms" the upper part of the shoulder, since it is the only muscle that lies upon this part. You should make the incision along the vein downwards and, of course, after having taken away the entire skin of this region, as well as all the membranes around the muscles. The incision in the upper part of the shoulder should be made by paying attention to the likeness and unlikeness of the fibres, whereby you will perceive that the whole circumference of the muscle reaches up to one apex and, like a triangle, is inserted into the upper arm. This muscle belongs to the shoulder joint, and it alone among the muscles moving the shoulder joint has now to be removed, in order that the double head of the foremost muscle of the upper arm become visible. . . .

The foremost muscle of the upper arm which, even before dissection, is clearly visible along the shoulder vein in all people, particularly in athletes, has two heads, one of which[6] is attached to the rim of the neck of the scapula, the other[7] to the apophysis which some people call ancyroeides [i.e. similar to an anchor], others coracoeides [i.e. similar to a raven's beak]. The ligament of each head is rather strong and not quite round. Now you should follow these heads as they extend downwards along the upper arm. For at the point where they unite, they produce this muscle which, however, is no longer lifted nor distant from the bones like its heads, but is forthwith grown to the bone of the upper arm, together with the other smaller muscle which lies underneath,[8] and on top of which it rides as far as the elbow joint. Here it forms its aponeurosis and brings forth a strong tendon by means of which it grows to the radius. Moreover, it takes a

---

[1] M. brachioradialis.
[2] M. extensor carpi radialis longus and M. extensor carpi radialis brevis.
[3] M. biceps brachii.
[4] Vena cephalica.
[5] M. deltoideus.
[6] Caput longum of M. biceps brachii.
[7] Caput breve of M. biceps brachii.
[8] M. brachialis.

share in the membranous ligament around the joint, by means of which it also flexes the joint, bending it slightly to the inner side.

If this muscle is taken away, you will find another one lying underneath. This muscle too clings to the upper arm from two fleshy origins, one in the back of the upper arm, the other one more in the front. The posterior head, however, reaches much higher. You will see that these heads also unite and form a muscle which becomes tendinous and, with the tendon produced, attaches to the ulnar bone, flexing the joint and bending it slightly outwards. But when both muscles[1] function exactly, the flexion of the joint is not inclined to either side. Thus there are two anterior muscles which, as has been said, flex the elbow joint, while three others, grown to one another, extend this articulation.[2] They should also be prepared as follows.

First one has to dissect the muscle which lies in the inner parts of the upper arm below the skin.[3] It has its head near the limit of the posterior muscle of the arm-pit[4] (whose nature will be discussed in the anatomy of the muscles moving the shoulder joint). Its end reaches the elbow joint near the inner condyle of the upper arm and is membranous and thin. After this muscle has been removed, study two origins of the muscles which extend the whole forearm. One of their origins[5] grows out from the inside of the shoulder blade, not from the whole but from approximately half of its upper part. The other one[6] grows out from the posterior part of the upper arm below its head. Advancing, these origins grow together at the upper arm and in their further advance they become attached to the bend of the ulna[7] by means of a broad tendon. If you follow the fibres from above in a longitudinal direction, this tendon will appear double to you, taking its external part from the first muscle mentioned before, and the internal part from the second. If you separate them from each other and endeavor to pull each part of the muscle, you will see that each of them extends the whole forearm, but that there is a difference regarding lateral inclination. For the first of the muscles mentioned effects an oblique inclination outwards, the second inwards. Beneath this one there is another muscle which surrounds the bone of the upper arm.[8] It is united with the second muscle and believed to be part of it by the anatomists. Thus it is

---

[1] M. biceps brachii and M. brachialis.

[2] M. triceps brachii.

[3] M. dorso-epitrochlearis; this muscle is normal for macaques and other monkeys, but is absent in man (except as a very rare anomaly).

[4] M. latissimus dorsi.

[5] Caput longum of M. triceps brachii.

[6] Caput laterale of M. triceps brachii.

[7] I.e., the olecranon.

[8] Caput mediale of M. triceps brachii.

if one considers this whole muscle as a double muscle. Nevertheless, it is possible to separate them from each other along the length of the fibres. And if you do this, you will find that this muscle remains entirely fleshy and grows into the posterior part of the elbow. If this muscle is pulled, sometimes a straight and direct tension around the elbow joint seemed to me to take place, sometimes it seemed to incline a little inwards.

Experiments in Physiology

DOES DRINK ENTER THE LUNGS?

Hippocratic Collection, *On the Heart* 2[1]

Now most of what we drink is taken into the stomach (for the gullet is like a funnel and receives the great bulk of what we take). But some part of what we drink is also taken into the windpipe, but only the small amount that can flow through the opening without being felt. For the epiglottis is a perfect cover and nothing except liquid could pass through. The proof is as follows. If you stain some water with blue or red and give it to an animal that is very thirsty, preferably a pig (an animal that is neither careful nor fastidious), and if, while it is still drinking, you cut its throat, you will find this stained with the drink.[2] But the operation is not one that can be performed by everybody. We must not therefore be disbelieved when we say that drink lubricates the windpipe. Then how is it that water, when it enters the throat without any restraint, causes distress and much coughing? Because, I hold, it enters in opposition to the respiration. For that which flows in through the opening, entering a very little at a time,[3] does not oppose the upward movement of the air; on the contrary the moisture makes the path of the air a smooth one. This liquid passes off from the lung together with the air.

---

[1] This treatise is generally considered to have been composed in the fourth century B.C. The question whether any part of what we drink passes into the lungs was widely discussed in antiquity.

Cf. Plato, *Timaeus* 70 C: "The lung is, to begin with, soft and bloodless. Furthermore it contains within it perforated openings like those of a sponge, so that, as it receives breath *and drink*, it may produce coolness and furnish rest and relief in the burning heat."

Plato's view is opposed by Aristotle, *On the Parts of Animals* III.3, 664*b*4-19.

That some drink enters the lungs was held by the author of the Hippocratic book *De Natura Ossium.* The authority of Hippocrates, Philistion, and Dioxippus was later invoked for this view against the criticism of Erasistratus. See Plutarch *Symposiaca* VII. 1 and Gellius XVII. 11.

In *De Morbis* IV. 56 (VII. 604–8 Littré) numerous "proofs" are adduced against the proposition that drink enters the lungs. It is noteworthy that while the experiment of *De Corde* 2 gave rise to an erroneous conclusion, the passage of the *De Morbis*, despite its largely irrelevant rhetoric and generally lower level of scientific method, happened to answer the question correctly.

[2] The liquid may have entered as the pig squealed.

[3] According to another reading, "along the wall [of the throat]."

PROOF OF EFFLUVIA FROM ANIMALS

Anonymus Londinensis, col. XXXIII[1]

Clearly, then, there are emanations from beasts. Erasistratus[2] tries to show this also as follows. If one should take an animal, for example a bird, and keep it in a vessel without food for a given period of time, and then weigh the animal along with the visible excreta, the weight will be found to be much less[3] because of the invisible passage of considerable effluvia.

THE IRREVERSIBILITY OF THE FLOW FROM KIDNEY TO BLADDER

Galen, *On the Natural Faculties* I. 13. Translation of A. J. Brock (London, 1916)[4]

Thus, one of our Sophists, who is a thoroughly hardened disputer and as skillful a master of language as there ever was, once got into a discussion with me on this subject;[5] so far from being put out of countenance by any of the above-mentioned considerations, he even expressed his surprise that I should try to overturn obvious facts by ridiculous arguments! "For," said he, "one may clearly observe any day in the case of any bladder, that, if one fills it with water or air and then ties up its neck and squeezes it all round, it does not let anything out at any point, but accurately retains all

---

[1] This papyrus, discovered in 1891, contains material of great interest in connection with medical history. Certain parts seem to be based on a history of medicine by Menon, a pupil of Aristotle.

[2] On Erasistratus see also pp. 468, 471. A similar experiment conducted in the seventeenth century by Sanctorius upon himself is generally considered the beginning of the modern study of metabolism.

[3] Than the original weight of the bird.

[4] Galen (*ca.* A.D. 130–*ca.* 200), the most outstanding figure of medical science in the Greco-Roman period, was born at Pergamum but traveled extensively and spent many years at Rome. He was a most prolific writer not merely in the field of medicine but in philosophy, philology, and rhetoric, as well. Much of his work is extant, including treatises on almost every branch of medicine as it was then constituted. His most extended works are *On the Use of the Parts* (17 books) and *Anatomical Procedures* (15 books of which, save for a few chapters, the last seven books are known only in an Arabic version). The treatises *On the Use of the Parts* and *On the Natural Faculties* reveal the author's teleological viewpoint. Every part of the body, he holds, is so constructed as best to perform a predetermined function. This doctrine was particularly congenial to medieval thought and partly accounts for Galen's influence in the Middle Ages. The treatise *Anatomical Procedures*, on the other hand, is a guide for the anatomist and is concerned with methods and results of the actual dissection (and vivisection) of animals.

Galen was an eclectic dogmatist, drawing on the various schools of medicine of his time, as well as on the Hippocratic writings, Plato, and Aristotle. Noteworthy is his adherence to the old doctrine of the humors, which had been rejected by Erasistratus and others.

The passages selected indicate something of Galen's experimental methods. [Edd.]

[5] I.e., the function of the kidneys. Galen held the function was to separate out the urine and pass the same into the bladder by way of the ureters. The followers of Asclepiades denied this. [Edd.]

its contents. And surely," said he, "if there were any large and perceptible channels coming into it from the kidneys the liquid would run out through these when the bladder was squeezed, in the same way that it entered."[1] Having abruptly made these and similar remarks in precise and clear tones, he concluded by jumping up and departing—leaving me as though I were quite incapable of finding any plausible answer!

The fact is that those who are enslaved to their sects are not merely devoid of all sound knowledge, but they will not even stop to learn! Instead of listening, as they ought, to the reason why liquid can enter the bladder through the ureters, but is unable to go back again the same way—instead of admiring Nature's artistic skill—they refuse to learn; they even go so far as to scoff, and maintain that the kidneys, as well as many other things, have been made by Nature *for no purpose!* And some of them who had allowed themselves to be shown the ureters coming from the kidneys and becoming implanted in the bladder even had the audacity to say that these also existed for no purpose, and others said that they were spermatic ducts, and that this was why they were inserted into the neck of the bladder and not into its cavity. When, therefore, we had demonstrated to them the real spermatic ducts[2] entering the neck of the bladder lower down than the ureters, we supposed that, if we had not done so before, we would now at least draw them away from their false assumptions, and convert them forthwith to the opposite view. But even this they presumed to dispute, and said that it was not to be wondered at that the semen should remain longer in these latter ducts, these being more constricted, and that it should flow quickly down the ducts which came from the kidneys, seeing that these were well dilated. We were, therefore, further compelled to show them in a still living animal the urine plainly running out through the ureters into the bladder; even thus we hardly hoped to check their nonsensical talk.

Now the method of demonstration is as follows. One has to divide the peritoneum in front of the ureters, then secure these with ligatures, and next, having bandaged up the animal, let him go (for he will not continue to urinate). After this one loosens the external bandages and shows the bladder empty and the ureters quite full and distended—in fact almost on the point of rupturing; on removing the ligature from them, one then plainly sees the bladder becoming filled with urine.

When this has been made quite clear, then, before the animal urinates, one has to tie a ligature round his penis and then to squeeze the bladder all over; still nothing goes back through the ureters to the kidneys. Here,

---

[1] Regurgitation, however, is prevented by the fact that the ureter runs for nearly one inch obliquely through the bladder wall before opening into its cavity, and thus an efficient *valve* is produced.

[2] The *vasa deferentia.*

then, it becomes obvious that not only in a dead animal, but in one which is still living, the ureters are prevented from receiving back the urine from the bladder. These observations having been made, one now loosens the ligature from the animal's penis and allows him to urinate, then again ligatures one of the ureters and leaves the other to discharge into the bladder. Allowing, then, some time to elapse, one now demonstrates that the ureter which was ligatured is obviously full and distended on the side next to the kidney, while the other one—that from which the ligature had been taken—is itself flaccid, but has filled the bladder with urine. Then, again, one must divide the full ureter, and demonstrate how the urine spurts out of it, like blood in the operation of venesection; and after this one cuts through the other also, and, both being thus divided, one bandages up the animal externally. Then when enough time seems to have elapsed, one takes off the bandages; the bladder will now be found empty, and the whole region between the intestines and the peritoneum full of urine, as if the animal were suffering from dropsy. Now, if anyone will but test this for himself on an animal, I think he will strongly condemn the rashness of Asclepiades, and if he also learns the reason why nothing regurgitates from the bladder into the ureters, I think he will be persuaded by this also of the forethought and art shown by Nature in relation to animals.

OBSERVATION OF DIGESTIVE PROCESSES BY VIVISECTION

Galen, *On the Natural Faculties* III. 4. Translation of A. J. Brock

Suppose you fill any animal whatsoever with liquid food—an experiment I have often carried out in pigs, to whom I give a sort of mess of wheaten flour and water, thereafter cutting them open after three or four hours; if you will do this yourself, you will find the food still in the stomach. For it is not *chylification*[1] which determines the length of its stay here—since this can also be effected outside the stomach; the determining factor is *digestion*, which is a different thing from chylification, as are blood-production and nutrition. For, just as it has been shown that these two processes depend upon a *change of qualities*, similarly also the digestion of food in the stomach involves a transmutation of it into the quality proper to that which is receiving nourishment. Then, when it is completely digested, the lower outlet opens and the food is quickly ejected through it, even if there should be amongst it abundance of stones, bones, grape-pips, or other things which cannot be reduced to chyle.[2] And you may observe this yourself in an animal, if you will try to hit upon the time at which the descent of food from the stomach takes place. But even if you should fail to discover the time, and nothing was yet passing down, and the food was

[1] I.e., not the mere mechanical breaking down of food, but a distinctly vital action of "alteration."

[2] I.e., an emulsion. [Edd.]

still undergoing digestion in the stomach, still even then you would find dissection not without its uses. You will observe, as we have just said, that the pylorus is accurately closed and that the whole stomach is in a state of contraction upon the food very much as the womb contracts upon the foetus. For it is never possible to find a vacant space in the uterus, the stomach, or in either of the two bladders—that is, either in that called bile-receiving or in the other; whether their contents be abundant or scanty, their cavities are seen to be replete and full, owing to the fact that their coats contract constantly upon the contents—so long, at least, as the animal is in a natural condition.

Now Erasistratus for some reason declares that it is the contractions of the stomach which are the cause of everything—that is to say, of the softening of the food,[1] the removal of waste matter, and the absorption of the food when chylified [emulsified].

Now I have personally, on countless occasions, divided the peritoneum of a still living animal and have always found all *the intestines* contracting peristaltically[2] upon their contents. The condition of *the stomach*, however, is found less simple; as regards the substance freshly swallowed, it had grasped these accurately both above and below, in fact at every point, and was as devoid of movement as though it had grown round and become united with the food. At the same time I found the pylorus persistently closed and accurately shut, like the os uteri on the foetus.

In the cases, however, where digestion had been completed the pylorus had opened, and the stomach was undergoing peristaltic movements, similar to those of the intestines.[3]

THE SPINAL CORD

Oribasius, *Medical Collection* XXIV. 3[4]

The spinal cord is covered by two membranes, the one thick, the other thin.[5] These arise from the membranes which cover the brain and it is, in fact, to these latter membranes that the spinal cord is united. And

[1] This is an instance of Erasistratus' quest for a mechanistic explanation. [Edd.]

[2] I.e., contracting and dilating; no longitudinal movements involved. . . .

[3] The passage indicates the interrelation of two of the "natural faculties," the retentive faculty and the assimilative or alterative faculty (here digestion). [Edd.]

[4] Oribasius (*ca.* A.D. 325–*ca.* 400), was the physician of the emperor Julian the Apostate. His largest work, the *Medical Collection* (70 books, of which about a third are extant), is an encyclopedia of Greek medical knowledge in which the teachings of Galen are prominent. Shorter works of Oribasius, *Synopsis* (9 books) and *Remedia* (4 books), are extant in complete form.

The present passage collects the results of experiments and observations, recorded by Galen in various works, on the effect of incision of the spinal cord at various levels. Some of the observations go back at least to Erasistratus.

[5] *Dura mater* and *pia mater*, respectively.

there is a third body[1] which, in turn, encases these membranes and acts as a sort of covering and protection for the spinal cord. This body arises as an outgrowth from the condyles of the head. Its nature, moreover, is like that of ligaments, in so far as it arises from a bone in the same way as do ligaments. This body in fact, which is double, seems to bind together the anterior surfaces of the vertebrae by penetrating into the spaces between the vertebrae. In thickness, color, and hardness it is like the thick membrane.

Now if a longitudinal or a latitudinal section of this ligament is made, or a section in both directions at once, no harm is done to the animal. The same is true if the hard membrane is cut, or even the cord itself in a longitudinal direction. For each of the nerves that originate in pairs from the cord at the juncture of the vertebrae extends laterally, the one on the right towards the right, the other towards the left. But if a transverse section of the cord is made, the result is a paralysis of motion of those parts of the animal, the nerves leading to which originate below the point where the cord has been cut.

It will, consequently, be quite easy to determine, by dissection of nerves, what lesions an animal will suffer from the section of each part of the spinal cord.

For the present I shall add only this to what I have just said:

1. The cutting of the spinal cord at a point between the head and the first vertebra, an incision which divides the membrane covering the extremity of the posterior ventricle of the brain, immediately renders the animal incapable of controlling any part of its body, and at the same time removes all sensation. It is at this point that bulls are cut by sacrificing priests on the occasion of religious ceremonies, as you may observe.

2. Incision just below the first vertebra produces the same symptoms not only because such incision touches the aforesaid ventricle but because it paralyzes the limbs of the animal and completely suppresses respiration.

3. The same result attends an incision below the second, or the third, or the fourth vertebra (provided that in making this last incision you are careful to cut the nerve that originates in the space between the fourth and fifth vertebrae). But animals upon which this incision has been made are able to move the upper parts of the neck.

4. Incision of the cord below the fifth vertebra paralyzes all parts of the thorax, but leaves the action of the diaphragm and of a small portion of the muscles in the uppermost part of the chest almost entirely unimpaired.

5. Incision of the cord below the sixth vertebra has the same effect on the muscles in the upper part of the chest as that just described, and has less effect upon the diaphragm.

6. Incision below the seventh vertebra (and *a fortiori* below the eighth) has no effect at all upon the motion of the diaphragm, and practically no

[1] The exterior covering of the *dura mater*.

effect upon the motion of the muscles of the upper part of the chest and those of the whole neck. But the effect upon the motion of the intercostal muscles is different. For the motion of these muscles is entirely lost when the incision is made at any of the vertebrae of the neck. Even when a complete section of the spinal cord is made just below the first vertebra of the chest, the activity of the intercostal muscles is completely suppressed. But when the incision is made at the level of the second intercostal space, a very little of this activity is preserved. Below that point, however, the amount of activity on the part of the intercostal muscles that is possible after such an incision depends on the place of the incision, for the intercostal muscles situated above the level of the incision continue to function, whereas those situated below are paralyzed.

NOTE ON GALEN'S PHYSIOLOGY

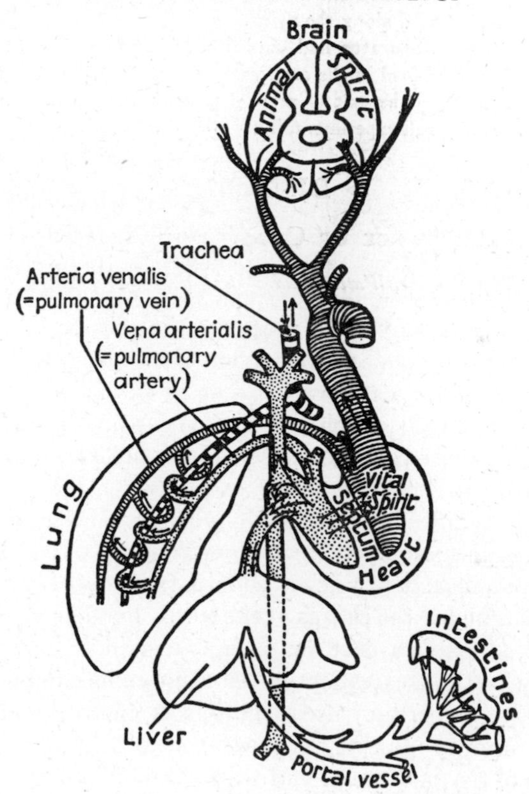

Galen's physiology. From Charles Singer, *A Short History of Science* (Oxford, Clarendon Press, 1941).

In keeping with his teleological philosophy, Galen assigned a purpose to every part of the organism and held that the purpose was carried out by the exercise of the faculties of attraction, retention, alteration, expulsion, and excretion, with which the various parts were endowed by nature.

Blood and pneuma are the basic substances in the life process. That part of the food which is not passed off as waste is elaborated into venous blood in the liver as the end product of a series of alterative processes. This blood is distributed to the parts of the body by a sort of ebbing and flowing within the venous system. Pneuma is obtained from the air we breathe, but is differentiated by the power of various organs. Blood reaching the right ventricle for the most part flows back to the general venous system after impurities are carried off through the arterial vein (i.e., our pulmonary artery) to the lung, whence they are exhaled. But a small part of the blood in the right ventricle trickles through invisible pores of the septum into the left ventricle, where it joins the pneuma that has come there from the lung by way of the arteria venalis (i.e., our pulmonary vein). Here is formed the zotic pneuma, called in later ages "vital spirit." Charged with this pneuma the bright-red arterial blood is carried throughout the body by the arterial system. The part of this blood that reaches the brain is used there in the formation of psychic pneuma (animal spirit). This pneuma is distributed by the nerve channels.

Later ages made the system symmetrical by adding a third type of pneuma, which Galen had mentioned but hardly incorporated into his system, the physical pneuma (natural spirit), centering in the liver and carried along the veins.

It may be noted that Erasistratus had denied that blood normally was carried in the arteries. He had considered blood in the arteries a pathological condition (plethora) caused, e.g., by suction resulting from the exit of arterial pneuma through a wound. Galen combated this view dialectically and experimentally.

DISEASE

Influence of Climate on Health

Hippocratic Collection, *On Airs, Waters, and Places* 1–3.[1] Translation of W. H. S. Jones

1. Whoever wishes to pursue properly the science of medicine must proceed thus. First we ought to consider what effects each season of the year can produce; for the seasons are not at all alike, but differ widely both in themselves and at their changes. The next point is the hot winds and the cold, especially those that are universal, but also those that are peculiar to each particular region. He must also consider the properties of the waters; for as these differ in taste and in weight, so the property of each is far different from that of any other. Therefore, on arrival at a town with which he is unfamiliar, a physician should examine its position with respect to the winds and to the risings of the sun. For a northern, a southern, an eastern, and a western aspect has each its own individual property. He must consider with the greatest care both these things and how the natives are off for water, whether they use marshy, soft waters, or such as are hard and come from rocky heights, or brackish and harsh. The soil, too, whether bare and dry or wooded and watered, hollow and hot or high and

[1] The first part of the treatise *On Airs, Waters, and Places* deals with the effect of climate on health. In the second part of the work the characteristics of various nations are considered in connection with their geographical situation. The usefulness of such a treatise for physicians who traveled from city to city is obvious; and it is to be noted that many Hippocratic books stress the importance of seasonal and geographical factors. [Edd.]

cold. The mode of life also of the inhabitants that is pleasing to them, whether they are heavy drinkers, taking lunch,[1] and inactive, or athletic, industrious, eating much and drinking little.

2. Using this evidence he must examine the several problems that arise. For if a physician know these things well, by preference all of them, but at any rate most, he will not, on arrival at a town with which he is unfamiliar, be ignorant of the local diseases, or of the nature of those that commonly prevail; so that he will not be at a loss in the treatment of diseases, or make blunders, as is likely to be the case if he have not this knowledge before he consider his several problems. As time and the year passes he will be able to tell what epidemic diseases will attack the city either in summer or in winter, as well as those peculiar to the individual which are likely to occur through change in mode of life. For knowing the changes of the seasons, and the risings and settings of the stars, with the circumstances of each of these phenomena, he will know beforehand the nature of the year that is coming. Through these considerations and by learning the times beforehand, he will have full knowledge of each particular case, will succeed best in securing health, and will achieve the greatest triumphs in the practice of his art. If it be thought that all this belongs to meteorology, he will find out, on second thoughts, that the contribution of astronomy to medicine is not a very small one but a very great one indeed. For with the seasons men's diseases, like their digestive organs, suffer change.

3. I will now set forth clearly how each of the foregoing questions ought to be investigated, and the tests to be applied. A city that lies exposed to the hot winds—these are those between the winter rising of the sun and its winter setting[2]—when subject to these and sheltered from the north winds, the waters here are plentiful and brackish, and must be near the surface, hot in summer and cold in winter. The heads of the inhabitants are moist and full of phlegm, and their digestive organs are frequently deranged from the phlegm that runs down into them from the head. Most of them have a rather flabby physique, and they are poor eaters and poor drinkers. For men with weak heads will be poor drinkers, as the after-effects are more distressing to them. The endemic diseases are these. In the first place, the women are unhealthy and subject to excessive fluxes. Then many are barren through disease and not by nature, while abortions are frequent. Children are liable to convulsions and asthma, and to what they think causes the disease of childhood, and to be a sacred disease.[3] Men suffer from dysentery, diarrhoea, ague, chronic fevers in winter, many

[1] That is, taking more than one full meal every day.

[2] That is, roughly between ESE and WSW (for Mediterranean latitudes). [Edd.]

[3] The reference is primarily to epilepsy (see p. 473, n. 2). [Edd.]

attacks of eczema, and from hemorrhoids. Cases of pleurisy, pneumonia, ardent fever, and of diseases considered acute rarely occur. These diseases cannot prevail where the bowels are loose. Inflammations of the eyes occur with running, but are not serious; they are of short duration, unless a general epidemic take place after a violent change. When they are more than fifty years old, they are paralyzed by catarrhs supervening from the brain, when the sun suddenly strikes their head or they are chilled. These are their endemic diseases, but besides, they are liable to any epidemic disease that prevails through the change of the seasons.

Influence of the Humors

Hippocratic Collection, *On the Nature of Man* 4–5.[1] Translation of W. H. S. Jones

4. The body of man has in itself blood, phlegm, yellow bile, and black bile; these make up the nature of his body, and through these he feels pain or enjoys health. Now he enjoys the most perfect health when these elements are duly proportioned to one another in respect of compounding, power and bulk, and when they are perfectly mingled. Pain is felt when one of these elements is in defect or excess, or is isolated in the body without being compounded with all the others. For when an element is isolated and stands by itself, not only must the place which it left become diseased, but the place where it stands in a flood must, because of the excess, cause pain and distress. In fact, when more of an element flows out of the body than is necessary to get rid of superfluity, the emptying causes pain. If, on the other hand, it be to an inward part that there takes place the emptying, the shifting, and the separation from other elements, the man certainly must, according to what has been said, suffer from a double pain, one in the place left, and another in the place flooded.

5. Now I promised to show that what are according to me the constituents of man remain always the same, according to both convention

[1] The lack of unity in the work *On the Nature of Man* has given rise to various hypotheses about diversity of authorship. The first part, that dealing with the humors, may be the work of Polybus, son-in-law of Hippocrates.

The theory of the four humors set forth in the treatise has had great influence in medical history. The development of the doctrine of humors in Greek medicine is closely linked to certain philosophical ideas prominent among the Presocratics. The notion of "opposite" qualities and of the equal balance between them came to be the basis for a doctrine of the constitution of man. In one form the number of humors was indeterminate, in another it was limited to four. The latter view is a counterpart of the Empedoclean theory of the elements (see p. 352).

Attacks on the doctrine of humors were made in antiquity notably by Erasistratus, Asclepiades, and the Methodists. Mainly through the influence of Galen a humoral as opposed to a "solidistic" system dominated medicine until comparatively recent times. For those who espoused a humoral doctrine, the basic motive in therapy was the restoration of the balance among the humors. [Edd.]

and nature. These constituents are, I hold, blood, phlegm, yellow bile, and black bile. First I assert that the names of these according to convention are separated, and that none of them has the same name as the others; furthermore, that according to nature their essential forms are separated, phlegm being quite unlike blood, blood being quite unlike bile, bile being quite unlike phlegm. How could they be like one another, when their colours appear not alike to the sight nor does their touch seem alike to the hand? For they are not equally warm, nor cold, nor dry, nor moist. Since they are so different from one another in essential form and in power, they cannot be one, if fire and water are not one.

Cf. Hippocratic Collection, *On Diseases* IV. 32–33[1]

All parts of the male and female body contribute seed for the generation of the human being.[2] This seed is deposited in the womb of the female and congeals. And after a lapse of time a human being is produced from it. Both male and female have four types of humor within the body. From these originate diseases except such as are caused by external violence. The four types of humor are phlegm, blood, bile, and water. It is not the least nor the weakest part of these that enters the seed. As a matter of fact when the living being is formed it has within itself humors healthy and unhealthy corresponding to those of the parents. I shall show how in the case of each of these types of humor there may be an excess or a deficiency in the body, a circumstance which produces disease, and that the crises of diseases come on odd numbered days. . . .

The source for the blood is the heart, for the phlegm the head, for the water the spleen, and for the bile the liver. These, apart from the alimentary tract, are the four sources for the humors. Of these sources the head and the spleen are most hollow, for the amount of open space in them is largest. But I shall speak of this more fully a little later.

Now the situation is as follows. All food and drink contains something of the nature of bile, something of the nature of water, something of the nature of blood, and something of the nature of phlegm, but in different proportions in the different cases. For this reason various foods and drinks differ from one another in point of healthfulness. That is all I shall say now on this point.

When a person has eaten or drunk something the body draws to itself from the alimentary tract something of the aforesaid humors. The sources draw from the alimentary tract through the veins, the like humor drawing the like. Thus the humors are distributed through the entire body, just as in the case of plants the like humor draws the like from the earth.

---

[1] In this treatise, of which the date, school, and authorship are uncertain, the humors are stated to be blood, phlegm, water, and bile. Cf. the treatise *On the Nature of Man*.

[2] See pp. 417ff.

Cf. Aëtius, *Placita* V. 30.1 (= Alcmaeon, Fragment 4 [Diels])

Alcmaeon held that health is a condition of equal rights (*isonomia*) among the qualities of moist, warm, dry, cold, bitter, sweet, and the rest. The excessive power (*monarchia*) of any one of these produces disease.

Descriptions of Disease and Clinical Records

Hippocratic Collection, *Epidemics* I. 1–3, Cases 2, 13; *Epidemics* III, Case 3 (Second Series).[1] Translation of W. H. S. Jones

First Constitution

1. In Thasos during autumn, about the time of the equinox to near the setting of the Pleiades,[2] there were many rains, gently continuous, with southerly winds. Winter southerly,[3] north winds light, droughts; on the whole, the winter was like a spring. Spring southerly and chilly; slight showers. Summer in general cloudy. No rain. Etesian winds few, light, and irregular.

The whole weather proved southerly, with droughts, but early in the spring, as the previous constitution had proved the opposite and northerly, a few patients suffered from ardent fevers, and these very mild, causing hemorrhage in few cases and no deaths. Many had swellings beside one ear, or both ears, in most cases unattended with fever,[4] so that confinement to bed was unnecessary. In some cases there was slight heat, but in all the swellings subsided, without causing harm; in no case was there suppuration such as attends swellings of other origin. This was the character of them:—flabby, big, spreading, with neither inflammation nor pain, in every case they disappeared without a sign.[5] The sufferers were youths, young men, and men in their prime, usually those who frequented the wrestling school and gymnasia. Few women were attacked. Many had dry coughs which brought up nothing when they coughed, but their voices were hoarse. Soon

[1] Though the title *Epidemiae* is traditionally rendered "Epidemics," the term "Visits" is employed by W. Jaeger in his *Paideia*. In his opinion the reference is to the travels of the physician to foreign cities. The first and third of the seven books are frequently, though doubtfully, ascribed to Hippocrates. They deal with various "constitutions" and record some 42 case histories. The "constitutions" are descriptions of specific climatic conditions prevailing at a particular place and time and the effect on the health of the people. Though some of the case histories have no connection with the constitutions, they are all of the highest interest as examples of painstaking observation of the course of disease. The concern here is not with therapeutics but with the recording of symptoms as they evolve. Opinions differ as to the diseases involved, though some of the cases seem to be instances of malarial fevers. Of the 42 cases, 25 end fatally. [Edd.]

[2] . . .The period is roughly from September 21 to November 8.

[3] That is, the winds were generally from the south, and such north winds as blew were light.

[4] Or [with different punctuation], "There were swellings beside the ears, in many cases on one side, but in most on both." The epidemic was obviously mumps.

[5] That is, with no symptoms indicative of a crisis.

after, though in some cases after some time, painful inflammations occurred either in one testicle or in both, sometimes accompanied with fever, in other cases not. Usually they caused much suffering. In other respects the people had no ailments requiring medical assistance.

2. Beginning early in the summer, throughout the summer and in winter many of those who had been ailing a long time took to their beds in a state of consumption, while many also who had hitherto been doubtful sufferers at this time showed undoubted symptoms. Some showed the symptoms now for the first time; these were those whose constitution inclined to be consumptive. Many, in fact most of these, died; of those who took to their beds I do not know one who survived even for a short time. Death came more promptly than is usual in consumption, and yet the other complaints, which will be described presently, though longer and attended with fever, were easily supported and did not prove fatal. For consumption was the worst of the diseases that occurred and alone was responsible for the great mortality.

In the majority of cases the symptoms were these. Fever with shivering, continuous, acute, not completely intermitting, but of the semitertian type; remitting during one day they were exacerbated on the next, becoming on the whole more acute. Sweats were continual, but not all over the body. Severe chill in the extremities, which with difficulty recovered their warmth. Bowels disordered, with bilious, scanty, unmixed, thin, smarting stools, causing the patient to get up often. Urine either thin, colourless,[1] unconcocted and scanty, or thick and with a slight deposit, not settling favourably, but with a crude and unfavourable deposit. The patients frequently coughed up small, concocted sputa, brought up little by little with difficulty. Those exhibiting the symptoms in their most violent form showed no concoction at all, but continued spitting crude sputa. In the majority of these cases the throat was throughout painful from the beginning, being red and inflamed. Fluxes slight, thin, pungent. Patients quickly wasted away and grew worse, being throughout averse to all food and experiencing no thirst. Delirium in many cases as death approached. Such were the symptoms of the consumption.

3. But when summer came, and during autumn occurred many continuous but not violent fevers, which attacked persons who were long ailing without suffering distress in any other particular manner; for the bowels were in most cases quite easy, and hurt to no appreciable extent. Urine in most cases of good colour and clear, but thin, and after a time near the crisis it grew concocted. Coughing was slight and caused no distress. No

[1] Throughout *Epidemics* ἄχρως may mean, not merely "without colour," but "of bad colour". . . .

lack of appetite; in fact it was quite possible even to give food. In general the patients did not sicken, as did the consumptives, with shivering fevers, but with slight sweats, the paroxysms being variable and irregular. The earliest crisis was about the twentieth day; in most cases the crisis was about the fortieth day, though in many it was about the eightieth. In some cases the illness did not end in this way, but in an irregular manner without a crisis. In the majority of these cases the fevers relapsed after a brief interval, and after the relapse a crisis occurred at the end of the same periods as before. The disease in many of these instances was so protracted that it even lasted during the winter.

Out of all those described in this constitution only the consumptives showed a high mortality-rate; for all the other patients bore up well, and the other fevers did not prove fatal.

Case II

Silenus lived on Broadway near the place of Eualcidas. After overexertion, drinking, and exercises at the wrong time he was attacked by fever. He began by having pains in the loins, with heaviness in the head and tightness of the neck. From the bowels on the first day there passed copious discharges of bilious matter, unmixed, frothy, and highly coloured. Urine black, with a black sediment; thirst; tongue dry; no sleep at night.

Second day. Acute fever, stools more copious, thinner, frothy; urine black; uncomfortable night; slightly out of his mind.

Third day. General exacerbation; oblong tightness of the hypochondrium, soft underneath, extending on both sides to the navel; stools thin, blackish; urine turbid, blackish; no sleep at night; much rambling, laughter, singing; no power of restraining himself.

Fourth day. Same symptoms.

Fifth day. Stools unmixed, bilious, smooth, greasy; urine thin, transparent; lucid intervals.

Sixth day. Slight sweats about the head; extremities cold and livid; much tossing; nothing passed from the bowels; urine suppressed; acute fever.

Seventh day. Speechless; extremities would no longer get warm; no urine.

Eighth day. Cold sweat all over; red spots with sweat, round, small like acne, which persisted without subsiding. From the bowels with slight stimulus there came a copious discharge of solid stools, thin,[1] as it were unconcocted, painful. Urine painful and irritating. Extremities grow a little warmer; fitful sleep; coma; speechlessness; thin, transparent urine.

Ninth day. Same symptoms.

---

[1] I take λεπτός here to mean "thinner than usual, than might have been expected."... It might also mean "consisting of small pieces."...

Tenth day. Took no drink; coma; fitful sleep. Discharges from the bowels similar; had a copious discharge of thickish urine, which on standing left a farinaceous, white deposit, extremities again cold.

Eleventh day. Death.

From the beginning the breath in this case was throughout rare and large. Continuous throbbing of the hypochondrium; age about twenty years.

Case XIII

A woman lying sick by the shore, who was three months gone with child, was seized with fever, and immediately began to feel pains in the loins.

Third day. Pain in the neck and in the head, and in the region of the right collar-bone. Quickly she lost her power of speech, the right arm was paralyzed, with a convulsion, after the manner of a stroke; completely delirious. An uncomfortable night, without sleep; bowels disordered with bilious, unmixed, scanty stools.

Fourth day. Her speech was recovered, but was indistinct; convulsions; pains of the same parts remained; painful swelling in the hypochondrium; no sleep; utter delirium; bowels disordered; urine thin, and not of good colour.

Fifth day. Acute fever; pain in the hypochondrium; utter delirium; bilious stools. At night sweated; was without fever.

Sixth day. Rational; general relief, but pain remained about the left collar-bone; thirst; urine thin; no sleep.

Seventh day. Trembling; some coma; slight delirium; pains in the region of the collar-bone and left upper arm remained; other symptoms relieved; quite rational. For three days there was an intermission of fever.

Eleventh day. Relapse; rigor; attack of fever. But about the fourteenth day the patient vomited bilious, yellow matter fairly frequently; sweated; a crisis took off the fever.

Case III (Second Series)

In Thasos Pythion, who lay sick above the shrine of Heracles, after labour, fatigue and careless living, was seized by violent rigor and acute fever. Tongue dry; thirst; bilious; no sleep; urine rather black, with a substance suspended in it, which formed no sediment.

Second day. About mid-day chill in the extremities, especially in the hands and head; could not speak or utter a sound; respiration short for a long time; recovered warmth; thirst; a quiet night; slight sweats about the head.

Third day. A quiet day, but later, about sunset, grew rather chilly; nausea; distress;[1] painful night without sleep; small, solid stools were passed.

[1] Probably bowel trouble. . . .

Fourth day. Early morning peaceful, but about mid-day all symptoms were exacerbated; chill; speechless and voiceless; grew worse; recovered warmth after a time; black urine with a substance floating in it; night peaceful; slept.

Fifth day. Seemed to be relieved, but there was heaviness in the bowels with pain; thirst; painful night.

Sixth day. Early morning peaceful; towards evening the pains were greater; exacerbation; but later a little clyster caused a good movement of the bowels. Slept at night.

Seventh day. Nausea; rather uneasy; urine oily; much distress at night; wandering; no sleep at all.

Eighth day. Early in the morning snatches of sleep; but quickly there was chill; loss of speech; respiration thin and weak; in the evening he recovered warmth again; was delirious; towards morning slightly better; stools uncompounded, small, bilious.

Ninth day. Comatose; nausea whenever he woke up. Not over-thirsty. About sunset was uncomfortable; wandered; a bad night.

Tenth day. In the early morning was speechless; great chill; acute fever; much sweat; death.

In this case the pains on even days.

INFLAMMATION OF THE WOMB

Soranus, *Gynaecia* III. 17 (Ilberg)[1]

The word "inflammation" (φλεγμονή) is derived from the verb "to inflame" (φλέγειν) and not, as Democritus said, from the notion that inflammation is caused by phlegm. Inflammation of the womb originates in many different ways, but more frequently with a chill, or blow, or miscarriage, or careless delivery, none of which requires a different treatment.

When the womb is inflamed not only are general symptoms present, but also special symptoms, indicative of the part of the womb affected. For sometimes the whole is inflamed, but at other times the mouth, the neck, the fundus, or the hollow part itself, either above, below, or at the sides, and at times some of these parts or most of them.

[1] Soranus of Ephesus, the outstanding physician of the Methodist School, practiced medicine in Rome in the reigns of Trajan and Hadrian. His writings were philosophical and philological as well as medical. His work on gynecology and obstetrics in four books has been largely preserved. A treatise on bandages and a brief biography of Hippocrates are also extant. Though his work on acute and chronic diseases has not been preserved, its contents may be reconstructed from a translation by the fifth-century Latin author, Caelius Aurelianus. Other medical works of Soranus in the fields of hygiene, pharmacology, pathology, therapeutics, ophthalmology, and surgery have not survived.

The passage here quoted describes symptoms that may, in some cases, be due to puerperal fever. Cf. Hippocratic Collection, *On Diseases of Women* I. 50.

The general accompanying symptoms, then, are fever, pain and throbbing in the part affected, distention and hardness of the epigastric region, burning heat, dryness, tenseness in the hip-joints or heaviness in the loins, flanks, lower abdomen, groins, and thighs, an attack of shivering, a pricking sensation, numbness of the feet and coldness of the knees, sweating all over, small and very rapid pulse, concomitant pain in the stomach, fainting, and weakness. As the disease progresses, the patient also suffers hiccups, pains in the neck, throat, jaws, the front part of the head, and the eyes, especially the bases; there is also obstruction of the urine or feces or both. When the inflammation becomes worse, the fever increases and swelling of the epigastric region and delirium ensue, along with clenching of the teeth and convulsions.[1]

TETANUS

Aretaeus of Cappadocia, *On the Causes and Symptoms of Acute Diseases* I. 6.[2] Translation of Francis Adams (London, 1856)

Tetanus, in all its varieties, is a spasm of an exceedingly painful nature, very swift to prove fatal, but neither easy to be removed. They are affections of the muscles and tendons about the jaws; but the illness is communicated to the whole frame, for all parts are affected sympathetically with the primary organs. There are three forms of the convulsion, namely, in a straight line, backwards, and forwards. Tetanus is in a direct line when the person labouring under the distention is stretched out straight and inflexible. The contractions forwards and backwards have their appellation from the tension and the place; for that backwards we call Opisthotonos; and that variety we call Emprosthotonos in which the patient is bent forwards by the anterior nerves. For the Greek word *τόνος* is applied both to a nerve, and to signify tension.

The causes of these complaints are many; for some are apt to supervene on the wound of a membrane, or of muscles, or of punctured nerves, when, for the most part, the patients die; for, "spasm from a wound is fatal."[3] And women also suffer from this spasm after abortion; and, in this case, they seldom recover. Others are attacked with the spasm owing to a severe blow in the neck. Severe cold also sometimes proves a cause; for this reason, winter of all the seasons most especially engenders these affections; next to it, spring and autumn, but least of all summer, unless when

[1] Soranus goes on to describe special symptoms associated with inflammation in various parts of the womb.

[2] Aretaeus of Cappadocia lived at about the same time as Galen. His extant writings deal with the causes, symptoms, and therapy of acute and chronic diseases. He also wrote on fevers, surgery, gynecology, and pharmacy. The descriptions are particularly good and the system of therapy employed is in the conservative tradition. [Edd.]

[3] Hippocratic Collection, *Aphorisms* V. 2. [Edd.]

preceded by a wound, or when any strange diseases prevail epidemically. Women are more disposed to tetanus than men, because they are of a cold temperament; but they more readily recover, because they are of a humid. With respect to the different ages, children are frequently affected, but do not often die, because the affection is familiar and akin to them; striplings are less liable to suffer, but more readily die; adults least of all, whereas old men are most subject to the disease, and most apt to die; the cause of this is the frigidity and dryness of old age, and the nature of the death.[1] But if the cold be along with humidity, these spasmodic diseases are more innocent, and attended with less danger.

In all these varieties, then, to speak generally, there is a pain and tension of the tendons and spine, and of the muscles connected with the jaws and the cheek; for they fasten the lower jaw to the upper, so that it could not easily be separated even with levers or a wedge. But if one, by forcibly separating the teeth, pour in some liquid, the patients do not drink it but squirt it out, or retain it in the mouth, or it regurgitates by the nostrils; for the isthmus faucium is strongly compressed, and the tonsils being hard and tense, do not coalesce so as to propel that which is swallowed. The face is ruddy, and of mixed colours, the eyes almost immovable, or are rolled about with difficulty; strong feeling of suffocation; respiration bad, distention of the arms and legs; subsultus of the muscles; the countenance variously distorted; the cheeks and lips tremulous; the jaw quivering, and the teeth rattling, and in certain rare cases even the ears are thus affected. I myself have beheld this and wondered. The urine is retained, so as to induce strong dysuria, or passes spontaneously from the contraction of the bladder. These symptoms occur in each variety of the spasms.

But there are peculiarities in each; in Tetanus there is tension in a straight line of the whole body, which is unbent and inflexible; the legs and arms are straight.

Opisthotonos bends the patient backward, like a bow, so that the reflected head is lodged between the shoulder-blades; the throat protrudes; the jaw sometimes gapes, but in some rare cases it is fixed in the upper one; respiration stertorous; the belly and chest prominent, and in these there is usually incontinence of urine; the abdomen stretched, and resonant if tapped; the arms strongly bent back in a state of extension; the legs and thighs are bent together, for the legs are bent in the opposite direction to the hams.

But if they are bent forwards, they are protuberant at the back, the loins being extruded in a line with the back, the whole of the spine being straight; the vertex prone, the head inclining towards the chest; the lower jaw fixed upon the breast bone; the hands clasped together, the lower ex-

[1] The Greek text is uncertain. The reference may be to the frigidity of death. [Edd.]

tremities extended; pains intense; the voice altogether dolorous; they groan, making deep moaning. Should the mischief then seize the chest and the respiratory organs, it readily frees the patient from life; a blessing this, to himself, as being a deliverance from pains, distortion, and deformity; and a contingency less than usual to be lamented by the spectators, were he a son or a father. But should the powers of life still stand out, the respiration, although bad, being still prolonged, the patient is not only bent up into an arch but rolled together like a ball, so that the head rests upon the knees, while the legs and back are bent forwards, so as to convey the impression of the articulation of the knee being dislocated backwards.

PARALYSIS AND APOPLEXY

Galen, *On the Parts Affected* III. 14 (VIII. 208–214 Kühn)

When the whole body suffers a lesion in the functioning of the nerves, it is an indication that their origin is affected. And this origin can be found only by dissection. When all the nerves have simultaneously lost their sensation and movement, the affection is called apoplexy. . . . Now paralysis sometimes attacks the whole arm, or the whole leg, and sometimes only the foot of a leg and the parts below the knee, or the corresponding parts of the arm. Dissection has shown us that in the case of all the parts of an animal below the neck that are moved voluntarily, the motor nerves originate in the so-called spinal marrow. . . . You have also seen in dissections that the nerves which move the thorax originate in the spinal cord at the neck, and again, it has been pointed out that transverse incisions which completely cut the cord destroy the sensation and mobility of all parts of the body situated below, since the spinal cord receives from the brain the faculties of sensation and of voluntary movement.[1] You have also seen in dissections that transverse incisions of the cord which stop at the center do not paralyze all the lower parts but only those situated on the same side as the incision, i.e., those parts on the right when it is the right portion of the cord which is cut, and those parts on the left when it is the left portion that is cut.

Clearly, then, when there is a condition at the origin of the cord which prevents the powers of the brain from reaching the cord, all the parts situated below, with the exception of the face, will be deprived of movement and sensation. Similarly, if only half of the cord is affected, there will be a paralysis not of all the parts situated below, but only of the right or left parts. Some paralyses of this type are also seen to attack the face, but it is the side of the face opposite the lesion that is paralyzed.

Dissection has taught us that the nerves which go to the parts of the face come from the brain itself. Hence when one of these parts is paralyzed along with the whole body, you may be sure that the seat of the paralysis

[1] See p. 484.

is in the brain itself; but when these facial parts are unaffected, the para
originates in the cord. . . .

Since apoplexy at one time destroys all psychic functions, it is c
that the brain itself is affected. An indication of the seriousness of
affection is given by the extent to which respiration is impaired. In
cases in which it departs considerably from its natural rhythm the in
to the brain must be considered serious, while in the cases where respira
is little impeded, the injury is not serious. . . .

Now just as paralysis in the whole body, the parts of the face remai
untouched, indicates that the affection is at the beginning of the sp
cord, so too if an apoplectic convulsion takes place in the whole bod
will indicate that the same region of the spinal cord is affected, provi
that the parts of the face are unaffected. But if these parts are also
fected, it will be an indication that the seat of the affection is in the br
If a single part is convulsed, the motor nerve of this part or the muscles must be affected. Therefore one who knows, by dissection, the origin of the nerves that pass to each part will be better able to cure each part deprived of sensation and movement. . . .

The sophist Pausanias, who was born in Syria and came to Rome, experienced what was at first a partial loss of sensation in his two little fingers and half of the middle finger of his left hand. And as his doctors did not treat the condition properly, the loss of sensation became complete. But when I saw him I asked him about everything that had happened to him previously and heard, among other things, that he had fallen from his carriage while traveling and had suffered a blow at the upper part of the back. This part had quickly become better but the loss of sensation in the fingers had gradually become worse. I ordered that the medicaments which his doctors had applied to his fingers be applied to the part where the blow had been received. In this way he was quickly cured. The doctors do not even know that there are special nerve roots which are distributed over the skin of the entire arm and to which the arm owes its sensation, and a different set of roots for the nerves which move the muscles.

Diagnosis and Prognosis[1]

PURPOSE OF PROGNOSIS: CRITERIA OF JUDGMENT

Hippocratic Collection, *Prognostic* 1–2.[2] Translation of W. H. S. Jones

1. I hold that it is an excellent thing for a physician to practise forecasting. For if he discover and declare unaided by the side of his patients

[1] From this point on the material in the section on medicine is chiefly concerned with the medical art as an application of medical science. We include this material for reasons that prompted us to include technological applications in dealing with other branches of Greek science. [Edd.]

[2] In addition to the obvious reasons for the importance of prognosis, there is the special

the present, the past and the future, and fill in the gaps in the account given by the sick, he will be the more believed to understand the cases, so that men will confidently entrust themselves to him for treatment. Furthermore, he will carry out the treatment best if he know beforehand from the present symptoms what will take place later. Now to restore every patient to health is impossible. To do so indeed would have been better ever than forecasting the future. But as a matter of fact men do die, some owing to the severity of the disease before they summon the physician, others expiring immediately after calling him in—living one day or a little longer—before the physician by his art can combat each disease. It is necessary, therefore, to learn the natures of such diseases, how much they exceed the strength of men's bodies, and to learn how to forecast them. For in this way you will justly win respect and be an able physician. For the longer time you plan to meet each emergency the greater your power to save those who have a chance of recovery, while you will be blameless if you learn and declare beforehand those who will die and those who will get better.

2. In acute diseases the physician must conduct his inquiries in the following way. First he must examine the face of the patient, and see whether it is like the faces of healthy people, and especially whether it is like its usual self. Such likeness will be the best sign, and the greatest unlikeness will be the most dangerous sign. The latter will be as follows. Nose sharp, eyes hollow, temples sunken, ears cold and contracted with their lobes turned outwards, the skin about the face hard and tense and parched, the colour of the face as a whole being yellow or black.[1] If at the beginning of the disease the face be like this, and if it be not yet possible with the other symptoms to make a complete prognosis, you must go on to inquire whether the patient has been sleepless, whether his bowels have been very loose, and whether he suffers at all from hunger. And if anything of the kind be confessed you must consider the danger to be less. The crisis comes after a day and a night if through these causes the face has such an appearance. But should no such confession be made, and should a recovery not take place within this period, know that it is a sign of death. If the disease be of longer standing than three days when the face has these characteristics, go on to make the same inquiries as I ordered in the previous

---

circumstance that physicians of the early period generally traveled from city to city in the practice of their art and had no official status. The present passage indicates how prognosis provided a ready means by which the physician might establish his competence.

The emphasis in the work *Prognostics* is on signs to be observed. With this should be compared Rufus of Ephesus, *On the Interrogation of the Patient*, where the emphasis is on the information that physicians may obtain from patients' answers to their questions and the manner in which these answers are given. A considerable portion of this interesting treatise is translated in A. J. Brock, *Greek Medicine*. [Edd.]

[1] This is the famous "Hippocratic face" describing the appearance of a dying person. [Edd.]

case, and also examine the other symptoms, both of the body generally and those of the eyes. For if they shun the light, or weep involuntarily, or are distorted, or if one becomes less than the other, if the whites be red or livid or have black veins in them, should rheum appear around the eyeballs, should they be restless or protruding, or very sunken, or if the complexion of the whole face be changed—all these symptoms must be considered bad, in fact fatal. You must also examine the partial appearance of the eyes in sleep. For if a part of the white appear when the lids are closed, should the cause not be diarrhoea or purging, or should the patient not be in the habit of so sleeping, it is an unfavourable, in fact a very deadly symptom. But if, along with one of the other symptoms, eyelid, lip or nose be bent or livid, you must know that death is close at hand. It is also a deadly sign when the lips are loose, hanging, cold and very white.

Hippocratic Collection, *Epidemics* I. 23. Translation of W. H. S. Jones

The following were the circumstances attending the diseases, from which I framed my judgments, learning from the common nature of all and the particular nature of the individual, from the disease, the patient, the regimen prescribed and the prescriber—for these make a diagnosis more favorable or less; from the constitution, both as a whole and with respect to the parts, of the weather and of each region; from the custom, mode of life, practices and ages of each patient; from talk, manner, silence, thoughts, sleep or absence of sleep, the nature and time of dreams, pluckings, scratchings, tears; from the exacerbations, stools, urine, sputa, vomit, the antecedents and consequents of each member in the successions of diseases, and the abscessions to a fatal issue or a crisis, sweat, rigor, chill, cough, sneezes, hiccoughs, breathings, belchings, flatulence, silent or noisy, hemorrhages, and hemorrhoids. From these things must we consider what their consequents also will be.

Fevers and Crises

Hippocratic Collection, *Epidemics* I. 24–26.[1] Translation of W. H. S. Jones

24. Some fevers are continuous, some have an access during the day and an intermission during the night, or an access during the night and an

[1] The phenomenon of a crisis in cases of acute fevers, often on a given day after the inception of the fever, is a matter of common observation to a physician. In the humoral pathology, which had many adherents when the Hippocratic treatises were written, crisis is viewed as the removal of the offending humor after its coction. Elaborate systems were developed concerning the precise day on which crisis might occur, and the treatment was so regulated as to enable the patient to cooperate with nature in bringing the disease to a successful crisis on such a critical day. While the observed periodicity of certain types of fevers common in the Greek world undoubtedly was the point of departure for these systems, there seems to be a latent element of astrology and number mysticism. [Edd.]

intermission during the day; there are semitertians, tertians, quartans, quintans, septans, nonans.[1] The most acute diseases, the most severe, difficult and fatal, belong to the continuous fevers. The least fatal and least difficult of all, but the longest of all, is the quartan. Not only is it such in itself, but it also ends other, and serious, diseases. In the fever called semitertian, which is more fatal than any other, there occur also acute diseases, while it especially precedes the illness of consumptives, and of those who suffer from other and longer diseases. The nocturnal is not very fatal, but it is long. The diurnal is longer still, and to some it also brings a tendency to consumption. The septan is long but not fatal. The nonan is longer still but not fatal. The exact tertian has a speedy crisis and is not fatal. But the quintan is the worst of all. For if it comes on before consumption or during consumption the patient dies

25. Each of these fevers has its modes, its constitutions and its exacerbations. For example, a continuous fever in some cases from the beginning is high and at its worst, leading up to the most severe stage, but about and at the crisis it moderates. In other cases it begins gently and in a suppressed manner, but rises and is exacerbated each day, bursting out violently near the crisis. In some cases it begins mildly, but increases and is exacerbated, reaching its height after a time; then it declines again until the crisis or near the crisis. These characteristics may show themselves in any fever and in any disease. It is necessary also to consider the patient's mode of life and to take it into account when prescribing. Many other important symptoms there are which are akin to these, some of which I have described, while others I shall describe later. These must be duly weighed when considering and deciding who is suffering from one of these diseases in an acute, fatal form, or whether the patient may recover; who has a chronic, fatal illness, or one from which he may recover; who is to be prescribed for or not, what the prescription is to be, the quantity to be given and the time to give it.

26. When the exacerbations are on even days, the crises are on even days. But the diseases exacerbated on odd days have their crises on odd days. The first period of diseases with crises on the even days is the fourth day, then the sixth, eighth, tenth, fourteenth, twentieth, twenty-fourth, thirtieth, fortieth, sixtieth, eightieth, hundred and twentieth. Of those with a crisis on the odd days the first period is the third, then the fifth, seventh, ninth, eleventh, seventeenth, twenty-first, twenty-seventh, thirty-

[1] I.e., fevers are continuous or intermittent; and if intermittent reach a maximum once a day (quotidian), or once every other day (tertian), or once every third day (quartan), etc. The semitertian shows characteristics sometimes of the quotidian and sometimes of the tertian. [Edd.]

first.[1] Further, one must know that, if the crises be on other days than the above, there will be relapses, and there may also be a fatal issue. So one must be attentive and know that at these times there will be the crises resulting in recovery, or death, or a tendency for better or worse. One must also consider in what periods the crises occur of irregular fevers, of quartans, of quintans, of septans, and of nonans.

Hippocratic Collection, *Prognostic* 20. Translation of W. H. S. Jones

Fevers come to a crisis on the same days, both those from which patients recover and those from which they die. The mildest fevers, with the most favorable symptoms, cease on the fourth day or earlier. The most malignant fevers, with the most dangerous symptoms, end fatally on the fourth day or earlier. The first assault of fevers ends at this time; the second lasts until the seventh day, the third until the eleventh, the fourth until the fourteenth, the fifth until the seventeenth, and the sixth until the twentieth day. So in the most acute diseases keep on adding periods of four days[2] up to twenty, to find the time when the attacks end. None of them, however, can be exactly calculated in whole days; neither can whole days be used to measure the solar year and the lunar month.

Afterwards, in the same manner and by the same increment, the first period is one of thirty-four days, the second of forty days and the third of sixty days.[3] At the commencement of these it is very difficult to forecast those which will come to a crisis after a protracted interval, for at the beginning they are very much alike. From the first day, however, you must pay attention, and consider the question at the end of every four days, and then the issue will not escape you. The constitution of quartans too is of this order. Those that will reach a crisis after the shortest interval are easier to determine, for their differences are very great from the commencement. Those who will recover breathe easily, are free from pain, sleep during the night, and show generally the most favorable symptoms; those who will die have difficulty in breathing, are sleepless and delirious, and show generally the worst symptoms. Learning these things beforehand you must make your conjectures at the end of each increment as the illness advances to the crisis. In the case of women too after delivery, the crises occur according to the same rules.

---

[1] This system of critical days is different from that set forth in the *Prognostic* (see below). In the Hippocratic treatise *Sevens* the critical days are 7, 14, 21, . . . 63. This system seems to reflect Pythagorean numerology. [Edd.]

[2] In the modern way of counting, three.

[3] The series apparently are these:—

1, 4, 7, 11, 14, 17, 20
[24, 27, 31,] 34
[37,] 40
[44, 47, 51, 54, 57,] 60.

The whole question, however, is involved in uncertainty. . . .

Supplementary Note on Pathology

Of prime importance in following ancient works on pathology and therapy is the identification of the disease, a task which in some cases is extremely difficult. The modern name of a disease may be the same as the ancient and yet refer to something quite different. Again, the Greeks did not generally give specific names to what we consider specific varieties of a disease. And our association of certain diseases with a specific germ makes the modern viewpoint quite different from the ancient.

The fact of contagion was, of course, observed in certain diseases, most strikingly in the plagues. For example, Thucydides tells us (II. 47) that in the plague at Athens the doctors died in great numbers because of their frequent contact with patients. But there is no understanding in antiquity of the underlying reason.

Something resembling a germ theory was the subject of speculation in antiquity. Thus Lucretius speaks of seeds wafted in the air as causing disease. And there is the well-known passage of Varro (*Res Rusticae* I. 12.2) in which the author speaking of the dangers of swamps says: "Certain minute animals which the eye cannot see are bred there and are carried by the air, enter the body through the mouth and nose, and cause serious diseases."

Speculation along these lines, however, could bear no fruit before the invention of the microscope. The basic conflict in antiquity was rather between the humoral and the so-called "solidistic" pathology. In the former the emphasis was on dyscrasia of the bodily humors as the cause of disease, in the latter the emphasis was on the individual organs of the body or on assumed bodily corpuscles. This conflict reappears in various forms throughout medical history. It should be added that an important school of ancient medicine, the Empiricists, refused to speculate on "hidden" causes of disease and were concerned chiefly with finding an effective treatment through experience.

THERAPEUTICS AND HYGIENE

Diet and Regimen

Dietetics in the broadest sense received great emphasis in Greek medicine. Among the extant Hippocratic and Galenic works are books devoted to diet and regimen, including baths, exercise, and massage, not merely for the sick but for the healthy. Though the tendency was generally sound, health faddism not infrequently manifested itself. Thus the first writer on gymnastics whose name is known to us, Herodicus of Selymbria, laid down such a multiplicity of rules about diet, exercise, baths, and massage that to follow them one would have to give up all other pursuits. Diocles' famous regimen of health (Frag. 141, Wellmann) is another instance. This type of extremism is criticized by Plato and Aristotle as well as by medical writers. Undoubtedly the early Greek physicians learned much from observation and practice at the gymnasium, and this fact may be the basis of the tradition that Herodicus was a teacher of Hippocrates. Physicians differed widely in antiquity as they have at all times on the relative importance of diet and drugs as therapeutic principles.

Hippocratic Collection, *On Ancient Medicine* 8–11.[1] Translation of W. H. S. Jones

8. A consideration of the diet of the sick, as compared with that of men in health, would show that the diet of wild beasts and of animals generally is not more harmful, as compared with that of men in health. Take a man

[1] The treatise *On Ancient Medicine* is a polemic against the intrusion of philosophical abstractions into medicine. It holds that medicine need not take account of such hypothetical entities as "the hot," "the cold," etc., but admits observable elements such as humors. The treatise is generally considered to be one of the earlier of the Hippocratic writings and may reflect such views as Hippocrates himself held. [Edd.]

sick of a disease which is neither severe and desperate nor yet altogether mild, but likely to be pronounced under wrong treatment, and suppose that he resolved to eat bread, and meat, or any other food that is beneficial to men in health, not much of it, but far less than he could have taken had he been well; take again a man in health, with a constitution neither altogether weak nor altogether strong, and suppose he were to eat one of the foods that would be beneficial and strength-giving to an ox or a horse, vetches or barley or something similar, not much of it, but far less than he could take. If the man in health did this he would suffer no less pain and danger than that sick man who took bread or barley-cake at a time when he ought not. . . .

9. If the matter were simple, as in these instances, and both sick and well were hurt by too strong foods, benefited and nourished by weaker foods, there would be no difficulty. For recourse to weaker food must have secured a great degree of safety. But as it is, if a man takes insufficient food, the mistake is as great as that of excess, and harms the man just as much. For abstinence has upon the human constitution a most powerful effect, to enervate, to weaken and to kill. Depletion produces many other evils, different from those of repletion, but just as severe. . . .

10. That the discomforts a man feels after unseasonable abstinence are no less than those of unseasonable repletion, it were well to learn by a reference to men in health. For some of them benefit by taking one meal only each day, and because of this benefit they make a rule of having only one meal; others again, because of the same reason, that they are benefited thereby, take lunch also. Moreover some have adopted one or other of these two practices for the sake of pleasure or for some other chance reason. For the great majority of men can follow indifferently either the one habit or the other, and can take lunch or only one daily meal. Others again, if they were to do anything outside what is beneficial, would not get off easily, but if they change their respective ways for a single day, nay, for a part of a single day, they suffer excessive discomfort. Some, who lunch although lunch does not suit them, forthwith become heavy and sluggish in body and in mind, a prey to yawning, drowsiness and thirst; while, if they go on to eat dinner as well, flatulence follows with colic and violent diarrhoea. Many have found such action to result in a serious illness, even if the quantity of food they take twice a day be no greater than that which they have grown accustomed to digest once a day. On the other hand, if a man who has grown accustomed, and has found it beneficial, to take lunch, should miss taking it, he suffers, as soon as the lunch-hour is passed, from prostrating weakness, trembling and faintness. Hollowness of the eyes follows; urine becomes paler and hotter, and the mouth bitter; his bowels seem to hang, there come dizziness, depression and listlessness. Besides all this,

when he attempts to dine, he has the following troubles: his food is less pleasant, and he cannot digest what formerly he used to dine on when he had lunch. The mere food, descending into the bowels with colic and noise, burns them, and disturbed sleep follows, accompanied by wild and troubled dreams. Many such sufferers also have found these symptoms the beginning of an illness.

11. It is necessary to inquire into the cause why such symptoms come to these men. The one who had grown accustomed to one meal suffered, I think, because he did not wait sufficient time, until his digestive organs had completely digested and assimilated the food taken the day before, and until they had become empty and quiet, but had taken fresh food while the organs were still in a state of hot turmoil and ferment. Such organs digest much more slowly than others, and need longer rest and quiet. The man accustomed to take lunch, since no fresh nourishment was given to him as soon as his body needed nourishment, when the previous meal was digested and there was nothing to sustain him, naturally wastes and pines away through want. For I put down to want all the symptoms which I have said such a man shows. And I assert furthermore that all other men besides, who when in good health fast for two or three days, will show the same symptoms as I have said those exhibit who do not take their lunch.

Hippocratic Collection, *On Regimen in Acute Diseases* 10–13.[1] Translation of W. H. S. Jones

10. Now I think that gruel made from barley has rightly been preferred over other cereal foods in acute diseases, and I commend those who preferred it; for the gluten of it is smooth, consistent, soothing, lubricant, moderately soft, thirst-quenching, easy of evacuation, should this property too be valuable, and it neither has astringency nor causes disturbance in the bowels or swells up in them. During the boiling, in fact, it has expanded to the utmost of its capacity.

11. Those who use this gruel in acute diseases must not fast, generally speaking, on any day, but they must use it without intermission unless

[1] Various books of the Hippocratic Collection deal especially with diet and regimen, e.g., *Regimen* (4 books), *Regimen in Health*, and *Regimen in Acute Diseases*. Practically all the other treatises consider diet and regimen incidentally. In the course of the second book of the treatise on *Regimen*, an attempt is made to give a systematic account of all human foods, based generally on the qualities of hot, cold, moist, and dry.

The acute diseases referred to in *Regimen in Acute Diseases* include chest complaints like pleurisy and pneumonia and fevers perhaps of a malarial type. The treatment is mild and conservative, a decoction of barley (ptisan) is prescribed as the patient's food during the continuance of the disease, and detailed directions are given for its preparation and administration. As drinks mixtures of honey and water (hydromel) and honey and vinegar (oxymel) are recommended. Fomentations, baths, purges, and venesection are other treatments referred to in the Hippocratic writings, but rest and diet are stressed above all. [Edd.]

some intermission be called for because of a purge or enema. Those who are wont to eat two meals a day should take gruel twice; those wont to have one meal only should have gruel once on the first day. Gradually, if it be thought that they need it, these also may take a second dose. At first it is sufficient to administer a small quantity, not overthick, just enough, in fact, to satisfy habit and to prevent severe pangs of hunger.

12. As to increasing the quantity of the gruel, if the disease be drier than one would wish, you ought not to increase the dose, but to give to drink before the gruel either hydromel or wine, whichever is suitable; it will be stated later what is suitable in each form of illness. Should the mouth be moist, and the sputa as they should be, increase as a general rule the quantity of the gruel; for early appearance of abundant moisture indicates an early crisis, while a later appearance of scanty moisture indicates a late crisis. In their essence the facts are on the whole as stated.

13. Many other important points have been passed over which must be used in prognosis; these will be discussed later. The more complete the purging of the bowels the more the quantity of gruel administered should be increased until the crisis. In particular, proceed thus for two days after the crisis, in such cases as lead you to suppose that the crisis will be on the fifth, seventh or ninth day, so as to make sure of both the even and the odd day. Afterwards you must administer gruel in the morning, but you may change to solid food in the evening.

Exercise and Massage

Hippocratic Collection, *Regimen* II. 62. Translation of W. H. S. Jones

Walking is a natural exercise, much more so than the other exercises, but there is something violent about it. The properties of the several kinds of walking are as follows. A walk after dinner dries the belly and body; it prevents the stomach becoming fat for the following reasons. As the man moves, the food and his body grow warm. So the flesh draws the moisture, and prevents it accumulating about the belly. So the body is filled while the belly grows thin. The drying is caused thus. As the body moves and grows warm, the finest part of the nourishment is either consumed by the innate heat, or secreted out with the breath or by the urine. What is left behind in the body is the driest part from the food, so that the belly and the flesh dry up. Early-morning walks too reduce [the body], and render the parts about the head light, bright and of good hearing, while they relax the bowels. They reduce because the body as it moves grows hot, and the moisture is thinned and purged, partly by the breath, partly when the nose is blown and the throat cleared, partly being consumed by the heat of the soul for the nourishment thereof. They relax the bowels because, cold breath rushing into them from above while they are hot, the heat gives way before the cold. It makes light the parts about the head for the following reasons. When the bowels have been emptied, being hot they draw to themselves the moisture from the body generally, and especially from the head; when the head is emptied sight and hearing are purged, and the man becomes bright. Walks after gymnastics render the body pure and thin, prevent the flesh melted by exercise from collecting together, and purge it away.

Hippocratic Collection, *On Joints* 9. Translation of E. T. Withington

The practitioner [of reduction of dislocations] must be skilled in many things and particularly in friction [massage]. Though called by one name it has not one and the same effect, for friction will make a joint firm when looser than it should be, and relax it when too stiff. But we shall define the rules for friction in another treatise. Now, for such a shoulder[1] the proper friction is that with soft hands, and always gently. Move the joint about, without force, but so far as it can be moved without pain.

Celsus, *On Medicine* II. 14.2–4. Translation of W. G. Spencer

It cannot be disputed that Asclepiades has taught when and how rubbing should be practised, with a wider application, and in a clearer way, although he has discovered nothing which had not been comprised in a few words by that most ancient writer Hippocrates, who said that rubbing, if strenuous, hardens the body, if gentle, relaxes; if much, it diminishes, if moderate, fills out. It follows, therefore, that in the following cases rubbing should be employed, when either a feeble body has to be toned up, or one indurated has to be softened, or a harmful superfluity is to be dispersed, or a thin and infirm body has to be nourished. Yet when examined with attention (although this no longer concerns the medical man) the various species of rubbing may be easily recognized as all dependent on causing one thing, depletion. For an object is toned up when that is removed which by its presence was the cause of the laxness; and is softened when that which has been producing induration is abstracted; and it is filled up, not by the rubbing itself, but by the nutriment, which subsequently penetrates by some sort of dispersal to the very skin itself after it has become relaxed.

Fomentations and Baths

Hippocratic Collection, *Regimen in Acute Diseases* 21, 65–68. Translation of W. H. S. Jones

21. When there is pain in the side, whether at the beginning or later, it is not amiss to try to dissipate it first by hot fomentations. The best

[1] The reference is to a dislocated shoulder. [Edd.]

fomentation is hot water in a skin, or bladder, or bronze or earthen vessel. Apply something soft to the side first to prevent discomfort. A good thing also to apply is a big, soft sponge dipped in hot water and squeezed out. You must, however, cover up the heat on the upper part,[1] for doing so will make it hold out and last for a longer time; besides, it will prevent the steam being carried towards the breath of the patient—unless indeed the patient's breathing it be considered an advantage, as in fact it occasionally is. Barley, too, or vetches: soak in vinegar that is slightly stronger than could be drunk, boil, sew up in bags and then apply. Bran may be used in like manner. For dry fomentations, salt or toasted millet in woolen bags is most suitable; millet is also light and soothing.

65. The bath will be beneficial to many patients, sometimes when used continuously, sometimes at intervals. Occasionally its use must be restricted, because the patients have not the necessary accommodation, for few houses have suitable apparatus and attendants to manage the bath properly. Now if the bath be not carried out thoroughly well, no little harm will be done. The necessary things include a covered place free from smoke, and an abundant supply of water, permitting bathings that are frequent but not violent, unless violence is necessary.

If rubbing with soap be avoided, so much the better; but if the patient be rubbed, let it be with soap that is warm, and many times greater in amount than is usual, while an abundant affusion should be used both at the time and immediately afterwards. A further necessity is that the passage to the basin should be short, and that the basin should be easy to enter and to leave. The bather must be quiet and silent; he should do nothing himself, but leave the pouring of water and the rubbing to others. Prepare a copious supply of tepid water, and let the affusions be rapidly made. Use sponges instead of a scraper, and anoint the body before it is quite dry. The head, however, should be rubbed with a sponge until it is as dry as possible. Keep chill from the extremities, and the head, as well as from the body generally. The bath must not be given soon after gruel or drink has been taken, nor must these be taken soon after a bath.

66. Let the habits of the patient carry great weight—whether he is very fond of his bath when in health and is in the habit of bathing. Such people feel the need of a bath more, are more benefited by its use and more harmed by its omission. On the whole, bathing suits pneumonia rather than ardent fevers, for it soothes pain in the sides, chest and back; besides, it concocts and brings up sputum, eases respiration, and removes fatigue, as it softens the joints and the surface of the skin. It is diuretic, relieves heaviness of the head, and moistens the nostrils.

[1] I.e., on the part of the sponge not next to the skin.

67. Such are the benefits from bathing, and they are all needed. If, however, one or more requisites be wanting, there is a danger that the bath will do no good, but rather harm. For each neglect of the attendants to make proper preparations brings great harm. It is a very bad time to bathe when the bowels are looser than they ought to be in acute diseases, likewise too when they are more costive than they ought to be, and have not previously been moved. Do not bathe the debilitated, those affected by nausea or vomiting, those who belch up bile, nor yet those who bleed from the nose, unless the hemorrhage be less than normal, and you know what the normal is. If the hemorrhage be less than normal, bathe either the whole body, if that be desirable for other considerations, or else the head only.

68. If the preparations be adequate, and the patient likely to benefit by the bath, bathe every day. Those who are fond of bathing will not be harmed even by two baths a day. Patients taking unstrained gruel are much more capable of using the bath than those taking the juice only, though these too can use it sometimes. Those taking nothing but drink are the least capable, though some even of these can bathe. Judge by means of the principles given above who are likely and who are unlikely to profit by the bath in each kind of regimen. Those who really need one of the benefits given by the bath you should bathe as far as they are profited by the bath. Those should not be bathed who have no need of these benefits, and who furthermore show one of the symptoms that bathing is not suitable.

Pharmacology and Materia Medica

The use of inorganic and organic substances, particularly plants, for medical purposes and for the preparation of poisons antedates recorded history; and there is an extensive literature in this field that has come down from Greco-Roman antiquity. Not only do the medical writings of every period abound in this material, but the authors of treatises on botany, e.g., Theophrastus in the ninth book of his *History of Plants*, take up the subject. Special works on pharmacology appear in the Alexandrian and subsequent periods. They generally set forth a traditional lore sometimes with an admixture of superstition. Note in this connection the Greek poems of Nicander of Colophon (third century B.C.), which treat of poisons derived from animals as well as plants, and their antidotes. Crateuas in the first century B.C. wrote on the medicinal properties of plants. He lived at the court of Mithridates of Pontus, whose interest in toxicology was shared by many kings of that and other periods. Mithridates' notes on the subject are said to have been ordered by his conqueror Pompey to be translated into Latin.

Among the larger treatments of pharmacology that have come to us from Greco-Roman antiquity are those of Pliny and Celsus in Latin and of Dioscorides and Galen in Greek.

Pliny's account takes up a large part of Books 20–27 of the *Natural History*; Celsus' account covers most of the fifth book of his treatise *On Medicine*. In addition to the *De Materia Medica* of Dioscorides (five books), there are other extant treatises doubtfully ascribed to this author, dealing with poisons derived from plant and animal substances and with the preparation of simple and compound remedies.

Galen devoted several special works to this subject, including among others, one *On Simples* (11 books) and one *On the Composition of Remedies* (17 books). While Galen drew heavily on his predecessors, especially Dioscorides, he also made independent investigations, traveled widely for the purpose of obtaining drugs, and even sought instruction in methods of adulteration in order that he might detect frauds. Professional sellers of drugs were notoriously fraudulent.

In the earlier period doctors compounded their own medicaments, but this practice had declined in Galen's time. There was, however, a widespread sale of special proprietary remedies for various diseases. These drugs were often compounded by well-known physicians and sold in containers stamped with the compounder's name. Many such boxes that had contained eye salves have been found.

The tendency to compounds with large numbers of ingredients developed in the earlier Alexandrian period and persisted throughout antiquity. Panaceas, universal antidotes, and the like were concocted. In these preparations superstition and magic frequently played a great part. The various Greek and Latin herbals of the later period of antiquity drew heavily on the earlier material and made practically no original contribution.

HISTORY AND GENERAL CONSIDERATIONS

Dioscorides, *De Materia Medica* I, Introduction.[1] Translation of John Goodyer (1655) in R. T. Gunther, *The Greek Herbal of Dioscorides* (Oxford, 1934)

Dearest Areius,

Although many writers of modern times, as well as of antiquity, have composed treatises on the preparation, power and testing of medicines, I will try to show you that I was not moved to this undertaking by any vain or senseless impulse. It was because some of these authors did not perfect their work, while others derived most of their account from histories. Iollas the Bithynian[2] and Heraclides the Tarentine[3] did indeed touch upon the subject, but they entirely omitted the Treatise on Herbs, and failed to record all metallics and spices. Crateuas, the rhizotomist, and Andreas,[4] the physician, appear to have been better versed in this part of the subject

[1] Pedanius Dioscorides of Anazarba (near Tarsus in Cilicia) made extensive observations on plants in the course of his travels as physician with the Roman armies. Where he uses literary sources in his treatise *De Materia Medica* (five books), they are in many cases the same as those of his contemporary, Pliny. Perhaps the best known is Sextius Niger, who wrote in the first century B.C. Two extant works on poisons are ascribed to Dioscorides but were probably not written by him. His authorship of the *Euporista* (two books) has also been questioned.

The work of Dioscorides is of fundamental importance in the history of botany because of the minute description of some six hundred plants. There is evidence that the drawings of plants in Crateuas's no longer extant treatise became the basis for those of subsequent authors and that many of the drawings in the famous Codex Aniciae Julianae (sixth century) of Dioscorides derive from Crateuas. [Edd.]

[2] A physician of the third century B.C. [Edd.]

[3] See p. 469. [Edd.]

[4] The reference is probably to the physician of Ptolemy IV (Philopator) in the third century B.C., an adherent of the school of Herophilus and author of a work on pharmacology of which fragments are extant. [Edd.]

than the others, but have passed over many very serviceable roots and have given insufficient descriptions of several herbs. Yet it must be confessed that though the matters, which they have transmitted, are few, yet the ancients have used great diligence in their work.

We may not be wholly in agreement with the modern writers, among whom are Julius Bassus, Niceratus and Petronius, Niger and Diodotus,[1] all Asclepiads.[2] They have in a manner deigned to describe familiar facts well-known to all, but they have transmitted the powers of Medicines and their examination cursorily, not estimating their efficacy by experience, but by vain prating about the cause, having lifted up each medicine to a heap of controversy: and besides this they have recorded one thing by mistake for another. Thus Niger, who seems to be a man of special note amongst them, states that Euphorbion is the juice of Chamelaia growing in Italy, that Androsaimon is the same as Hypericon, and that Aloe is a mineral growing in Judaea, and in the face of plain evidence he sets down many more such falsehoods, which are tokens that he acquired his information not by his own observation, but had it only from the false relation of hearsay. Moreover, they have offended in the classification of medicines: some couple together those of quite contrary faculties, others follow an alphabetical arrangement in their writing, and have separated both the kinds and the operations of things that are closely related, so that thereby they come to be harder to remember.

But we, as I may say from our first growth, having an unceasing desire to acquire knowledge of this matter, and having travelled much (for you know that I led a soldier's life), have by your advice gathered together all that I have commented hereupon, and have committed it into five books. This compilation I dedicate to you, thus fulfilling my grateful affection for the goodwill you have towards us. It is your nature to show yourself a familiar friend to all who are led by learning, but especially to those who follow the same profession, and yet more particularly to myself. But the great affection that that most excellent man, Licinius Bassus (doth bear) unto thee, is no small token of the loveable goodness that is in thee, which I know, beholding when I lived with you, the goodwill worthy of emulation, that passed between you both.

But I beg that you, and all that may peruse these Commentaries, will not pay attention so much to the force of our words, as to the industry and experience that I have brought to bear in the matter. For with very accurate diligence, knowing most herbs with mine own eyes, others by his-

[1] The writers here mentioned were used, for the most part, as sources by Pliny, and fragments of their work are in some cases extant. Pliny seems to have erred in identifying Petronius with Diodotus. [Edd.]

[2] Dioscorides probably meant "followers of the physician Asclepiades" (first century B.C.). [Edd.]

torical relation agreeable to all, and by questioning, diligently enquiring of the inhabitants of each sort, we will endeavor both to make use of another arrangement, and also to describe the kinds and forces of every one of them.

Now it is obvious to everybody that a Treatise on Medicines is necessary, for it is conjoined to the whole Art of Healing, and by itself yields a mighty assistance to every part. And because its scope may be enlarged both in the direction of methods of preparation, and of mixtures, and of experiments on diseases and because a knowledge of each separate medicine contributes much hereunto, we will include matter that is familiar and closely allied, that the book may be complete.

Before all else it is proper to use care both in the storing up and in the gathering of herbs each at its due season, for it is according to this that medicines either do their work, or become quite ineffectual. We ought to gather herbs when the weather is clear, for there is a great difference whether it be dry or rainy when the gathering is made. The place also makes a difference: whether the localities be mountainous and high, whether they lie open to the wind, whether they be cold and dry; upon this the stronger forces of drugs depend. Medicinal plants found growing on plains, in plashy and shady localities, where the wind cannot blow through, are for the most part the weaker; and especially those that are not gathered at the right season, or else are decayed through weakness. It must also not be forgotten that herbs frequently ripen earlier or later according to the characteristics of the country and the temperature of the year, and that while some of them by an innate property bear flowers and leaves in winter, others flower twice in a year. Now it behoves anyone who desires to be a skillful herbalist, to be present when the plants first shoot out of the earth, when they are fully grown, and when they begin to fade. For he who is only present at the budding of the herb cannot know it when full-grown, nor can he who hath examined a full-grown herb recognize it when it has only just appeared above ground. Owing to changes in the shape of leaves and the size of stalks, and of the flowers and fruits, and of certain other known characteristics, a great mistake has been made by some who have not paid proper attention to them in this manner. For this very reason, some authors have blundered when they have written of some plants that they bear neither flowers, not stalk, nor fruit, citing Gramen, Tussilago and Quinquefolium. Therefore the man who will observe his herbs oftentimes and in divers places will acquire the greatest knowledge of them. We must likewise be aware that only those Medicinal Herbs, the White and Black Hellebore, *Veratrum album*[1] and *nigrum*, retain their

[1] See p. 515. European white hellebore is similar to the so-called "false" hellebore, *veratrum viride*, native to the United States. [Edd.]

power for many years; the others, for the most part, will only keep good for use for three years. But herbs which are full of branches, like Stoechas, Chamaidrus, Polion, Abrotonum, Seriphium, Absinthium, Hyssopum, and the like should be gathered whilst they are great with seed; flowers ought to be gathered before they fall; fruits when they are ripe, and seeds when they begin to be dry, and before they fall out. To extract the juice of herbs, take their stalks when they have newly sprouted; and so too with the leaves. But for taking juices and tears, the stems should be cut while yet in their ripeness. Roots for storing or for the extraction of juices and the peeling of barks should be collected when the herbs are beginning to lose their leaves, when those which are clean may be set to dry forthwith in dry places, but those which have earth or clay sticking to them must be washed with water. Flowers and sweet-scented things should be laid up in dry boxes of Lime-wood; but there are some herbs which do well enough if wrapped up in papers or leaves for the preservation of their seeds. For moist medicines some thicker material such as silver, or glass, or horn will agree best. Yes, and earthenware if it be not thin is fitting enough, and so is wood, particularly if it is box-wood. Vessels of bronze will be suitable for eye medicines and for liquids and for all that are compounded of vinegar or of liquid pitch or of Cedria, but fats and marrows ought to be put up in vessels of tin.

Celsus, *On Medicine* V, Introduction.[1] Translation of W. G. Spencer

[Medicaments] were held of high value by ancient writers, both by Erasistratus and those who styled themselves Empirics, especially however by Herophilus and his school, insomuch that they treated no kind of disease without them. A great deal has also been recorded concerning the powers of medicaments, as in the works of Zeno[2] or of Andreas or of Apollonius, surnamed Mys.[3] On the other hand, Asclepiades dispensed with the use of these for the most part, not without reason; and since nearly all medicaments harm the stomach and contain bad juices, he transferred all his treatment rather to the management of the actual diet. But while in most diseases that is the more useful method, yet very many illnesses attack our bodies which cannot be cured without medicaments. This before all things it is well to recognize, that all branches of medicine are so connected together, that it is impossible to separate off any one part completely, but each gets its name from the treatment which it uses most. Therefore, both that which treats by dieting has recourse at times to medicaments,

[1] The extent to which drugs should be used in medicine has been the subject of endless controversy, tradition recording an opposition even between Coan and Cnidian physicians on this question. [Edd.]

[2] An important anatomist of about 200 B.C. [Edd.]

[3] A pharmacologist contemporary with Celsus. [Edd.]

and that which combats disease mainly by medicaments ought also to regulate diet, which produces a good deal of effect in all maladies of the body.

PREPARATION AND USE OF DRUGS

Theophrastus, *History of Plants* IX. 8, 10, 17, 19. Translation of Arthur Hort

[Collection of Plant Juices]

8. Now we must endeavour to speak in like manner of those juices which have not been mentioned already, I mean, such as are medicinal or have other properties; and at the same time we must speak of roots; for some of the juices are derived from roots, and apart from that roots have in themselves divers properties of all kinds; and in general we must discuss medicinal things of all kinds, as fruit, extracted juice, leaves, roots, "herbs"; for the herb-diggers call some medicinal things by this name.

The properties of "roots"[1] are numerous and they have numerous uses; but those which have medicinal virtues are especially sought after, as being the most useful; and they differ in not all being applied to the same purposes and in not all having their virtue in the same parts of them. To speak generally, most "roots" have it in themselves; or else it is found in the fruits or the juices of the plant; and in some cases in the leaves as well, and it is to the virtues of the leaves in most cases that the herb-diggers refer, when they speak, as has just been said, of "herbs."

The collection of the juice from plants from which it is collected is mostly done in summer, in some cases at the beginning of that season, in others when it is well advanced. The digging of roots is done in some cases at the time of wheat-harvest or a little earlier, but the greater part of it in autumn after the rising of Arcturus when the plants have shed their leaves, and, in the case of those whose fruit is serviceable, when they have lost their fruit. The collection of juice is made either from the stalks, as with *tithymallos* (spurge), wild lettuce, and the majority of plants, or from the roots, or thirdly from the head, as in the case of the poppy; for this is the only plant which is so treated and this is its peculiarity. In some plants the juice collects of its own accord in the form of a sort of gum, as with tragacanth; for incision of this plant cannot be made; but in most it is obtained by incision. In some cases the juice is collected straight into vessels, for instance, that of *tithymallos* (spurge) or *mekonion* (for the plant has both names) and in general the juice of specially juicy plants is so collected. But that of those which do not yield abundant juice is taken with a piece of wool, as also that of wild lettuce.

In some cases there can be no collection of juice, but there is a sort of extraction of it, for instance, in the case of plants which are cut down or

---

[1] "Root" is used in this translation to refer to a medicinal plant in general; *root* to refer to the part of a plant. [Edd.]

bruised; they then pour water over them and strain off the fluid, keeping the sediment; but it is plain that in these cases the juice obtained is dry and less copious. In most "roots" the juice thus extracted is less powerful than that of the fruit, but in hemlock it is stronger and it causes an easier and speedier death even when administered in a quite small pill; and it is also more effective for other uses. That of *thapsia* is also powerful, while all the rest are less so. Such then is a general account of the various ways of obtaining the juices of plants. . . .

[Hellebore]

10. The white and the black hellebore appear to have nothing in common except the name. But accounts differ as to the appearance of the plants; some say that the two are alike and differ only in colour, the root of the one being white, of the other black; some, however, say that the leaf of the "black" is like that of bay, that of the white like that of the leek, but that the roots are alike except for their respective colours. Now those who say that the two plants are alike describe the appearance[1] as follows: the stem is like that of asphodel and very short; the leaf has broad divisions, and is extremely like that of ferula, but is long; it is closely attached to the root and creeps on the ground; the plant has numerous roots, to wit, the slender roots which are serviceable.

Also they say that the black is fatal to horses, oxen and pigs, wherefore none of these animals eat it; while the white is eaten by sheep, and from this circumstance the virtue of the plant was first observed, since it purges them; it is at its prime in autumn, and past its prime when spring comes. However the people of Mount Oeta gather it for the meetings[2] of the Amphictyons; for it grows there in greatest abundance and best, though at only one place in the district of Oeta, namely, about Pyra.

(The seed of rupture-wort is mixed with the potion given to promote easy vomiting; this plant is a small herb.)

The black kind of hellebore grows everywhere; it is found in Boeotia, in Euboea and in many other places; but best is that from Mount Helicon, which mountain is in general rich in medicinal herbs. The white occurs in few places; the best and that which is most used comes from one of four places, Oeta, Pontus, Elea, and Malea.[3] They say that that of Elea grows in the vineyards and makes the wine so diuretic that those who drink it become quite emaciated.

[1] I.e., of the two plants regarded as one; but the text of the following description seems to be hopelessly confused.

[2] Which were held apparently at Thermopylae regularly in autumn and sometimes in spring: the meeting would give opportunities for sale. . . .

[3] . . . Pliny (Natural History XXV, 49) gives Parnassus as the fourth locality. Cf. the following paragraph.

But best of all these and better than that found anywhere else is that of Mount Oeta, while that of Parnassus and that of Aetolia (for the plant is common in these parts too and men buy and sell it, not knowing the difference) are tough and exceeding harsh. These plants then, while resembling the best form in appearance, differ in their virtues.

Some call the black the "hellebore of Melampus," saying that he first cut and discovered it. Men also purify horses and sheep with it, at the same time chanting an incantation; and they put it to several other uses.

[Effect of Drugs]

17. The virtues of all drugs become weaker to those who are accustomed to them, and in some cases become entirely ineffective. Thus some eat enough hellebore to consume whole bundles and yet suffer no hurt; this is what Thrasyas did, who, as it appeared, was very cunning in the use of herbs. And it appears that shepherds sometimes do the like; wherefore the shepherd who came before the vendor of drugs (at whom men marvelled because he ate one or two roots) and himself consumed the whole bundle, destroyed the vendor's reputation: it was said that both this man and others did this every day.

For it seems that some poisons became poisonous because they are unfamiliar, or perhaps it is a more accurate way of putting it to say that familiarity makes poisons non-poisonous; for, when the constitution has accepted them and prevails over them, they cease to be poisons, as Thrasyas also remarked; for he said that the same thing was a poison to one and not to another; thus he distinguished between different constitutions, as he thought was right; and he was clever at observing the differences. Also, besides the constitution it is plain that use has something to do with it. At least Eudemus, the vendor of drugs, who had a high reputation in his business, after making a wager that he would experience no effect before sunset, drank a quite moderate dose, and it proved too strong for his power of resistance: while the Chian Eudemus took a draught of hellebore and was not purged. And on one occasion he said that in a single day he took two and twenty draughts in the market-place as he sat at his stall, and did not leave the place till it was evening, and then he went home and had a bath and dined, and was not sick. However, this man was able to hold out because he had provided himself with an antidote; for he said that after the seventh dose he took a draught of tart vinegar with pumice-stone dust in it, and later on took a draught of the same in wine in like manner; and that the virtue of the pumice-stone dust is so great that, if one puts it into a boiling pot of wine, it causes it to cease to boil, not merely for the moment, but altogether, clearly because it has a drying effect and it catches the vapour and passes it off. It was then by this antidote that Eudemus was able to contain himself in spite of the large quantity of hellebore which he took.

However, many things go to show that use makes much difference; thus some say that the sheep of some places do not eat wormwood; yet those of Pontus not only eat it but become fatter and fairer and, as some say, have no bile. But these things may be said to belong to a different enquiry.

[The Problem of Specific Cause]

19. . . . Now since the natural qualities of roots, fruits and juices have many virtues of all sorts, some having the same virtue and causing the same result, while others have opposite virtues, one might raise a question which is perhaps equally perplexing in regard to other matters, to wit, whether those that produce the same effect do so in virtue of some single virtue which is common to them all, or whether the same result may not come about also from different causes.

Dioscorides, *De Materia Medica* IV. 75 (Wellmann)

Mandragora[1]

Others call it antimimon, others bombochylon, others circaea, [others dircaea,] for the root is held to be useful in making philtres. There is a female variety which is black. This is called thridacias. Its leaves are narrower and smaller than those of lettuce, have a foul and heavy odor, and spread out over the ground. Along the leaves there is fruit which resembles the service-berry, being yellow and of pleasant smell. In it is the seed, which is like that of the pear. The roots are quite large, numbering two or three, intertwined with one another, black on the surface, white within. The bark is thick. There is no stalk.

The white male variety, which some have called morion, has white leaves, large, flat and smooth like those of the beet. Its fruit is twice as large, saffron colored, and of good but somewhat strong odor. Shepherds eat it and become somewhat drowsy. The root is like that just described, but larger and whiter. This variety is also without stalk.

Juice is obtained from the bark of the root by beating it when it is tender and placing it under a press. The juice must then be warmed in the sun

[1] The plant is generally known as *Atropa mandragora* and the "female" and "male" varieties referred to by Dioscorides are sometimes known as *mandragora autumnalis* and *mandragora vernalis*, respectively. The root owes its narcotic properties to hyoscyamine, hyoscine, and atropine, and, as our passage indicates, was used as an anesthetic. A third variety referred to later by Dioscorides may be *atropa belladonna*, deadly nightshade, which also contains atropine. The mandrake root seems to have been used among various peoples of antiquity in compounding aphrodisiacs and other potions. The alleged resemblance of the root to the human form added to the superstitions regarding the plant.

and, after being condensed, must be stored in an earthen vessel. Juice is obtained also from the fruit in a similar way but this juice is weak. The bark of the root is stripped off, pierced with thread, hung up and thus stored. Some boil down the roots in wine to a third their original size and, straining the juice, store it up. They administer a cyathus[1] of it to those who cannot sleep or who suffer severe pain, or those whom they do not wish to feel pain during an operation of cutting or cauterizing.

If a weight of two obols[2] of the juice is drunk with melicraton[3] it brings up the phlegm and bile, as does hellebore. Drunk in larger quantities it is fatal. It is used as a component in eye medicines, anodynes, and soothing pessaries. But an application of half an obol of it unmixed causes menstrual flow and expels the embryo. Inserted in the anus as a suppository it causes sleep. And the root, it is said, softens ivory, when boiled together with it for six hours, and makes it easy to mold into any desired shape.

Fresh leaves of this plant when spread on with barley are good for inflammations of the eyes and those upon ulcers. It removes all hardness and abscesses, glandular swellings, and tumors. If rubbed on gently for five or six days it causes branded marks to disappear without ulceration. The leaves are preserved in brine and stored for the same purposes. The root rubbed down fine with vinegar heals erysipelas, but for snake bites is used with honey or oil. It removes glandular swellings and tumors when used with water, and in combination with barley gives relief from pains of the joints.

Wine from the bark of the root is also prepared without boiling. One must place three minas into a metretes[4] of sweet wine. Three cyathi of this preparation should be given to those who are about to be operated upon with knife or cautery, as we have said. For they do not feel the pain, being overcome by sleep. Again, when the fruit is eaten or smelled it causes drowsiness, as does also the juice of the fruit. Taken in excessive quantities it causes loss of speech.

The seed of the fruit when drunk purges the womb; applied as a pessary with native sulphur it stops the red discharge. Liquor is obtained when the root is grooved around, and the flow caught in the hollow. But the [pressed out] juice is stronger than this liquor. Roots do not yield liquor at every point, as experience shows.

Some report another variety[5] called morion growing in shady places and around caves and having leaves like those of white mandragora but

---

[1] About 1/12 of a pint.

[2] An obol was about 1/40 of an ounce in the Attic system; about 1/30 in the Aeginetan system.

[3] A mixture of honey and milk or water.

[4] The mina was about a pound in the Attic system, the metretes about 9 gallons.

[5] See p. 517, n. 1.

smaller, about a span long, white, falling in a circle around the root. The root is tender and white, a little longer than a span, of the thickness of a thumb. They say that when only so much as a drachma[1] of this is drunk or eaten with barley on a cake or in a sauce it causes a deep sleep. For the person remains unconscious for three or four hours from the time he takes it and in the same position. Physicians use this too when they are about to cut or cauterize.

They say that the root when drunk with the so-called strychnus manicus[2] is an antidote.

Celsus, *On Medicine* V. 1–4, 17–20, 23, 25.[3] Translation of W. G. Spencer

Since all medicaments have special powers, and afford relief, often when simple, often when mixed, it does not seem amiss beforehand to state both their names and their virtues and how to compound them, that there may be less delay when we are describing the treatment itself.

1. The following suppress bleeding: Blacking which the Greeks call chalcanthon, copper ore, acacia, lycium with water, frankincense, lign aloe, gums, lead sulphide, leek, polygonum; Cimolian chalk or potter's clay, antimony sulphide; cold water, wine, vinegar; alum from Melos, iron and copper scales. . . .

2. The following agglutinate a wound: myrrh, frankincense, gums, especially gum arabic, fleawort, tragacanth, cardamon, bulbs, linseed, nasturtium; white of egg, glue, isinglass; white vine, snails pounded with their shells, cooked honey, a sponge squeezed out of cold water or out of wine or out of vinegar; unscoured wool squeezed out of the same; if the wound is slight, even cobwebs.

The following subdue inflammation: alum, both split alum called schiston, and alum brine; quince oil, orpiment, verdigris, copper ore, blacking.

3. The following mature abscessions and promote suppuration: nard, myrrh, costmary, balsam, galbanum, propolis, storax, frankincense, both the soot and the bark, bitumen, pitch, sulphur, resin, suet, fat, oil.

4. The following open, as it were, mouths[4] in our bodies, called in Greek *στομοῦν*: cinnamon, balsam, all-heal; rush-root, pennyroyal, white violet flowers, bdellium, galbanum, turpentine and pine-resin, propolis, old

---

[1] About ⅙ of an ounce.

[2] I.e., thorn apple (*Datura stramonium,*

[3] Celsus lists hundreds of simple drugs and compounds and classifies them according to their effect on the body. Both organic (vegetable and animal) and inorganic substances are prescribed in the form of emollients, pastils, plasters, pessaries, and pills. Also in describing various diseases Celsus indicates the type of drug, if any, that is effective in each case. [Edd.]

[4] These drugs were intended to open the pores (*stomata* of Erasistratus) at the ends of veins, and so to relieve congestion. . . .

olive-oil; pepper, pyrethrum, ground pine thistle, black bryony berries, sulphur, alum, rue seed.[1]

17. The powers of medicaments when unmixed having been set out, we have to say how they may be mixed together, and what are the compositions so made. Now they are mixed in various ways and there is no limit to this, since some simples may be omitted, others added, and when the same ingredients are used the proportion of their weights may be changed. Hence though there are not so very many substances having medicinal powers, there are innumerable kinds of mixtures; and, even if all of them could be included, yet this would be needless. For the same effects are produced by but a few compositions, and to vary these is easy to anyone who knows their powers. Therefore I shall content myself with those I have heard of as the best known. . . .

18. . . . I will speak first of emollients, almost all of which were invented, not for the purpose of cooling but for heating. There is, however, one kind which can cool, being suitable for hot podagras. It is a cupful of oak-galls, unripe or otherwise, coriander seed, hemlock, dried poppy-tears, and gum, of each 63 c.cm.; of washed cerate called by the Greeks πεπλύμενον, 168 grms.

Almost all the rest are heating. But some disperse the diseased matter, some extract it and are called epispastic; most are designed rather for particular parts of the body. . . .

19. Among the plasters none renders greater service than those for immediate application to bleeding wounds, which the Greeks call enhaema. For these repress inflammation, unless a severe cause excites it, and even then they lessen its attack; further, they agglutinate wounds which allow of it, and induce a scar in them. But as the plasters consist of medicaments which are not greasy, they are named alipe.[2]

The best of these is the plaster called barbarum. It contains scraped verdigris 48 grms., litharge 80 grms., alum, dried pitch, dried pine-resin, 4 grms. each, to which is added oil and vinegar 250 c.cm. each. . . .

20. Pastils have also divers faculties. For some are suitable for agglutinating and making the scar upon recent wounds: such as that containing copper ore, antimony sulphide, soda-scum, flowers of copper, oak-galls, split alum moderately boiled, each 4 grms., calcined copper and pomegranate-heads, each 12 grms. It should be dissolved with vinegar, and so smeared on for agglutinating a wound. But if the part wounded involves sinews or muscles, it is better to mix the pastil with a cerate, eight parts of the former to nine of the latter.

[1] The author goes on to list, *inter alia*, drugs, that are caustic, scab-inducing, scab-loosening, soothing, emollient, and cleansing. [Edd.]

[2] I.e., without grease.

Another for the same purpose is composed of bitumen and split alum, each 4 grms., calcined copper 16 grms., litharge 44 grms., oil half a litre.

23. . . . [Antidotes] are chiefly necessary against poisons introduced into our bodies through bites or food or drink. . . .

The most famous antidote is that of Mithridates,[1] which that king is said to have taken daily and by it to have rendered his body safe against danger from poison. It contains costmary 1.66 grms., sweet flag 20 grms., hypericum, gum, sagapenum, acacia juice, Illyrian iris, cardamon, 8 grms. each, anise 12 grms., Gallic nard, gentian root and dried rose-leaves, 16 grms. each, poppy-tears and parsley, 17 grms. each, casia, saxifrage, darnel, long pepper, 20.66 grms. each, storax 21 grms., castoreum, frankincense, hypocistis juice, myrrh and opopanax, 24 grms. each, malabathrum leaves, 24 grms., flower of round rush, turpentine-resin, galbanum, Cretan carrot seeds, 24.66 grms. each, nard and opobalsam, 25 grms. each, shepherd's purse 25 grms., rhubarb root 28 grms., saffron, ginger, cinnamon, 29 grms. each. These are pounded and taken up in honey. Against poisoning, a piece the size of an almond is given in wine. In other affections an amount corresponding in size to an Egyptian bean is sufficient. . . .

25. Pills are also numerous, and are made for various purposes. Those which relieve pain through sleep are called anodynes; unless there is overwhelming necessity, it is improper to use them; for they are composed of medicaments which are very active and alien to the stomach. There is one, however, which actually promotes digestion; it is composed of poppy-tears and galbanum, 4 grms. each, myrrh, castory, and pepper, 8 grms. each. Of this it is enough to swallow an amount the size of a vetch.

Another, worse for the stomach, but more soporific, consists of mandragora 1 grm., celery-seed and hyoscyamus seed, 16 grms. each, which are rubbed up after soaking in wine. One of the same size mentioned above is quite enough to take.

Surgery

Numerous writings on the subject of surgery as well as many surgical instruments have been preserved from ancient Greece and Rome. The special works devoted to surgery in the Hippocratic Collection are the treatises *On Wounds in the Head*, *On Joints*, *In the Surgery*, and *Instruments of Reduction*.

The systematic practice of dissection of human corpses in the Alexandrian Age and until about the end of the first century A.D. and the refinement of mechanical instruments in that period aided the development of surgery. Though the works of the great surgeons, e.g., Hero (the Alexandrian surgeon), Meges of Sidon, Philoxenus, Leonidas, Heliodorus, Archigenes, and Antyllus, have not come down to us, their methods are preserved in Celsus (especially Books 7 and 8), Galen, Oribasius, Aëtius of Amida, and Paul of Aegina. Note also the Galenic treatise *On Bandages* and that of Soranus on the same subject. There is a manuscript (ninth or tenth century) of the commentary by Apollonius of Citium (first century

[1] See p. 509. [Edd.]

B.C.) on the Hippocratic treatise *On Joints*, with drawings illustrating various methods of reducing dislocations. The use of mandragora as an anesthetic has already been noted.

It is to be noted that Greek *cheirurgia* (from *cheir*, "hand," and the root of *ergon*, "work") denotes all healing by the operation of the hand, as opposed to medicaments.

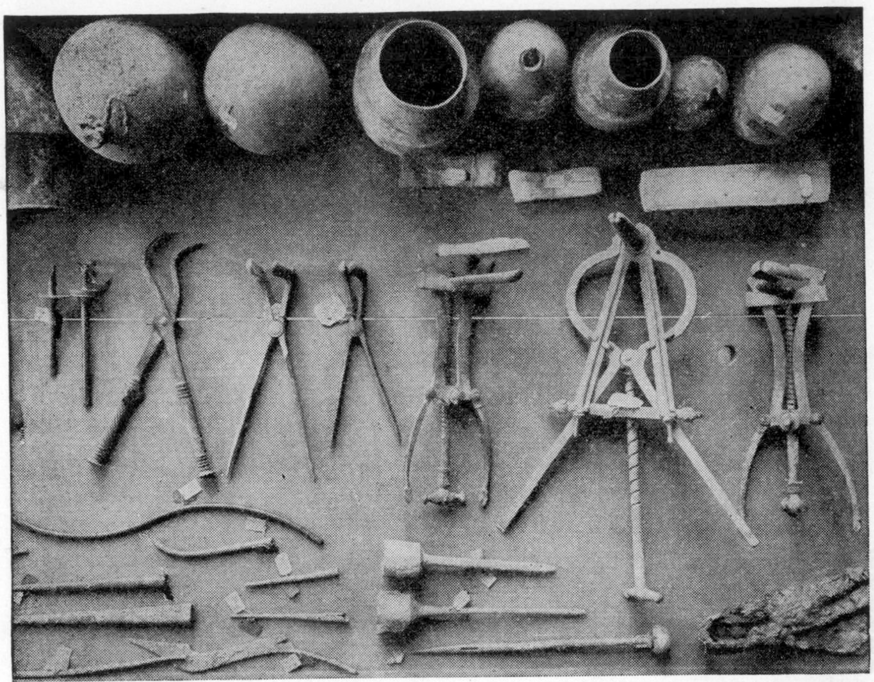

Ancient surgical instruments from Pompeii, now in the National Museum at Naples. Among them are bleeding cups, forceps, cannulae, dilators, catheters, and rectal and vaginal specula.

WOUNDS

Hippocratic Collection, *On Ulcers* 1–2.[1] Translation of Francis Adams (London, 1849)

We must avoid wetting all sorts of ulcers except with wine, unless the ulcer be situated in a joint. For the dry is nearer to the sound, and the wet to the unsound, since an ulcer is wet, but a sound part is dry. And it is better to leave the part without a bandage unless a cataplasm be applied. Neither do certain ulcers admit of cataplasms, and this is the case with the recent rather than the old, and with those situated in joints. A spare diet and water agree with all ulcers, and with the more recent rather than the older; and with an ulcer which either is inflamed or is about to be so; and where there is danger of gangrene; and with the ulcers and inflammations

[1] The Hippocratic treatise περὶ ἑλκῶν gives rules for the treatment of wounds and sores of various types, a subject which is fundamental for surgery. It is to be noted that the translator often uses the term "ulcer" to include both wounds and sores, as he does in the title of the work. [Edd.]

in joints; and where there is danger of convulsion; and in wounds of the belly; but most especially in fractures of the head and thigh, or any other member in which a fracture may have occurred. In the case of an ulcer, it is not expedient to stand; more especially if the ulcer be situated in the leg; but neither, also, is it proper to sit or walk. But quiet and rest are particularly expedient. Recent ulcers, both the ulcers themselves and the surrounding parts, will be least exposed to inflammation, if one shall bring them to a suppuration as expeditiously as possible, and if the matter is not prevented from escaping by the mouth of the sore, or, if one should restrain the suppuration, so that only a small and necessary quantity of pus may be formed, and the sore may be kept dry by a medicine which does not create irritation. For the part becomes inflamed when rigor and throbbing supervene; for ulcers then get inflamed when suppuration is about to form. A sore suppurates when the blood is changed and becomes heated; so that, becoming putrid, it constitutes the pus of such ulcers. When you seem to require a cataplasm, it is not the ulcer itself to which you must apply the cataplasm, but to the surrounding parts, so that the pus may escape and the hardened parts may become soft. Ulcers formed either from the parts having been cut through by a sharp instrument, or excised, admit of medicaments for bloody wounds, and which will prevent suppuration by being desiccant to a certain degree.[1] But when the flesh has been contused and roughly cut by the weapon, it is to be so treated that it may suppurate as quickly as possible; for thus the inflammation is less, and it is necessary that the pieces of flesh which are bruised and cut should melt away by becoming putrid, being converted into pus, and that new flesh should then grow up.[2] In every recent ulcer, except in the belly, it is expedient to cause blood to flow from it abundantly, and as may seem seasonable; for thus will the wound and the adjacent parts be less attacked with inflammation. . . .

TREPHINING

Hippocratic Collection, *On Wounds in the Head* 21.[3] Translation of E. T. Withington

As to trephining, when it is necessary to trephine a patient, keep the following in mind. If you operate after taking on the treatment from the beginning, you should not, in trephining, remove the bone at once down to the membrane, for it is not good for the membrane to be denuded of bone and exposed to morbid influences for a long time, or it may end by

[1] I.e., permit of healing "by the first intention." [Edd.]

[2] I.e., heal "by the second intention." [Edd.]

[3] After describing the skull and its sutures the author discusses the various types of skull fracture and the systemic complications that may ensue. Trephining is recommended in certain cases as an early measure. The portion of the skull is removed with a circular saw rotated by a cord. This operation is also common among many primitive peoples, in connection with ritual medicine, even when there has been no head wound. [Edd.]

becoming macerated. There is also another danger that, if you immediately remove the bone by trephining down to the membrane, you may, in operating, wound the membrane with the trephine. You should rather stop the operation when there is very little left to be sawn through, and the bone is movable; and allow it to separate of its own accord. For no harm will supervene in the trephined bone, or in the part left unsawn, since what remains is thin enough. For the rest the treatment should be such as may seem beneficial to the lesion.

While trephining, you should frequently take out the saw and plunge it into cold water to avoid heating the bone, for the saw gets heated by rotation, and by heating and drying the bone cauterises it and makes more of the bone around the trephined part come away than was going to do. If you want to trephine down to the membrane at once, and then remove the bone, the trephine should in like manner be often taken out and plunged in cold water.

If you do not take on the cure from the beginning, but receive it from another, coming late to the treatment, trephine the bone at once down to the membrane with a sharp-toothed trephine, taking it out frequently for inspection, and also examining with a probe around the track of the saw. For the bone is much more quickly sawn through if you perate when it is already suppurating and full of pus; and the skull is often found to have no depth, especially if the wound happens to be in the part of the head where the bone inclines to be thin rather than thick. You must be careful not to be heedless in placing the trephine, but always to fix it where the bone seems thickest. Examine often, and try by to-and-fro movements to lift up the bone; and, after removing it, treat the rest as may seem beneficial to the lesion [having regard to what has happened].[1]

If you take on the case from the beginning, and want to trephine the bone at once completely and remove it from the membrane, you should likewise often examine the circular track of the saw with the probe, always fixing the trephine in the thickest part of the bone, and aim at getting it away by to-and-fro movements. If you use a perforating trepan, do not go down to the membrane, if you perforate on taking the case from the beginning; but leave a thin layer of bone, as was directed in trephining.

CONGENITAL CLUBFOOT

Hippocratic Collection, *On Joints* 62.[2] Translation of E. T. Withington

There are certain congenital displacements which, when they are slight, can be reduced to their natural position, especially those at the foot-joints.

---

[1] Some editors have considered this phrase an interpolation. [Edd.]

[2] The treatise *On Joints* deals with various types of dislocation and fracture, e.g., of the shoulder joint, collarbone, jaw, nose, spine, ribs, and hip. Methods of setting, reduction,

Cases of congenital clubfoot are, for the most part, curable, if the deviation is not very great or the children advanced in growth. It is therefore best to treat such cases as soon as possible, before there is any very great deficiency in the bones of the foot, and before the like occurs in the tissues of the leg. Now the mode of clubfoot is not one, but manifold; and most cases are not the result of complete dislocation, but are deformities due to the constant retention of the foot in a contracted position. The things to bear in mind in treatment are the following: push back and adjust the bone of the leg at the ankle from without inwards, making counter-pressure outwards on the bone of the heel where it comes in line with the leg, so as to bring together the bones which project at the middle and side of the foot; at the same time, bend inwards and rotate the toes all together, including the big toe. Dress with cerate well stiffened with resin, pads and soft bandages, sufficiently numerous, but without too much compression. Bring round the turns of the bandaging in a way corresponding with the manual adjustment of the foot, so that the latter has an inclination somewhat towards splay-footedness.[1] A sole should be made of not too stiff leather or of lead, and should be bound on as well, not immediately on to the skin, but just when you are going to apply the last dressings. When the dressing is completed, the end of one of the bandages used should be sewn on to the under side of the foot dressings, in a line with the little toe; then, making such tension upwards as may seem suitable, pass it round the calf-muscle at the top, so as to keep it firm and on the stretch.[2] In a word, as in wax modelling, one should bring the parts into their true natural position, both those that are twisted and those that are abnormally contracted, adjusting them in this way both with the hands and by bandaging in like manner; but draw them into position by gentle means, and not violently. Sew on the bandages so as to give the appropriate support; for different forms of lameness require different kinds of support. A leaden shoe shaped as the Chian boots used to be, might be made, and fastened on outside the dressing; but this is quite unnecessary if the manual adjustment, the dressing with bandages, and the contrivance for drawing up are properly done. This then is the treatment, and there is no need for incision, cautery, or complicated methods; for such cases yield to treatment more rapidly than one would think. Still, time is required for complete success, till the part has acquired growth in its proper position. When the time has come for footwear, the most suitable are the so-called "mud-shoes," for this kind of boot yields least to the foot; indeed, the foot rather yields to it. The Cretan form[3] of footwear is also suitable.

---

and amputation are described. The chapter on clubfoot, here quoted, has evoked the admiration of modern surgeons. [Edd.]

[1] I.e., *valgus* (outward distortion).

[2] I.e., so as to hold up the outer side of the foot.

[3] "Reaching to the middle of the leg." Galen.

PRICKING OF CATARACT BY A NEEDLE

Celsus, *On Medicine* VII. 7. 13–14.[1] Translation of W. G. Spencer

13. . . . When it [cataract] is more chronic it requires treatment by surgery, and this is one of the most delicate operations. . . .[2]

14. . . . We must wait until it is no longer fluid, but appears to have coalesced to some sort of hardness. Before treatment the patient should eat in moderation, and for three days beforehand drink water, for the day before abstain from everything. Then he is to be seated opposite the surgeon in a light room, facing the light, while the surgeon sits on a slightly higher seat; the assistant from behind holds the head so that the patient does not move: for vision can be destroyed permanently by a slight movement. In order also that the eye to be treated may be held more still, wool is put over the opposite eye and bandaged on: further the left eye should be operated upon with the right hand, and the right eye with the left hand. Thereupon a needle is to be taken pointed enough to penetrate, yet not too fine; and this is to be inserted straight through the two outer tunics[3] at a spot intermediate between the pupil of the eye and the angle adjacent to the temple, away from the middle of the cataract, in such a way that no vein is wounded. The needle should not be, however, entered timidly, for it passes into the empty space; and when this is reached even a man of moderate experience cannot be mistaken, for there is then no resistance to pressure. When the spot is reached, the needle is to be sloped against the suffusion itself and should gently rotate there and little by little guide it below the region of the pupil; when the cataract has passed below the pupil it is pressed upon more firmly in order that it may settle below. If it sticks there the cure is accomplished; if it returns to some extent, it is to be cut up with the same needle and separated into several pieces, which can be the more easily stowed away singly, and form smaller obstacles to vision. After this the needle is drawn straight out; and soft wool soaked in white of egg is to be put on, and above this something to check inflammation; and then bandages. Subsequently the patient must have rest, abstinence, and

---

[1] Ophthalmology was one of the most highly developed branches of Greek medicine, perhaps because of the prevalence of eye affections in the Mediterranean world. Numerous diseases and operations are recorded. The operation here described seeks to clear the vision not by extracting the cataract but by bringing it below the pupil, or, if that cannot be done, by breaking the cataract up into small pieces. The translator points out that though the operation of extraction has been widely used in recent times, the old operation is still used for certain soft forms of cataract. It may be added that the extraction of cataract was discussed at least as early as Antyllus (second century A.D.). [Edd.]

[2] In the part here omitted Celsus describes the eye, noting the cornea, pupil, chorioid membrane, arachnoid membrane (retina), vitreous humor, and crystalline humor. He then speaks of the development of a cataract from a blow or disease. [Edd.]

[3] The cornea and the chorioid membrane. [Edd.]

inunction with soothing medicaments; the day following will be soon enough for food, which at first should be liquid to avoid the use of the jaws; then, when the inflammation is over, such as has been prescribed for wounds, and in addition to these directions it is necessary that water should for some time be the only drink.

ANEURYSMS

Antyllus in Oribasius, *Medical Collection* XLV. 24[1]

There are two kinds of aneurysm. The first results when the artery undergoes a local dilatation, and from this it gets the name of aneurysm.[2] In the second the artery is ruptured and discharges blood into the surrounding flesh. Aneurysms which result from dilatation of an artery are rather elongated; those which result from rupture are more rounded. Several layers of tissue cover the aneurysm caused by dilatation of the artery. When the aneurysm caused by rupture is pressed by the fingers, a kind of creaking noise is heard, but no noise comes from pressing the other type of aneurysm. Now to refuse to treat any aneurysm, as the surgeons of old advised, is foolish. But it is also dangerous to operate on every aneurysm. And so we should avoid treating aneurysms in the armpit, the groin, and the neck because of the size of the vessels and the impossibility and danger of isolating and binding them. We should also refuse to operate on a very large aneurysm, even when it is in some other part of the body. Aneurysms in the extremities, the limbs, and the head are operated upon in the following manner. In the case of an aneurysm caused by dilatation, make a straight incision in the skin along the length of the vessel; then, separating the lips of the wound with hooks, divide carefully all the membranes between the skin and the artery. Pushing aside with blunt hooks the vein which lies alongside the artery, expose completely the dilated part of the artery. Insert the round head of a probe from below and, raising the aneurysm, move a needle with a double thread along the probe in such a way that it is brought under the artery; clip the thread near the needle so that there are two threads and four ends. Then, taking the two ends of one thread, bring them gently to one end of the aneurysm

---

[1] Antyllus, a leading physician among the so-called "Pneumatists," lived in the second century A.D. Most of our knowledge about him is derived from Oribasius, who tells us of his investigations on climatology, diet and regimen, general therapeutics, and surgery. In this latter branch he seems to have been a follower of Heliodorus, who lived at the time of Trajan. There are references in Oribasius and others to Antyllus's operations for abscesses, fistulas, various types of bone removal, and the extraction of cataract, in addition to the passage given here on the operation for aneurysm.

Aneurysms are soft, pulsating tumors containing blood and formed by arterial dilatation or rupture. With the passage here given may be compared the description of aneurysms and of an operation for aneurysm in the hollow of the elbow in Aëtius XV. 10.

[2] I.e., from *aneurynō*, "to dilate."

and tie them carefully. Similarly, bring the other thread to the opposite extremity, and tie the artery carefully at that point. In this way the entire aneurysm is enclosed within two ligatures. Then make a small incision in the middle of the aneurysm; by this means all the matter contained therein will be evacuated and there will be no danger of hemorrhage.

Those who tie the artery on both sides as we have indicated, but proceed to cut out the dilated part in the middle, perform a dangerous operation. For the force and intensity of the pneuma frequently push off the ligatures.

If the aneurysm is the result of rupture of an artery, it is necessary to separate with the fingers as much of the tumor, along with the skin, as is possible. Then insert a needle with a double thread of linen or sinew under the isolated part; and after passing the needle through, cut the thread at the end so that there are two threads. Then take both ends of one thread, bring them over to the right, and tie them firmly so that the thread will not slip. Similarly bring the other ends to the opposite side, i.e., to the left. If the surgeon fears that the thread will slip, he must pass another needle, also with a double thread, through the same place, crossing the place where the first needle passed so as to form the letter *X*. The threads must be cut and fastened just as those of the first needle were, so that four threads make up the ligature. Then open the swelling at the top; after the matter in it is evacuated, remove the superfluous skin, leaving in place the part which is bound by the threads. In this way the operation is conducted without hemorrhage.

DENTAL SURGERY

Celsus, *On Medicine* VII. 12.1.[1] Translation of W. G. Spencer

In the mouth too some conditions are treated by surgery. In the first place, teeth sometimes become loose, either from weakness of the roots, or from disease drying up the gums. In either case the cautery should be applied so as to touch the gums lightly without pressure. The gums so cauterized are smeared with honey, and swilled with honey wine. When the ulcerations have begun to clean, dry medicaments, acting as repressants, are dusted on. But if a tooth gives pain and it is decided to extract it because medicaments afford no relief, the tooth should be scraped round in order that the gum may become separated from it; then the tooth is to be shaken. And this is to be done until it is quite moveable; for it is very dangerous to extract a tooth that is tight, and sometimes the jaw is dis-

[1] There are numerous references in Greek and Roman medical literature to the subject of dentistry. Not only are surgical aspects discussed, as here, but also the hygiene of the teeth and the preparation and use of dentifrices. Gold and lead were used to fill teeth and gold wire to bind loose teeth together. False teeth furnished satirists material for jest then as now. [Edd.]

located. With the upper teeth there is even greater danger, for the temples or eyes may be concussed. Then the tooth is to be extracted, by hand, if possible, failing that with forceps. But if the tooth is decayed, the cavity should be neatly filled first, whether with lint or with lead, so that the tooth does not break in pieces under the forceps. The forceps is to be pulled straight upwards, lest if the roots are bent, the thin bone to which the tooth is attached should break at some part. And this procedure is not altogether free from danger, especially in the case of the short teeth, which generally have shorter roots, for often when the forceps cannot grip the tooth, or does not do so properly, it grips and breaks the bone under the gum. But as soon as there is a large flow of blood it is clear that something has been broken off the bone. It is necessary therefore to search with a probe for the scale of bone which has been separated, and to extract it with a small forceps. If this does not succeed the gum must be cut into until the loose scale is found. And if this has not been done at once, the jaw outside the tooth hardens, so that the patient cannot open his mouth. But a hot poultice made of flour and a fig is then to be put on until pus is formed there: then the gum should be cut into. A free flow of pus also indicates a fragment of bone; so then too it is proper to extract the fragment; sometimes also when the bone is injured a fistula is formed which has to be scraped out. But a rough tooth is to be scraped in the part which has become black, and smeared with crushed rose-petals to which a fourth part of oak galls and the same amount of myrrh has been added; and at frequent intervals undiluted wine is to be held in the mouth; and in this case the head is to be wrapped up, and the patient should have much walking exercise, massage of his head and food which is not too bitter. But if teeth become loosened by a blow, or any other accident, they are to be tied by gold wire to firmly fixed teeth, and repressants must be held in the mouth, such as wine in which some pomegranate rind has been cooked, or into which burning oak galls have been thrown. In children too if a second tooth is growing up before the first one has fallen out, the tooth which ought to come out must be freed all round and extracted; the tooth which has grown up in the place of the former one is to be pressed upwards with a finger every day until it has reached its proper height. And whenever, after extraction, a root has been left behind, this too must be at once removed by the forceps made for the purpose which the Greeks called *rhizagra*.[1]

[1] I.e., "root catcher." The translator notes that such an instrument has been found in a collection of Roman surgical instruments. [Edd.]

PHYSIOLOGICAL PSYCHOLOGY

The sharp separation of psychology as an empirical or natural science from the part of metaphysics known as "rational psychology," "epistemology," or "theory of mind" is a relatively recent event. Yet it would be a mistake to ignore what Hellenic thinkers did to lay the foundation for verifiable or scientific knowledge about human nature. They assumed from the beginning that mental activity depended on bodily conditions, and to this day we continue to use their distinctions between the different psychologic functions such as perception, imagination, reason, emotion, and volition. In regard to scientific method in psychology the Aristotelian distinction between the physiologic and the dialectic or analytic point of view is still valid.

Even in the restricted field of empirical psychology abundant material is extant. For the period from the earliest Greek philosophers to the Peripatetics our chief sources are Plato's dialogues, Aristotle's *De Anima* and *De Sensu*, as well as considerable material in other parts of the *Parva Naturalia* and in the zoological treatises, and the fragments of the *De Sensibus* and *De Odoribus* of Theophrastus. The later doxographical collections are also important sources for the psychological theories of the philosophers.

Extensive discussion of human nature and the psychological aspects of health and disease is to be found in the medical writers of all periods. It may be noted, however, that the early Alexandrian discoveries in the anatomy and physiology of muscles and nerves do not seem to have produced any fruitful development in experimental psychology. In connection with vision and hearing reference may also be made to the sections on Optics and Acoustics.

The material on psychology in Stoic, Epicurean, Skeptic, Neoplatonic, and Patristic authors is generally based on dialectical and ethical rather than on physiologic considerations.

While we have grouped the selections on the mental traits of animals separately from those on human psychology, the method of treatment is identical in both. It is true, however, that post-Aristotelian writers, such as Pliny in the first century A.D. and Aelian at the end of the second, are far less concerned with general theory and are far less critical in sifting the stories they read and hear. They deal chiefly with the anecdotal and the miraculous. This tendency begins perhaps with early Alexandrian writers on animals, among whom are the famous librarians, Callimachus of Cyrene and Aristophanes of Byzantium, followed, in the first century B.C., by Alexander of Myndos. The importance of Pliny, Aelian, and the anonymous Physiologus of the second century A.D. for medieval animal lore and the beast fable is unquestionably great, but from the scientific viewpoint the extant Greek works of highest importance in animal psychology are those of Aristotle. Note also the selections given above under "Zoology: Habits of Animals."

INTRODUCTION TO THE STUDY OF PSYCHOLOGY

Aristotle, *On the Soul* I. 1; II. 1–2. Translation of R. D. Hicks

I. 1. . . . The first thing necessary is no doubt to determine under which of the summa genera soul[1] comes and what it is; I mean, whether it is a

[1] The word "soul" has many associations and connotations that are quite foreign to the sense in which Aristotle uses *psyche*, and much may be said for those who have translated the

particular thing, i.e., substance, or is quality or is quantity, or falls under any other of the categories already determined. We must further ask whether it is amongst things potentially existent or is rather a sort of actuality, the distinction being all-important. Again, we must consider whether it is divisible or indivisible; whether, again, all and every soul is homogeneous or not; and, if not, whether the difference between the various souls is a difference of species or a difference of genus: for at present discussions and investigations about soul would appear to be restricted to the human soul. We must take care not to overlook the question whether there is a single definition of soul answering to a single definition of animal;[1] or whether there is a different definition for each separate soul, as for horse and dog, man and god: animal, as the universal, being regarded either as non-existent or, if existent, as logically posterior. . . .

The starting point of every demonstration is a definition of what something is.[2] Hence the definitions which lead to no information about attributes and do not facilitate even conjecture respecting them have clearly been framed for dialectic and are void of content, one and all.

A further difficulty arises as to whether all attributes of the soul are also shared by that which contains the soul or whether any of them are peculiar to the soul itself: a question which it is indispensable, and yet by no means easy, to decide. It would appear that in most cases soul neither acts nor is acted upon apart from the body:[3] as, e.g., in anger, confidence, desire and sensation in general. Thought, if anything, would seem to be peculiar to the soul. Yet, if thought is a sort of imagination, or not independent of imagination, it will follow that even thought cannot be independent of the body. If, then, there be any of the functions or affections of the soul peculiar to it, it will be possible for the soul to be separated from the body: if, on the other hand, there is nothing of the sort peculiar to it, the soul will not be capable of separate existence. As with the straight

---

latter by "vital principle." Certainly what Aristotle says about the vegetative and animal soul shows that he conceives psychology as the study of the functions of a living organism. The *psyche* is not separable from the material body, and in its lower phases it is identical with such unconscious activity as organic growth. [Edd.]

[1] Cf. Aristotle, *On the Soul* 414*b*19–24 (translation of R. D. Hicks); "From this it is clear that there is one definition of soul exactly as there is one definition of figure: for there is in the one case no figure excepting the triangle, quadrilateral and the rest, nor is there any species of soul apart from those above mentioned." [Edd.]

[2] The function of a definition is to enable us to deduce the properties of the object defined. Although the preliminary study of properties contributes to the framing of the definition, the ultimate aim of the sciences is to approach as closely as possible to the deductive character of mathematics. [Edd.]

[3] Aristotle calls upon experience to confirm the inseparability of soul, in its actions and affections, from body, and goes on to say that the soul differs from the objects of mathematics in being inseparable, *even in thought*, from material body. [Edd.]

line, so with it. The line, *quâ* straight, has many properties; for instance, it touches the brazen sphere at a point; but it by no means follows that it will so touch it if separated. In fact it is inseparable, since it is always conjoined with body of some sort. So, too, the attributes of the soul appear to be all conjoined with body: such attributes, viz., as anger, mildness, fear, pity, courage; also joy, love and hate; all of which are attended by some particular affection of the body. This indeed is shown by the fact that sometimes violent and palpable incentives occur without producing in us exasperation or fear, while at other times we are moved by slight and scarcely perceptible causes, when the blood is up and the bodily condition that of anger. Still more is this evident from the fact that sometimes even without the occurrence of anything terrible men exhibit all the symptoms of terror. If this be so, the attributes are evidently forms or notions realised in matter. Hence they must be defined accordingly: anger, for instance, as a certain movement in a body of a given kind, or some part or faculty of it, produced by such and such a cause and for such and such an end. These facts at once bring the investigation of soul, whether in its entirety or in the particular aspect described, within the province of the natural philosopher. But every such attribute would be differently defined by the physicist and the dialectician or philosopher. Anger, for instance, would be defined by the dialectician as desire for retaliation or the like, by the physicist as a ferment of the blood or heat which is about the heart: the one of them gives the matter, the other the form or notion.

II. 1. . . . It must follow, then, that soul is substance in the sense that it is the form of a natural body having in it the capacity of life. . . .

Further, we must view our statement in the light of the parts of the body. For, if the eye were an animal, eyesight would be its soul, this being the substance as notion or form of the eye. The eye is the matter of eyesight, and in default of eyesight it is no longer an eye, except equivocally, like an eye in stone or in a picture. What has been said of the part must be understood to apply to the whole living body; for, as the sensation of a part of the body is to that part, so is sensation as a whole to the whole sentient body as such. By that which has in it the capacity of life is meant not the body which has lost its soul, but that which possesses it. Now the seed in animals, like the fruit in plants, is that which is potentially such and such a body. As, then, the cutting of the axe or the seeing of the eye is full actuality, so, too, is the waking state; while the soul is actuality in the same sense as eyesight and the capacity of the instrument. The body, on the other hand, is simply that which is potentially existent. But, just as in the one case the eye means the pupil in conjunction with the eyesight, so in the other soul and body together constitute the animal. . . .

II. 2. . . . We take, then, as our starting-point for discussion that it is life which distinguishes the animate from the inanimate. But the term life is used in various senses; and, if life is present in but a single one of these senses, we speak of a thing as living. Thus there is intellect, sensation, motion from place to place and rest, the motion concerned with nutrition and, further, decay and growth. Hence it is that all plants are supposed to have life. For apparently they have within themselves a faculty and principle whereby they grow and decay in opposite directions. For plants do not grow upwards without growing downwards; they grow in both directions equally, in fact in all directions, as many as are constantly nourished and therefore continue to live, so long as they are capable of absorbing nutriment. This form of life can be separated from the others, though in mortal creatures the others cannot be separated from it. In the case of plants the fact is manifest: for they have no other faculty of soul at all.

It is, then, in virtue of this principle that all living things live, whether animals or plants. But it is sensation primarily which constitutes the animal. For, provided they have sensation, even those creatures which are devoid of movement and do not change their place are called animals and are not merely said to be alive. . . .

Just as in the case of plants some of them are found to live when divided and separated from each other (which implies that the soul in each plant, though actually one, is potentially several souls), so, too, when insects or annelida are cut up, we see the same thing happen with other varieties of soul: I mean, each of the segments has sensation and moves from place to place, and, if it has sensation, it has also imagination and appetency.

ANIMAL PSYCHOLOGY

Traits and Character of Animals

GENERAL OBSERVATIONS

Aristotle, *History of Animals* VIII. 1. Translation of D. W. Thompson

In the great majority of animals there are traces of psychical qualities or attitudes, which qualities are more markedly differentiated in the case of human beings. For just as we pointed out resemblances in the physical organs, so in a number of animals we observe gentleness or fierceness, mildness or cross temper, courage or timidity, fear or confidence, high spirit or low cunning, and, with regard to intelligence, something equivalent to sagacity. Some of these qualities in man, as compared with the corresponding qualities in animals, differ only quantitatively: that is to say, a man has more or less of this quality, and an animal has more or less of some other; other qualities in man are represented by analogous and not identical qualities: for instance, just as in man we find knowledge, wisdom and sagacity, so in certain animals there exists some other natural potentiality akin to these. The truth of this statement will be the more clearly apprehended if we have regard to the phenomena of childhood: for in children may be observed the traces and seeds of what will one day be settled psychological habits, though psychologically a child hardly differs for the time being from an animal; so that one is quite justified in saying that, as regards man and animals, certain psychical qualities are identical with one another, whilst others resemble, and others are analogous to, each other.

Aristotle, *History of Animals* I. 1, 488*b*12–26. Translation of D. W. Thompson

Animals also differ from one another in regard to character in the following respects. Some are good-tempered, sluggish, and little prone to ferocity, as the ox; others are quick-tempered, ferocious and unteachable, as the wild boar; some are intelligent and timid, as the stag and the hare; others are mean and treacherous, as the snake; others are noble and courageous and high-bred, as the lion; others are thorough-bred and wild and treacherous, as the wolf: for, by the way, an animal is high-bred if it comes from a noble stock, and an animal is thorough-bred if it does not deflect from its racial characteristics.

Further, some are crafty and mischievous, as the fox; some are spirited and affectionate and fawning, as the dog; others are easy-tempered and easily domesticated, as the elephant; others are cautious and watchful, as the goose; others are jealous and self-conceited, as the peacock. But of all animals man alone is capable of deliberation.

Many animals have memory, and are capable of instruction; but no other creature except man can recall the past at will.

EFFECT OF THE COMPOSITION OF THE BLOOD

Aristotle, *On the Parts of Animals* II. 2 (648*a*2–13); II. 4. Translation of William Ogle

2. The thicker and the hotter blood is, the more conducive is it to strength, while in proportion to its thinness and its coldness is its suitability for sensation and intelligence. A like distinction exists also in the fluid which is analogous to blood. This explains how it is that bees and other similar creatures are of a more intelligent nature than many sanguineous animals; and that, of sanguineous animals, those are the most intelligent whose blood is thin and cold. Noblest of all are those whose blood is hot, and at the same time thin and clear. For such are suited alike for the development of courage and of intelligence. Accordingly, the upper parts are superior in these respects to the lower, the male superior to the female, and the right side to the left.

4. Some at any rate of the animals with watery blood have a keener intellect than those whose blood is of an earthier nature. This is due not to the coldness of their blood, but rather to its thinness and purity; neither of

which qualities belongs to the earthy matter. For the thinner and purer its fluid is, the more easily affected is an animal's sensibility. Thus it is that some bloodless animals, notwithstanding their want of blood, are yet more intelligent than some among the sanguineous kinds. Such, for instance, as already said, is the case with the bee and the tribe of ants, and whatever other animals there may be of a like nature. At the same time too great an excess of water makes animals timorous. For fear chills the body; so that in animals whose heart contains so watery a mixture the way is prepared for the operation of this emotion. For water is congealed by cold. This also explains why bloodless animals are, as general rule, more timorous than such as have blood, so that they remain motionless, when frightened, and discharge their excretions, and in some instances change colour.[1] Such animals, on the other hand, as have thick and abundant fibres in their blood are of a more earthy nature, and of a choleric temperament, and liable to bursts of passion. For anger is productive of heat; and solids, when they have been made hot, give off more heat than fluids. The fibres therefore, being earthy and solid, are turned into so many hot embers in the blood, like the embers in a vapour-bath, and cause ebullition in the fits of passion.

This explains why bulls and boars are so choleric and so passionate. For their blood is exceedingly rich in fibres,[2] and the bull's at any rate coagulates more rapidly than that of any other animal. If these fibres, that is to say, if the earthy constituents of which we are speaking, are taken out of the blood, the fluid that remains behind will no longer coagulate; just as the watery residue of mud will not coagulate after removal of the earth. But if the fibres are left the fluid coagulates, as also does mud, under the influence of cold. For when the heat is expelled by the cold, the fluid, as has been already stated, passes off with it by evaporation, and the residue is dried up and solidified, not by heat but by cold. So long, however, as the blood is in the body, it is kept fluid by animal heat.

The character of the blood affects both the temperament and the sensory faculties of animals in many ways. This is indeed what might reasonably be expected, seeing that the blood is the material of which the whole body is made. For nutriment supplies the material, and the blood is the ultimate nutriment. It makes then a considerable difference whether the blood be hot or cold, thin or thick, turbid or clear.

---

[1] Aristotle tells us elsewhere that certain beetles remain motionless when frightened. It is, of course, the cephalopods, and most notably the sepia, that discharge their ink because of fright. These cephalopods can all change color, but the poulp is particularly mentioned by Aristotle as having this faculty. [Edd.]

[2] The translator refers to some experimental evidence for Aristotle's view of the relation between ferocity and the amount of fibrin in the blood, and also to some evidence against Aristotle's view about the relative rapidity of coagulation. [Edd.]

DIFFERENCES BETWEEN THE SEXES

Aristotle, *History of Animals* IX. 1. Translation of D. W. Thompson

In all genera in which the distinction of male and female is found, Nature makes a similar differentiation in the mental characteristics of the two sexes. This differentiation is the most obvious in the case of human kind and in that of the larger animals and the viviparous quadrupeds. In the case of these latter the female is softer in character, is the sooner tamed, admits more readily of caressing, is more apt in the way of learning; as, for instance, in the Laconian breed of dogs the female is cleverer than the male. . . .

In all cases, excepting those of the bear and leopard, the female is less spirited than the male; in regard to the two exceptional cases, the superiority in courage rests with the female. With all other animals the female is softer in disposition than the male, is more mischievous, less simple, more impulsive, and more attentive to the nurture of the young; the male, on the other hand, is more spirited than the female, more savage, more simple, and less cunning. The traces of these differentiated characters are more or less visible everywhere, but they are especially visible where character is the more developed, and most of all in man.

The fact is, the nature of man is the most rounded off and complete, and consequently in man the qualities or capacities above referred to are found in their perfection. Hence woman is more compassionate than man, more easily moved to tears, at the same time is more jealous, more querulous, more apt to scold and to strike. She is, furthermore, more prone to despondency and less hopeful than the man, more void of shame or self-respect, more false of speech, more deceptive, and of more retentive memory. She is also more wakeful, more shrinking, more difficult to rouse to action, and requires a smaller quantity of nutriment.

As was previously stated, the male is more courageous than the female, and more sympathetic in the way of standing by to help. Even in the case of molluscs, when the cuttle-fish is struck with the trident the male stands by to help the female; but when the male is struck the female runs away.

EFFECT OF SEXUAL EXCITEMENT

Aristotle, *History of Animals* VI. 18. Translation of D. W. Thompson

It is common to all animals to be most excited by the desire of one sex for the other and by the pleasure derived from copulation. The female is most cross-tempered just after parturition, the male during the time of pairing; for instance, stallions at this period bite one another, throw their riders, and chase them. Wild boars, though usually enfeebled at this time as the result of copulation, are now unusually fierce, and fight with one another in an extraordinary way, clothing themselves with defensive armor, or in other words deliberately thickening their hide by rubbing against trees or by coating themselves repeatedly all over with mud and then drying themselves in the sun. They drive one another away from the swine pastures, and fight with such fury that very often both combatants succumb. The case is similar with bulls, rams, and he-goats; for, though at ordinary times they herd together, at breeding time they hold aloof from and quarrel with one another. The male camel also is cross-tempered at pairing time if either a man or a camel comes near him; as for a horse, a camel is ready to fight him at any time. It is the same with wild animals. The bear, the wolf, and the lion are all at this time ferocious towards such as come in their way, but the males of these animals are less given to fight with one another from the fact that they are at no time gregarious. The she-bear is fierce after cubbing, and the bitch after pupping.

Male elephants get savage about pairing time, and for this reason it is stated that men who have charge of elephants in India never allow the males to have intercourse with the females; on the ground that the males go wild at this time and turn topsy-turvy the dwellings of their keepers lightly constructed as they are, and commit all kinds of havoc. They also state that abundancy of food has a tendency to tame the males. They further introduce other elephants amongst the wild ones, and punish and break them in by setting on the new-comers to chastise the others.

Animals that pair frequently and not at a single specific season, as for instance animals domesticated by man, such as swine and dogs, are found to indulge in such freaks to a lesser degree owing to the frequency of their sexual intercourse.

The Senses of Animals

Aristotle, *History of Animals* IV. 8. Translation of D. W. Thompson

We now proceed to treat of the senses; for there are diversities in animals with regard to the senses, seeing that some animals have the use of all the senses, and others the use of a limited number of them. The total number of the senses (for we have no experience of any special sense not here included), is five: sight, hearing, smell, taste, and touch.

Man, then, and all vivipara that have feet, and, further, all red-blooded ovipara, appear to have the use of all five senses, except where some isolated species has been subjected to mutilation, as in the case of the mole. For this animal is deprived of sight; it has no eyes visible, but if the skin—a thick one, by the way—be stripped off the head, about the place in the exterior where eyes usually are, the eyes are found inside in a stunted condition, furnished with all the parts found in ordinary eyes; that is to say, we find there the black rim, and the fatty part surrounding it; but all these parts are smaller than the same parts in ordinary visible eyes. There is no external sign of the existence of these organs in the mole, owing to the thickness of the skin drawn over them, so that it would seem that the natural course of development were congenitally arrested; [for extending from the brain at its junction with the marrow are two strong sinewy ducts running past the sockets of the eyes, and terminating at the upper eye-teeth.][1] All the other animals of the kinds above mentioned have a perception of colour and of sound, and the senses of smell and taste; the fifth sense, that, namely, of touch, is common to all animals whatsoever.

In some animals the organs of sense are plainly discernible; and this is especially the case with the eyes. For animals have a special locality for the eyes, and also a special locality for hearing: that is to say, some animals have ears, while others have the passage for sound discernible. It is the same with the sense of smell; that is to say, some animals have nostrils, and others have only the passages for smell, such as birds. It is the same also with the organ of taste, the tongue. Of aquatic red-blooded animals, fishes possess the organ of taste, namely, the tongue, but it is in an imperfect and amorphous form, in other words it is osseous and undetached. In some fish the palate is fleshy, as in the fresh-water carp, so that by an inattentive observer it might be mistaken for a tongue.

There is no doubt but that fishes have the sense of taste, for a great number of them delight in special flavours; and fishes freely take the hook if it be baited with a piece of flesh from a tunny or from any fat fish, obviously enjoying the taste and the eating of food of this kind. Fishes have no visible organs for hearing or for smell; for what might appear to indicate an organ for smell in the region of the nostril has no communication with the brain. These indications, in fact, in some cases lead nowhere, like blind alleys, and in other cases lead only to the gills; but for all this fishes undoubtedly hear and smell. For they are observed to run away from any loud noise, such as would be made by the rowing of a galley, so as to become easy of capture in their holes; for, by the way, though a sound be very slight in the open air, it has a loud and alarming resonance to creatures that hear under water. And this is shown in the capture of the dolphin; for when the hunters have enclosed a shoal of these fishes with a ring of their canoes, they set up from inside the canoes a loud splashing in the water, and by so doing induce the creatures to run in a shoal high and dry up on the beach, and so capture them while stupefied with the noise. And yet, for all this, the dolphin has no organ of hearing discernible. Furthermore, when engaged in their craft, fishermen are particularly careful to make no noise with oar or net; and after they have spied a shoal, they let down their nets at a spot so far off that they count upon no noise being likely to reach the shoal, occasioned either by oar or by the surging of their

[1] In this selection material within brackets is, in the opinion of the editors Aubert and Wimmer, a later interpolation. Here the reference can hardly be to the mole. [Edd.]

boats through the water; and the crews are strictly enjoined to preserve silence until the shoal has been surrounded. And, at times, when they want the fish to crowd together, they adopt the stratagem of the dolphin-hunter; in other words they clatter stones together, that the fish may, in their fright, gather close into one spot, and so they envelop them with their nets. . . .

The case is similar in regard to the sense of smell. Thus, as a rule, fishes will not touch a bait that is not fresh, neither are they all caught by one and the same bait, but they are severally caught by baits suited to their several likings, and these baits they distinguish by their sense of smell; and, by the way, some fishes are attracted by malodorous baits, as the saupe, for instance, is attracted by excrement. Again, a number of fishes live in caves; and accordingly fishermen when they want to entice them out, smear the mouth of a cave with strong-smelling pickles, and the fish are soon attracted to the smell. And the eel is caught in a similar way; for the fisherman lays down an earthen pot that has held pickles, after inserting a "weel" in the neck thereof. As a general rule, fishes are especially attracted by savoury smells. For this reason, fishermen roast the fleshy parts of the cuttle-fish and use it as bait on account of its smell, for fish are peculiarly attracted by it; they also bake the octopus and bait their fish-baskets or weels with it, entirely, as they say, on account of its smell. Furthermore, gregarious fishes, if fish-washings or bilge-water be thrown overboard, are observed to scud off to a distance from apparent dislike of the smell. And it is asserted that they can at once detect by smell the presence of their own blood; and this faculty is manifested by their hurrying off to a great distance whenever fish-blood is spilt in the sea. And, as a general rule, if you bait your weel with a stinking bait, the fish refuse to enter the weel or even to draw near; but if you bait the weel with a fresh and savoury bait, they come at once from long distances and swim into it. [And all this is particularly manifest in the dolphin; for, as was stated, it has no visible organ of hearing, and yet it is captured when stupefied with noise; and so, while it has no visible organ for smell, it has the sense of smell remarkably keen.] It is manifest, then, that the animals above mentioned are in possession of all the five senses.

All other animals may, with very few exceptions, be comprehended within four genera: to wit, molluscs, crustaceans, testaceans, and insects. Of these four genera, the mollusc, the crustacean, and the insect have all the senses: at all events, they have sight, smell, and taste. As for insects, both winged and wingless, they can detect the presence of scented objects afar off, as for instance bees and cnipes detect the presence of honey at a distance; and they do so recognizing it by smell. Many insects are killed by the smell of brimstone; ants, if the apertures to their dwellings be smeared with powdered origanum and brimstone, quit their nests; and most insects may be banished with burnt hart's horn, or better still by the burning of the gum styrax. The cuttle-fish, the octopus, and the crawfish may be caught by bait. The octopus, in fact, clings so tightly to the rocks that it cannot be pulled off, but remains attached even when the knife is employed to sever it; and yet, if you apply fleabane to the creature, it drops off at the very smell of it.[1] The facts are similar in regard to taste. For the food that insects go in quest of is of diverse kinds, and they do not all delight in the same flavours: for instance, the bee never settles on a withered or wilted flower, but on fresh and sweet ones; and the conops or gnat settles only on acrid substances and not on sweet. The sense of touch, by the way, as has been remarked, is common to all animals. Testaceans have the senses of smell and taste. With regard to their possession of the sense of smell, that is proved by the use of baits, e.g., in the case of the purple-fish; for this creature is enticed by baits of rancid meat, which it perceives and is attracted to from a great distance. The proof that it possesses a sense of taste hangs by the proof of its sense of smell; for whenever an animal is attracted to a thing by perceiving its smell, it is sure to like the taste of it. Further, all animals furnished with a mouth derive pleasure or pain from the touch of sapid juices.

With regard to sight and hearing, we cannot make statements with thorough confidence or on irrefutable evidence. However, the solen or razor-fish, if you make a noise, appears to burrow in the sand, and to hide himself deeper when he hears the approach of the iron rod[2] (for the animal, be it observed, juts a little out of its hole, while the greater part of the body remains within), and scallops if you present your finger near their open valves, close them tight again as though they could see what you were doing. Furthermore, when fishermen are laying bait for neritae, they always get to leeward of them, and never speak a word while so engaged, under the firm impression that the animal can smell and hear; and they assure us that, if any one speaks aloud, the creatures make efforts to escape. With regard to testaceans, of the walking or creeping species the urchin appears to have the least developed sense of smell; and, of the stationary species, the ascidian and the barnacle.

So much for the organs of sense in the general run of animals.

Aristotle, *History of Animals* I. 15 (494*b*11–18). Translation of D. W. Thompson

As for the senses and for the organs of sensation, the eyes, the nostrils, and the tongue, all alike are situated frontwards; the sense of hearing, and

[1] The translator notes that this procedure is still common in Greece, either fleabane or tobacco being used. [Edd.]

[2] The translator notes that an iron rod with a conical knob at the head is still used in the Adriatic for the capture of razor fish. The rod is let down into the burrow between the creature's valves. These close upon the iron and the fish is thus drawn up. [Edd.]

the organ of hearing, the ear, is situated sideways, on the same horizontal plane with the eyes. The eyes in man are, in proportion to his size, nearer to one another than in any other animal.

Of the senses man has the sense of touch more refined than any animal, and so also, but in less degree, the sense of taste; in the development of the other senses he is surpassed by a great number of animals.

Aristotle, *History of Animals* I. 3 (489*a*17–19); I. 4 (489*a*23–26). Translation of D. W. Thompson

One sense, and one alone, is common to all animals—the sense of touch. Consequently, there is no special name for the organ in which it has its seat; for in some groups of animals the organ is identical, in others it is only analogous.

Touch has its seat in a part uniform and homogeneous, as in the flesh or something of the kind, and generally, with animals supplied with blood, in the parts charged with blood. In other animals it has its seat in parts analogous to the parts charged with blood; but in all cases it is seated in parts that in their texture are homogeneous.

Cf. Aristotle, *On the Soul* III. 13 (435*b*4–7)

Clearly, then, animals deprived of this sense alone [the sense of touch] must die. For it is impossible that that which is not an animal should have this sense, and, again, it is unnecessary that an animal possess any other sense but this.

HUMAN PSYCHOLOGY

Sensation and Perception

GENERAL CONSIDERATIONS

Theophrastus, *On the Senses* 1–2, 19.[1] Translation of G. M. Stratton (London, 1917)

1. The various opinions concerning sense perception, when regarded broadly, fall into two groups. By some investigators it is ascribed to similarity, while by others it is ascribed to contrast: Parmenides, Empedocles, and Plato attribute it to similarity; Anaxagoras and Heraclitus attribute it to contrast.[2]

---

[1] Theophrastus' work *On the Senses* is of the greatest value for its account and criticism of Greek psychology before Aristotle and is concerned not merely with perception but with such psychological subjects as pleasure and pain, temperaments and emotion, and the relation of bodily to psychic states.

Material within [ ] is added by the translator for the sake of clearness. [Edd.]

[2] Besides those philosophers mentioned there are others, e.g., Alcmaeon, Diogenes of Apollonia, and Democritus, whom Theophrastus finds it more difficult to fit into this classification. The justice of citing Plato as an adherent of the "likeness" theory may be questioned. [Edd.]

The one party is persuaded by the thought that other things are, for the most part, best interpreted in the light of what is like them; that it is a native endowment of all creatures to know their kin; and furthermore, that sense perception takes place by means of an effluence, and like is borne toward like

2. The rival party assumes that perception comes to pass by an alteration; that the like is unaffected by the like, whereas opposites are affected by each other. So they give their verdict for this [idea of opposition]. And to their mind further evidence is given by what occurs in connection with touch, since a degree of heat or cold the same as that of our flesh arouses no sensation.

19. . . . Although it is a fairly difficult task to explain the facts of vision, yet how could we by *likeness* discern the objects with which the other senses deal? For the word "likeness" is quite vague. [We do] not [discern] sound by sound, nor smell by smell, nor other objects by what is kindred to them; but rather, we may say, by their opposites. To these objects it is necessary to offer the sense organ in a passive state. If we have a ringing in the ears, or a taste in the tongue, or a smell in the nostrils, these organs all become blunted; and the more so, the fuller they are of what is like them, unless there be a further distinction of these terms.

Aristotle, *On the Soul* II. 6, 12. Translation of R. D. Hicks

6. In considering each separate sense we must first treat of their objects. By the sensible object may be meant any one of three things, two of which we say are perceived in themselves or directly, while the third is perceived *per accidens* or indirectly. Of the first two the one is the special object of a particular sense, the other an object common to all the senses. By a special object of a particular sense I mean that which cannot be perceived by any other sense and in respect to which deception is impossible, for example, sight is of colour, hearing of sound and taste of flavour, while touch no doubt has for its object several varieties. But at any rate each single sense judges of its proper objects and is not deceived as to the fact that there is a colour or a sound; though as to what or where the coloured object is or what or where the object is which produces the sound, mistake is possible. Such, then, are the special objects of the several senses. By common sensibles are meant motion, rest, number, figure, size:[1] for such qualities are not the special objects of any single sense, but are common to all.

[1] In addition to the five senses Aristotle recognizes a synthetic faculty. The perception of the "common sensibles" here mentioned is one of the functions of this synthetic faculty. With the latter is also connected the power of discriminating between and comparing the data of the special senses, as well as the faculties of consciousness, imagination, memory, and reminiscence; sleeping and dreaming are affections of this same synthetic faculty. [Edd.]

For example, a particular motion can be perceived by touch as well as by sight. What is meant by the indirect object of sense may be illustrated if we suppose that the white thing before you is Diares' son. You perceive Diares' son, but indirectly, for that which you perceive is accessory to the whiteness.[1] Hence you are not affected by the indirect sensible as such. Of the two classes of sensibles directly perceived it is the objects special to the different senses which are properly perceptible: and it is to these that the essential character of each sense is naturally adapted.

12. In regard to all sense generally we must understand that sense is that which is receptive of sensible forms apart from their matter, as wax receives the imprint of the signet-ring apart from the iron or gold of which it is made: it takes the imprint which is of gold or bronze, but not *qua* gold or bronze. And similarly sense as relative to each sensible is acted upon by that which possesses colour, flavour or sound, not in so far as each of those sensibles is called a particular thing, but in so far as it possesses a particular quality and in respect of its character or form. The primary sense-organ is that in which such a power resides, the power to receive sensible forms. Thus the organ is one and the same with the power, but logically distinct from it. For that which perceives must be an extended magnitude. Sensitivity, however, is not an extended magnitude, nor is the sense: they are rather a certain character or power of the organ. From this it is evident why excesses in the sensible objects destroy the sense-organs. For if the motion is too violent for the sense-organ, the character or form (and this, as we saw, constitutes the sense) is annulled, just as the harmony and the pitch of the lyre suffer by too violent jangling of the strings. It is evident, again, why plants have no sensation, although they have one part of soul and are in some degree affected by the things themselves which are tangible: for example, they become cold and hot. The reason is that they have in them no mean, no principle capable of receiving the forms of sensible objects without their matter, but on the contrary, when they are acted upon, the matter acts upon them as well.

VISION

Alcmaeon and Anaxagoras

Theophrastus, *On the Senses* 26, 36. Translation of G. M. Stratton

26. [Alcmaeon[2] states that] eyes see through the water round about. And the eye obviously has fire within, for when one is struck [this fire]

---

[1] For the special object of vision is color. [Edd.]

[2] Alcmaeon of Croton seems to have been influenced by Pythagorean views, though the tradition of his having been a pupil of Pythagoras is doubtful. According to Chalcidius (*Commentary on the Timaeus* 246), "he was the first who ventured to dissect the eye." His theory of health and disease may be a foreshadowing of the humoral view (see p. 490). [Edd.]

flashes out.[1] Vision is due to the gleaming—that is to say, the transparent—character of that which [in the eye] reflects to the object; and sight is the more perfect, the greater the purity of this substance. All the senses are connected in some way with the brain;[2] consequently they are incapable of action if [the brain] is disturbed or shifts its position, for [this organ] stops up the passages through which the senses act.

36. Anaxagoras' doctrine of the visual image is one somewhat commonly held; for nearly everyone assumes that seeing is occasioned by the reflection in the eyes.[3] They took no account of the fact, however, that the size of objects seen is incommensurate with the size of their reflection; and that it is impossible to have many contrasting objects reflected at the same time; and farther, that motion, distance, and size are visual objects and yet produce no image. And with some animals nothing whatever is reflected—for example, with those that have horny eyes, or that live in the water. Moreover, according to this theory many *lifeless* things would possess the power of sight; for there is a reflection certainly in water, in bronze, and in many other things.

Criticism of Democritus by Theophrastus

Theophrastus, *On the Senses* 50–53. Translation of G. M. Stratton

Vision he [Democritus] explains by the reflection [in the eye], of which he gives a unique account. For the reflection does not arise immediately in the pupil. On the contrary, the air between the eye and the object of sight is compressed by the object and the visual organ, and thus becomes imprinted; since there is always an effluence of some kind arising from everything. Thereupon this imprinted air, because it is solid and is of a hue contrasting [with the pupil], is reflected in the eyes, which are moist. A dense substance does not receive [this reflection], but what is moist gives it admission. Moist eyes accordingly have a better power of vision than have hard eyes; provided their outer tunic be exceedingly fine and close-knit, and the inner [tissues] be to the last degree spongy and free from dense and stubborn flesh, and free too, from thick oily moisture; and provided, also, the ducts connected with the eyes be straight and dry that they may

[1] In the absence of knowledge of the mechanism of the retina and optic nerve the flash "seen" when the eyeball is struck was attributed to an inner fire. This played an important role in ancient optical theory. [Edd.]

[2] Aristotle criticizes this view and holds the region of the heart to be the sensory center (e.g., *On the Parts of Animals* II. 10). [Edd.]

[3] Observation of the image seen in the pupils of the eyes by one who looks closely into a mirror may have given rise to conjectures of this kind. The structure of the eye and optic nerve were quite accurately demonstrated in the early Alexandrian period, but the physiology of vision was never understood in antiquity. [Edd.]

"perfectly conform" to the entering imprints. For each knows best its kindred.

Now in the first place this imprint upon the air is an absurdity. For the substance receiving such an imprint must have a certain consistence and not be "fragile"; even as Democritus himself, in illustrating the character of the "impression," says that "it is as if one were to take a mould in wax." In the second place, an object could make a better imprint upon water [than upon air], since water is denser. While the theory would require us to see more distinctly [an object in water], we actually see it less so. In general, why should Democritus assume this *imprint*, when in his discussion of forms he has supposed an *emanation* that conveys the object's form? For these images [due to emanation] would be reflected.

But if such an imprint actually occurs and the air is moulded like wax that is squeezed and pressed, how does the reflection [in the eye] come into existence, and what is its character? For the imprint here as in other cases will evidently face the object seen. But since this is so, it is impossible for a reflection facing us to arise unless this imprint is turned around. What would cause this reversal, and what the manner of its operation, ought, however, to be shown; for in no other way could vision come to pass. Moreover, when several objects are seen in one and the same place, how can so many imprints be made upon the self-same air? And again, how could we possibly see each other? For the imprints would inevitably clash, since each of them would be facing [the person] from whom it sprung. All of which gives us pause.

Furthermore, why does not each person see himself? For the imprints [from ourselves] would be reflected in our own eyes quite as they are in the eyes of our companions, especially if these imprints directly face us and if the effect here is the same as with an echo—since Democritus says that [in the case of the echo] the vocal sound is reflected back to him who utters it. Indeed the whole idea of imprints made on the air is extravagant.[1]

Aristotle

Aristotle, *On the Soul* II. 7.[2] Translation of R. D. Hicks

The object seen in light is colour, and this is why it is not seen without light. For the very quiddity of colour is, as we saw, just this, that it is capable of exciting change in the operantly transparent medium: and the

[1] The atomistic theory of effluences or idols is considered at length in Epicurus's *Letter to Herodotus* and in Lucretius IV. [Edd.]

[2] In the explanation of sight Aristotle proceeds from the cardinal facts that by this sense we distinguish objects (1) at a distance, (2) as coloured. Hence he assumes a medium upon which colour can act. The medium, in itself neutral, has two determinations, a positive state when it is illuminated and we say there is light, a negative state when we say there is darkness. . . .

activity of the transparent is light. There is clear evidence of this. If you lay the coloured object upon your eye, you will not see it. On the contrary, what the colour excites is the transparent medium, say, the air, and by this, which is continuous, the sense-organ is stimulated. For it was a mistake in Democritus to suppose that if the intervening space became a void, even an ant would be distinctly seen, supposing there were one in the sky.[1] That is impossible. For sight takes place through an affection of the sensitive faculty. Now it cannot be affected by that which is seen, the colour itself: therefore it can only be by the intervening medium: hence the existence of some medium is necessary. But, if the intermediate space became a void, so far from being seen distinctly, an object would not be visible at all.[2]

Cf. Aristotle, *De Sensu* 438*b*2–16. Translation of J. I. Beare (Oxford, 1908)

That without light vision is impossible has been stated elsewhere; but, whether the medium between the eye and its objects is air or light, vision is caused by a process through this medium.

Accordingly, that the inner part of the eye consists of water is easily intelligible, water being translucent.

Now, as vision outwardly is impossible without [extra-organic] light, so also it is impossible inwardly [without light within the organ]. There must, therefore, be some translucent medium within the eye, and, as this is not air, it must be water. The soul or its perceptive part is not situated at the external surface of the eye, but obviously somewhere within: whence the necessity of the interior of the eye being translucent, i.e., capable of admitting light. And that it is so is plain from actual occurrences. It is matter of experience that soldiers wounded in battle by a sword slash on the temple, so inflicted as to sever the passages of [i.e., inward from] the eye, feel a sudden onset of darkness, as if a lamp had gone out; because what is called the pupil, i.e., the translucent, which is a sort of inner lamp, is then cut off [from its connexion with the soul].

HEARING[3]

Theophrastus, *On the Senses* 25, 9. Translation of G. M. Stratton

25 . . . . Hearing is by means of the ears, he [Alcmaeon] says, because within them is an empty space, and this empty space resounds. A kind of

[1] The films that fly from objects and cause vision, according to Democritus, are prevented by the intervening air from reaching our eyes. Instead they impress their form on the air; but since this process entails progressive distortion the clearness of perception decreases as the distance increases. If, however, the intervening air were removed, such distortion would not take place and the image would reach the eye unaltered. [Edd.]

[2] On Aristotle's theory of light, see p. 285. [Edd.]

[3] The Greeks attacked the problem of the nature of sound more successfully than that of

noise is produced by the cavity, and the internal air re-echoes this sound.

9. He [Empedocles] says that hearing results from sounds within [the head], whenever the air, set in motion by a voice, resounds within. For the organ of hearing, which he calls a "fleshy off-shoot," acts as the "bell" of a trumpet, ringing with sounds like [those it receives]. When set in motion [this organ] drives the air against the solid parts and produces there a sound.

Plato, *Timaeus* 67 B. Translation of R. G. Bury

The third organ of perception within us which we have to describe in our survey is that of hearing, and the causes whereby its affections are produced. In general, then, let us lay it down that sound is a stroke transmitted through the ears, by the action of the air upon the brain and the blood, and reaching to the soul; and that the motion caused thereby, which begins in the head and ends about the seat of the liver, is "hearing"; and that every rapid motion produces a "shrill" sound, and every slower motion a more "deep" sound; and that uniform motion produces an "even" and smooth sound and the opposite kind of motion a "harsh" sound; and that large motion produces "loud" sound, and motion of the opposite kind "soft" sound.

Aristotle, *On the Soul* II. 8 (420*a*4–5)

Now there is air naturally connected with the organ of hearing. And because this organ is in air, motion of the external air produces motion of the air within the organ.

SMELL AND TASTE

Empedocles

Theophrastus, *On the Senses* 9. 21–22. Translation of G. M. Stratton

9. . . . Smell, according to Empedocles, is due to the act of breathing. As a consequence, those have keenest smell in whom the movement of the breath is most vigorous. The intensest odour emanates from bodies that are subtile and light. Of taste and touch severally he offers no precise account, telling us neither the manner nor the means of their operation, save the [assertion he makes with regard to all the senses in] common, that perception arises because emanations fit into the passages of sense.[1] Pleasure is excited by things that are similar [to our organs], both in their constituent parts and in the manner of their composition; pain, by things opposed.

---

the nature of hearing (see pp. 286ff.). Here we have merely suggested a few viewpoints of the latter problem. [Edd.]

[1] The reference is to the pores, which played a part in all sense perceptions in the theory of Empedocles. [Edd.]

21. . . . It is silly to assert [as does Empedocles] that those have the keenest sense of smell who inhale most; for if the organ is not in health or is, for any cause, not unobstructed, mere breathing is of no avail. It often happens that a man has suffered injury [to the organ] and has no sensation at all. Furthermore, persons "short of breath" or at hard labour or asleep—since they inhale most air—should be most sensitive to odours. Yet the reverse is the fact. For in all likelihood respiration is not of itself the cause of smell, but is connected with it incidentally.

Democritus

Theophrastus, *On the Causes of Plants* VI. 1.6. Translation of J. I. Beare, *Greek Theories of Elementary Cognition from Alcmaeon to Aristotle*, p. 164 (Oxford, 1906)

Democritus investing each taste with its characteristic figure makes the *sweet* that which is round and large in its atoms; the astringently *sour* that which is large in its atoms, but rough, angular, and not spherical; the *acid*, as its name imports, that which is sharp in its bodily shape, angular, and curving, thin, and not spherical; the *pungent* that which is spherical, thin, angular, and curving; the *saline*, that of which the atoms are angular, and large, and crooked and isosceles; the *bitter*, that which is spherical, smooth, scalene, and small. The *succulent* is that which is thin, spherical, and small.[1]

Aristotle

Aristotle, *On the Soul* II. 9–10. Translation of R. D. Hicks

9. . . . As with flavours, so with odours: some are sweet, some bitter. (But in some objects smell and flavour correspond; for example, they have sweet odour and sweet flavour: in other things the opposite is the case.) Similarly, too, an odour may be pungent, irritant, acid or oily. But because, as we said above, odours are not as clearly defined as the corresponding flavours, it is from these latter that the odours have taken their names, in virtue of the resemblance in the things. Thus the odour of saffron and honey is sweet, while the odour of thyme and the like is pungent; and so in all the other cases. Again, smell corresponds to hearing and to each of the other senses in that, as hearing is of the audible and inaudible, and sight of the visible and invisible, so smell is of the odorous and inodorous. By inodorous may be meant either that which is wholly incapable of having odour or that which has a slight or faint odour. The term tasteless involves a similar ambiguity.

---

[1] The dependence of sensations on atomic sizes, shapes, arrangements, and motions is cardinal in the philosophy of the Greek atomists. Thus Democritus formulated the distinction between what came to be called primary and secondary qualities when he said, "By convention we speak of color, and of sweet and bitter, but in reality there are atoms and void." (Frag. 125, Diels.) [Edd.]

Further, smell also operates through a medium, namely, air or water. For water animals too, whether they are, or are not, possessed of blood, seem to perceive odour as much as the creatures in the air: since some of them also come from a great distance to seek their food, guided by the scent. . . .

The inability to perceive what is placed immediately on the sense organ man shares with all animals: what is peculiar to him is that he cannot smell without inhaling. . . .[1]

10. The object of taste is a species of tangible. And this is the reason why it is not perceived through a foreign body as medium: for touch employs no such medium either. The body, too, in which the flavour resides, the proper object of taste, has the moist, which is something tangible, for its matter or vehicle. Hence, even if we lived in water, we should still perceive anything sweet thrown into the water, but our perception would not have come through the medium, but by the admixture of sweetness with the fluid, as is the case with what we drink. But it is not in this way, namely, by admixture, that colour is perceived, nor yet by emanations. Nothing, then, corresponds to the medium; but to colour, which is the object of sight, corresponds the flavour, which is the object of taste. But nothing produces perception of flavour in the absence of moisture, but either actually or potentially the producing cause must have liquid in it: salt, for instance, for that is easily dissolved and acts as a dissolvent upon the tongue. . . .

The organ of taste, then, which needs to be moistened, must have the capacity of absorbing moisture without being dissolved, while at the same time it must not be actually moist. A proof of this is the fact that the tongue has no perception either when very dry or very moist. In the latter case the contact is with the moisture originally in the tongue, just as when a man first makes trial of a strong flavour and then tastes some other flavour; or as with the sick, to whom all things appear bitter because they perceive them with their tongue full of bitter moisture.

As with the colours, so with the species of flavour, there are, first, simple flavours, which are opposites, the sweet and the bitter; next to these on one side the succulent, on the other the salt; and, thirdly, intermediate between these, the pungent, the rough, the astringent, and the acid. These seem to be practically all the varieties of flavour. Consequently, while the faculty of taste has potentially the qualities just described, the object of taste converts the potentiality into actuality.[2]

---

[1] Taken literally, this is not true. Other air-breathing animals also smell while inhaling breath. . . .

[2] Cf. the treatment in Plato, *Timaeus* 65C–66C. Theophrastus in his discussion of plant saps (*On the Causes of Plants* VI) has much to say about flavors. [Edd.]

[Aristotle], *Problemata* XIII. 2. Translation of E. S. Forster

Why is it that things of unpleasant odour do not seem to have an odour to those who have eaten them? Is it because, owing to the fact that the sense penetrates to the mouth through the palate, the sense of smell soon becomes satiated and so it no longer perceives the odour inside the mouth to the same extent—for at first every one perceives the odour, but, when they are in actual contact with it, they no longer do so, as though it had become part of themselves—and the similar odour from without is overpowered by the odour within?

Theophrastus

Theophrastus, *On Odors* 1, 5. Translation of Arthur Hort

1. Odours in general, like tastes, are due to mixture: for anything which is uncompounded has no smell, just as it has no taste: wherefore simple substances have no smell, such as water, air, and fire: on the other hand earth is the only elementary substance which has a smell, or at least it has one to a greater extent than the others, because it is of a more composite character than they.

Of odours some are, as it were, indistinct and insipid, as is the case with tastes, while some have a distinct character. And these characters appear to correspond to those of tastes, yet they have not in all cases the same names, as we said in a former treatise; nor in general are they marked off from one another by such specific differences as are tastes: rather the differences are, one may say, in generic character, some things having a good, some an evil odour. But the various kinds of good or evil odour, although they exhibit considerable differences, have not received further distinguishing names, marking off one particular kind of sweetness or of bitterness from another: we speak of an odour as pungent, powerful, faint, sweet, or heavy, though some of these descriptions apply to evil-smelling things as well as to those which have a good odour.

5. Now the odour of some things which have a good odour resides in things which are used for food, for instance that of stone-fruits, pears, and apples, the smell of which is sweet even if one does not eat them; indeed it may be said to be sweeter in that case. However, to make a general distinction, some odours exist independently, while others are incidental;[1] those of juices and things used for food are incidental, those of flowers exist independently. And, as was said above, things which have a good odour are generally of unpleasant, astringent, or somewhat bitter taste. Again some things which have a good taste have also an evil odour, such as the carob, which is sweet (this is true of some regions, if not of all). Again

[1] I.e., the smell is a kind of "accident" or by-product of the taste.

the Phoenician cedar, though it is sweet to the taste, when chewed produces a sort of evil odour, though it makes the water fragrant.

TOUCH

Aristotle, *On the Soul* II. 11.[1] Translation of R. D. Hicks

If touch is not a single sense but includes more senses than one, there must be a plurality of tangible objects also. It is a question whether touch is several senses or only one. What, moreover, is the sense-organ for the faculty of touch? Is it the flesh or what is analogous to this in creatures that have not flesh? Or is flesh, on the contrary, the medium, while the primary sense-organ is something different, something internal? We may argue thus: every sense seems to deal with a single pair of opposites, sight with white and black, hearing with high and low pitch, taste with bitter and sweet; but under the tangible are included several pairs of opposites, hot and cold, dry and moist, hard and soft and the like. A partial solution of this difficulty lies in the consideration that the other senses also apprehend more than one pair of opposites. Thus in vocal sound there is not only high and low pitch, but also loudness and faintness, smoothness and roughness, and so on. In regard to colour also there are other similar varieties. But what the one thing is which is subordinated to touch as sound is to hearing is not clear.

But is the organ of sense internal or is the flesh the immediate organ? No inference can be drawn, seemingly, from the fact that the sensation occurs simultaneously with contact. For even under present conditions, if a sort of membrane were constructed and stretched over the flesh, this would immediately on contact transmit the sensation as before. And yet it is clear that the organ of sense is not in this membrane; although, if by growth it became united to the flesh, the sensation would be transmitted even more quickly. Hence it appears that the part of the body in question, that is, the flesh, is related to us as the air would be if it were united to us all round by natural growth. We should then have thought we were perceiving sound, colour and smell by one and the same instrument: in fact, sight, hearing and smell would have seemed to us in a manner to constitute a single sense. But as it is, owing to the media, by which the various motions are transmitted, being separated from us, the difference of the organs of these three senses is manifest. But in regard to touch this point is at present obscure. . . .

It is, then, the distinctive qualities of body as body which are the objects of touch: I mean those qualities which determine the elements, hot or cold, dry or moist, of which we have previously given an account in our discussion of the elements. And their sense-organ, the tactile organ, that

[1] Cf. p. 541, above. [Edd.]

is, in which the sense called touch primarily resides, is the part which has potentially the qualities of the tangible object. For perceiving is a sort of suffering or being acted upon: so that when the object makes the organ in actuality like itself it does so because that organ is potentially like it. Hence it is that we do not perceive what is just as hot or cold, hard or soft, as we are, but only the excesses of these qualities: which implies that the sense is a kind of mean between the opposite extremes in the sensibles. This is why it passes judgment on the things of sense. For the mean is capable of judging, becoming to each extreme in turn its opposite. And, as that which is to perceive white and black must not be actually either, though potentially both, and similarly for the other senses also, so in the case of touch the organ must be neither hot nor cold. Further, sight is in a manner, as we say, of the invisible as well as the visible, and in the same way the remaining senses deal with opposites. So, too, touch is of the tangible and the intangible: where by intangible is meant, first, that which has the distinguishing quality of things tangible in quite a faint degree, as is the case with the air; and, secondly, tangibles which are in excess, such as those which are positively destructive.

Lucretius, *On the Nature of Things* III. 374–395

Not only are the atoms of the soul much smaller than those of which our body and flesh are composed, but they are also fewer in number and are scattered only here and there over our body. Thus one may say that the intervals between the atoms of the soul are equal to the size of the smallest bodies that produce sensation by impact upon our body.[1]

For sometimes we do not feel the adhesion of dust to our body, or the settling of powdered chalk on our limbs, or a mist at night, or a spider's fine threads when we become entangled in them, or its shrivelled web falling on our head, or the feathers of birds, or the seeds of thistle-down as they are wafted to us and fall so slowly because they are so light, or the approach of every crawling creature, or every footstep that gnats and their like place on our body. So true is it that many particles must be moved in us before the atoms of the soul, interspersed through all the parts of our bodies, can perceive the impact and move in these intervals to buffet others, and meeting them leap apart in turn.

The Afterimage

Aristotle, *On Dreams* 2 (459*a*24–*b*23). Translation of J. I. Beare (Oxford, 1908)

The objects of sense-perception corresponding to each sensory organ produce sense-perception in us, and the affection due to their operation is present in the organs of sense not only when the perceptions are actualized, but even when they have departed.

[1] I.e., a smaller body may fail to touch any atoms of the soul.

What happens in these cases may be compared with what happens in the case of projectiles moving in space. For in the case of these the movement continues even when that which set up the movement is no longer in contact [with the things that are moved]. For that which set them in motion moves a certain portion of air, and this, in turn, being moved excites motion in another portion; and so, accordingly, it is in this way that [the bodies], whether in air or in liquids, continue moving, until they come to a standstill.[1]

This we must likewise assume to happen in the case of qualitative change; for that part which [for example] has been heated by something hot, heats [in turn] the part next to it, and this propagates the affection continuously onwards until the process has come round to its point of origination. This must also happen in the organ wherein the exercise of sense-perception takes place, since sense-perception, as realized in actual perceiving, is a mode of qualitative change. This explains why the affection continues in the sensory organs, both in their deeper and in their more superficial parts, not merely while they are actually engaged in perceiving, but even after they have ceased to do so. That they do this, indeed, is obvious in cases where we continue for some time engaged in a particular form of perception, for then, when we shift the scene of our perceptive activity, the previous affection remains; for instance, when we have turned our gaze from sunlight into darkness. For the result of this is that one sees nothing, owing to the motion excited by the light still subsisting in our eyes. Also, when we have looked steadily for a long while at one colour, e.g., at white or green, that to which we next transfer our gaze appears to be of the same colour. Again if, after having looked at the sun or some other brilliant object, we close the eyes, then, if we watch carefully, it appears in a right line with the direction of vision (whatever this may be), at first in its own colour; then it changes to crimson, next to purple, until it becomes black and disappears. And also when persons turn away from looking at objects in motion, e.g., rivers, and especially those which flow very rapidly, they find that the visual stimulations still present themselves, for the things really at rest are then seen moving: persons become very deaf after hearing loud noises, and after smelling very strong odours their power of smelling is impaired; and similarly in other cases. These phenomena manifestly take place in the way above described.

Association of Ideas

Aristotle, *On Memory and Recollection* 2. Translation of J. I. Beare (Oxford, 1908)

Acts of recollection, as they occur in experience, are due to the fact that one movement has by nature another that succeeds it in regular order. . . .

[1] A reference to the doctrine of *antiperistasis*. See p. 221. [Edd.]

Whenever, therefore, we are recollecting, we are experiencing certain of the antecedent movements until finally we experience the one after which customarily comes that which we seek. This explains why we hunt up the series having started in thought either from a present intuition or some other, and from something either similar, or contrary, to what we seek, or else from that which is contiguous with it. Such is the empirical ground of the process of recollection; for the mnemonic movements involved in these starting-points are in some cases identical, in others, again, simultaneous, with those of the idea we seek, while in others they comprise a portion of them, so that the remnant which one experienced after that portion [and which still requires to be excited in memory] is comparatively small.

Thus, then, it is that persons seek to recollect, and thus, too, it is that they recollect even without the effort of seeking to do so, viz., when the movement implied in recollection has supervened on some other which is its condition. For, as a rule, it is when antecedent movements of the classes here described have first been excited, that the particular movement implied in recollection follows. . . . Accordingly, therefore, when one wishes to recollect, this is what he will do: he will try to obtain a beginning of movement whose sequel shall be the movement which he desires to reawaken. This explains why attempts at recollection succeed soonest and best when they start from a beginning [of some objective series]. For, in order of succession, the mnemonic movements are to one another as the objective facts [from which they are derived]. Accordingly, things arranged in a fixed order, like the successive demonstrations in geometry, are easy to remember [or recollect], while badly arranged subjects are remembered with difficulty. . . .

Hence it is that [from the same starting-point] the mind receives an impulse to move sometimes in the required direction, and at other times otherwise, [doing the latter] particularly when something else somehow deflects the mind from the right direction and attracts it to itself. This last consideration explains too how it happens that, when we want to remember a name, we remember one somewhat like it, indeed, but blunder in reference to the one we intended.

Thus, then, recollection takes place.

The Interrelation of Bodily and Mental States

TEMPERATURE AND EMOTIONS

Aristotle, *On the Motion of Animals*, 7–8. Translation of A. S. L. Farquharson

Sensations are obviously a form of change of quality, and imagination and conception have the same effect as the objects so imagined and conceived. For in a measure the form conceived be it of hot or cold or pleasant or fearful is like what the actual objects would be, and so we shudder and are frightened at a mere idea. Now all these affections involve changes of quality, and with those changes some parts of the body enlarge, others grow smaller. And it is not hard to see that a small change occurring at the centre makes great and numerous changes at the circumference, just as by shifting the rudder a hair's breadth you get a wide deviation at the prow. And further, when by reason of heat or cold or some kindred affection a change is set up in the region of the heart, or even in an imperceptibly small part of the heart, it produces a vast difference in the periphery of the body—blushing, let us say, or turning white, goose-skin and shivers and their opposites.

But to return, the object we pursue or avoid in the field of action is, as has been explained, the original of movement, and upon the conception and imagination of this there necessarily follows a change in the temperature of the body. For what is painful we avoid, what is pleasing we pursue. We are, however, unconscious of what happens in the minute parts; still anything painful or pleasing is generally speaking accompanied by a definite change of temperature in the body. One may see this by considering the affections. Blind courage and panic fears, erotic motions, and the rest of the corporeal affections, pleasant and painful, are all accompanied by a change of temperature, some in a particular member, others in the body generally. So, memories and anticipations, using as it were the reflected images of these pleasures and pains, are now more and now less causes of the same changes of temperature.

Aberrations of the Senses

Aristotle, *On Dreams* 2. Translation of J. I. Beare

We are easily deceived respecting the operations of sense-perception when we are excited by emotions, and different persons according to their different emotions; for example, the coward when excited by fear, the amorous person by amorous desire; so that, with but little resemblance to go upon, the former thinks he sees his foes approaching, the latter, that he sees the object of his desire; and the more deeply one is under the influence of the emotion, the less similarity is required to give rise to these illusory impressions. Thus too, both in fits of anger, and also in all states of appetite, all men become easily deceived, and more so the more their emotions are excited. This is the reason too why persons in the delirium of fever sometimes think they see animals on their chamber walls, an illusion arising from the faint resemblance to animals of the markings thereon when put together in patterns; and this sometimes corresponds with the emotional states of the sufferers, in such a way that, if the latter be not very ill, they know well enough that it is an illusion; but if the illness is more severe they actually move according to the appearances. The cause of these occurrences is that the faculty in virtue of which the controlling sense judges is not identical with that in virtue of which presentations come before the mind. A proof of this is that the sun presents itself as only a foot in diameter, though often something else gainsays the presentation.[1] Again, when the fingers are crossed, the one object [placed between them] is felt [by the touch] as two;[2] but yet we deny that it is two; for sight is more authoritative than touch. Yet, if touch stood alone, we should actually have pronounced the one object to be two. The ground of such false judgments is that any appearances whatever present themselves, not only when its object stimulates a sense, but also when the sense by itself alone is stimulated, provided only it be stimulated in the same manner as it is by the object.

PSYCHOPATHOLOGY[3]

Plato, *Timaeus* 86 B–87 A. Translation of R. G. Bury

Such is the manner in which diseases of the body come about; and those of the soul which are due to the condition of the body arise in the following way. We must agree that folly is a disease of the soul; and of folly there are two kinds, the one of which is madness, the other ignorance. Whatever affection a man suffers from, if it involves either of these conditions it must be termed "disease"; and we must maintain that pleasures and pains in excess are the greatest of the soul's diseases. For when a man is overjoyed or contrariwise suffering excessively from pain, being in haste to seize on the one and avoid the other beyond measure, he is unable either to see or to hear anything correctly, and he is at such a time distraught and wholly incapable of exercising reason. . . . And again, in respect of pains likewise the soul acquires much evil because of the body.

For whenever the humours which arise from acid and saline phlegms, and all humours that are bitter and bilious wander through the body and find no external vent but are confined within, and mingle their vapour with the movement of the soul and are blended therewith, they implant diseases of the soul of all kinds, varying in intensity and in extent; and as these humours penetrate to the three regions of the soul,[4] according to the region which they severally attack, they give rise to all varieties of bad temper and

---

[1] The Epicureans, it will be recalled, maintained that the sun and moon were actually about a foot in diameter. [Edd.]

[2] This experiment of crossing the fingers is often referred to by Aristotle. [Edd.]

[3] Numerous passages in biological, medical, philosophical, and purely literary works illustrate this topic, and in this connection we may note references to the effect of age on character (e.g., Aristotle, *Rhetoric* II. 12–14), the effect of drugs on the mind (e.g., Theophrastus, *History of Plants* IX. 19), and the effect of music (Plato *Republic* 398–400, Aristotle, *Politics* VIII. 7, Theophrastus, *Fragments* 87, 88). [Edd.]

[4] I.e., the brain, the heart, and the liver, which are the centers, respectively, of the rational, "spirited," and appetitive aspects of soul. [Edd.]

bad spirits, and they give rise to all manner of rashness and cowardice, and of forgetfulness also, as well as of stupidity.[1]

Aretaeus, *On the Causes and Symptoms of Chronic Diseases* I.5. Translation of Francis Adams

Melancholy

But if it [black bile] be determined upwards to the stomach and diaphragm, it forms melancholy;[2] for it produced flatulence and eructations of a fetid and fishy nature, and it sends rumbling wind downwards, and disturbs the understanding. On this account, in former days, they were called melancholics and flatulent persons. And yet, in certain of these cases there is neither flatulence nor black bile, but mere anger and grief, and sad dejection of mind. . . .

It is a lowness of spirits from a single phantasy, without fever; and it appears to me that melancholy is the commencement and a part of mania. For in those who are mad, the understanding is turned sometimes to anger and sometimes to joy, but in the melancholics to sorrow and despondency only. But they who are mad are so for the greater part of life, becoming silly, and doing dreadful and disgraceful things; but those affected with melancholy are not every one of them affected according to one particular form; but they are either suspicious of poisoning, or flee to the desert from misanthropy, or turn superstitious, or contract a hatred of life. Or if at any time a relaxation takes place, in most cases hilarity supervenes, but these persons go mad. . . .

But if it also affects the head from sympathy, and the abnormal irritability of temper change to laughter and joy for the greater part of their life, these become mad rather from the increase of the disease than from change of the affection.

Dryness is the cause of both. Adult men, therefore, are subject to mania and melancholy, or persons of less age than adults. Women are worse affected with mania than men. As to age, towards manhood, and those actually in the prime of life. The seasons of summer and of autumn engender, and spring brings it to a crisis.

The characteristic appearances, then, are not obscure; for the patients are dull or stern, dejected or unreasonably torpid, without any manifest cause: such is the commencement of melancholy. And they also become peevish, dispirited, sleepless, and start up from a disturbed sleep.

Unreasonable fear also seizes them, if the disease tend to increase, when

[1] The connection here alluded to between the humors and psychological states has persisted in language and in folklore, as when we speak of sanguine, phlegmatic, melancholic, and choleric types. There is a classic discussion of the melancholic type in [Aristotle], *Problemata* XXX. 1. [Edd.]

[2] Cf. pp. 488–490. [Edd.]

their dreams are true, terrifying, and clear: for whatever, when awake, they have an aversion to, as being an evil, rushes upon their visions in sleep. They are prone to change their mind readily; to become base, mean-spirited, illiberal, and in a little time, perhaps, simple, extravagant, munificent, not from any virtue of the soul, but from the changeableness of the disease. But if the illness become more urgent, hatred, avoidance of the haunts of men, vain lamentations; they complain of life, and desire to die. In many, the understanding so leads to insensibility and fatuousness, that they become ignorant of all things, or forgetful of themselves, and live the life of the inferior animals. The habit of the body also becomes perverted. . . . Therefore the bowels are dried up, and discharge nothing; or, if they do, the dejections are dried, round, with a black and bilious fluid, in which they float; urine scanty, acrid, tinged with bile. They are flatulent about the hypochondriac region; the eructations fetid, virulent, like brine from salt; and sometimes an acrid fluid, mixed with bile, floats in the stomach. Pulse for the most part small, torpid, feeble, dense, like that from cold.

A story is told, that a certain person, incurably affected, fell in love with a girl; and when the physicians could bring him no relief, love cured him. But I think that he was originally in love, and that he was dejected and spiritless from being unsuccessful with the girl, and appeared to the common people to be melancholic.[1] He then did not know that it was love; but when he imparted the love to the girl, he ceased from his dejection, and dispelled his passion and sorrow; and with joy he awoke from his lowness of spirits, and he became restored to understanding, love being his physician.

[1] Lovesickness is found to be the cause of insomnia in a case described by Galen, XIV. 631 (Kühn). The quickening of the pulse at the mention of the name of the beloved gives the clue. [Edd.]

SOME IMPORTANT BOOKS ON GREEK SCIENCE

The books here cited do not constitute an exhaustive bibliography of the subject; for such an undertaking a separate volume would be necessary. There are, however, certain outstanding contributions that should be made known to the student. We also wish here to indicate the range of primary sources, for which more precise bibliographical data may be found in the body of this volume.

Much of the literature on Greek science is dispersed among philological books, periodicals, and encyclopedias. We mention first the two great encyclopedias devoted to classical antiquity:

Paulys Real-Encyclopädie der classischen Altertumswissenschaft. Ed. by G. Wissowa, W. Kroll, and K. Mittelhaus. Stuttgart, publication begun in 1893; not yet complete. Hereafter abbreviated RE.

Daremberg, C., and E. Saglio. *Dictionnaire des antiquités grecques et romaines.* 5 volumes, Paris, 1873–1917.

Older than these, but occasionally very helpful:

A Dictionary of Greek and Roman Biography and Mythology. Ed. by William Smith. 3 volumes, London, 1876.

A Dictionary of Greek and Roman Antiquities. Ed. by W. Smith, W. Wayte, and G. E. Marindin. 3d ed., 2 volumes, London, 1890–1891.

Note also:

A Companion to Latin Studies. Ed. by J. E. Sandys, Cambridge, 1910.

A Companion to Greek Studies. Ed. by L. Whibley, Cambridge, 1905.

In many types of investigation the student of Greek science will find it profitable to employ the bibliographical resources of the classicist, e.g.:

Marouzeau, J. *Dix Années de bibliographie classique* (covering the period 1914–1924). Paris, 1927–1928.

L'Année philologique (Paris).

Bibliotheca philologica classica (Leipzig).

Klassieke bibliographie (Utrecht).

Bursian's Jahresbericht über die Fortschritte der klassischen Altertumswissenschaft. Note the classified list of articles (1873–1923) in McFayden, D., "Fifty Years of Bursian's Jahresbericht." *Washington University Studies. Humanistic Series* 12 (1924) 111.

Of treatises dealing wholly or in large measure with Greek science note the following:

Sarton, George. *Introduction to the History of Science.* Vol. I, *From Homer to Omar Khayyam.* Baltimore, 1927. This book includes rich bibliographies which are supplemented in the issues of the periodical *Isis.*

Brunet, P., and A. Mieli. *Histoire des sciences: Antiquité.* Paris, 1935 (a history and an anthology, with extensive bibliographies).

Mieli, A. *Manuale di storia della scienza. Storia, antologia, bibliografia.* Rome, 1925.

Enriques, F., and G. De Santillana. *Storia del pensiero scientifico.* Vol. I, *Il Mondo antico.* Bologna, 1932.

Heiberg, J. L. *Geschichte der Mathematik und Naturwissenschaften im Altertum.* Munich, 1925.

——— *Mathematics and Physical Science in Classical Antiquity.* London, 1922. A translation by D. C. Macgregor of *Naturwissenschaften Mathematik und Medizin im klassischen Altertum*, Leipzig, 1912.

Rehm, A., and K. Vogel. *Exakte Wissenschaften.* Leipzig, 1933. (*Einleitung in die Altertumswissenschaft* II. 2, ed. by A. Gercke and E. Norden.)

Rey, Abel. *La Science orientale avant les Grecs.* Paris, 1930.

——— *La Jeunesse de la science grecque.* Paris, 1933.

——— *La Maturité de la pensée scientifique en Grèce.* Paris, 1939.

——— *L'Apogée de la science technique grecque.* Paris, 1946. The emphasis in these works in on the philosophy of science.

Milhaud, G. *Leçons sur les origines de la science grecque.* Paris, 1894.

——— *Etudes sur la pensée scientifique chez les Grecs et chez les modernes.* Paris, 1906.

——— *Les Philosophes géomètres de la Grèce.* Paris, 1906.

——— *Nouvelles Etudes sur l'histoire de la pensée scientifique.* Paris, 1912.

Reymond, A. *Histoire des sciences exactes et naturelles dans l'antiquité gréco-romaine.* Paris, 1924. English translation by R. G. de Bray, New York [1927].

Heidel, W. A. *The Heroic Age of Science: the Conception, Ideals, and Methods of Science among the Ancient Greeks.* Baltimore, 1933.

Farrington, B. *Science in Antiquity.* London, 1936.

——— *Science and Politics in the Ancient World.* London, 1939.

Thorndike, Lynn. *A History of Magic and Experimental Science during the First Thirteen Centuries of Our Era.* Vol. I (New York, 1923) begins with the Roman Empire.

Thompson, D. W. *Science and the Classics.* Oxford, 1940.

Tannery, Paul. *Mémoires scientifiques.* 14 volumes, 1912–1937. Volumes I–III are especially devoted to ancient science, but there is extensive material on antiquity in the other volumes.

The intimate relation of philosophy and science in the period covered by our volume makes it advisable to mention here, for its bibliographical richness:

Friedrich Ueberwegs Grundriss der Geschichte der Philosophie. Erster Teil. 12th ed., ed. by K. Praechter. Berlin, 1926.

Note also:

Zeller, E. *Die Philosophie der Griechen in ihrer geschichtlichen Entwicklung.* English translation of various parts by Alleyne, Alleyne and Goodwin, Costelloe and Muirhead, Reichel.

——— *Grundriss der Geschichte der griechischen Philosophie.* English translation by L. R. Palmer. New York, 1931.

Gomperz, T. *Griechische Denker.* 3 volumes, Leipzig, 1896–1909. English translation, 4 volumes, London, 1911–1912.

Burnet, John. *Greek Philosophy.* Part I, *Thales to Plato.* London, 1914.

——— *Early Greek Philosophy.* 3d ed., London, 1920.

Tannery, P. *Pour l'Histoire de la science hellène.* 2d ed., Paris, 1930.

Robin, Léon. *La Pensée grecque et les origines de l'esprit scientifique.* Paris, 1923. English translation by M. R. Dobie, New York, 1928.

Mieli, Aldo. *La Scienza greca prearistotelica.* Florence, 1916.

Frank, E. *Platon und die sogenannten Pythagoreer.* Halle, 1923.

Jaeger, W. *Aristoteles.* English translation by R. Robinson. Oxford, 1934.

——— *Paideia.* English translation by G. Highet. 3 volumes, Oxford, 1939–1944.

In addition to the extant philosophical works, such as those of Plato, Aristotle, and Lucretius, the collections of fragments furnish many of our sources. Note, e.g.:

Diels, H. *Die Fragmente der Vorsokratiker.* 5th ed., ed. by W. Kranz, 3 volumes, Berlin, 1934–1937.

Diels, H. *Doxographi graeci.* Berlin, 1879.

Usener, H. *Epicurea.* Leipzig, 1887.

Arnim, H. von. *Stoicorum veterum fragmenta.* Leipzig, 1903–1924.

Useful English translations are contained in:

Fairbanks, A. *The First Philosophers of Greece.* London, 1898.

Selections from Early Greek Philosophy. Ed. by Milton C. Nahm. New York, 1935.

Selections from Hellenistic Philosophy. Ed. by G. H. Clark. New York, 1940.

Periodicals

Among those periodicals dealing generally with the history of science, including the ancient period:

Archeion: archivio di storia della scienza. Founded in 1919 by A. Mieli.

Isis: International Review Devoted to the History of Science and Civilization. Founded in 1913 by George Sarton. Includes critical bibliographies of current literature in the history of science and thus forms a continuing supplement to Sarton's *Introduction to the History of Science.*

Osiris. Founded in 1936 by George Sarton for the publication of longer papers and monographs.

Mitteilungen zur Geschichte der Medizin, der Naturwissenschaften und der Technik. Founded in 1902. Especially important for its bibliographies.

Archiv für Geschichte der Naturwissenschaften und der Technik. Issued 1908–1922, 1927–1931.

Thalès. Recueil annuel des traveaux de l'institut d'histoire des sciences et des techniques de l'université de Paris. Founded in 1934.

Lychnos. Organ of the Swedish History of Science Society. Founded in 1936.

The series of translations, *Klassiker der exacten Naturwissenschaften*, founded by Wilhelm Ostwald, contains translations of various works of Greek mathematics, astronomy, and physics.

Many historical articles are published in periodicals dealing with special sciences (some will be noted below) and in such general scientific periodicals as *Science* (New York), *Nature* (London), and *Scientia* (Bologna).

Mathematics

An extensive bibliography of the history of mathematics (including the ancient period) is contained in George Sarton, *The Study of the History of Mathematics*, Cambridge, Mass., 1936, and in Gino Loria, *Guida allo studio della storia delle matematiche*, 2d. ed., Milan, 1946.

The standard histories of mathematics take up the Greek period quite fully, e.g.:

Cantor, Moritz. *Vorlesungen über Geschichte der Mathematik.* 4 volumes. Volume I (4th ed., Leipzig, 1922) includes the ancient period.

Smith, D. E. *History of Mathematics.* 2 volumes, Boston, 1923–1925.

Loria, Gino. *Storia delle matematiche.* 3 volumes, Turin, 1929–1933.

The following are devoted to Greek mathematics in particular.

Heath, T. L. *A History of Greek Mathematics.* 2 volumes, Oxford, 1921.

——— *A Manual of Greek Mathematics.* Oxford, 1931.

Thomas, Ivor. *Selections Illustrating the History of Greek Mathematics.* With English translation. Loeb Classical Library, 2 volumes, Cambridge, Mass., and London, 1939, 1941. An excellent anthology, which closely follows Heath's treatment of the subject.

Loria, Gino. *Le Scienze esatte nell' antica Grecia.* 5 volumes, Modena, 1893–1902.

——— *Histoire des sciences mathématiques dans l'antiquité hellénique.* Paris, 1929.

Among the older books note:

Gow, James. *A Short History of Greek Mathematics.* Cambridge, 1884.

Tannery, P. *La Géométrie grecque.* Paris, 1887.

Zeuthen, H. G. *Die Lehre von den Kegelschnitten im Altertum.* Copenhagen, 1886.

On pre-Greek mathematics see:

Neugebauer, O. *Vorgriechische Mathematik.* Berlin, 1934.

Chase, A. B., H. P. Manning, and R. C. Archibald. *The Rhind Mathematical Papyrus.* 2 volumes, Oberlin, 1927–1929.

Vogel, Kurt. *Die Grundlagen der ägyptischen Arithmetik.* Munich, 1929.

The knowledge of pre-Greek mathematics has developed rapidly in recent years and may best be followed in the books and papers of O. Neugebauer, F. Thureau-Dangin, K. Vogel, S. Gandz, and others. See *Isis* 31 (1940) 399ff.

Periodicals

Many recent papers of great importance on Greek and pre-Greek mathematics are to be found in *Quellen und Studien zur Geschichte der Mathematik, Astronomie, und Physik.* Begun in 1930.

Scripta Mathematica. Ed. by J. Ginsburg. Begun in 1932.

Of the earlier periodicals and series containing important material for the student of Greek mathematics note especially:

Bibliotheca Mathematica. Zeitschrift für Geschichte der mathematischen Wissenschaften. Ed. by Gustaf Eneström. 1884–1886; 1887–1899; 1900–1914.

Abhandlungen zur Geschichte der mathematischen Wissenschaften. Founded by Moritz Cantor. 1877–1913.

Bullettino di bibliografia e storia delle scienze matematiche e fisiche. Ed. by B. Boncompagni. 1868–1887.

Bollettino di bibliografia e storia delle scienze matematiche. Ed. by G. Loria, 1898–1917. (Later an appendix to *Bollettino di matematica.*)

Sources

See p. 1. Among the more important Greek writers on mathematics (and their modern editors) we may mention:

Euclid (Heiberg and Menge); Archimedes (Heiberg); Apollonius (Heiberg: this edition does not contain *Conics* V–VII, extant only in Arabic); Diophantus (Tannery); Pappus (Hultsch); Autolycus (Hultsch); Hero of Alexandria (Schöne and Heiberg); Theodosius, *Sphaerica* (Heiberg); Menelaus, *Sphaerica* (Arabic text ed. by Krause); Theon of Smyrna (Hiller); Nicomachus (Hoche); Serenus (Heiberg); Iamblichus, *Introduction to Arithmetic* (Pistelli); and Proclus, *Commentary on Euclid's Elements I* (Friedlein). Many of these editions are accompanied by Latin translations.

The English reader is fortunate in having T. L. Heath's series of translations, paraphrases, and commentaries:

The Thirteen Books of Euclid's Elements. 3 volumes, Cambridge, 1908.

The Works of Archimedes. Cambridge, 1897.

The "Method" of Archimedes. Cambridge, 1912.

Apollonius of Perga. Treatise on Conic Sections. Cambridge, 1896.
Diophantus of Alexandria. 2d ed., Cambridge, 1910.

Note also Nicomachus of Gerasa, *Introduction to Arithmetic.* English translation by M. L. D'Ooge with studies in Greek arithmetic by F. E. Robbins and L. C. Karpinski. New York, 1926.

Paul Ver Eecke has in recent years published French translations of a series of Greek mathematical authors, including works of Archimedes, Apollonius, Pappus, Diophantus, Theodosius (*Sphaerica*), Euclid (*Optics* and *Catoptrics*), and Serenus.

Important mathematical passages sometimes occur in non-mathematical works, e.g., passages from Eudemus preserved in Simplicius' commentaries on Aristotle's *Physics* and *De Caelo*, passages in Plato, etc.

Astronomy and Mathematical Geography

Many of the books noted under Mathematics contain material on Greek astronomy and mathematical geography. Apart from general histories of astronomy, the following contain important treatments of Greek astronomy.

Heath, T. L. *Aristarchus of Samos.* Oxford, 1913. The text and translation of Aristarchus's treatise *On the Sizes and Distances of the Sun and Moon* are preceded by a study of early Greek astronomy.

——— *Greek Astronomy.* London, 1932. A historical survey followed by an anthology.

Dreyer, J. L. E. *History of the Planetary Systems from Thales to Kepler.* Cambridge, 1906.

Delambre, J. B. J. *Histoire de l'astronomie ancienne.* 2 volumes, Paris, 1817.

Duhem, P. *Le Système du monde. Histoire des doctrines cosmologiques de Platon à Copernic.* 5 volumes, Paris, 1913–1917.

Tannery, P. *Recherches sur l'histoire de l'astronomie ancienne.* Paris, 1893.

Schiaparelli, G. *Scritti sulla storia dell' astronomia antica.* 3 volumes, Bologna, 1925–1927.

The papers of J. K. Fotheringham and O. Neugebauer (some of them in *Quellen und Studien zur Geschichte der Mathematik, Astronomie, und Physik*) include important recent contributions to Babylonian and Greek astronomy. For a general summary see O. Neugebauer, "The History of Ancient Astronomy: Problems and Methods." *Journal of Near Eastern Studies* 4 (1945) 1–38: reprinted, with some amplification, in *Publication of the Astronomical Society of the Pacific* 58 (1946) 17–43, 104–142.

The books dealing with mathematical geography are substantially the same as those on astronomy. In addition to the latter, note Wilhelm Kubitschek, articles "Karten" (RE, vol. X) and "Erdmessung" (RE Supplement VI), and F. Gisinger, article "Geographie" (RE, vol. IV).

Bunbury, E. H. *A History of Ancient Geography among the Greeks and Romans from the Earliest Ages till the Fall of the Roman Empire.* 2d ed., 2 volumes, London, 1883.

Wright, J. K. *The Geographical Lore of the Time of the Crusades.* New York, 1925. Has extensive bibliographies and numerous references to antiquity.

Berger, Hugo. *Geschichte der wissenschaftlichen Erdkunde der Griechen.* 2d ed., Leipzig, 1903.

Heidel, W. A. *The Frame of the Ancient Greek Maps.* New York, 1937.

Warmington, E. H. *Greek Geography.* London and New York, 1934. An anthology with a considerable section on mathematical geography.

Sources

See p. 89. The basic texts include Euclid, *Phaenomena* (Menge); Autolycus (Hultsch: German translation by Czwalina); Aristarchus (Heath, with English translation); Geminus,

Elements of Astronomy (Manitius); Hipparchus, *Commentaries on the "Phaenomena" of Aratus and Eudoxus* (Manitius, with German translation); Ptolemy, *Syntaxis Mathematica* (=*Almagest*) and *Opera Astronomica Minora* (Heiberg: German translation of *Syntaxis* by Manitius, French translation by Halma), *Catalogue of Stars* [from *Syntaxis* VII and VIII] (Peters and Knobel), commentaries on *Syntaxis* by Theon of Alexandria and Pappus (A. Rome); Cleomedes (Ziegler: German translation by Czwalina); Theodosius, *De Diebus et Noctibus* (Heiberg), *De Habitationibus* (Fecht); Proclus, *Hypotyposis* (Manitius, with German translation); Ptolemy, *Geography* (Müller, incomplete, Nobbe: German translation and commentary on Book I—the part relating to mathematical geography—by H. v. Mžik); Strabo (Meinecke: English translation by H. L. Jones); Philoponus, *On the Astrolabe* (Hase: French translation by P. Tannery, English translation by H. W. Greene in R. T. Gunther, *The Astrolabes of the World*, vol. I).

In addition there is extensive source material on astronomy and mathematical geography in the philosophical and doxographical literature and in such Latin writers as Pliny the Elder, Macrobius, Martianus Capella, and Censorinus.

Physics

Of the general histories of physics that of Ernst Gerland, *Geschichte aer Physik* (Munich, 1913), has a good treatment of the ancient period.

Mechanics

Duhem, P. *L'Evolution de la mécanique.* Paris, 1903.

——— *Les Origines de la statique.* 2 volumes, Paris, 1905–1906.

——— Σώζειν τὰ φαινόμενα. *Essai sur la notion de théorie physique de Platon à Galilée.* Paris, 1908.

——— *Etudes sur Léonard de Vinci.* 3 volumes, Paris, 1906–1913.

——— *La Théorie physique: son objet, sa structure.* 2d ed., Paris, 1914.

Vailati, Giovanni. *Scritti.* Leipzig, 1911.

Mach, Ernst. *Die Mechanik in ihrer Entwicklung historisch-kritisch dargestellt.* 7th ed., Leipzig, 1912. English translation (*The Science of Mechanics*) by T. J. McCormack, 4th ed., Chicago, 1919.

Atomism

Mabilleau, L. *Histoire de la philosophie atomistique.* Paris, 1895.

Lasswitz, K. *Geschichte der Atomistik im Mittelalter bis Newton.* 2 volumes, Hamburg, 1890.

Bailey, C. *The Greek Atomists and Epicurus.* Oxford, 1928.

Technology

Enciclopedia delle scienze e delle loro applicazioni. 2 volumes, Milan, 1941–1943. Contains much material pertaining to antiquity.

Feldhaus, F. M. *Die Technik der Antike und des Mittelalters.* Potsdam, 1931.

Diels, Hermann. *Antike Technik.* 3d ed., Leipzig, 1924.

Sources

Mechanics: Archimedes, various works, especially *On the Equilibrium of Planes* and the *Method* (ed. by Heiberg: English translation by Heath); the Aristotelian *Mechanics* (text and English translation by W. S. Hett, English translation by E. S. Forster); Hero of Alexandria, *Mechanics* (ed. by L. Nix, with German translation of the Arabic; B. Carra de Vaux, with French translation); Pappus, *Mathematical Collection*, Book VIII (Hultsch: see under Mathematics). For basic questions of dynamics, the philosophers and commentators (e.g., Simplicius and Philoponus) must be consulted.

Hydrostatics: Archimedes, *On Floating Bodies* (ed. by Heiberg: English translation by Heath).

Optics and catoptrics: Euclid (Heiberg: English translation by H. E. Burton, French translation by Ver Eecke).

Ptolemy (?) (Govi), Hero of Alexandria (W. Schmidt), Damianus (Schöne).

Acoustics and musical theory: Euclid (Menge); Ptolemy, *Harmonics* (Düring, also German translation); Porphyrius, *Commentary on Ptolemy's Harmonics* (Düring); Aristoxenus (Macran, with English translation); the acoustical and musical problems in the Aristotelian *Problemata* Books 11 and 19 (English translations by W. S. Hett [Loeb Classical Library] and E. S. Forster [Oxford]); Theon of Smyrna (Hiller); Plutarch, *On Music* (Volkmann: French translation by Weil and T. Reinach); Boethius, *On Music* (Friedlein). Works of Nicomachus, Bacchius, Gaudentius, and Alypius in addition to some of the others mentioned above are included in K. von Jan, *Musici Scriptores Graeci*, Leipzig, 1895. In this branch, as in so many others of ancient science, the philosophers (notably in this case Plato and Aristotle) and their commentators (e.g., Alexander of Aphrodisias, Themistius, Proclus, Porphyrius, Simplicius, Philoponus, etc.) must be consulted.

Pneumatics: Hero of Alexandria, *Pneumatics* (Schmidt, with German translation; English translation by J. G. Greenwood); Philo of Byzantium, *Pneumatics* (Arabic ed. by Carra de Vaux, with French translation; Latin ed. by Schmidt; French translation by A. de Rochas).

For applied mechanics, see p. 314. Among the literary sources for our knowledge of ancient machinery—agricultural, industrial, military, theatrical, etc.—are passages in the Aristotelian *Problemata* and *Mechanica*, Hero of Alexandria, Philo of Byzantium, Biton, Vitruvius, Pappus, *Mathematical Collection* VIII, Cato, Varro, and the minor writers on geoponics.

Chemistry and Chemical Technology

Kopp, Hermann. *Geschichte der Chemie.* 4 volumes. Braunschweig, 1843–1847.

——— *Beiträge zur Geschichte der Chemie.* Part I. Braunschweig, 1869.

Meyer, Ernst von. *Geschichte der Chemie von den ältesten Zeiten bis zur Gegenwart.* 4th ed., Leipzig, 1914. English translation of 3d ed. by G. McGowan, New York, 1906.

Stillman, J. M. *The Story of Early Chemistry.* New York, 1924.

Berthelot, M. *Les Origines de l'alchimie.* Paris, 1885.

——— *Introduction à l'étude de la chimie des anciens et du moyen âge.* Paris, 1889.

Lippmann, E. O. von. *Abhandlungen und Vorträge zur Geschichte der Naturwissenschaften.* Leipzig, 1906.

——— *Entstehung und Ausbreitung der Alchemie.* Berlin, 1919.

Periodicals

Ambix, Journal of the Society for the Study of Alchemy and Early Chemistry, was published in 1937–1938 under the editorship of F. S. Taylor. Publication resumed in 1946.

Chymia: Studies in the History of Chemistry. University of Pennsylvania Press. An annual begun in 1948.

Sources

See p. 352. Important source material is to be found in:

Berthelot, M., and C. E. Ruelle. *Collection des alchimistes grecs.* 3 volumes, Paris, 1885–1888.

Lagercrantz, O. *Papyrus graecus Holmiensis.* Uppsala, 1913.

Leemans, C. *Papyri graeci musei antiquarii publici Lugduni-Batavi.* Vol. 2, Leyden, 1885 (containing Leyden Papyrus X).

Mieli, Aldo. *Pagine di storia della chimica.* Rome, 1922.

Bailey, K. C. *The Elder Pliny's Chapters on Chemical Subjects.* 2 volumes, London, 1929, 1932.

Medical and pharmacological authors contribute important evidence and, of course, on the basic questions of matter and its transformations the philosophical literature (e.g., Aristotle (?), *Meteorologica* Bk. IV) must be consulted.

Geology, Physiography, and Meteorology

Adams, F. D. *Birth and Development of the Geological Sciences.* Baltimore, 1938.

Gilbert, Otto. *Die meteorologischen Theorien des griechischen Altertums.* Leipzig, 1907.

Capelle, Wilhelm. Article "Meteorologie" in RE, Supplement VI.

Lones, T. E. (see under Biology).

Lenz, H. O. *Mineralogie der alten Griechen und Römer.* Gotha, 1861. An anthology (German translation of sources).

Sources

See p. 374.

Biology

Singer, Charles. *A Short History of Biology.* Oxford, 1931.

Nordenskiöld, E. *History of Biology.* English translation, New York, 1928.

Locy, W. A. *The Growth of Biology.* New York, 1925.

Singer, Charles. "Greek Biology and Its Relations to the Rise of Modern Biology." *Studies in the History and Method of Science*, ed. by C. Singer, vol. 2, Oxford, 1921.

Senn, Gustav. *Die Entwicklung der biologischen Forschungsmethode in der Antike.* Aarau, 1933.

Thompson, D'Arcy W. *On Aristotle as a Biologist.* Oxford, 1916.

Lones, T. E. *Aristotle's Researches in Natural Science.* London, 1913. The larger part is concerned with Aristotelian biology, though the non-biological sciences are also treated.

Botany

Meyer, Ernst H. F. *Geschichte der Botanik.* 4 volumes. Königsberg, 1854–1857.

Reed, Howard Sprague. *A Short History of the Plant Sciences.* Waltham, Mass., 1942.

Greene, Edward L. *Landmarks of Botanical History.* Washington, 1909.

Strömberg, Reinhold. *Theophrastea. Studien zur botanischen Begriffsbildung.* Göteborg, 1937.

Bretzl, Hugo. *Botanische Forschungen des Alexanderzuges.* Leipzig, 1903.

Lenz, H. O. *Botanik der alten Griechen und Römer.* Gotha, 1859. An anthology (German translation of sources).

Periodical

Chronica Botanica. Founded by F. Verdoorn, 1933. In recent years has stressed historical studies.

Zoology

Keller, Otto. *Die antike Tierwelt.* Leipzig, 1909.

Thompson, D. W. *A Glossary of Greek Birds.* 2d ed., London, 1936.

——— *A Glossary of Greek Fishes.* London, 1947.

Lenz, H. O. *Zoologie der alten Griechen und Römer.* Gotha, 1850. An anthology (German translation of sources).

Sources

See pp. 394, 400, 438.

Botany. Theophrastus, *History of Plants* and *Causes of Plants* (ed. by F. Wimmer: English translation of the *History* by A. Hort and of Book I of the *Causes* by R. E. Dengler). Treatise *On Plants* of the Aristotelian *Corpus* (ed. by E. H. F. Meyer, English translations by E. S. Forster and W. S. Hett); Dioscorides (ed. by M. Wellmann, English translation by J. Goodyer, German translation by J. Berendes); Pliny, *Natural History*, especially Books 12–27. Note in addition Vergil (*Georgics*), Varro, Columella, Ps.-Democritus, the minor geoponic authors, and the medical authors who discuss pharmacology.

Zoology. Aristotle's *History of Animals*, *Parts of Animals*, *Motion of Animals*, *Progression of Animals*, and *Generation of Animals*; Pliny's *Natural History*, Books 8–11, Aelian's treatise *On Animals*, the poem *Halieutica* attributed to Ovid, and the anonymous *Physiologus* are the chief extant works. The veterinary literature may also be noted in this connection. (See *Corpus Hippiatricorum Graecorum*, ed. by E. Oder and C. Hoppe, 2 volumes, Leipzig, 1924, 1927; *Mulomedicina Chironis*, ed. by E. Oder, Leipzig, 1901; and Vegetius, *Digesta Artis Mulomedicinae*, ed. by E. Lommatzsch, Leipzig, 1903.)

In addition to botanical and zoological works, many philosophical works, e.g., Aristotle's *De Anima*, deal with general biology.

Medicine

Drabkin, Miriam. "A Select Bibliography of Greek and Roman Medicine." *Bulletin of the History of Medicine* 11 (1942) 399–408. This bibliography lists the best editions and translations of Greek and Roman medical writings as well as general works on Greek and Roman medicine and works on the special branches of medicine. The literature on Greek and Roman medicine is so vast that only a few selected titles may be noted here.

Singer, Charles. *Greek Biology and Greek Medicine* Oxford, 1922.

Taylor, H. O. *Greek Biology and Medicine.* Boston, 1922.

Puschmann, T. *A History of Medical Education.* English translation by E. H. Hare. London, 1891.

Allbutt, Clifford. *Greek Medicine in Rome.* New York, 1921.

Brock, A. J. *Greek Medicine.* New York, 1929. An anthology preceded by a general treatment of the subject.

Singer, Charles. *The Evolution of Anatomy.* New York, 1925.

Needham, Joseph. *A History of Embryology.* Cambridge, 1934.

Kremers, E. and G. Urdang. *History of Pharmacy.* Philadelphia, 1940.

Schmidt, A. *Drogen und Drogenhandel in Altertum.* Leipzig, 1924.

Gurlt, E. *Geschichte der Chirurgie.* 3 volumes, Berlin, 1898.

Milne, J. S. *Surgical Instruments in Greek and Roman Times.* Oxford, 1907.

Because of their rather ample treatment of antiquity, the following general works should be noted.

Daremberg, C. *Histoire des sciences médicales.* 2 volumes, Paris, 1870.

Neuburger, M., and J. Pagel. *Handbuch der Geschichte der Medizin.* 3 volumes. The sections on Greek medicine by Robert Fuchs and on Roman medicine by Iwan Bloch are in vol. I (Jena, 1902).

Singer, Charles. *A Short History of Medicine.* Oxford, 1928.

Diepgen, P. *Geschichte der Medizin.* 5 volumes, Berlin, 1923–1928.

Periodicals. See also p. 561.

Janus, Zeitschrift für Geschichte und Literatur der Medizin. Ed. by A. Henschel. 1846–1848.

——— *Central-Magazin für Geschichte und Literaturgeschichte der Medizin.* Ed. by H. Bretschneider, A. Henschel, C. Heusinger, J. G. Thierfelder. 1851–1853.

——— *Archives internationales pour l'histoire de la médecine et la géographie médicale.* Founded by H. F. A. Peypers in 1896.

Archiv für Geschichte der Medizin. Founded by K. Sudhoff, 1907.

Bulletin of the History of Medicine. Founded by H. E. Sigerist, 1933.

Journal of the History of Medicine and Allied Sciences. Ed. by G. Rosen. Begun in 1946.

Quellen und Studien zur Geschichte der Naturwissenschaften und der Medizin. Ed. by P. Diepgen and J. Ruska. Begun in 1931.

Index Medicus. A quarterly bibliography on all phases of medicine. Begun in 1916.

Sources

See p. 467 and bibliography of M. Drabkin. A series of definitive texts is in course of being published in the *Corpus Medicorum Graecorum* and *Corpus Medicorum Latinorum.*

Among the more important Greek and Roman writers on medicine (and their modern editors) we may mention Aëtius of Amida (Olivieri, still incomplete, separate books by Hirschberg, Costomiris, Zervos, others); Aretaeus (Adams, Hude); Caelius Aurelianus (Amman); Cassius Felix (Rose); Celsus (Marx); Diocles (Wellmann); Galen (Diels, Marquardt, Helmreich, Kalbfleisch, Müller, Simon, others; the edition by C. G. Kühn is not completely superseded); Hippocratic Collection (Littré, Kuehlewein, Heiberg, Jones, Withington, others); Oribasius (Bussemaker and Daremberg, Raeder); Paul of Aegina (Heiberg); Rufus of Ephesus (Daremberg and Ruelle); Soranus (Ilberg); Theodore Priscian (Rose); Fragments of the Empiric School (Deichgräber); of the Pneumatic School (Wellmann).

Note the following important English translations.

The Genuine Works of Hippocrates. Trans. by F. Adams. London, 1848. (Reprinted New York, 1929.)

Hippocrates. Ed. and trans. by W. H. S. Jones and E. T. Withington. 4 volumes, New York, 1923–1931. (Loeb Classical Library.)

——— *On Ancient Medicine.* Ed. and trans. by W. H. S. Jones, in *Philosophy and Medicine in Ancient Greece.* Bulletin of the History of Medicine, Suppl. 8, 1946.

The Medical Writings of Anonymus Londinensis. Ed. and trans. by W. H. S. Jones. Cambridge, 1947.

Galen. *On the Natural Faculties.* Ed. and trans. by A. J. Brock. London, 1916. (Loeb Classical Library.)

Galen. *On Medical Experience.* Ed. and trans. by R. Walzer. Oxford, 1944.

Celsus. *On Medicine.* Ed. and trans. by W. G. Spencer. 3 volumes, London, 1935–1938. (Loeb Classical Library.)

The Extant Works of Aretaeus. Ed. and trans. by F. Adams. London, 1856.

The Seven Books of Paulus Aegineta. Trans. by F. Adams. 3 volumes, London, 1844–1846.

Physiological Psychology

See p. 530, where the chief primary sources are indicated. J. I. Beare, *Greek Theories of Elementary Cognition from Alcmaeon to Aristotle* (Oxford, 1906) and G. M. Stratton, *Theophrastus and the Greek Physiological Psychology before Aristotle* (London, 1917) are important works in this field.

ADDENDA

P. 80 (last line before footnotes). I.e., Euclid's *Elements*. We adopt Halley's emendation of the Greek text.

P. 80, n. 2. Although Guldin's Theorem is now sometimes called Pappus's Theorem, the latter term is usually employed to designate a quite different theorem of Pappus (*Mathematical Collection* VII. 139): *If A, B, C are any three distinct points on a line l, and A', B', C' any three distinct points on a line l' intersecting l, the points of intersection of BC' and $B'C$, of CA' and $C'A$, and of AB' and $A'B$ are collinear.* This theorem and outgrowths of it have played an important part in projective geometry and in modern discussions of the foundations of geometry.

P. 82, n. 1. In modern times isoperimetric problems and their generalizations have had a central role in the development of the Calculus of Variations.

P. 128. Although Ptolemy in his mathematical exposition is generally concerned with the geometry and trigonometry of circles, his system, like that of his predecessors from the time of Eudoxus (and his successors until Kepler), is fundamentally a system of spheres. We have alluded (p. 102) to the question whether the spheres have physical reality, a question about which there was considerable discussion, particularly in medieval times. It was not until after Brahe's observations of planetary orbits (which would require that certain of the spheres intersect) and of the paths of comets (which would penetrate planetary spheres) that the idea of the physical reality of the spheres was seriously shaken; Kepler's work definitely ended the controversy.

The order of the heavenly bodies in Ptolemy's system, proceeding outward from the earth, was: Moon, Mercury, Venus, Sun, Mars, Jupiter, Saturn, fixed stars. But because he did not observe any planetary parallax, Ptolemy was unable to determine the distances of the planets from the earth; his system merely gave a ratio between the perigee and the apogee distances of any particular planet. But from first beginnings in the fifth century or earlier, the idea developed, primarily among Arab astronomers, that the apogee distance of any planet was just exceeded by the perigee distance of the next outer planet. Since the distance from the earth to the first heavenly body, the moon, could be found with some accuracy, the distance of every other heavenly body could also, on this theory, be found.

The history of this theory, which is linked with the philosophic concept of a full universe and with the controversy about the physical reality of the spheres, is traced by Edward Rosen in *Scientific Monthly* 63 (1946) 213–217.

P. 130 (top). Actually a planetary orbit is not an absolutely perfect ellipse; this is due to the presence of bodies other than the planet and the sun. Now planetary motion could theoretically be described, to any required degree of approximation, by sufficiently complex combinations of circular motions. But the great advantages of Kepler's system over Ptolemy's lay in the relative simplicity of its structure and in its making possible the unification of celestial and terrestrial mechanics under the same dynamical principles.

P. 130 (Star Catalogue). In classifying stars into six magnitudes, from the brightest stars (first magnitude) to those barely visible to the unaided eye (sixth magnitude), Ptolemy followed what seems to have been a traditional system of comparison. With the development of the telescope and photometric techniques in modern times, the number of magnitudes has been greatly extended, and more objective methods of distinguishing the magnitudes have been evolved.

Pp. 135–138. See the article "Horologium," RE vol. VIII. 2416–2433.

P. 245, n. 3. The clepsydra used for measuring speaking time in the law courts was merely a vessel with an orifice near the bottom, through which water flowed out. An example from the fifth century B.C., recently found at Athens, is described by S. Young in *Hesperia* 8 (1939) 274. The vessel is, of course, quite different from the device described on pp. 246 and 327, and from the clock described on p. 137.

P. 295, n. 1. The reference here is probably to Glaucus of Rhegium. But the origin of the proverbial "art of Glaucus" was uncertain even in antiquity. Some connected the expression with Glaucus of Chios and his discovery, in the sixth century B.C., of the technique of soldering iron (cf. Herodotus I. 25). Others mentioned a Glaucus of Samos and a supposed innovation in the art of writing.

P. 332. To accord with the text the diagram should show a pivot linking *TU* and *QU* at *U* rather than an aperture.

P. 336. See the article "Dioptra," RE Supplement VI. 1287–1290.

P. 376, line 8. Some have identified the Heraclean with the Lydian stone and have transferred the examples to stones that attract (cf. pp. 311–312, above); others have taken the Heraclean stone as an example of an attracting stone, and the Lydian (a black flinty jasper?) as an example of a testing stone. But any dislocation or misunderstanding of Theophrastus' text on this point was already present in antiquity, for Pliny (*Natural History* 33. 126), quoting Theophrastus, speaks of the Heraclean stone (with which Pliny here identifies the Lydian) as an example of a testing stone.

Pp. 401–403. Aristotle's emphasis here and elsewhere (see pp. 405–409) on continuity between non-living and living matter and throughout the *scala naturae* is discussed by A. O. Lovejoy in connection with the history of what he calls "the principle of plenitude" (*The Great Chain of Being*, pp. 55–59).

P. 469, n. 8. In his mechanical principles Erasistratus may have been greatly influenced by Strato of Lampsacus (p. 211, n.). This view is set forth by H. Diels in a study of Strato's physical system, *Sitzungsberichte der Preussischen Akademie der Wissenschaften*, 1893, pp. 109, 117.

P. 559. A good treatment of scientific and technical writings is contained in the volumes on Greek and Roman literature in Müller's *Handbuch der klassischen Altertumswissenschaft*, and in F. Susemihl's *Geschichte der griechischen Literatur in der Alexandrinerzeit.*

P. 562. Thaer, Clemens, "Antike Mathematik," *Bursian's Jahresbericht* 283 (1943) 1–144. Report on the literature of the subject from 1906 to 1930.

P. 563 (top). Edition and English translation, by W. Thomson, of an Arabic version of what is probably Pappus's Commentary on Book X of Euclid's *Elements.* Introduction, notes, and glossary by W. Thomson and G. Junge. Cambridge, Mass., 1930.

P. 564 (top). Ptolemy, *Almagest.* English translation of Books I–V by R. C. Taliaferro, Books VII–IX.6 by G. A. Bingley. Annapolis, 1938–1939.

INDEX OF NAMES

Biographical and bibliographical references are placed first and separated from other references by a semicolon.

The following list of subjects will serve to indicate the general field in which a given reference lies (see also the Table of Contents):

Mathematics	1–88
Astronomy	89–143
Mathematical Geography	143–181
Physics	182–351
Chemistry and Chemical Technology	352–373
Geology and Meteorology	374–393
Biology	394–466
Zoology	400–437
Botany	437–466
Medicine	467–529
Physiological Psychology	530–558
Bibliography	559–568
Addenda	569–570

B

C

I

J

K

L

P

R